MATHEMATICAL INSTRUMENTS

THEIR CONSTRUCTION, ADJUSTMENT, TESTING, AND USE

COMPRISING

DRAWING, MEASURING, OPTICAL, SURVEYING, AND ASTRONOMICAL INSTRUMENTS

Owing to the antiquity and rarity of this book, certain pages are regrettably absent, and readability is compromised due to the blurring of the original text. If you possess any insights regarding the availability of these missing sections, kindly inform us. Your assistance is crucial in ensuring the completion of this significant work for the benefit of future generations.

MATHEMATICAL INSTRUMENTS

THEIR CONSTRUCTION, ADJUSTMENT, TESTING, AND USE

COMPRISING

DRAWING, MEASURING, OPTICAL, SURVEYING, AND ASTRONOMICAL INSTRUMENTS

BY

J. F. HEATHER

GYAN

Published by

GYAN

Gyan Publishing House
5, Ansari Road
Daryaganj, New Delhi-110002
Phone: 011-47034999, 9811692060
E-mail: books@gyanbooks.com

Distribution Network
gyanbooks.com
India, USA, Canada, UK, Australia

ISBN: 978-93-6433-032-9 (PB)
First Published, 1884

2nd Impression 2024

Printed at: Gyan Press, Delhi.

MATHEMATICAL INSTRUMENTS: THEIR CONSTRUCTION, ADJUSTMENT, TESTING, AND USE COMPRISING DRAWING, MEASURING, OPTICAL, SURVEYING, AND ASTRONOMICAL INSTRUMENTS
Author: J. F. HEATHER

Plate I

Frontispiece.

1

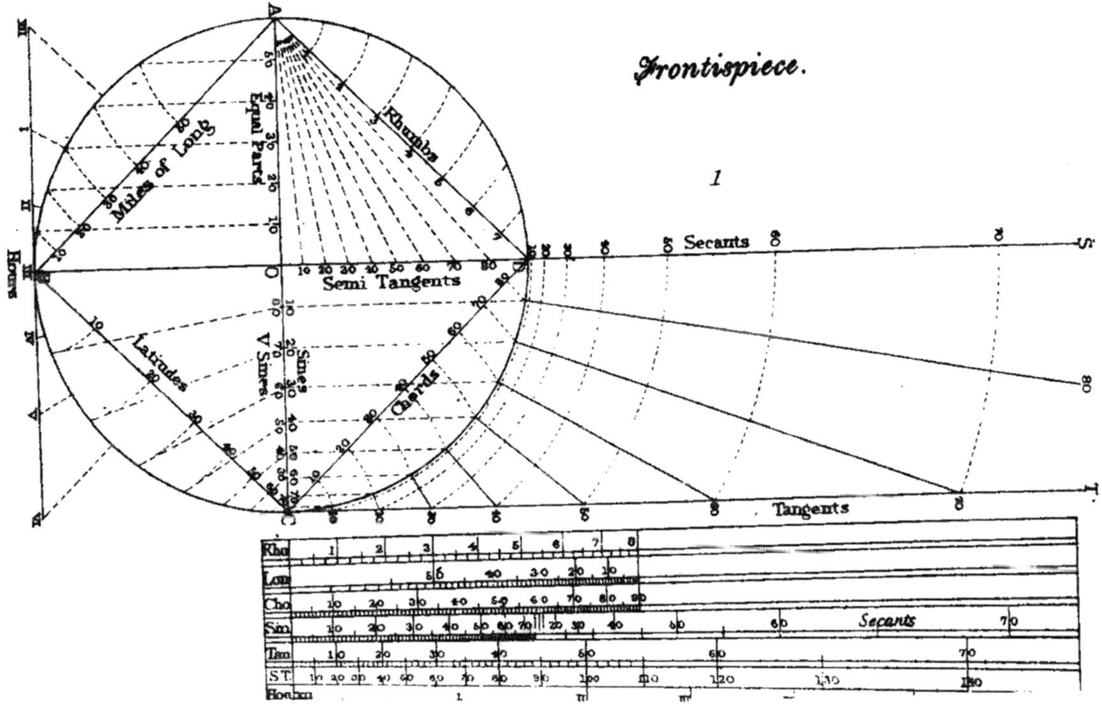

MATHEMATICAL INSTRUMENTS

THEIR CONSTRUCTION, ADJUSTMENT, TESTING, AND USE

COMPRISING

DRAWING, MEASURING, OPTICAL, SURVEYING, AND ASTRONOMICAL INSTRUMENTS

By J. F. HEATHER, M.A.

LATE OF THE ROYAL MILITARY ACADEMY, WOOLWICH; AUTHOR OF "PRACTICAL PLANE GEOMETRY," "DESCRIPTIVE GEOMETRY," ETC.

ENLARGED EDITION, FOR THE MOST PART ENTIRELY RE-WRITTEN

With Numerous Illustrations

Capio Lumen

LONDON
CROSBY LOCKWOOD AND CO.
7, STATIONERS' HALL COURT, LUDGATE HILL
1884

ADVERTISEMENT.

This popular treatise on the "Construction, Adjustment, Testing, and Use of Mathematical Instruments" appeared originally in 1849, and has since been used continuously in Government military and naval schools; and forms, by authority, part of a midshipman's kit. In presenting it for the tenth time to the public it has been entirely re-written, and full development given to the description, with their most recent improvements, of such mathematical instruments as come within the classes mentioned on either of its title-pages. This has necessitated the division of the subjects into three distinct portions, the second volume being devoted to Optical Instruments, and the third to Surveying and Astronomical Instruments.* To meet, however, the convenience of those schools which have introduced the original work as a text-book, the NINTH EDITION has been stereotyped, and copies of that edition will continue to be supplied as long as the heads of such establishments may require them.

* Mathematical Instruments Part III., Optical Instruments, by J. F. H. Mathematical Instruments: Parts IV. and V., Surveying and Astronomical Instruments, by J. F. H.

MATHEMATICAL INSTRUMENTS

VOL. I.

DRAWING AND MEASURING INSTRUMENTS

INCLUDING

INSTRUMENTS EMPLOYED IN GEOMETRICAL DRAWING, AND IN THE MEASUREMENT OF MAPS AND PLANS

PREFACE TO VOL. I.

UPON the publication of the original edition of this treatise the author was very greatly and very agreeably surprised by the large number of complimentary letters that he received from persons engaged in a variety of pursuits, many of whom were entirely unknown to him. The praises, thus liberally bestowed on his humble work, were supplemented by its introduction, as a text-book, into numerous educational establishments, and its selection by the Admiralty to form part of a midshipman's kit.

It was manifest, then, that his little book had supplied a want that was strongly felt; and the writer, fully aware of its deficiencies, has long been anxious for the opportunity to improve upon his first essay, which has been afforded him by the liberality of the present proprietors of the WEALE SERIES.

The subjects of Optical Instruments, and of Surveying and Astronomical Instruments, having been removed, to be treated of more fully in separate volumes, sufficient space has been obtained in this volume for a considerable development of that portion of the work, entitled Part I., allotted to the description of Drawing Instruments; while in Part II., now devoted to the consideration of Instruments for Measuring, and for the mechanical performance of arithmetical computations, the principle of the slide-rule has been, it is hoped, clearly explained, the comparative merits of different modes of construction have been discussed, and plain and simple rules for the numeration of the results obtained have been given.

The want of a method of so using the slide-rule, that it should of itself indicate the numeration of the results obtained, and thus become its own interpreter, has been so serious a drawback on its value as a mechanical calculator, that the small extent to which it has hitherto come into general use is not to be wondered at. Not only has no such method been heretofore published, but the accomplishment of so desirable an object has been deemed impracticable. Thus the eminent writer* of the article on the slide-rule in Knight's "English Encyclopædia" remarks:—"Attempts are made in works professing to explain the slide-rule to give rules for the denomination of the character of the figures in the answer, but without any success. It is all very well for a few chosen examples, but an attempt to do without the book soon shows the insufficiency of rules. If on a large scale 653 should be the figures of an answer, common sense applied to the problem must say whether it is ·0653, ·653, 6·53, 65·3, 653, 6530, 65300, &c., which is meant."

The principal obstacle to the discovery of the very simple rules for numeration on the A and B lines, laid down in the text (p. 102), appears to have arisen from the unnecessary repetition of the line of numbers, from 1 to 10, upon *both* these lines.

At all events, the Author was led to the discovery of these rules by the examination of the rule designed by Mr. S. Coulson, supplied to him by the manufacturers, Messrs. Dring and Fage, on which the line of numbers, from 1 to 10, is laid down on the A line *once only* in the entire length of the rule, while it is laid down on the B line *twice*, on the slide, formed in two pieces, numbered on from one to the other. This rule, being used so that the results sought shall be found on the B line, or slide, the correct denomination of its figures is determined by the section of the slide upon which it appears.

To apply, then, the same method of numeration to the ordinary form of the rule, in which the line of numbers, from 1 to 10, is laid down *twice* on the A line, it is manifestly necessary only to disregard one-half of this line, and work with a single section of it.

It is evident, however, that, as regards the A lines, the ordinary rules are made twice as long as they need be, only to be encumbered with an useless repetition of the numbers laid down on them.

* Professor de Morgan.

As regards the C and D lines, there is a very serious defect in the ordinary construction; for, except in the case in which the slide is laid evenly in the groove, so that the 1 on the first section of the C line and the 10 on its second section coincide with the 1 and 10 on the D line, there are always portions of the D line which lie beyond the C line at one end, and fall short of it at the other, so that numerous problems cannot be *directly* solved. Their direct solution would require, in fact, an additional section to be added to the C line. With the diminished length of the rule in the *Coulson* form of its construction, the D line being laid down, half above and half below the groove for the slide, this defect is supplied, as a third section is by this means practically added to the C line, as will be seen in the text (p. 116).

The rules for the numeration of results have been extended to the C and D line (pp. 114—117) with but little addition to their complexity, and the Author flatters himself that he has thus added greatly to the practical utility of the instrument, and at the same time clearly pointed out the advantages of the form devised by Mr. Coulson.

It is remarkable that this form appears to have been designed solely with a view to the reduction of the length of the rule, its designer only claiming for it that with a nine-inch rule all the advantages are obtained of a rule of the ordinary construction of double the length.

In the preparation of this new edition the Author has received valuable assistance from several of the most eminent instrument-makers, and from personal examination and experience of the instruments treated of is enabled to hold himself responsible for the correctness of the description of their construction, and of the modes of testing, adjusting, and using them, herein laid down.

Messrs. Dring and Fage supplied the Author with numerous rules and gauges of their manufacture, and he originally contemplated including in the second part of the volume an account of the instruments used in gauging and measuring casks; but he found that to treat this subject, which is somewhat complex, in such a manner as he would have wished, would have required considerable space, and, in fact, that it would not be possible to do justice to it within the limits which he could allow himself.

Messrs. Elliott have not only afforded him every facility

for the inspection and study of numerous instruments constructed by them, but have also kindly made drawings for him of the spherometer, extensometer, and German-silver wire gauge, &c., and liberally allowed him the use of numerous woodcuts for the illustration of the work.

The Author cannot here forbear from calling particular attention to a little instrument called a polar planimeter (pp. 80—87), manufactured under a Swiss patent, and for which Messrs. Elliott are the sole agents in this country. He knows not whether he was agreeably surprised more by the remarkable geometrical property on which its construction is founded, the neatness and facility with which this principle jumps, as it were, from the theorem proved in Euclid, Bk. ii. props. 12 and 13, the simple manner in which it has been applied to the production of the planimeter, the beautiful manufacture of the little instrument itself, or the wonderful accuracy with which it performs its duty.

Messrs. Parkes and Son, of Birmingham, have assisted the Author by forwarding him a large number of instruments of their manufacture, and he will again have occasion to acknowledge their assistance in the succeeding volumes.

In returning his sincerest thanks to these gentlemen, and others who have assisted him in his task, the Author is compelled to acknowledge that he has been most liberally supplied with the best material, and can only hope that the skill of the workman may not appear to have been altogether inadequate to its manipulation.

J. F. H.

WATFORD,
November, 1870.

PREFACE TO THE FIRST EDITION.

An attempt has been made in the following pages to put within the reach of all a short and compendious treatise upon some of the ingenious instruments by which the scientific practitioner is aided in his observations, and in the delineation of the results obtained from them.

The instruments treated of have been divided into five classes, to each of which a part of the work has been devoted. The first part treats of Mathematical Drawing Instruments; the second, of Optical Instruments; the third, of Surveying Instruments; the fourth, of Astronomical Instruments; and the fifth, and last, of Goniometrical Instruments, for measuring the angles of crystals.

The greater part of the Wood Engravings, and some parts of the Text, of Simms's Mathematical Drawing Instruments, have been pressed into the service of the present work; and the works of the best writers upon the several parts of the subject have been consulted, and much valuable matter has been extracted from them, particularly from Pearson's Astronomy.

The limits of the bulk and cost of the work have forbidden any extensive excursion into the sciences in which the instruments are used; but it is hoped that a large mass of information has here been placed in a small compass without sacrificing perspicuity to undue compression.

J. F. H.

Royal Military Academy, Woolwich.
March, 1849.

CONTENTS OF VOL. I.

PART I.

ON MATHEMATICAL DRAWING INSTRUMENTS.

CHAPTER I.

THE DRAWING BOARD, PAPER, AND PENS.

PAG

Mounting Paper . . . .
Dimensions of Drawing Paper . . . .
Drawing Boards . . . .
Drawing Pens and Pricking Point . . . . . .
Stencilled Letters . . . .

CHAPTER II.

INSTRUMENTS FOR DRAWING STRAIGHT LINES.

Straight Edge 5
Bow Line 6
Parallel Rules 7
The T-square 7
The Triangle or Set Square 8
The Centrolinead 8

CHAPTER III.

INSTRUMENTS FOR DRAWING CURVED LINES.

Drawing Compasses 13
Compasses with Movable Points 14
Bow Compasses 15
Horn Centres 16
Tubular Compasses 16
Beam Compasses 17
Templets 19
The Ellipse 20
The Hyperbola 21
The Parabola 22

CHAPTER IV.

INSTRUMENTS FOR TAKING OFF AND LAYING DOWN DISTANCES AND ANGLES.

	PAGE
Scales of Equal Parts	22
Plotting Scales	23
Diagonal Scales	24
Protracting Scales	26
Projection of the Sphere	30
Construction of Dials	36
The Protractor or Plain Scale	38
The Military Protractor	39
The Sector	42
Marquois's Scales	52
The Vernier	54
The Circular Protractor	57
The Semicircular Protractor	59
The Station Pointer	60

CHAPTER V.

COPYING INSTRUMENTS.

Triangular Compasses	62
Wholes and Halves	62
Proportional Compasses	63
The Pantagraph	65
Copying-glass	68
Method of Squares	68
The Eidograph	70

CHAPTER VI.

INSTRUMENTS FOR MEASURING LINES AND AREAS ON PLANS.

The Opisometer	75
The Computer	77
The Polar Planimeter	80

PART II.

ON RULES AND GAUGES.

INSTRUMENTS FOR ARITHMETICAL COMPUTATION.

Gunter's Lines	88
Gunter's Scale	98
The Slide-Rule	99
Uses of the A and B lines	101

	PA
Numeration on the A and B lines	1
Problems on the A and B lines	1
Different modes of construction of the C and D lines	1
Squaring Numbers by the lines C and D	1
Extraction of Square Root	1
General Rules for Numeration upon the C and D lines	1
Comparative Merits of the different modes of construction	11
Problems on the C and D lines	11
Uses of the A, B, C, and D lines conjointly	12
The Inverted Slide	12
The Inverted Line	12
Examples of Inverted Lines	12
Numeration with Inverted Line	13
The Trigonometrical Slide	13
Ratios	13
Divisors and Gauge Points	13
The Routledge Slide-Rule	14
Bayley's Slide-Rule	14
The Kentish Slide-Rule	14
Coulson's Slide-Rule	14
The Cubing Rule	14
The Timber-contenting Rule	149

MEASURING INSTRUMENTS.

Instruments for measuring Timber	151
The Spherometer	153
The Extensometer	154
The German-silver Wire Guage	154

MATHEMATICAL INSTRUMENTS.

PART I.

ON MATHEMATICAL DRAWING INSTRUMENTS.

CHAPTER I.

THE DRAWING-BOARD, PAPER, AND PENS.

THE DRAWING BOARD.

The first thing to be done, preparatory to the commencement of a drawing, is to stretch the paper evenly upon the smooth and flat surface of a drawing board. The edges of the paper should first be cut straight, and, as nearly as possible, at right angles with each other; also the sheet should be so much larger than the intended drawing and its margin, as to admit of being afterwards cut from the board, leaving the border by which it is attached thereto by glue or paste, as we shall next explain.

The paper must first be thoroughly and equally damped with a sponge and clean water, on the opposite side from that on which the drawing is to be made. When the paper absorbs the water, which may be seen by the wetted side becoming dim, as its surface is viewed slantwise against the light, it is to be laid on the drawing board with the wetted side downwards, and placed so that its edges may be nearly parallel with those of the board; otherwise, in using a T-square, an inconvenience may be experienced. This done, lay a straight flat ruler on the paper, with its edge parallel to, and about half an inch from one of its edges. The ruler

n

must now be held firm, while the said projecting half inch of paper is turned up along its edge; then, a piece of solid glue (common glue will answer the purpose), having its edge partially dissolved by holding it in boiling water for a few seconds, must be passed once or twice along the turned edge of the paper, after which, this glued border must be again laid flat by sliding the rule over it, and, the ruler being pressed down upon it, the edge of the paper will adhere to the board. If sufficient glue has been applied, the ruler may be removed directly, and the edge finally rubbed down by an ivory book-knife, or any clean polished substance at hand, which will then firmly cement the paper to the board. Another but *adjoining* edge of the paper must, next, be acted upon in like manner, and then the remaining edges in succession; we say the adjoining edges, because we have occasionally observed that, when the opposite and parallel edges have been laid down first, without continuing the process progressively round the board, a greater degree of care is required to prevent undulations in the paper as it dries.

Sometimes strong paste is used instead of glue; but, as this takes a longer time to set, it is usual to wet the paper also on the upper surface to within an inch of the paste mark, care being taken not to rub or injure the surface in the process. The wetting of the paper in either case is for the purpose of expanding it; and the edges being fixed to the board in its enlarged state, act as stretchers upon the paper while it contracts in drying, which it should be allowed to do gradually. All creases or undulations by this means disappear from the surface, and it forms a smooth plane to receive the drawing.

TABLE OF DIMENSIONS OF DRAWING PAPER.

Demy . . .	20	in. by	15½ in.		Columbier .	35	in. by	23½ in.
Medium . .	22¾	,,	17½ ,,		Atlas . . .	34	,,	26
Royal . . .	24	,,	19¼ ,,		Dbl. Elephant	40	,,	27
Super Royal .	27¼	,,	19¼ ,,		Antiquarian .	53	,,	31
Imperial . .	30	,,	22 ,,		Emperor . .	68	,,	48
Elephant . .	28	,,	23 ,,					

The requisites of a good drawing board are (1) that its surface should be perfectly level, and should remain so under all changes of temperature; (2) that its edges should be straight and well squared; (3) that it should be as light as it can be conveniently made to fulfil the preceding conditions. Unfortunately, all woods will expand and contract laterally under the influence of variations in temperature, and as drawing boards are constantly being wetted and then dried by the fire, no attempt to control the alternate expansions and contractions can prevent the warping of the board. The only effective construction, then, must be such as will allow of any possible amount of expansion and contraction, without destroying the even surface of the board. The desired result is perfectly accomplished by the following mode of construction, the invention of Mr. Howlett, late chief draughtsman at the War Office.

The board is made of well-seasoned pine, three-quarters of an inch thick, without knots, well planed and smoothed, the edges being either simply glued together, or tongued and glued. At the back of the board are the cross pieces, or clamps, consisting each of two pieces, A, B, three quarters of an inch thick, passing through the iron straps C C, into which they fit accurately, but through which they can slide stiffly. The width of the board is divided into three parts at the points A and B, and at these points the clamps are screwed firmly to the board. The clamps are cut away at C to a depth equal to or slightly in excess of the thickness of the iron strap, to prevent the latter from injuring the table on which the board is placed for use. This board then may contract and expand laterally without any strain either upon the joints, which consequently remain close, or upon the clamps, which slide through the straps, and they have been found to stand without injury the severest tests to which they could be exposed.

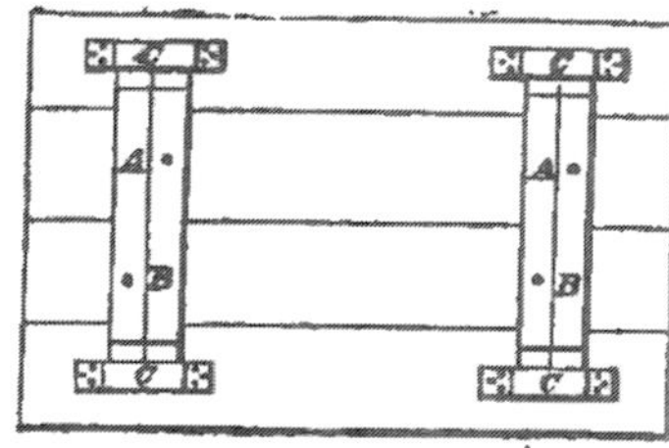

The accompanying illustration represents another mode of construction, which also sets at defiance the evils arising from alternate expansions and contractions. The screws here pass through slots in the clamps, so that while they hold the pieces forming the board firmly against these clamps, they leave them free to expand and

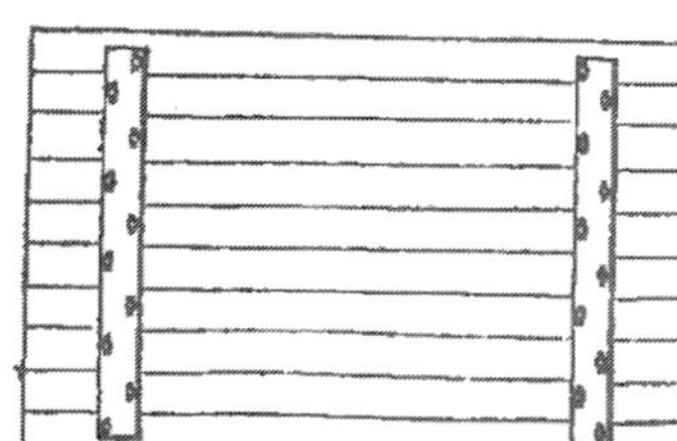

contract laterally, carrying the screws with them through the amount of play allowed them in the slots.

THE DRAWING PEN.

This instrument (Fig. 1) consists of two blades with steel points fixed to a handle; and they are so bent that a sufficient cavity is left between them for the ink, when the ends of the steel points meet close together, or nearly so. In the best construction the blades are set with the points more or less open by means of a mill-headed screw, so as to draw lines of any required fineness or thickness. One of the blades is framed with a joint, so that by taking out the screw the blades may be completely opened, and the points effectively cleaned after use. The ink is to be put between the blades by a common pen, and in using the pen it should be slightly inclined in the direction of the line drawn, and care should be taken that both points touch the paper. The drawing-pen should be kept close to the straight or other guiding edge (see STRAIGHT-EDGE), and held in the same direction during the whole operation of drawing the line.

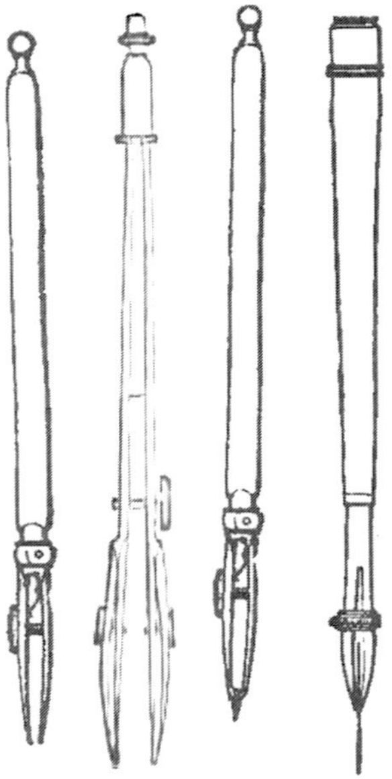

Fig. 1. Fig. 2. Fig. 3. Fig. 4.

For drawing close parallel lines in mechanical and architectural drawings, or to represent canals or roads, a double pen (Fig. 2) is frequently used, with an adjusting screw, to set the pen to any required small distance. This is usually called the road pen. The wheel or dotting pen (Fig. 3), is used for producing dotted lines of different patterns to distinguish the parts of a geometrical construction, or boundaries, back portions of solids, &c. This pen, when the wheel is removed, forms an excellent broad-pointed pen for drawing thick marginal lines. The best pricking point is a fine needle held in a pair of forceps (Fig. 4). It is used to mark the intersections of lines, or to set off divisions from the plotting scale and protractor (pp. 23 and 38). This point may also be used to prick through a drawing upon an intended copy, or, the needle being reversed, the eye end forms a good tracing point.

Letters used to give a description of a drawing, or for the purpose of reference to a written explanation, may be put in very expeditiously by means of stencil plates. These are thin plates of metal having the letters very neatly cut out from them; by laying them on the paper, and brushing the ink over them with a hard brush, it is painted on the paper through the openings, forming a very elegant typography.

CHAPTER II.

INSTRUMENTS FOR DRAWING STRAIGHT LINES.

A STRAIGHT-EDGE.

ALL instruments used for drawing straight lines must be provided with straight edges, and long flat rulers, intended solely for this purpose, are termed straight-edges. They are made of plain mahogany, pear wood, mahogany edged with ebony, or of the very best of steel. It is important in the first place to test the accuracy of such edges. This may be done by placing two such edges in contact and sliding them along each other, while held up between the eye and the light: if the edges fit close in some parts, so as to exclude the light, but admit it to pass between them at other parts, the edges are not true: if, however, the edges appear, as far as the test has now proceeded, to be true, still this may arise from a curvature in one edge fitting into an opposite curvature in the other; the final step then is to take a third edge, and try it in the same manner with each of the other two, and if in each case the contact be close throughout the whole extent of the edges, then they are all three good.*

"To draw a straight line between two points upon a plane, we lay a rule so that the straight edge thereof may just pass by the two points; then moving a fine-pointed needle, or drawing-pen, along this edge, we draw a line from one point to the other, which, for common purposes, is sufficiently exact; but, where great accuracy is required, it will be found extremely difficult to lay the rule equally with respect to both the points, so as not to be nearer to one point than the other. It is difficult also so to carry the needle, or pen, that it shall

* Euclid bk. i. axiom 10. Peacock's Algebra, 1st edition, art. 532. p. 429.

incline neither towards the rule nor from it; and thirdly, it, is very difficult to find a rule that shall be perfectly straight."*

A right line should never be obtained by the prolongation of a very short one, unless some point in that prolongation is first found by some other means, especially in any essential part of a problem.

All lines, perpendicular or parallel to another, should be drawn long enough at first.

Whenever a line is required to be drawn to a point, in order to insure the coincidence of them, it is better to commence the line from the point; and if the line is to pass through two points, before drawing it the pencil should be moved along the rule, so as to ascertain whether the line will, when drawn, pass through them both. If several radii to a circle are required to pass through any number of points respectively, the lines should be begun from the centre of the circle; any error being more obvious when several lines meet in a point.

"If the two points be very far distant, it is almost impossible to draw the line with accuracy and exactness; a circular line may be described more easily, and more exactly, than a straight or any other line, though even then many difficulties occur, when the circle is required to be of a large radius.

"And let no one consider these reflections as the effect of too scrupulous exactness, or as an unnecessary aim at precision; for, as the foundation of all our knowledge in geography, navigation, and astronomy, is built on observations, and all observations are made with instruments, it follows that the truth of the observations, and the accuracy of the deductions therefrom, will principally depend on the exactness with which the instruments are made and divided, and that those sciences will advance in proportion as these are less difficult in their use, and more perfect in the performance of their respective operations."*

BOW LINE.

This is merely a slip of wood which has a small piece of hard wood fastened upon one side of it at each end, across which a fine wire is drawn; a plain key at the back of the

* Geometrical and Graphical Essays, by George Adams, jun., edited by William Jones.

lath, similar to the key of a violin, will tighten the wire to any degree. It is used when a line considerably longer than the straight-edge is required. To produce a long line with it, marks must be made exactly under the line at distances apart that a straight edge will reach; the bow line is then removed, and the straight edge is used to produce the line.

PARALLEL RULES.

Lines are drawn parallel to each other with the greatest facility and accuracy, either by means of a triangle and straight-edge, or by the T-square; but rolling parallel rules are also much used by professional draughtsmen. They will produce parallel lines in any direction; they move freely and admit of rapid use. The accuracy of the work can also be readily tested by running the rule back to the original line, and again by turning the rule round end for end and repeating the operation.

THE T-SQUARE.

This instrument consists of a blade A, which forms the straight-edge, and the stock B, which guides its motion and has its guiding edge set accurately at right angles to the straight edge of the blade. The following are the two most useful forms of this instrument.

In Fig. (1) the stock is on one side flush with the blade, and on the other projects beyond it, forming a rabbet which passes over the edge of the drawing board, and being gently pressed against it, guides the straight-edge, and enables us to draw lines *c*, *c*, *d*, *d*, throughout the whole extent of the paper, at right angles, to the edges of the drawing board, and consequently parallel and perpendicular to each other.

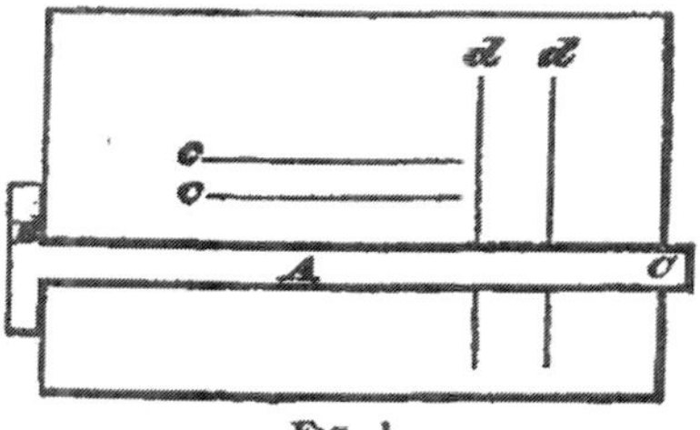

Fig. 1.

In the absence of a *good* drawing board, this T-square may be turned over and worked against a straight-edge clamped to a drawing board or a table.

In Fig. (2) the square is merely a blade screwed upon a stock without being either sunk or mortised into it. This form can be used only with a *good* drawing board; but it allows the set square to pass over the head, so as to be user for drawing lines close up to the edge of the drawing board The instrument makers also supply movable head squares to

Fig. 2.

form bevels for drawing oblique parallel lines. One of these has a graduated arm attached to the blade, for setting the stock at any desired angle with it.

THE TRIANGLE OR SET SQUARE.

For drawing parallel lines of short lengths, and in any directions the best instrument is a triangle worked against a straight edge. They are made with angles of various sizes, but the most useful are those with angles of 45° and 30°,

and the triangle of the Marquois's scales. The angle of 30° enables us at once to draw all the principal lines in an isometrical drawing.

THE CENTROLINEAD.

In the perspective representation of objects, all parallel lines in the object represented, which are not at the same time parallel to the plane of the picture, upon which they are to be represented, must be so drawn as to meet in a point, which is denominated the vanishing point. This point is generally so distant as to require a very long straight-edge and drawing board to produce the lines, and often to render even this mode impracticable, and necessitate a troublesome geometrical construction for each line to be drawn. A convenient instrument for drawing lines which shall all converge to the same point, no matter how distant, is, therefore, a very great desideratum, and in the centrolinead we have an instrument of very simple construction, which perfectly accomplishes this object. It consists of a rule A, with a straight edge, to which a circular head B is so fitted that the prolongation of the straight edge would pass exactly through its centre. To this head two arms, C C, of equal length are

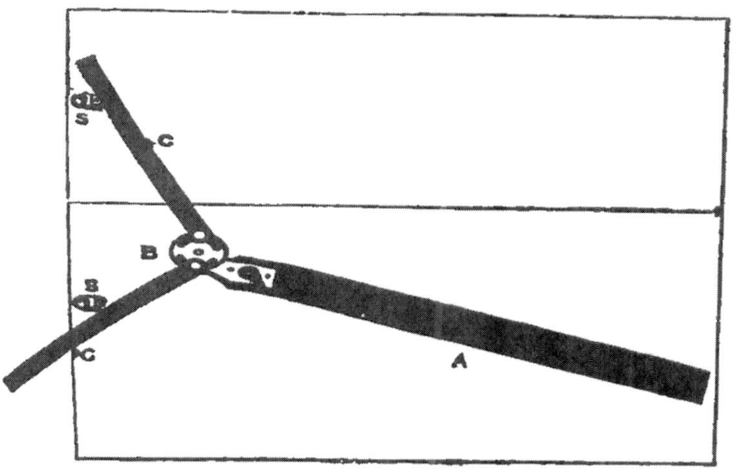

jointed so as to have their working edges, which are the edges farthest from the rule, also each in line with the centre of the head. These arms can be moved into any desired position, and then securely clamped. They work against two studs, S S, which are to be fixed to the drawing board.

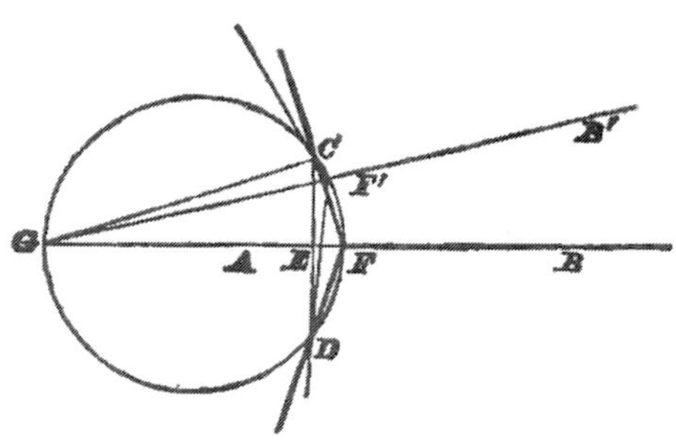

Now if A B and C D be two lines at right angles to each other, intersecting in E, and if the instrument be set so that its straight edge B F may coincide with A B, while the arms lie against the studs placed at the points C and D equally distant from A B, and the arms be then clamped, as the instrument is moved, keeping the arms in contact with the studs, the centre of the head will describe a circular arc* C F D, whose diameter F G will be found by drawing C G at right angles to C F, intersecting B A produced in G,† through which point the prolongation of the straight edge, when moved into any new position, B′ F′, will always pass.

The point E in a picture to which the eye is supposed to be directly opposite is called the centre of the picture; and a line A B drawn through this point parallel to the base of the picture is called *the horizontal line*, and the vanishing points for all horizontal lines lie upon this line or its production.

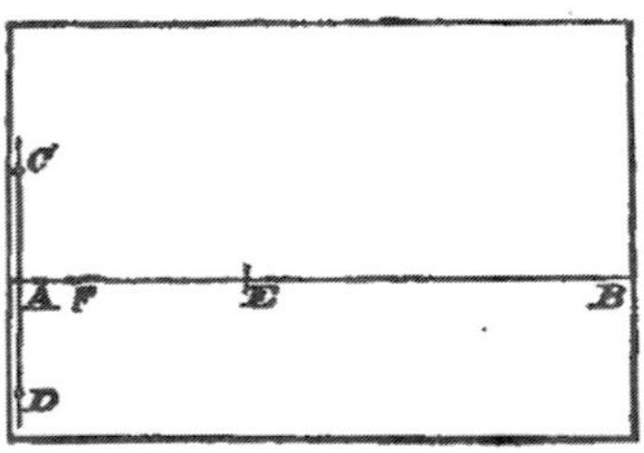

To set the centrolinead so as to draw a series of horizontal lines converging to a given vanishing point. The horizontal line A B must be first drawn, and then a line C A D at right angles to it, as near as possible to the edge of the drawing board on that side beyond which the vanishing point is to lie. Upon C D set off the points C, D, upon which to place the studs, equally distant from A, and as far apart as the dimensions of the instrument and the drawing board will permit, and as will be convenient for the next step. Set off A F the same part of A C or A D that these distances are of the required distance from A of the vanishing point, and the arms being unclamped, place the straight edge in coincidence with the horizontal line A B, so that the centre of the head may coincide with the point F, and bringing the arms to bear against the studs at C and D, clamp them in this position, and the instrument will be set as desired.

Ex.—To make the lines converge to a vanishing point 4 feet distant from A, make the distances A C and A D each 8 inches for the stud points, and 8 inches being the sixth part of 4 feet, set off A F equal to 1⅓ inches, which is the sixth part of 8 inches, for the centre of the head.

A Line to represent one Horizontal Line in a picture being already drawn, to set the instrument so as to draw all other Lines in the picture which are Parallel to it.—Let L M be a

* Euclid iii. 21 † Euclid iii. 31.

line drawn to represent one horizontal line in a drawing, the top of a building suppose; draw C D at right angles to the horizontal line A B, and place the centrolinead so as to have its straight edge coincident with L M, and one of its arms coincident with C D; clamp this arm, and fix the studs, one at C and the other at D, making A D equal to A C; now move the instrument so as to bring the straight edge into coincidence with A B, while the clamped arm bears against the stud at D, and clamping the other arm so as to bear against the stud at C, the instrument will be set as desired.

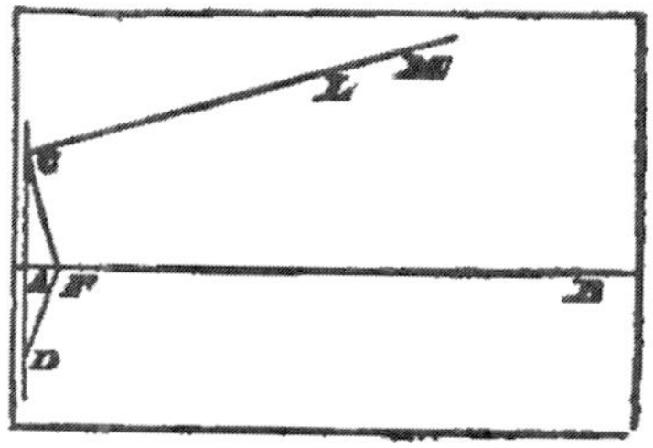

If the prolongation of the line L M to C should make A C inconveniently short, this distance may be multiplied by any quantity, provided A F be multiplied by the square of the same quantity: thus if A C be doubled A F must be four times as great as before, and if A C be trebled A F must be made nine time as great as before, and the instrument will still produce lines converging to the same point.

The instrument might also be set by placing the straight edge in coincidence with A B, and one of the arms with a line at right angles to L M, upon which line the studs would have to be fixed; but this would generally place the studs in an inconvenient position. In a similar manner the instrument may be set to draw any other series of lines to represent oblique parallels, when two lines of the series are determined.

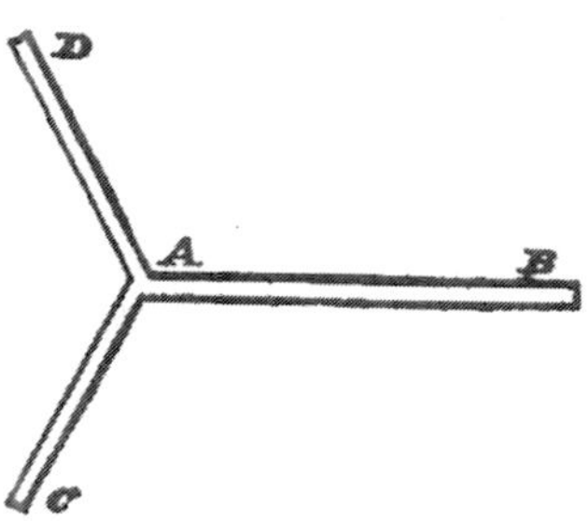

There is a simpler form of centrolinead in which the arms A C, A D, and the straight edge A B are all formed in one piece. The mode of setting this instrument will be seen to be extremely simple; but as the angle of the arms is invariable, the distance of the vanishing point can only be regulated by the distance the pins are placed apart. This distance will for a near vanishing point be inconveniently short, while for a distant vanishing point it will become inconveniently great, and, in fact, may not come within the limits of the drawing board.

Two Lines to represent Parallel Lines being already drawn in a picture, to set the pins for this instrument so as to draw any Number of Lines to represent Lines Parallel to them.—Let F G and H K be two lines in the picture, representing parallels. Bring the straight edge first into coincidence with F G, and draw pencil lines A C, A D, along the guiding edges of the two arms; again bring the straight edge into coincidence with H K, and draw pencil lines A′ C, and A′ D, along the guiding edges of the arms in their new position; then fixing the pins at the points C and D, in which the pencil lines intersect, they will be correctly set for drawing lines to represent parallels to F G and H K.

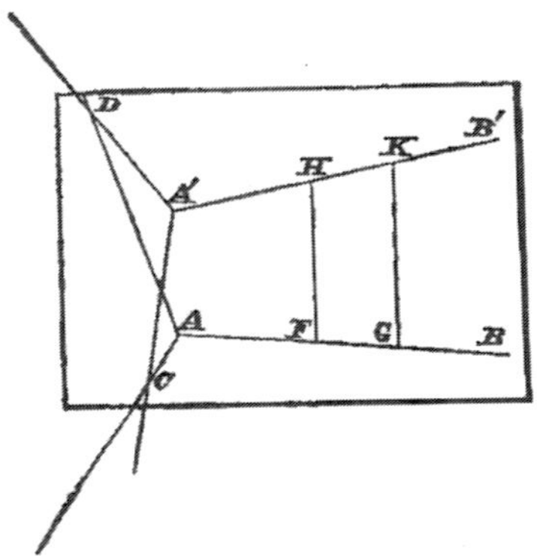

If it be desired to make the lines have a given vanishing point two lines must first be drawn vanishing at the point proposed, and the pins be then set by means of these lines in the manner explained above. The two lines required may be drawn thus:—Draw a line F G, the prolongation of which would pass through the desired vanishing point, and make the distance F G an aliquot part of the desired distance of the vanishing point from F; through F and G draw F H and G K parallel to one another, and of such length that the excess of G K above F H, may be the same aliquot part of F H that F G is of the distance of the vanishing point from F; and joining H K, H K and F G will be such lines as were required.

A ruler similar in appearance to the rolling parallel ruler, but rolling upon wheels at each end, which have different diameters, also forms a simple centrolinead.

CHAPTER III.

INSTRUMENTS FOR DRAWING CURVED LINES.

DRAWING COMPASSES.

This instrument consists of two legs movable about a joint, so that the points at the extremities of the legs may be set at any required distance from one another; it is used to take and transfer distances, and to describe arcs and circles. When used for the former purpose they are termed dividers.

The points of the compasses Figs. (1 and 2) should be formed of well-tempered steel, that cannot be easily bent or blunted, the upper part being formed of brass or silver. The joint is formed of two substances; one side being of the same material as the upper part of the compasses, either brass or silver, and the other of steel. This arrangement diminishes the wear of the parts, and promotes uniformity in their motion. If this uniformity be wanting, it is extremely difficult to set the compasses at any desired distance, for, being opened or closed by the pressure of the finger, if the joint be not good, they will move by fits and starts, and either stop short of, or go beyond the distance required; but, when they move evenly, the pressure may be regulated so as to open the legs to the desired extent, and the joint should be stiff enough to hold them in this position, and not to permit them to deviate from it in consequence of the small amount of pressure which is inseparable from their use.

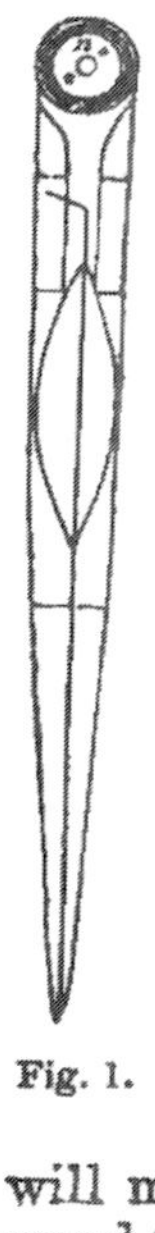

Fig. 1.

Fig. 2.

Fig. 3.

Fig. (3) represents a key of which the two points fit into the two holes seen in the nut *n*, of the joint; and by turning this nut the joint is made stiffer or easier at pleasure.

In the cases of drawing instruments usually sold by the mathematical instrument maker are inserted what are called *hair compasses*, intended by means of a second adjustment to afford greater accuracy in the setting of the points than can be effected by the joint alone; but the precision supposed to be derived from this arrangement is fallacious, while the weakness of the point on which the additional adjustment acts is a positive disadvantage.

COMPASSES WITH MOVABLE POINTS.

If an arc or circle is to be described faintly, merely as a guide for the terminating points of other lines, the steel points are generally sufficient for the purpose, and are susceptible of adjustment with greater accuracy than a pencil point; but, in order to draw arcs or circles with ink or black lead, compasses with a movable point are used. In the best description of these compasses the end of the shank is formed into a strong spring, which holds firmly the movable point, or a pencil or ink point, as may be required. A lengthening bar may also be attached between the shank and the movable point, so as to strike larger circles, and measure greater distances. The movable point to be attached to the lengthening bar, as also the pen point and pencil point, are furnished with a joint, that they may be set nearly perpendicular to the paper.

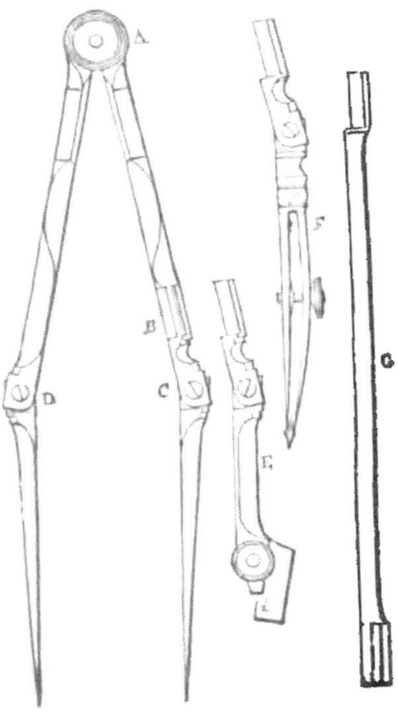

A, the compasses, with a movable point at B.

C and D, the joints to set each point perpendicular to the paper.

E, the pencil point.

F, the pen point. (This is represented with a dotting

wheel, the pen point and the dotting point being similar in shape to each other.)

e, the lengthening bar.

(1.) In using the compasses either as dividers or for drawing arcs or circles, care should be taken so to hold them, that the distance between the points should not be altered by the pressure of the fingers. A good position is given by placing the forefinger upon the head, and the second finger and thumb on the flat sides of the joint. In dividing a line it is better to move the dividers alternately above and below the line from each point of division, than to roll them over continually in one direction; as it saves the shifting of the fingers on the head of the dividers. In taking off distances with dividers it is always better first to open them a little too wide, and afterwards close them to the point required, than to set them by opening.

(2.) Arcs of circles, or right lines by which an important point is to be found, should never intersect each other very obliquely, or at an angle of less than 15 or 20 degrees; and, if this cannot be avoided, some other process should be had recourse to, to define the point more precisely.

(3.) When one arc of a circle is described, and a point in it is to be determined by the intersection of another arc, this latter need not be drawn at all, but only the point marked off on the first, as it is always desirable to avoid the drawing of unnecessary lines. The same observation applies to a point to be determined on one straight line by the intersection of another.

(4.) Whenever the compasses can be used in any part of a construction, or to construct the whole problem, they are to be preferred to the straight edge, unless the process is much more circuitous, or unless the second rule (above) forbids.

(5.) The larger the scale on which any problem, or any part of one, is constructed, the less is the result liable to error; hence all angles should be set off on the largest circles which circumstances will admit of being described, and the largest radius should be taken to describe the arcs by which to find a point through which a right line is to be drawn; and the greater attention is to be paid to this rule, in proportion as that step of the problem under consideration is conducive to the correctness of the final result.

To describe small arcs or circles a small pair of compasses, called *bow compasses*, with a permanent ink or pencil point, are used. They are formed with a round head, which rolls with ease between the fingers. The adjoining figures represent two constructions of pen bows, Fig. 1 being well adapted to describe arcs of not more than one inch radius, and Fig. 2 to describe arcs of small radii with exactness by means of the adjusting screw c.

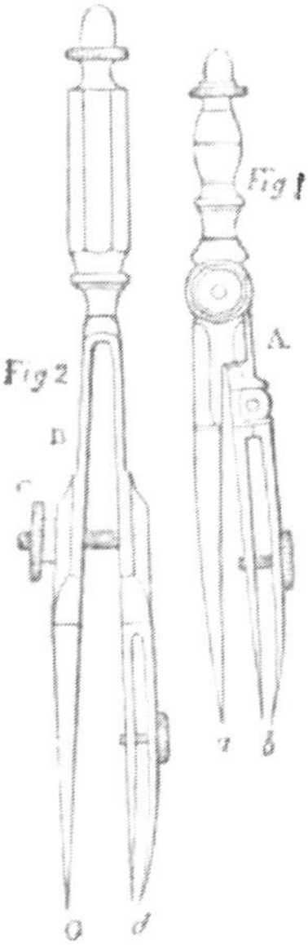

HORN CENTRES.

These are useful when many circles or arcs are to be struck from one centre, as they enable us to do so without injuring the paper beneath them by the continual working of the compass point upon the same spot. They consist merely of a small circular piece of thin horn, with small pins in the under side to prevent them from slipping when placed on the paper.

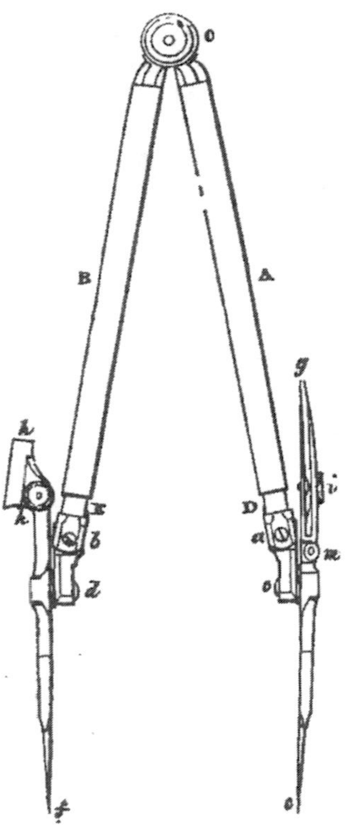

TUBULAR COMPASSES.

When the lengthening bar is used to describe large arcs, or take off long distances, the instrument is wanting in firmness and steadiness, and the object is better accomplished by the use of the tubular compasses, in which the legs consist of two tubes, each containing an inner tube or bar, which slides out like a telescope tube. These compasses, however, are not much used, the beam compasses which we proceed to describe being far preferable for taking off any but comparatively small distances, or describing arcs with any but comparatively small radii.

BEAM COMPASSES.

This instrument consists of a beam A A, of any length required, sometimes as much as six feet, the most usual lengths, however, being two, three, and four feet. Upon its face is sometimes inlaid throughout its whole length a slip of holly or boxwood, upon which are engraved the divisions of a scale.

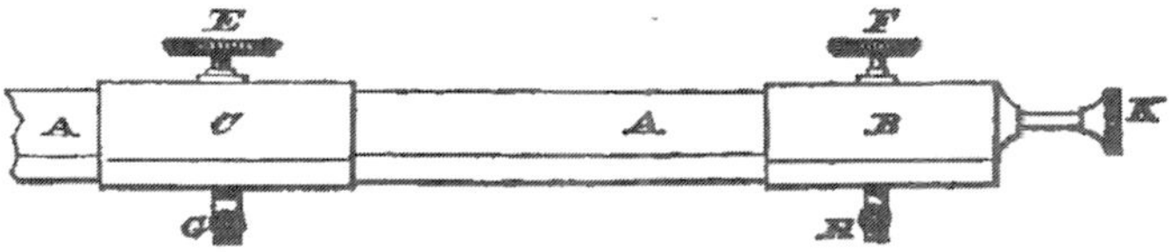

The advantages of such a scale are, however, doubtful for any purpose, since if the points be not perfectly parallel, and accurate in all respects, the distance between them will not accord with that indicated on the scale. For the draughtsman such a scale is actually a disadvantage, as he is apt to make use of it in filling in his work instead of setting his points by scales previously laid down on his paper, so as to expand and contract with it; for it must be borne in mind that paper is liable to expansion and contraction, as the air happens to be moist or dry, and the larger the lines to be laid down, the greater the errors that will arise from setting them off by an independent scale. Two scales should therefore be first laid down upon the paper, parallel to two of its adjacent edges. These scales expand and contract equally with the rest of the paper, according to the hydrometric state of the atmosphere; and the distances set off by their means will therefore always have their proper relative values. In laying down these scales the distances may be set off by applying the points of the compass to a good independent scale. To carry these points the instrument is furnished with two brass boxes, B and C, the first of which is fixed to one end of the beam, while the latter may be moved by sliding to any part of its length, and then fixed in position by turning the clamp screw E. To each of these boxes is attached a tube for holding one of the needle points, or a pen or pencil point as may be desired, and these points are firmly fixed in position by clamping screws, G and H. K is a tangent screw for giving a slow motion to the box B, after it has been clamped by the screw F, and thus producing a more accurate adjustment.

In tubular beam compasses, a series of tubes, as A, B, C, slide one within the other, like the tubes of a telescope, and when pulled out so as to allow the boxes holding the points to be set at the desired distance, are fixed in this position by the clamping screws D and E. One of the boxes, F, is fixed at the

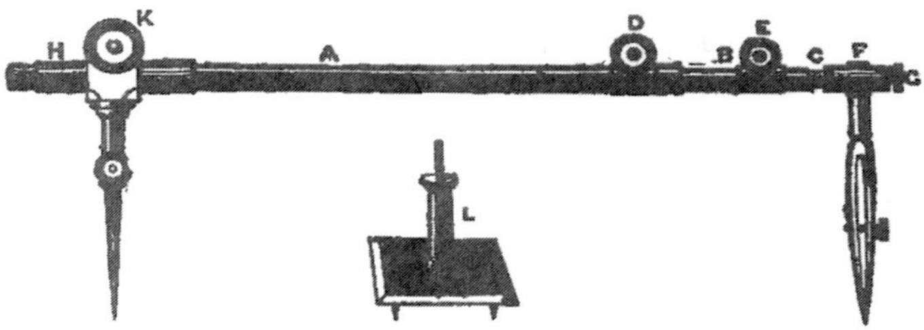

smaller end of the beam, and adjusted accurately by a tangent screw G, while the other box H slides upon the outside tube, which forms the larger end of the beam, and is fixed in any desired position by the clamping screw K. A universal centre is supplied with the instrument, the top of which fits into one of the boxes in place of the point, and thus enables the draughtsman to draw circles without making large holes in the paper. It in fact performs the same duty as the horn centre does for the ordinary compasses.

A very useful, if not the most useful, form of beam com-

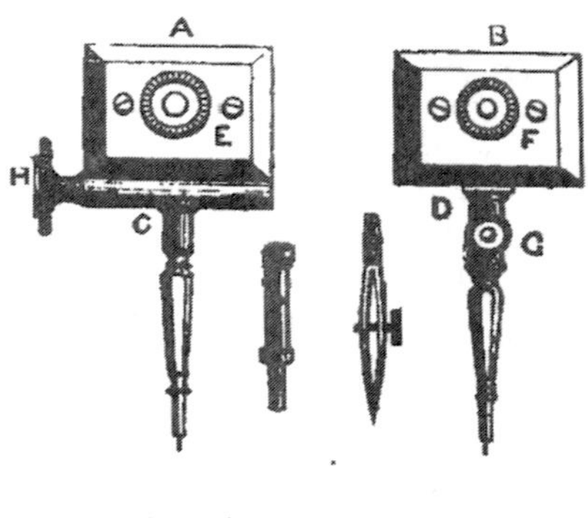

pass, called the portable beam compass, consists merely of boxes A, B, with tubes, C, D, for holding the points, and clamps, E, F, by which they may be fixed upon a lath or brass beam of any length. The tube D of the box B has a clamping screw G, by means of which the point inserted into this tube may be fixed when pushed more or less into the tube, so as to adjust it in accordance with the point inserted in the tube C of the box A. To the box A is also attached a tangent screw H, for giving a small motion to the tube C, and the point it carries, after the clamping of the boxes upon the lath or beam. The beam compasses above described form the most perfect instrument for describing arcs of circles of the largest radii that can ever be required by the draughtsman; but we proceed to explain a manner in which the same object may be accomplished without travelling beyond the limits of the drawing board.

By Euclid iii. 36, the angles in the same segment of a circle are equal to one another, and the property demonstrated in this theorem has already been applied in the construction of the centrolinead, in which the centre of the head describes a circular arc, the centre of which is on the production of the horizontal or other line to which the straight-edge is adjusted for setting the instrument. By an application of the same property we may produce an arc of any radius, or which will pass through any three points. Thus if A, B, C be the three points through which it is desired to describe an arc; join C A and C B and produce them to D and E until C D and C E each exceed A B in length; then cutting a triangle of thin

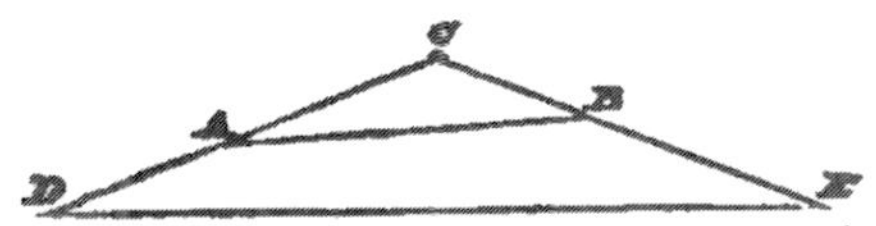

wood or card-board of the same dimensions as D, C, E, and moving it between two pins fixed at A and B, a pencil held at the apex C of the triangle will describe the desired arc. If in cutting the triangle a small circular head be left at the apex, and the point of the pencil be protruded through a pinhole at the vertex C, a very good instrument for the desired purpose will be produced.

Mr. Stanley has attempted to produce an instrument which should form a universal triangle for describing arcs of large radii; and his instrument, which he calls a centograph, does give an approximation to a circular arc; but the ink or pencil point not being placed at the vertex of the angle, the curve described is not really an arc of any circle whatever.

Instead of drawing arcs of large radii upon the paper by the immediate use of the beam compass, templets may first be cut out in wood by means of the compass, and then used as guiding edges for drawing the arcs.

Templets may also be procured from the instrument makers, cut into the form of almost every curve required by the draughtsman either for architectural ornament, or the lines of ship-building.

We proceed to make a few remarks on the mode of describing those of the curves having known geometrical properties, which are the most important to the draughtsman.

THE ELLIPSE.

This curve can be described by what is called the gardener's method. Two pins are fixed at the points A and B, which are called the foci of the ellipse, and a string A C B is attached

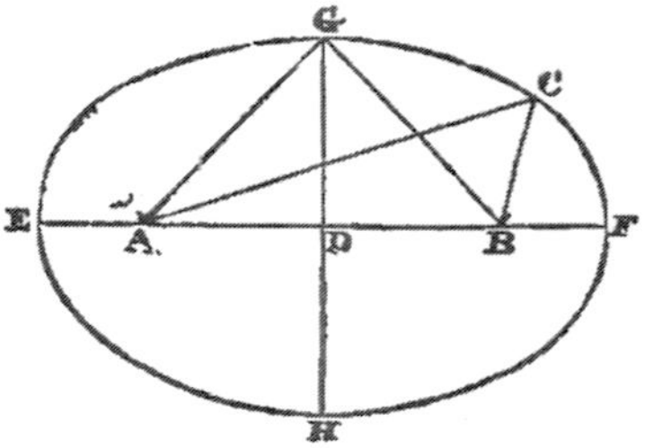

to these pins; a pencil is then moved round, keeping it always pressed against the string so that the sum of the distance A C and C B is constant.

If the lines E F and G H be drawn so as to intersect at right angles and mutually bisect each other in D, E F is called the major axis of the ellipse, G H its minor axis, and D its centre.

To find the foci when the axes are given, we have only to set off from one extremity, G, of the minor axis, the distances G A and G B, each equal the semi-major axis D E or D F, and the points A and B thus found will be the foci required.

The best mode, however, to be employed by the draughtsman for drawing an ellipse is the following:—

The axes E F and G H being first drawn, take a strip of paper with a straight edge M N P, and on it mark off M P and N P equal to the semi-axes D F and D G respectively; then if this strip of paper be moved round into different positions but so that the point M shall always lie on G H, and the point N on E F, the extremity P of the slip will determine a series of points in the ellipse, and the curve must then be drawn through these points. Upon the principle of the mode just explained, instruments called elliptographs have been constructed, in which a rod, filling the office of our piece of paper, has pins attached to it at the points M and N, which slide through grooves in a cross, which represents the axes. There are two constructions of the instrument, one for drawing large ellipses in which the describing point P necessarily passes outside the arms of the cross, and the other, called Farey's Elliptograph, in which the rod is placed beneath the cross between it and the paper, so that the describing point works within the cross, and very small ellipses can be described with it. Such instruments, however, do not work satisfactorily, and we cannot recommend them in preference to the modes above described.

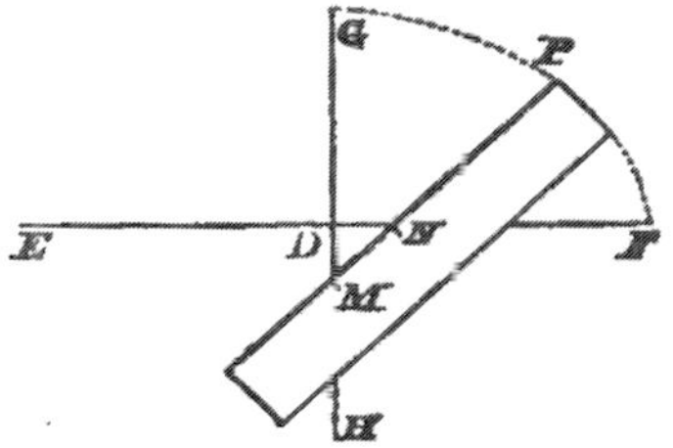

THE HYPERBOLA.

The hyperbola is the curve traced out by a point that moves so that the difference between the focal distances H P and S P is a constant quantity; it may therefore be thus described. Take a straight-edge and fix a string S P B to it at some point B, as remote as may be desired for the portion of the curve to be described; fix the other end of the string to a pin S, which point will form one of the foci of the curve; let a needle point be passed through the rod at H, and placed upon the paper, the point at which it meets the paper will be the other focus of the curve; then a pencil moved along the rod B P, so as to keep the portion S P of the string always stretched straight, and the portion B P in contact with the rod, while the rod itself moves round H, will trace out the required curve.

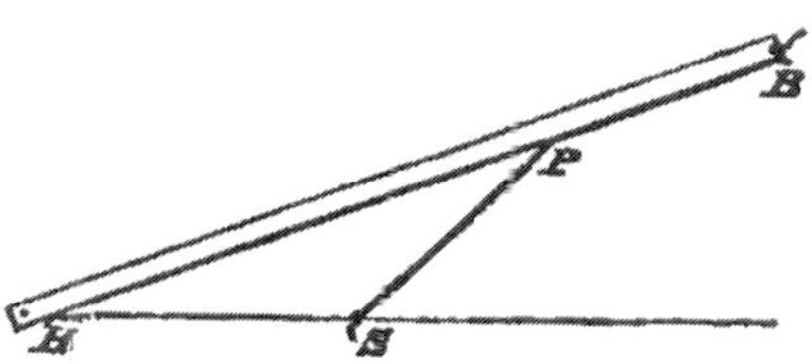

THE PARABOLA.

This curve may be traced from the property it possesses, that every point in it is at equal distances from a fixed point, and a line fixed in position. Let S be the fixed point, which is called the focus, and K K the line fixed in position. Place a straight-edge in coincidence with K K, and clamp it; place a set square A B C against the straight-edge, as represented in the figure. Fix a string S P B to a pin at S and to the point B of the set square, so that the length S P B of the string is equal to the length A B of the edge of the square; then a pencil moved so as to keep the portion S P of the string always straight, and the portion P B close against the edge A B of the set square, while the edge A C of this square is kept against the straight edge K K, will describe the curve required.

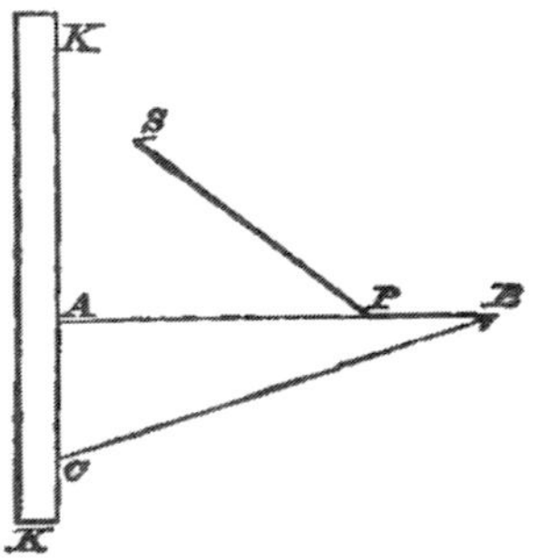

CHAPTER IV.

INSTRUMENTS FOR TAKING OFF AND LAYING DOWN DISTANCES AND ANGLES.

Scales of equal parts are used for measuring straight lines, and laying down distances, each part answering for one foot, one yard, one chain, &c., as may be convenient, and the plan will be larger or smaller as the scale contains a smaller or a greater number of parts in an inch.

Scales of equal parts may be divided into two kinds; simply-divided scales, and diagonal scales.

Simply-divided Scales.—Simply-divided scales consist of any extent of equal divisions, which are numbered 1, 2, 3, &c., beginning from the second division on the left hand. The first of these primary divisions is subdivided into ten equal parts, and from these last divisions the scale is named.

Thus it is called a scale of 30, when thirty of these small parts are equal to one inch. If, then, these subdivisions be taken as units, each to represent one mile, for instance, or one chain, or one foot, &c., the primary divisions will be so many tens of miles, or of chains, or of feet, &c.; if the subdivisions are taken as tens, the primary divisions will be hundreds; and, if the primary divisions be units, the subdivisions will be tenths.

The accompanying drawing represents six simply-divided scales, which are generally placed upon the plain scale (p. 38). To adapt them to feet and inches, the first primary division is divided duodecimally upon an upper line. To lay down 360, or 36, or 3·6, &c., from any one of these scales, extend the compasses from the primary division numbered 3 to the 6th lower sub-division, reckoning backwards, or towards the left hand. To take off any number of feet and inches, 6 feet 7 inches for instance, extend the compasses from the primary division numbered 6, to the 7th upper subdivision, reckoning backwards as before.

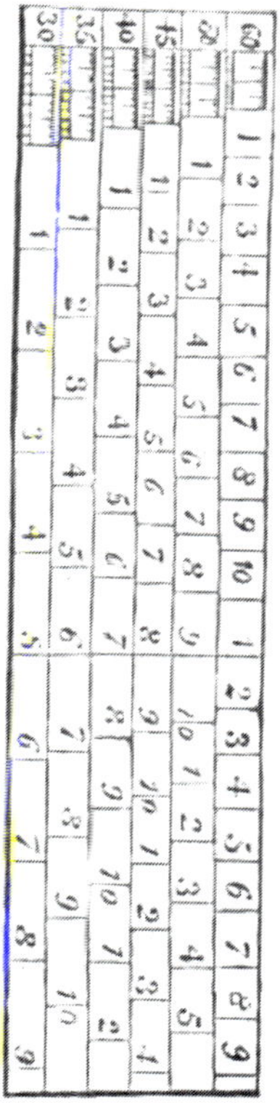

Such simply-divided scales as we have just described are usually placed upon one of the faces of the protractor or plain scale (p. 38) always supplied in the case of instruments; while on the other side are placed scales of inches and parts of inches, as the $\frac{3}{4}$-inch, $\frac{1}{2}$-inch, $\frac{1}{4}$-inch, $\frac{1}{8}$-inch, to represent feet, each of these scales having the first of the primary divisions divided into twelve equal subdivisions to represent inches. These scales have the disadvantage that in using them the divisional marks are liable to injury from the constant application of the points of the dividers.

Plotting scales, also called feather-edged scales, are straight rulers, usually about ten or twelve inches long. Each ruler has scales of equal parts, decimally divided, placed upon its edges, which are made sloping, so that the extremities of the strokes marking the divisions lie close to the paper. The primary divisions represent chains, and the subdivisions, consequently, ten links each, as there are 100 links on the surveying chain. Plotting scales may be procured in sets, each with a different number of chains to the inch

The advantages of this arrangement are, that the distances required can be transferred with great expedition from the scale to the paper by the aid of the pricking-point alone, and the marks denoting the divisions are in no danger of becoming defaced, as upon the plain scale, by the frequent application of the compasses.

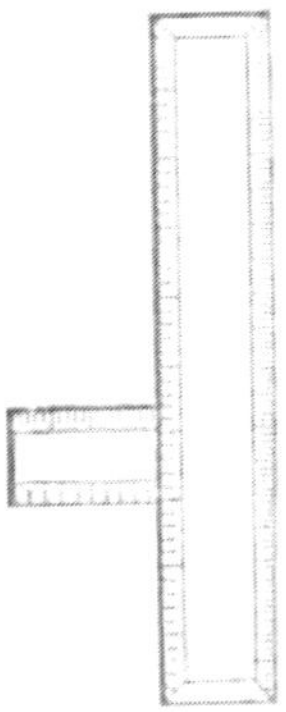

With each plotting scale, a small scale called an offsetting scale is generally supplied, which is worked along the edge of the plotting scale, as represented in the accompanying illustration, to set off distances from the main lines laid down by the plotting scale.

Diagonal Scales.—In the simply-divided scales one of the primary divisions is subdivided only into ten equal parts, and the parts of any distance which are less than tenths of a primary division cannot be accurately taken off from them; but, by means of a diagonal scale, the parts of any distance which are the hundredths of the primary divisions are indicated, as will easily be understood from its construction, which we proceed to describe.

Draw eleven parallel equidistant lines; divide the upper of these lines into equal parts of the intended length of the primary divisions; and through each of these divisions draw perpendicular lines, cutting all the eleven parallels, and number these primary divisions, 1, 2, 3, &c., beginning from the second.

Subdivide the first of these primary divisions into ten equal parts, both upon the highest and lowest of the eleven parallel lines, and let these subdivisions be reckoned in the opposite direction to the primary divisions, as in the simply-divided scales.

Draw the diagonal lines from the tenth subdivision below to the ninth above; from the ninth below to the eighth above; and so on; till we come to a line from the first below to the zero point above. Then, since these diagonal lines are all parallel, and consequently everywhere equidistant, the distance between any two of them in succession, measured upon any of the eleven parallel lines which they intersect, is the same as this distance measured upon the highest or lowest of these lines, that is, as one of the subdivisions before mentioned; but the distance between the perpendicular, which passes through the zero point, and the diagonal through the same point, being nothing on the highest line, and equal to one of the subdivisions on the lowest line, is equal (Euclid vi. prop. 4) to one-tenth of a subdivision on the second line, two-tenths of a subdivision on the third, and so on; so that this, and consequently each of the other diagonal lines, as it reaches each successive parallel, separates further from the perpendicular through the zero point by one-tenth of the extent of a subdivision, or one-hundredth of the extent of a primary division. Our figure represents the two diagonal scales which are usually placed upon the plane scale of six inches in length. In one, the distances between the primary divisions are each half an inch, and in the other a quarter of an inch. The parallel next to the figures numbering these divisions must be considered the highest or first parallel in each of these scales to accord with the above description.

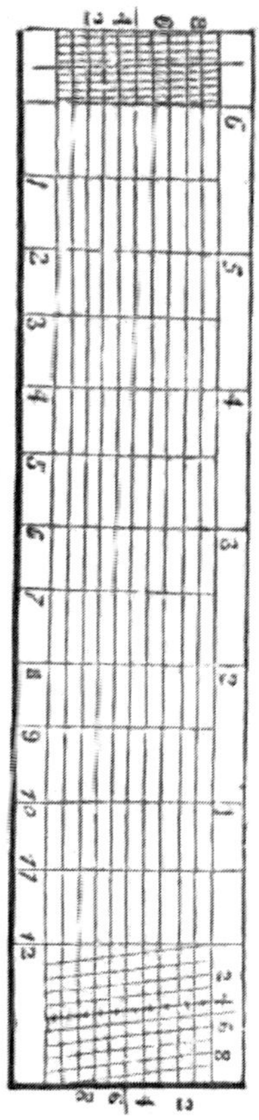

The primary divisions being taken for units, to set off the numbers 5·74 by the diagonal scale. Set one foot of the compasses on the point where the fifth parallel cuts the eighth diagonal line, and extend the other foot to the point where the same parallel cuts the sixth vertical line.

The primary divisions being reckoned as tens, to take off the number 46·7. Extend the compasses from the point where the eighth parallel cuts the seventh diagonal to the point where it cuts the fifth vertical.

The primary divisions being hundreds, to take off the number 253. Extend the compasses from the point where the fourth parallel cuts the sixth diagonal to the point where it cuts the third vertical.

Now, since the first of the parallels, of the diagonals, and of the verticals, indicate the zero points for the third, second, and first figures respectively, the second of each of them stands for, and is marked, 1, the third 2, and so on, and we have the following.

General Rule.—To take off any Number to Three Places of figures upon a Diagonal Scale.—On the parallel indicated by the third figure, measure from the diagonal indicated by the second figure to the vertical indicated by the first.

ON THE PROTRACTING SCALES.

The nature of these scales will be understood from the following construction (Plate 1, Fig. 1):—

With centre O, and radius O A, describe the circle A B C D; and through the centre O draw the diameters A C, and B D, at right angles to each other, which will divide the circle into four quadrants, A B, B C, C D, and D A.

Divide the quadrant C D into nine equal parts, each of which will contain ten degrees, and these parts may again be subdivided into degrees, and, if the circle be sufficiently large, into minutes.

Set one foot of the compasses upon C, and transfer the divisions in the quadrant C D, to the right line C D, and we shall have a scale of chords.*

From the divisions in the quadrant C D draw right lines parallel to D O, to cut the radius O C, and, numbering the divisions from O, towards C, we shall have a scale of sines.

If the same divisions be numbered from C, and continued to A, we shall have a scale of versed sines.

From the centre O draw right lines through the divisions of the quadrant C D, to meet the line C T, touching the circle at C, and, numbering from C, towards T, we shall have a scale of tangents.

Set one foot of the compasses upon the centre O, and transfer

* We give the constructions in the text to show the nature of the scales; but in practice a scale of chords is most accurately constructed by values computed from tabulated arithmetical values of sines, which computed values are set off from a scale of equal parts; and the circle is divided most accurately by means of such computed chords. The limits of our work forbid our entering further upon this interesting subject. All the other scales will also be most accurately constructed from computed arithmetical values, taken off by means of the beam compasses from an accurately constructed scale of equal parts.

the divisions in C T into the right line O S, and we shall have a scale of secants.

Right lines, drawn from A to the several divisions in the quadrant C D, will divide the radius O D into a line of semi-tangents, or tangents of half the angles indicated by the numbers; and the scale may be continued by continuing the divisions from the quadrant C D through the quadrant D A, and drawing right lines from A, through these divisions, to meet the radius O D, produced.

Divide the quadrant A D into eight equal parts, subdivide each of these into four equal parts, and, setting one foot of the compasses upon A, transfer these divisions to the right line A D, and we shall have a scale of rhumbs.

Divide the radius A O into 60 equal parts, and number them from O towards A; through these divisions draw right lines parallel to the radius OB, to meet the quadrant A B; and, with one foot of the compasses upon A, transfer these divisions from the quadrant to the line A B, and we shall have a scale of longitudes.

*Place the chord of 60°, or radius,† between the radii O C and OB, meeting them at equal distances from the centre; divide the quadrant C B into six equal parts, for intervals of hours, subdividing each of these parts into twelve for intervals of five minutes, and further subdividing for single minutes if the circle be large enough; and from the centre O draw right lines to the divisions and subdivisions of the quadrant, and we shall have a scale of hours.

Prolong the touching line T C to L; set off the scale of sines from C to L; draw right lines from the centre O to the divisions upon C L, and from the intersections of these lines with the quadrant C B draw right lines parallel to the radius O C, to meet the radius O B, and we shall have a scale of latitudes.‡

* The constructions in this and the following paragraphs are not given in the figure, being superseded by the constructions given on the next page.

† Chord of 60° is equal to radius. Euclid iv. 15, Cor.

‡ The line of latitudes is a line of sines (to radius equal the whole length of the line of hours) of the angles, of which the tangents are equal to the sines of the latitudes. The middle of the hour line being numbered for three o'clock, the divisions for the other hours are found by setting off both ways from the middle the tangents of n. 15°, n. being the number of hours from three o'clock, that is, one for two o'clock and four o'clock, two for one o'clock and five o'clock, and three for twelve o'clock and six o'clock.

Corresponding lines of hours and latitudes may also be constructed (as represented in our figure) more simply, and on a scale twice as large as by the preceding method, as follows:—

With the chord of 45° set off from B to E, and again from B to F, we obtain a quadrant E F bisected in B; and, the chord of 60° or radius being set off from A, C, F, and E, this quadrant is divided into six equal parts. From the centre O draw straight lines through these divisions to meet the line touching the circle at B, and we shall have the line of hours.

From the point D draw right lines through the divisions upon the line of sines O C, to meet the circumference B C, and transferring these divisions from B, as a centre to the chord B C, we shall have the corresponding line of latitudes.

It is not necessary that these scales should all be projected to the same radius; but those which are used together, as the rhumbs and chords, the chords and longitudes, the sines, tangents, secants, and semitangents, and lastly, the hours and latitudes, must be so constructed necessarily. In the accompanying diagram (Plate 1, Fig. 2) we have laid down the hours and latitudes to a radius equal to the whole length of the scale, the other lines being laid down to the radius used in the foregoing construction.

The Line of Chords is used to set off an angle, or to measure an angle already laid down.

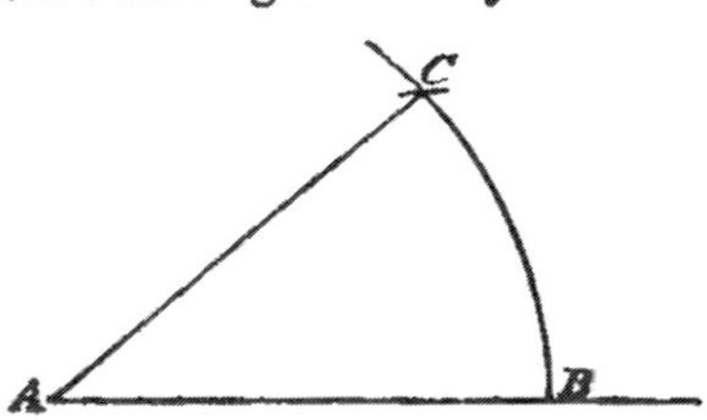

1st. *To set off an angle,* which shall contain D° from the point A, in the straight line A B. Open the compasses to the extent of 60° upon the line of chords, which equals the radius to which this line has been laid down (Euclid iv. prop. 15, Cor.), and, setting one foot upon A, with this extent describe an arc cutting A B in B; then, taking the extent of D° from the same line of chords, set it off from B to C; and, joining A C, B A C is the angle required. Thus, to set off an angle of 41°, having described the arc B C, as directed, with one foot of the compasses on B, and the extent of 41° on the line of chords, intersect B C in C, and join A C.

2nd. *To measure the angle contained by the straight lines* A B *and* A C *already laid down.*—Open the compasses to the extent of 60° on the line of chords, as before, and with this radius describe the arc B C, cutting A B and A C, produced, if necessary, in the points B and C; then, extending the compasses from B to C, place one point of the compasses on the beginning, or zero point, of the line of chords, and the other point will extend to the number upon this line, indicating the degrees in the angle B A C. If, for instance, this point fall on the 41st division, or the first division beyond that marked 40 in the figure (Plate 1, Fig. 2), the angle B A C will contain 41°.

The Line of Rhumbs is a scale of the chords of the angles of deviation from the meridian denoted by the several points and quarter points of the compass, enabling the navigator, without computation, to lay down or measure a ship's course upon a chart. Thus, supposing the ship's course to be N.N.E. ¾ E. Through the point A, representing the ship's place upon the chart, draw the meridian A B, and with centre A and distance equal to the extent of 60° upon the line of chords describe an arc cutting A B in B; then on the line of rhumbs take the extent to the third subdivision beyond the division marked 2, because N.N.E. is the second point of the compass from the north, and with one foot of the compasses on B describe an arc intersecting B C in C: join A C, and the angle B A C will represent the ship's course. On the other hand, if a ship is to be sailed from the point A to a point on the line A C on a chart, draw the meridian A B, describe the arc B C with radius equal to chord of 60°, as before, and the extent from B to C, applied to the line of rhumbs, will give 2 pts. 3 qrs., denoting that the ship must be sailed by the compass N.N.E. ¾ E.

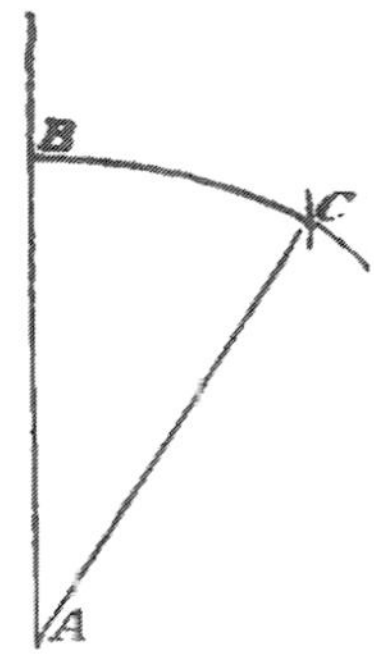

The Line of Longitudes shows the number of equatorial miles in a degree of longitude on the parallels of latitude indicated by the degrees on the corresponding points of the line of chords. *Ex.*—A ship in latitude 60° N. sailing E. 79 miles, required the difference of longitude between the beginning and end of her course. Opposite 60 on the line of chords stands 30 on the line of longitudes, which is, therefore, the number of equatorial miles in a degree of longitude at that latitude. Hence, as 30° : 79 : : 60 : 158 miles, the required difference of longitude.

The Lines of Sines, Secants, Tangents and Semitangents are principally used for the several projections, or perspective representations, of the circles of the sphere, by means of which maps are constructed. Thus, the meridians and parallels of latitude being projected, the countries intended to be represented are traced out according to their respective situations and extent, the position of every point being determined by the intersection of its given meridian and parallel of latitude.

The plane upon which the circles are to be delineated is called the primitive, and the circumference of a circle, described with a radius, representing, upon the reduced scale of the drawing, the radius of the sphere, is called the circumference of the primitive. Lines, drawn from all the points of the circles to the eye, by their intersection with the primitive form the projection.

When the eye is supposed to be infinitely distant, so that the lines of vision are parallel to one another and perpendicular to the primitive, the projection is called orthographic. When the primitive is a tangent plane to the sphere, and the eye is supposed to be at the centre of the sphere, the projection is called gnomonic. When the eye is supposed to be at the surface of the sphere, and the primitive to pass through the centre, so as to have the eye in its pole, the projection is called stereographic.

The projection is further termed equatorial, meridional, or horizontal, according as the primitive coincides with, or is parallel to, the equator, or the meridian or horizon of any place.

To delineate the Orthographic Projection of the Circles of the Terrestrial Sphere upon the Plane of the Meridian of any Place.—With a radius according to the contemplated scale of the projection, describe the circle W N E S for the circumference of the primitive, and draw the vertical and horizontal diameters N S and W E, which will be the projections of a meridian perpendicular to the primitive, and of the equator, respectively. Take out from the line of sines the sines of the latitudes through which the parallels are to be drawn, and, reducing these sines to the radius of the primitive,*

* If the proportional compasses, described hereafter (page 60), be set in the proportion of the sine 90° on the line of sines to the radius

set off these reduced distances both ways from the centre upon the line N S; and also both ways from the centre upon the line W E, for the sines of the angles which the meridians, to be drawn at the same intervals as the parallels, make with the meridian N S. Through the divisions thus set off, upon the line N S draw straight lines parallel to W E, and such straight lines will be the projections of the several parallels of latitude, which are to be numbered 0 to 90, from the equator to either pole for the latitudes. With distances from the centre to the divisions set off upon W E as semi-minor axes, and the distance from C to N or S, equal to radius of primitive, as a common semi-major axis, describe semi-ellipses, and they will be the projections of the several meridians, which are to be numbered either way from the first meridian for the longitudes. In the figure the primitive coincides with the plane of the meridian of a place in 30° west longitude, or 150° east longitude, the sum of these two being 180°, as must always be the case.

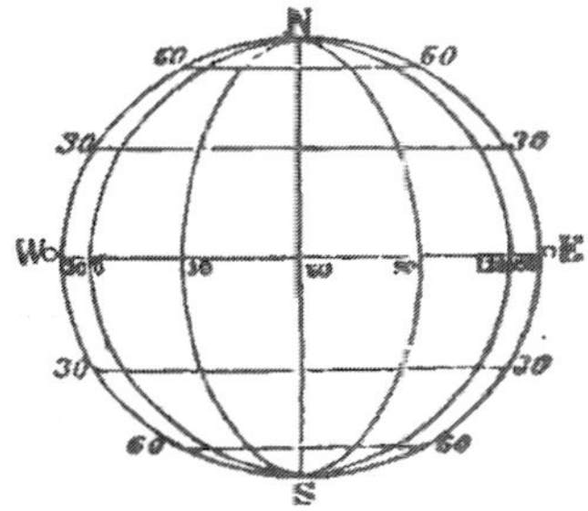

To delineate the Gnomonic Projection of the Circles of the Terrestrial Sphere upon a Plane parallel to the Equator.—In this case the meridians will all be projected into straight lines, making the same angles one with another that their originals do on the surface of the sphere; the projection of the pole will be the centre of the primitive, and the projections of the parallels of latitude will be circles described from the projection of the pole, as centre, with distances equal to the tangents of the respective colatitudes reduced to the radius of the primitive. The parallel of 45° will, therefore, coincide with the circumference of the primitive; the parallels of latitudes greater than 45° will lie within the primitive; and for latitudes less than 45° the parallels will fall without the primitive, the radii of their projections increasing as the latitude decreases, until the

of the primitive, one pair of points will give, reduced to this radius, the sines taken off by the other pair of points. The manner of taking from the sector a sine to any radius will also be hereafter pointed out (page 47).

radius for projecting the equator becomes infinite. Describe, then, a circle for the primitive; draw straight lines radiating from its centre, and equally inclined one to another for the projections of equidistant meridians; and number them 0 to 180 both ways from the first meridian for the longitudes. With the tangents of the colatitudes, taken at intervals equal to the angle between two successive meridians, and reduced to the radius of the primitive, as distances, describe from the centre of the primitive concentric circles; and number them 90 to 45 from the pole to the primitive for the latitudes, continuing the graduation beyond for the lower latitudes.

The gnomic projection affords a good representation of the polar regions, but all places in latitudes lower than 60° appear

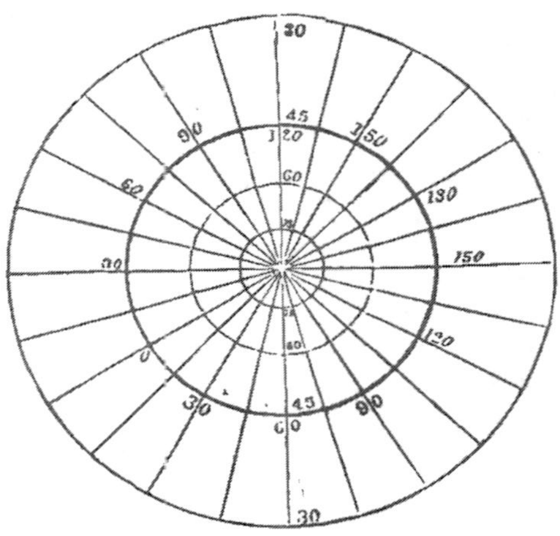

greatly distorted. The gnomonic projection enlarges the representations of places at a distance from the centre of projection.

To delineate the Stereographic Projection of the Circles of the Terrestrial Sphere upon the Horizon of any Place.—With radius determined upon describe a circle for the primitive, and draw its vertical and horizontal diameters, N S and W E, which will be the projections of the meridian of the place and of the prime vertical respectively. From the centre C set off upon the radius C S, produced, if necessary, the distance C A, equal to the tangent of the latitude of the place reduced to the radius of the primitive; and with centre A and distance A W or A E describe the circle W N E, which will be the projection of the meridian at right angles to N S, the meridian of the place; and, consequently, N will be the projection of the pole. Through A draw the right line A B at right angles to

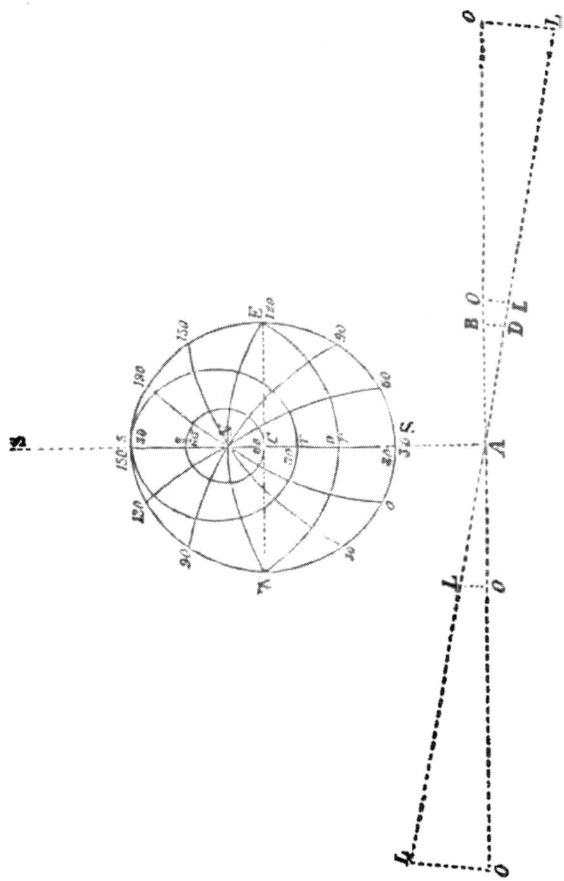

A C, and another line A D making any convenient angle with A B, and, setting off A B equal to the radius of the primitive, and A D equal to the sine of the colatitude, taken from the line of sines, join B D. Now take from the line of tangents the angles which the other meridians to be drawn are

to make with the meridian W N E, or the complements of the angles which they are to make with N S, and set them off both ways from A upon the line A D; through each of the divisions L, thus found, draw L O, parallels to B D, and we have at O the centres of the circles for describing the meridians.* With centres O and distances O N, describe the meridians, and number them 0 to 180, both ways, from the first meridian, for the longitudes. For a parallel through any given latitude, take the difference of the complement of the given latitude and of the colatitude of the place from the line of semitangents, and having reduced it to the radius of the primitive, set it off at *r* from C towards N for latitudes greater than the latitude of the place, and from C towards S for latitudes less than the latitude of the place:—again, take the sum of the complement of the given latitude and of the colatitude of the place from the line of semitangents, and set it off at *s* from C upon C N produced: then the circle described upon *r s* † as diameter will be the parallel required. Draw these parallels for intervals of latitude equal to the angles made by two successive meridians, and number them 90 to 0 from the pole N for the north latitudes, and again increasing from 0 on the other side of the equator for the south latitudes, if the place be in north latitude—or the converse if the place be in south latitude.

The practical application of the preceding methods of projection is usually confined to the representation of an entire hemisphere, or at least of a considerable portion of a sphere; but for laying down smaller portions of the sphere the method of development may be advantageously adopted. In this method the portion of the sphere to be represented is considered as coincident with a portion of a cone, touching the sphere in a circle which is the middle parallel of latitude of the country to be represented, and this portion of the cone when developed forms a portion of a sector of a circle.

To lay down the meridians and parallels of latitude for this development. 1. Take a straight line, B C A, for the middle meridian of the intended map, and divide it into equal parts, to represent degrees and minutes of latitude according to the

* The distance $A\,O = \dfrac{r \cot. L}{\cos. l}$ where r represents the radius of the primitive, l the latitude of the place, and L the angle at which the meridian is inclined to the meridian of the place.

† Diameter of parallel $= r \tan. \frac{1}{2}(c - \delta) + r \tan. \frac{1}{2}(c + \delta)$ where c = colatitude of place, and δ = colatitude of parallel.

scale determined upon for the map. 2. From one of these divisions, A, which is conveniently situated to form the centre of the map, set off from A to C the cotangent of the middle latitude, reduced to a radius equal to 57·3 of the divisions previously marked off as degrees, or to 3,438 of those marked off as minutes. 3. With C as a centre and radius C A, describe the arc D A E for the middle parallel of latitude, and divide it into equal parts to represent degrees and minutes of longitude, the lengths of these parts having, to the lengths of the parts previously set off on the meridian for degrees and minutes of latitude, the ratio cosine of middle latitude : radius. 4. With C as centre, describe concentric arcs, through the divisions on C A, for the parallels of latitude; and draw straight lines, radiating from C, through the divisions on D A E for the meridians.

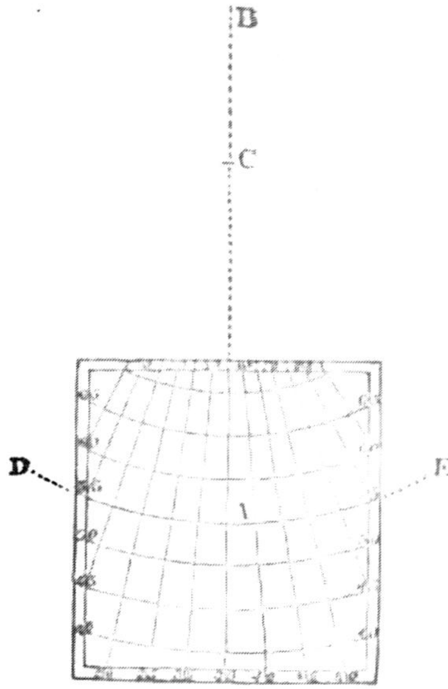

In our figure the middle latitude is 55°; A B is equal to the length of 57·3°, or the radius of the sphere; A C is equal to the cotangent of 55, or the tangent of 35 reduced to this radius; and C, consequently, is the centre for describing the parallels, and the radiating point for the meridians.

In drawing a map of small extent, it is usual to make all the meridians and parallels of latitude straight lines; and to make the extreme parallels, and the meridian passing through the centre of the map, proportional to their real magnitude.

Another and more exact method is to make the meridian passing through the centre of the map, and all the parallels of latitude, straight lines, as in the last method. Then all the degrees on each of the parallels are made proportional to their magnitude, and the lines passing through the corresponding points of division on the parallels will represent the meridians. These will be curved lines, and not straight, as in the last method. This is usually called *Flamstead's* Projection, as it was first used by that astronomer in constructing his "Celestial Atlas;" and it is extremely useful in geographical maps for countries lying on both sides of the equator.

A considerable improvement of this method, for countries of large extent, is to represent all the parallels of latitude by concentric circles, according to the principles of the conical development; and then to lay off the degrees on each parallel, proportional to their magnitude,* and draw lines through the corresponding divisions of these parallels to represent the meridians. This delineation, perhaps, will give the different parts of a map of some extent in as nearly their due proportions as the nature of the case will admit.

We will now briefly explain the manner of constructing some of the simplest dials by means of the dialling scales.

To construct a Horizontal Dial.—Draw on your dial plate two parallel lines, *a b*, *c d*, as a double meridian line, at a distance apart equal to the thickness of the intended style, or

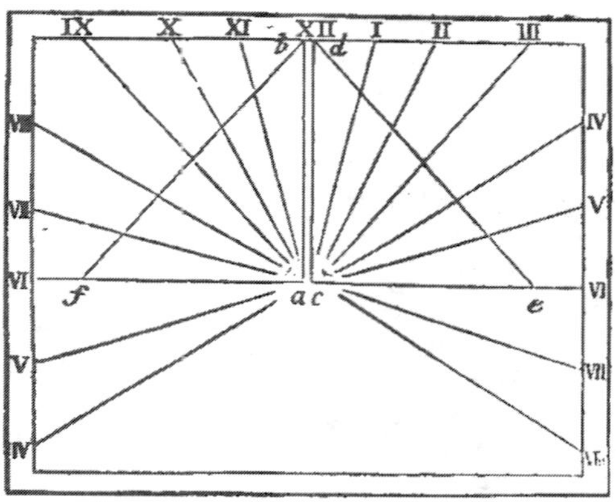

gnomon. Intersect them at right angles by another line, *e f*, called the six-o'clock line. From the scale of latitudes take the latitude of the place with the compasses, and set that extent from *c* to *e* and from *a* to *f* on the six-o'clock line, and then, taking the whole of six hours between the points of the compasses from the scale, with this extent set one foot in the point *e*, and with the other intersect the meridian line *c d* at *d*. Do the same from *f* to *b*, and draw the right lines *e d* and *f b*, which are of the same lengths as the scale of hours. Place one foot of the compasses on the beginning of the scale,

* That is, the degrees on each parallel must have to a degree of latitude the ratio of radius : cosine of the latitude of the parallel.

and, extending the other to any hour on the scale, lay these extents off from *d* to *e* for the afternoon hours, and from *b* to *f* for the forenoon. In the same manner the quarters or minutes may be laid down, if required. The edge of a ruler being now placed on the point *c*, draw the first five afternoon hours from that point through the marks on the line *d e*, and continue the lines of IV and V through the centre *c* to the other side of the dial for the like hours of the morning. Lastly, lay a ruler on the point *a*, and draw the last five forenoon hours through the marks on the line *f b*, continuing the hour lines of VII and VIII through the centre *a* to the outer side of the dial for the evening hours, and figure the hours to the respective lines.

To make the Gnomon.—From the line of chords, always placed on the same dialling scale, take the extent of 60°, and describe from the centre *a* the arc *g n*. Then with the extent of the latitude of the place, suppose London, 51½°, taken from the same line of chords, set one foot in *n*, and cross the arc with the other at *g*. From the centre at *a* draw the line *a g* for the axis of the gnomon *a g i*, and from *g* let fall the perpendicular *g i* upon the horizontal meridian line *a n*, and there will be formed a triangle *a g i*. A plate or triangular frame similar to this triangle, and of the thickness of the interval of the parallel lines *a c* and *b d*, being now made and set upright between them, touching at *a* and *b*, its hypothenuse or axis *a g* will be parallel to the axis of the earth when the dial is fixed truly, and will cast its shadow on the hour of the day.

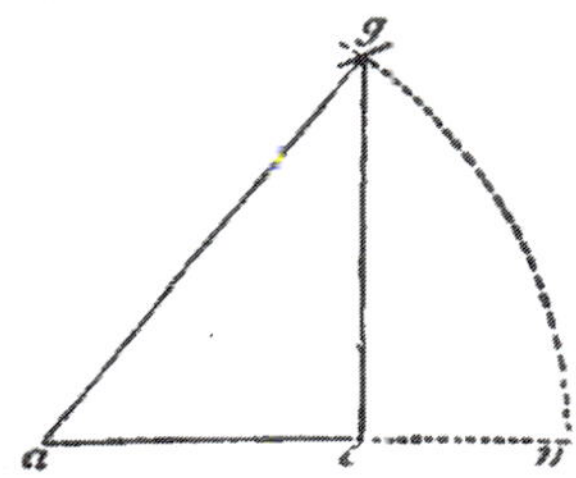

To make an erect South Dial.—Take the complement of the latitude of the place, which for London is 90° less 51½ = 38½, from the scale of latitudes, and proceed in all other respects for the hour lines as above for the horizontal dial; only reversing the hours, and limiting them to the VII, and for the gnomon making the angle of the style's height equal to the colatitude 38½.

To construct an East or West Dial.—Draw the two meridian lines as before, and intersect them at right angles by another line, upon which set off, from the meridian lines, the tangents of 15°, 30°, 45°, &c., for every 15°, reduced to a radius equal to the intended height of the style. The hour lines are to be drawn through the divisions thus marked, parallel to the meridian lines, and the meridian lines themselves are six-o'clock hour lines. The gnomon is a plate in the form of a parallelogram, the breadth of which forms the height of the style or gnomon, and must be equal to the radius to which the tangents have been set off on the dial plate. It is set up between the meridian lines, perpendicular to the dial plate; and the dial is set up so that the meridian lines, and consequently the edge of the gnomon, may *be parallel* to the earth's axis. As the sun only shines on the dial during half the day, if the dial fronts the east, it points out the time from sunrise to noon, or, if the dial fronts the west, from noon to night.

THE PROTRACTOR, OR PLAIN SCALE.

One of these instruments is represented in the annexed figure, being such a one as is usually supplied with a pocket case of instruments. It is made of ivory, six inches long, and one inch and three-quarters broad. On the face of the instrument represented in the engraving, a protractor is formed round three of its edges for readily setting off angles. In using this protractor, the fourth edge or base is to be made to coincide with the line from which the angle is to be set off, and the stroke in the middle of the base with the point in this line at which the angle is to be set off; a mark is then to be made with the pricking point at the point of the paper which coincides with the stroke on the protractor marked with the number of degrees in the angle required to be drawn; and, the protractor being now removed, a straight line is to be

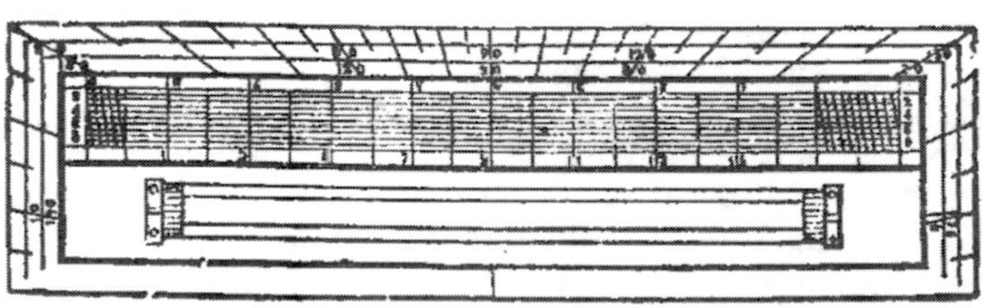

drawn through the given point in the given line and the point thus pricked off. The instrument has on the same face the two diagonal scales already described (p. 24), and on the opposite face scales of equal parts and several of the protracting scales already described (pp. 26—30), according to the purposes to which the scale is to be applied. The plain scale is sometimes fitted with roller, as represented in our engraving, making it at the same time a convenient small parallel ruler.

THE MILITARY PROTRACTOR.

The distinguishing feature of this instrument is a series of equidistant parallel lines across it, forming a scale of equal parts. They are ruled perpendicular to the longer sides of the instrument, the zero of the scale leading from the centre of the base to 90° on the protractor. These lines present a peculiar advantage for making sketches of field-work, the readings of which have been taken with the prismatic compass, as they may be made to represent lines perpendicular to the magnetic meridian, or magnetic east and west lines. The paper on which the sketch is to be made is also ruled with a series of lines parallel to each other, but at unequal distances, which may also be taken to represent lines from east to west, so that the protractor can be laid upon it, and some one of the lines ruled across it be made to coincide with one of the ruled lines upon the paper, at the same time that the zero of the scale is kept in coincidence with any desired point, and, the base of the scale when in this position coinciding with a magnetic meridian, the angular bearings of other points from this one, as taken by the compass, can be at once pricked off.

For the purpose of setting off the angles taken with the prismatic compass, it is a matter of indifference what scale the equidistant parallel lines may represent, or at what distances the lines may be ruled on the paper, provided that these be so arranged, that some one of the lines on the scale can always be made coincident with one of the lines on the paper, at whatever point on the latter the zero of the scale may be placed. By making, however, the scale such a one as has been found convenient for military sketches, it may be used for pricking off the distances at once, as well as for setting of the angles. The scale selected, therefore, is 4 inches to a mile; and 4 inches are divided into 176 parts, or each inch into 44 parts: the parallel lines are then ruled through each division, the zero line and every tenth line on either side of it being red, and the intermediate lines black; thus the distance between any two consecutive red lines represents 100 yards, and that between each two consecutive black lines represents 10 yards. The lines on the paper are ruled at distances varying from one-quarter to three-eighths of an inch.

Fig. 1 represents the instrument, and the method of using it explained above: thus, if A represents a point on the plan, and it is required to lay down three other points, B, C, D, on

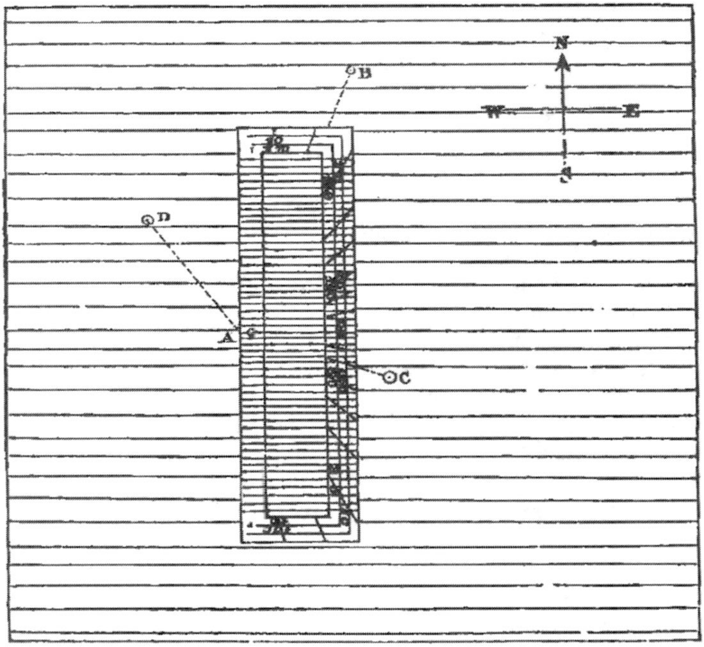

Fig. 1.

it, whose distances from A are 1,600, 800, and 970 yards, and their bearings, as taken by the prismatic compass, N. 25° E., N. 110° E., and N. 50° W., lay the instrument with the zero of the scale upon the point A, and turn it round until the lines on the scale and on the paper are all parallel, a position which can be well determined, since some one of the lines on the paper is certain to come sensibly into coincidence with a line on the scale. The base will now represent a magnetic meridian, and pricking off the angles 25°, and 110° from the north end of the protractor, lines through A and the points thus determined would bear from A, N. 25° E., and N. 110° E. respectively, and would pass through B and C. To prick off the third bearing without changing the position of the instrument, it is only necessary to consider that the same line which points N. 50° W. at one end, points S. 50° E. at the other; prick off, therefore, 50° from the south end of protector, and the production from A towards the left of the line passing through A, and the point thus determined would bear N. 50° W., and pass through D. Now lay the instrument so that its base coincides with the direction A B, above determined, and the zero of the scale coincides with A, and the extremity of the sixteenth red line will denote the position of B, and in like manner the positions of C and D will be denoted by the extremities of the eighth red line, and of the seventh black line beyond the ninth red line respectively, when the instrument is arranged in the proper directions with the point A at the zero of the scale.

Fig. 2 represents a mode of setting off the angles taken by the prismatic compass either with the ordinary or the military protractor; and when there is a scale marked on the bevelled edge of the base, the distances can be pricked off at the same time as the angles, and a sketch be very rapidly made. In this method the lines on the paper are taken as magnetic meridians, and the index at the centre of the base and the division on the protracting edge of the instrument denoting the angular bearing are both made coincident with one of these lines, and the instrument is slipped up or down, preserving this coincidence until the base passes exactly through the starting point. The direction of the base then gives the proper bearing, and the distance being set off along the base, the position of the second point is determined; thus, to lay down on the sketch the point B 1,000 yards from the point A already laid down, and bearing from it N. 30° E., place the instrument so that one of the lines, *n n*, on the paper coincides both with the extremity of the index and with the division marked 30° from the *south* end of the protractor, and slip it up or down until the point A touches the base; then the distance of 1,000 yards being taken along the base from A, gives the position of B. If the base has a scale marked on it, like the military protractor, this distance is at once indicated by it;

but if not, it must be taken from an independent scale, an set off on a line ruled from A to B.

If the distance be measured in paces instead of yards, th

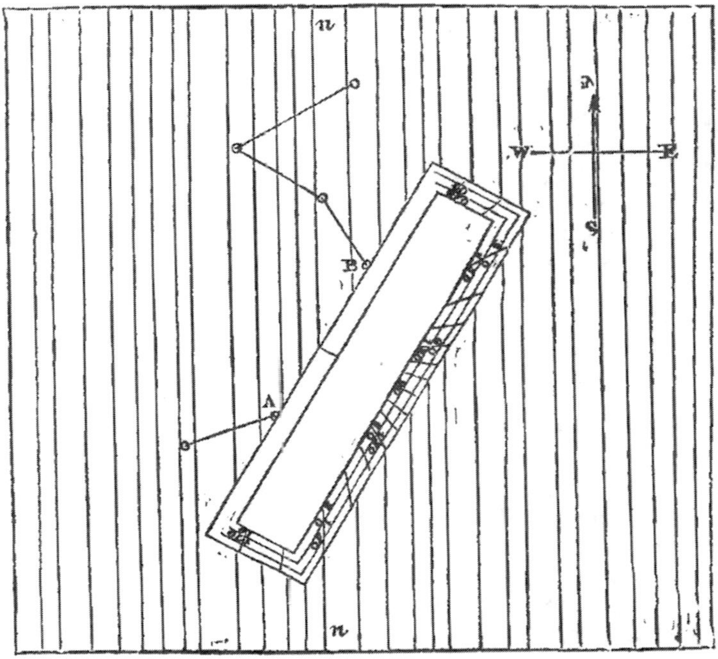

Fig. 2.

same scale may be used for laying them off; but the plan will now be one of as many paces to the inch, as the scale used is of yards to the inch.

THE SECTOR.

This instrument may well be called a universal scale. By its aid all questions in proportion may be solved; lines may be divided either equally or unequally into any number of parts that may be desired; the angular functions, viz., chords, sines, tangents, &c., may be set off or measured to any radius whatever; plans and drawings may be reduced or enlarged in any required proportion; and, in short, every operation in geometrical drawing may be performed by the aid of this instrument and the compasses only.

The name sector is derived from the tenth definition of the third Book of Euclid, in which this name is given to the figure contained by two radii of a circle, and the circumference between them. The instrument consists of two equal rulers, called legs, which represent the two radii, movable about the centre of a joint, which centre represents the centre of the circle. The legs can consequently be opened so as to contain any angle whatever, or completely opened out until their edges come into the same straight line.

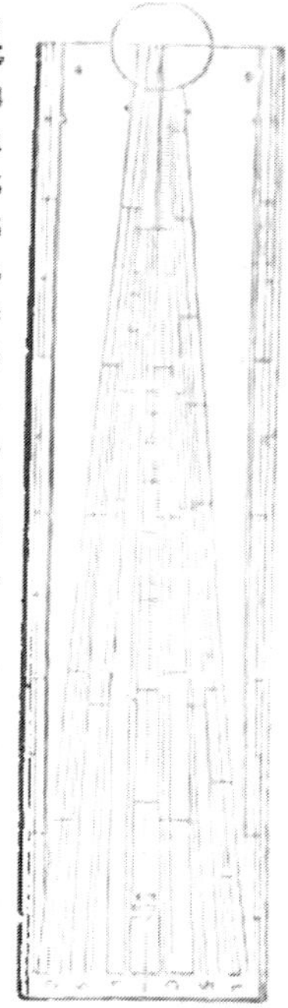

Sectors are made of different sizes, and their length is usually denominated from that of the legs when shut together. Thus, a sector of 6 inches, such as is supplied in the common pocket cases of instruments, forms a rule of 12 inches when opened; and this circumstance is taken advantage of, by filling up the spaces not occupied by the sectorial lines with such lines as it is most important to lay down upon a greater length than the 6-inch plain scale will admit. Among these the most usual are (1) the lines of logarithmic numbers, sines, and tangents, described in Part II.; (2) a scale of 12 inches, in which each inch is divided into ten equal parts; and (3) a foot divided into ten equal primary divisions, each of which is subdivided into ten equal parts, so that the whole is divided into 100 equal parts. The last-mentioned is called the decimal scale, and is placed on the edge of the instrument.

The *sectoral lines* proceed in pairs from the centre, one line of each pair on either leg, and are, upon one face of the instrument, a pair of scales of equal parts, called the *line of lines*, and marked L; a pair of lines of chords, marked C; a pair of lines of secants, marked S; a pair of lines of polygons, marked POL. Upon the other face, the sectoral lines are—a pair of lines of sines, marked S; a pair of lines of tangents up to 45°, marked T; and a second line of tangents to a lesser radius, extending from 45° to 75°.

Each pair of sectoral lines, except the line of polygons,

should be so adjusted as to make equal angles at the centre, so that the distances from the centre to the corresponding divisions of any pair of lines, and the transverse distance between these divisions, may always form similar triangles. On many instruments, however, the pairs of lines of secants, and of tangents from 45° to 75°, make angles at the centre equal to one another, but unequal to the angle made by all the other pairs of lines.

The solution of questions on the sector is said to be *simple* when the work is begun and ended upon the same pair of lines; *compound*, when the operation is begun upon one pair of lines and finished upon another.

In a compound solution the two pairs of lines used must make equal angles at the centre, and, consequently, in the exceptional case mentioned above, the lines of secants and of tangents above 45° cannot be used in connection with the other sectoral lines.*

When a measure is taken on any of the sectoral lines beginning at the centre, it is called a *lateral* distance; but when a measure is taken from any point on one line to its corresponding point on the line of the same denomination on the other leg, it is called a *transverse* or *parallel* distance.

The divisions of each sectoral line are contained within three parallel lines, *the innermost* being the line on which the points of the compasses are to be placed, because this is the only line of the three which goes to the centre, and is therefore the sectoral line.

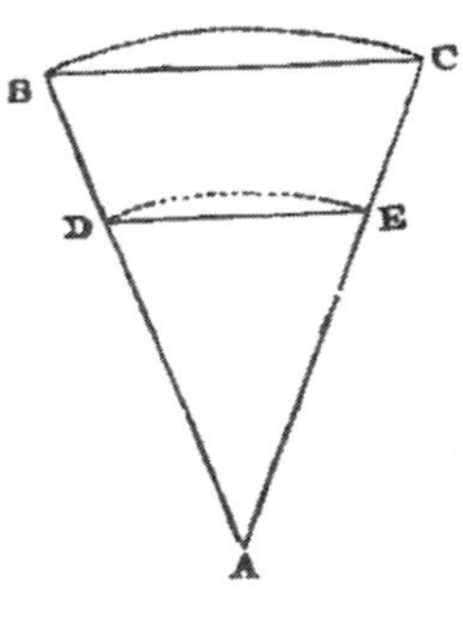

On the principle of the Use of the Sectoral Lines.—1st. *In the Case of a Simple Solution.*—Let the lines A B, A C represent a pair of sectoral lines, and B C, D E any two transverse distances taken on this pair of lines; then, from the construction of the instrument, we

* Since, however, secant : rad. : : rad. : cosine :
and tangent : rad. : : rad. : cotangent;
the line of sines may be used with the other sectoral lines in place of the line of secants, and the line of tangents less than 45° in place of the line of tangents greater than 45°, the complements of the angles being taken upon these lines, in either case, instead of the angles themselves. See page 51.

have A B equal to A C, and A D equal to A E, so that A B : A C :: A D : A E, and the triangles A B C and A D E have the angle at A common, and the sides about this common angle proportional; they are, therefore, similar (Euc. vi. prop. 6) and—

A B : B C :: A D : D E.

In the case of a compound solution, the angles at A are equal, but not common, and the reasoning is, in all other respects, exactly the same.

USES OF THE LINE OF LINES.

To find a Fourth Proportional to Three given Lines.—Set off from the centre a lateral distance equal to the first term, and open the sector till the transverse distance at the division thus found, expressing the first term, is equal to the second term; again, extend to a point whose lateral distance from the centre is equal to the third term, and the transverse distance at this point will be the fourth term required.

If the legs of the sector will not open far enough to make the lateral distance of the second term a transverse distance at the division expressing the first term, take any aliquot part of the second term which can conveniently be made such transverse distance, and the transverse distance at the third term will be the same aliquot part of the fourth proportional required.

A third proportional to two given lines is found by taking a third line equal to the second, and finding the fourth proportional to the three lines.

Example.—To find a fourth proportional to the numbers 2, 5, and 10. Open the sector till the lateral distance of the second term 5 becomes the transverse distance at 2, the first term; then the transverse distance at 10 will extend, as a lateral distance, from the centre to 25, the fourth proportional required.

To bisect a given Straight Line.—Take the extent of the line in the compasses, and open the sector till this extent is a transverse between 10 and 10 on the line of lines: then the transverse distance from 5 to 5 on the same pair of sectoral lines gives the half of the line, and this extent set off from either end will bisect it.

To divide a Straight Line into any Number of equal Parts.—1. When the number of parts is a power of 2, the operations are best performed by continual bisection. Thus, let i be required to divide the line A B into sixteen equal parts (1.) Make A B a tranverse distance between 10 and 10 on the

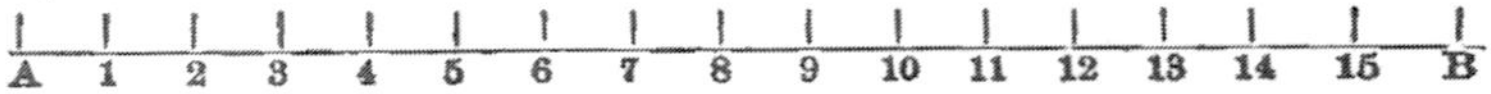

line of lines; then take off the transverse distance of 5 and 5 and set it off from A or* B to 8, and A B will be divided into two equal parts at the division 8. (2.) Make A 8 a transverse distance at 10, and then the transverse distance at 5, set off from A or* 8 at 4, and from B or* 8 at 12, will divide the line into four equal parts at the divisions 4, 8, and 12. (3.) Make the extent A 4 a transverse distance at 10, and the transverse distance at 5 will again bisect each of the parts last set off, and divide the whole line into eight equal parts at the divisions 2, 4, 6, 8, 10, 12, and 14. Each of these may be again bisected by taking the transverse distance at $2\frac{1}{2}$ or 2·5, that is, at the middle division between the 2 and the 3 upon the line of lines, and the line will be divided as required.

When the divisions become smaller than can be conveniently bisected by the method just explained, the operation may still be continued to any required extent by taking the extent of an odd number of the divisions already obtained as the transverse distance of 10 and 10; and setting off the half of this extent, or the transverse distance at 5, from the several divisions already obtained. Thus, in the preceding example, by making the extent of three of the divisions, or five, or seven, a transverse distance at 10, the transverse distance at 5, set off from the several divisions already obtained, will divide A B into 32 equal parts.

2. When the number of parts is not a power of 2, the operations cannot all be performed by bisections; but still, by a judicious selection of the parts into which the line is first divided, many of the after-operations may be performed by bisections. *Example.*—Let it be required to divide the line A B into seven equal parts. (1.) Make the whole extent,

* Greater accuracy is obtained by setting off the distance from both ends of the extent to be bisected, and then, in case the two points so found do not accurately coincide, taking the middle point between them, as near as the eye can judge, for the true point of bisection.

A B, a transverse distance between 7 and 7 on the line of lines; then take off the transverse distance of 4 and 4, and set it off from A and B to 4 and 3. (2.) Make this extent from A to 4 a transverse distance at 10; then the transverse distance at 5 bisects A 4 and 3 B in 2 and 5; set off from 3, gives 1, and from 4 gives 6; and thus the line A B is divided into seven equal parts as required.

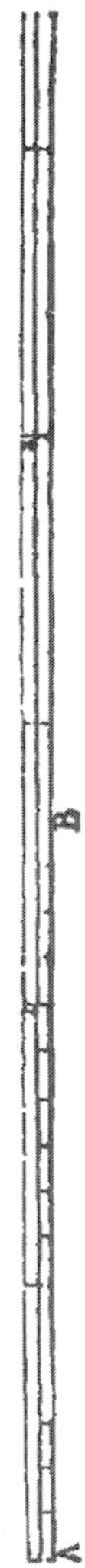

To open the Sector so that the Line of Lines may answer for any required Scale of equal Parts.—Take one inch in the compasses, and open the sector till this extent becomes a transverse distance at the division indicating the number of parts in an inch of the required scale; or, if there be not an integral number of parts in one inch, it will be better to take such a number of inches as will contain an integral number of parts, and make the extent of this number of inches, if it be not too great, a transverse distance at the division indicating the number of parts of the required scale in this extent.

Example.—To adjust the Sector as a Scale of One Inch to Four Chains.—Make one inch the transverse distance of 4 and 4; then the transverse distances of the other corresponding divisions and subdivisions will represent the number of chains and links indicated by these divisions: thus, the transverse distance from 3 to 3 will represent three chains; the transverse distance at 4·7, or the seventh principal subdivision after the primary division marked 4, will represent 4 chains 70 links, and so on.

To construct a Scale of Feet and Inches in such a manner that an extent of Three Inches shall represent Twenty Inches.—1. Make three inches a transverse distance between 10 and 10, and the transverse distance of 8 and 8 will represent 16 inches. 2. Set off this extent from A to B, divide it by continual bisection into 16 equal parts, and place permanent strokes to mark the first 12 of these divisions, which will represent inches. 3. Place the figure 1 at the twelfth stroke, and set off again the extent of the whole 12 parts, from 1 to 2, 2 to 3, &c., to represent the feet.

As an example of the Use of the Line of Lines in reducing Drawings, let it be required to reduce the Linear Dimensions of a

Drawing in the Proportion of 8 *to* 5.—1. Take in the compasses the distance between two points of the drawing, and make it a transverse distance at 8 and 8; then the transverse distance of 5 and 5 will be the distance between the two corresponding points of the copy. 2. These two points having been laid down, make the distance between one of them and a third point a tranverse distance at 8, and with the transverse distance at 5 describe, from that point as centre, a small arc. 3. Repeat the operation with the other point, and the intersection of the two small arcs will give the required position of the third point in the copy. In the same manner all the other points of the reduced copy may be set off, each one from two points previously laid down.

LINE OF CHORDS.

The double scales of chords upon the sector are more generally useful than the single line of chords laid down on the plain scale; for, on the sector, the radius with which the arc is to be described may be of any length between the transverse distance of 60 and 60 when the legs are closed, and that of the transverse of 60 and 60 when the legs are opened as far as the instrument will admit of; but, with the chords on the plain scale, the arc described must be always of the same radius.

To protract or lay down a right-lined Angle B A C, *which shall contain a given Number of Degrees, suppose* 46°.—*Case* 1.

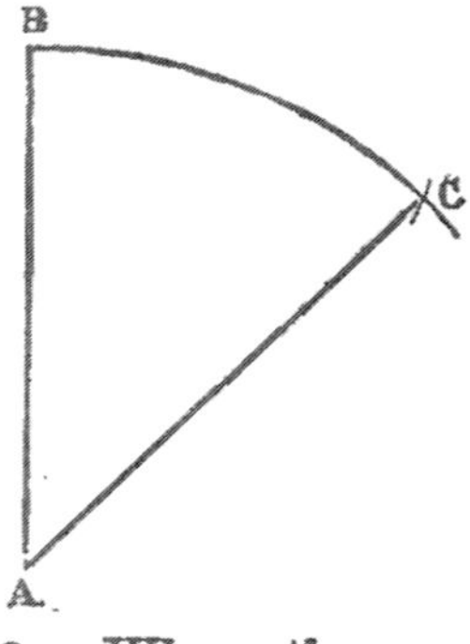

When the angle contains less than 60°, make the transverse distance of 60 and 60 equal to the length of the radius of the circle, and with that opening describe the arc B C. Take the transverse distance of the given degrees 46, and lay this distance on the arc from the point B to C. From the centre A of the arc draw two lines A C, A B, each passing through one extremity of the distance B C laid on the arc, and these two lines will contain the angle required. *Case* 2. When the angle contains more than 60°. Suppose, for example, we wish to form an angle containing 138°. Describe the arc B C D, and make the transverse distance of 60 and 60 equal the radius as before. Take the transverse distance of ½ or ⅓, &c., of the given number of degrees, and lay this distance on the arc twice or thrice, as from B to *a*, *a* to *b*, and from *b* to D. Draw two lines connecting B to A, and A to D, and they will form the angle required.

When the required angle contains less than 5°, suppose 3½, it will be better to proceed thus. With the given radius, and from the centre A, describe the arc D G; and from some point, D, lay off the chord of 60°, which suppose to give the point G, and also from the same point D lay off in the same direction the chord of 56½° (= 60° — 3½°), which would give the point E. Then through these two points E and G, draw lines to the point A, and they will represent the angle of 3½° as required.

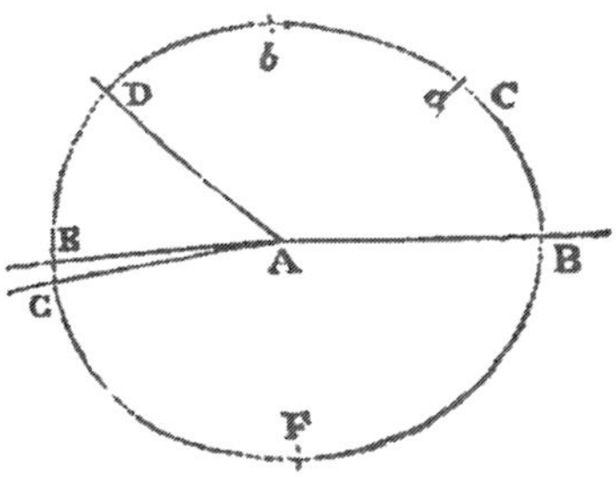

From what has been said about the protracting of an angle to contain a given number of degrees, it will easily be seen how to find the degrees (or measure) of an angle already laid down.

LINE OF POLYGONS.

The line of polygons is chiefly useful for the ready division of the circumference of a circle into any number of equal parts from 4 to 12; that is, as a ready means to inscribe regular polygons of any given number of sides, from 4 to 12, within a given circle; to do which, set off the radius of the given circle (which is always equal to the side of an inscribed hexagon) as the transverse distance of 6 and 6 upon the line of polygons. Then the transverse distance of 4 and 4 will be the side of a square; the transverse distance between 5 and 5, the side of a pentagon; between 7 and 7, the side of a heptagon; between 8 and 8, the side of an octagon; between 9 and 9, the side of a nonagon, &c., all of which is too plain to require an example.

If it be required to form a polygon, upon a given right line, set off the extent of the given line, as a transverse distance, between the points upon the line of polygons answering to the number of sides of which the polygon is to consist; as for a pentagon between 5 and 5; or for an octagon between 8 and 8: then the transverse distance between 6 and 6 will be the radius of a circle whose circumference would be divided by the given line into the number of sides required.

The line of polygons may likewise be used in describing, upon a given line, an isosceles triangle, whose angles at the base are each double that at the vertex. For, taking the given line between the compasses, open the sector till that extent becomes the transverse distance of 10 and 10, then the transverse distance of 6 and 6 will be the length of each of the two equal sides of the isosceles triangle.

All regular polygons whose number of sides will exactly divide 360 (the number of degrees into which all circles are supposed to be divided) without a remainder may likewise be set off upon the circumference of a circle by the line of chords. Thus, take the radius of the circle between the compasses, and open the sector till that extent becomes the transverse distance between 60 and 60 upon the line of chords; then, having divided 360 by the required number of sides, the transverse distance between the numbers of the quotient will be the side of the polygon required. Thus, for an octagon, take the distance between 45 and 45; and for a polygon of 36 sides take the distance between 10 and 10, &c.

LINES OF SINES, TANGENTS, AND SECANTS.

Given the Radius of a Circle (suppose equal to two inches), required the Sine and Tangent of 28° 30′ *to that Radius.*—Open the sector so that the transverse distance of 90 and 90 on the sines, or of 45 and 45 on the tangents, may be equal to the given radius, viz., two inches; then will the transverse distance of 28° 30′, taken from the sines, be the length of that sine to the given radius, or, if taken from the tangents, will be the length of that tangent to the given radius.

But if the Secant of 28° 30′ *is required*, make the given radius of two inches a transverse distance of 0 and 0 at the beginning of the line of secants, and then take the transverse distance of the degrees wanted, viz., 28° 30′.

A Tangent greater than 45° (*suppose* 60°) *is thus found*:—Make the given radius, suppose two inches, a transverse distance to 45 and 45 at the beginning of the line of upper tangents, and then the required degrees (60) may be taken from the scale.

The tangent to a given radius of any number of degrees

greater than 45° can also be taken from the line of lower tangents, if the radius can be made a transverse distance to the complement of those degrees on this line (see note, p. 44).

Example.—To find the tangent of 78° to a radius of two inches. Make two inches a transverse distance at 12 on the lower tangents, then the transverse distance of 45 will be the tangent of 78°.

In like manner the secant of any number of degrees may be taken from the sines, if the radius of the circle can be made a transverse distance to the complement of those degrees upon this line. Thus, making two inches a transverse distance to the sine of 12°, the transverse distance of 90 and 90 will be the secant of 78°.

To find, by means of the lower tangents and sines, the degrees answering to a given line, greater than a given radius which expresses the length of a tangent or secant to this radius. For a tangent, make the given line a transverse distance at 45 on the lower tangents; then take the extent of the given radius, and apply it to the lower tangents; and the complement of the degrees at which it becomes a transverse distance will be the number of degrees required. For a secant make the given line a transverse distance at 90 on the sines; then the extent of the radius will be a transverse distance at the complement of the number of degrees required.

Given the Length of the Sine, Tangent, or Secant of any Degrees, to find the Length of the Radius to that Sine, Tangent, or Secant.—Make the given length a transverse distance to its given degrees on its respective scale. Then,

If a sine	the transverse distance of	90 and 90 on the sines	will be the radius sought.
If a tangent under 45°		45 and 45 on the tangents	
If a tangent above 45°		45 and 45 on the upper tangents	
If a secant		0 and 0 on the secants	

To find the Length of a versed Sine to a given Number of Degrees and a given Radius.—1. Make the transverse distance of 90 and 90 on the sines equal to the given radius. 2. Take the transverse distance of the sine of the complement of the given number of degrees. 3. If the given number of degrees be less than 90, subtract the distance just taken, viz., the sine of the complement, from the radius, and the remainder will be the versed sine: but, if the given number of degrees is more than 90, add the complement of the sine to the radius, and the sum will be the versed sine.

To open the legs of a Sector, so that the corresponding double Scales of Lines, Chords, Sines, and Tangents may make each e Right Angle.—*On the line of lines* make the lateral distance 10, a transverse distance between 8 on one leg and 6 on the other leg.

On the line of sines make the lateral distance 90, a transverse distance from 45 to 45; or from 40 to 50; or from 30 to 60; or from any degree to its complement.

On the line of sines make the lateral distance of 45 a transverse distance between 30 and 30.

MARQUOIS'S SCALES. (Plate II. Fig. 3.)

These scales consist of a right-angled triangle, of which the hypothenuse or longest side is three times the length of the shortest, and two rectangular rules. Our figure, which is drawn one-third the actual size of the instruments from which it is taken, represents the triangle and one of the rules, as being used to draw a series of parallel lines. Either rule is one foot long, and has, parallel to each of its edges, two scales, one placed close to the edge and the other immediately within this, the outer being termed the artificial, and the inner the natural scale. The divisions upon the outer scale are three times the length of those upon the inner scale, so as to bear the same proportion to each other that the longest side of the triangle bears to the shortest. Each inner, or natural scale, is, in fact, a simply-divided scale of equal parts (see p. 22), having the primary divisions numbered from the left hand to the right throughout the whole extent of the rule. The first primary division on the left hand is subdivided into ten equal parts, and the number of these subdivisions in an inch is marked underneath the scale, and gives it its name. On one of the pair of Marquois's scales now before us, we have, on one face, scales of 30 and 60, on the obverse scales of 25 and 50, and on the other we have on one face scales of 35 and 45, and on the obverse scales of 20 and 40. In the artificial scales the zero point is placed in the middle of the edge of the rule, and the primary divisions are numbered both ways from this point to the two ends of the rule, and are every one subdivided into ten equal parts, each of which is, consequently, three times the length of a subdivision of the corresponding natural scale.

Plate II

3

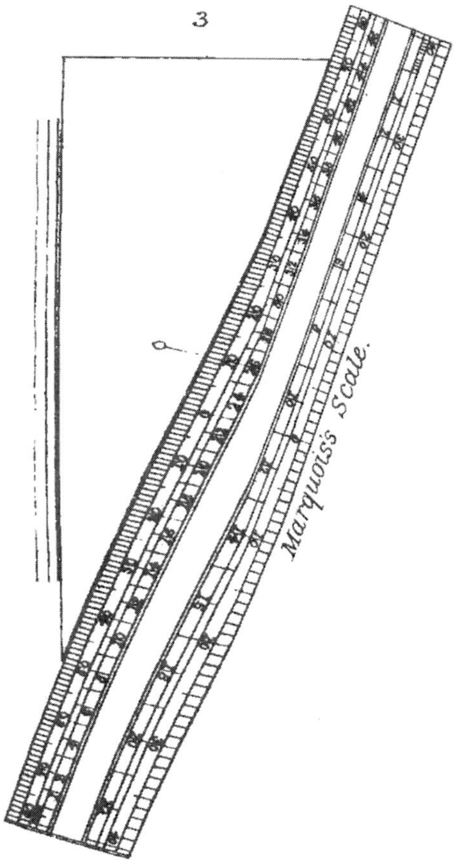

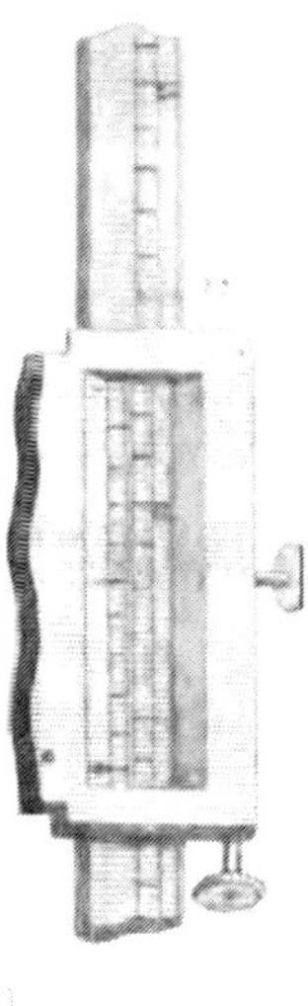

Vernier to straight limb.

Vernier to arched limb.

The triangle has a short line drawn perpendicular to the hypothenuse near the middle of it, to serve as an index or pointer; and the longest of the other two sides has a sloped edge.

To draw a Line parallel to a given Line, at a given Distance from it.—1. Having applied the given distance to the one of the natural scales which is found to measure it most conveniently, place the triangle with its sloped edge coincident with the given line, or rather, at such small distance from it that the pen or pencil passes directly over it when drawn along this edge. 2. Set the rule closely against the hypothenuse, making the zero point of the corresponding artificial scale coincide with the index upon the triangle. 3. Move the triangle along the rule to the left or right, according as the required line is to be above or below the given line, until the index coincides with the division or subdivision corresponding to the number of divisions or subdivisions of the natural scale, which measures the given distance; and the line drawn along the sloped edge in its new position will be the line required.*

NOTE.—The natural scale may be used advantageously in setting off the distances in a drawing, and the corresponding artificial scale in drawing parallels at required distances.

To draw a Line perpendicular to a given Line from a given Point in it.—1. Make the shortest side of the triangle coincide with the given line, and apply the rule closely against the hypothenuse. 2. Slide the triangle along the rule until a line drawn along the sloped edge passes through the given point, and the line so drawn will be the line required.

* If A B C represent the triangle in its new position, and the dotted

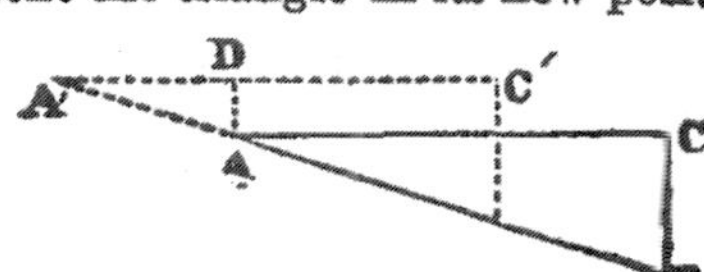

lines represent its original position, we have, by similar triangles A B C, A′ A D,

A D : A A′ : : B C : B A
: : 1 : 3;

and therefore A D contains as many divisions of the natural as A A′ contains of the artificial scale.

The advantages of Marquois's scales are: 1st, that the sight is greatly assisted by the divisions on the artificial scale being so much larger than those of the natural scale to which the drawing is constructed; 2nd, that any error in the setting of the index produces an error of but one-third the amount in the drawing.

If the triangle be accurately constructed, these scales may be advantageously used for dividing lines with accuracy and despatch; our figure of the Marquois's scales was drawn by the aid of themselves alone.

We proceed to explain a method which we have found to answer well for dividing lines with accuracy and despatch, and which is altogether independent of any error in the construction of the triangle.

Let *a b* represent the line to be divided into any number, *n*, of equal parts; select one of the natural scales, of which *n* divisions or some multiple *p n* of *n* divisions, are nearly equal to *a b*, but less than it; then setting the sloped edge of the triangle perpendicular to *a b*, so that a line drawn along it passes through *a*, place the rule closely against the hypothenuse, making a division of the artificial scale corresponding to the selected natural scale, coincide with the index upon the triangle; and, moving the triangle along the rule to the right, until the index coincides with the *p n*th division from that to which it was set, draw a line along the sloped edge intersecting *a b* in *c*. Upon *a b* as hypothenuse describe the right-angled triangle *a b d*, having the side *a d* equal to *a c*, and place the sloped edge of the triangle to coincide with *b d*, which touches the circular arc *c d* described from the centre *a*, and is consequently perpendicular to *a d*; then, sliding the triangle along the rule to the right or left, and drawing a line as the index comes into contact with every *p*th division of the artificial scale, the line *a b* will be divided as required.

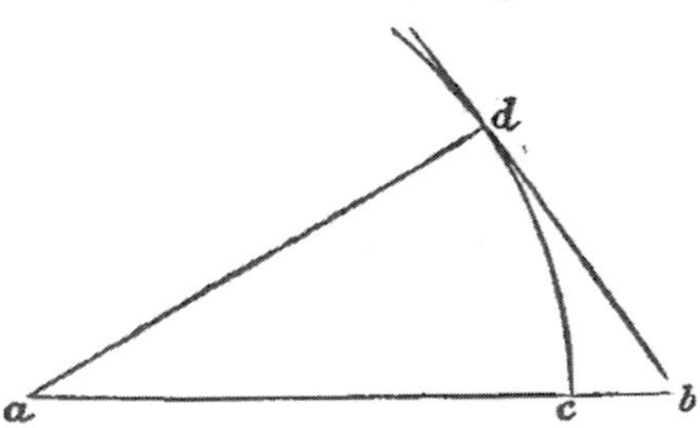

THE VERNIER.

Before proceeding to the description of the circular protractor, we will describe a subsidiary instrument called the

vernier, which is applied to all instruments in which great accuracy of measurement is required, in order to measure portions of the space between the equidistant divisions of a graduated scale.

This ingenious little instrument is so constructed as to slide evenly along the graduated limb of an instrument, and enables us to measure distances, or read off observations, with remarkable nicety. It is in fact a scale in which the divisions differ slightly in length from those of the graduated limb to which it is applied. The divisions are usually shorter than those upon the limb to which it is attached, the length of the graduated scale of the vernier being exactly equal to the length of a certain number $(n-1)$ of the divisions upon the limb, and the number (n) of divisions upon the vernier being one more than the number upon the same length of the limb.

Let, then, L represent the length of a division upon the limb, and V, the length of a division upon the vernier: so that $(n-1)\,\text{L} = n\,\text{V}$;

$$\text{and therefore } \text{L} - \text{V} = \text{L} - \frac{n-1}{n}\,\text{L} = \frac{1}{n}\,\text{L};$$

or the defect of a division upon the vernier from a division upon the limb is equal to the nth part of a division upon the limb, n being the number of divisions upon the vernier.

If n divisions of the vernier were equal to $(n+1)$ divisions of the limb, or $(n+1)\,\text{L} = n\,\text{V}$,

$$\text{then would } \text{V} - \text{L} = \frac{n+1}{n}\,\text{L} - \text{L} = \frac{1}{n}\,\text{L};$$

or the excess of a division upon the vernier above a division upon the limb would be equal to the nth part of a division upon the limb. With this arrangement, however, we should have the vernier reading backwards, an arrangement which is the most convenient in the case of the mountain barometer.

In fig. 1, six divisions of the vernier are equal to five divisions of the limb, and, consequently, the above defect, or L — V, is equal to a sixth part of a division upon the limb, or to 20′, since a division of the limb is equal to 2°.

In fig. 2, ten divisions of the vernier are equal to nine divisions of the limb, and, consequently, L — V is equal to a tenth part of a division upon the limb, or to the hundredth part of an inch, a division of the limb being equal to the tenth part of an inch.

In reading off we must first look to the lozenge, as pointing out the exact place upon the limb at which the required measurement is indicated. If, then, the stroke upon the vernier at the lozenge exactly coincides with a stroke upon the limb, the reading at this stroke gives the measurement required; but if the stroke at the lozenge be a distance beyond a stroke upon the limb, then will this distance be equal to once, or twice, or thrice, &c., the difference of a division upon the limb and upon the vernier, according as the stroke at the end of the first, or second, or third, &c., division upon the vernier coincides with a stroke upon the limb.

In Fig. 1 the stroke upon the vernier at the lozenge falls beyond the stroke indicating 22° upon the limb, and the stroke at the end of the second division upon the vernier coincides with a stroke upon the limb; the reading therefore is 22° 40′.

In Fig. 2 the stroke upon the vernier at the lozenge falls beyond the stroke indicating one inch and three-tenths upon the limb, and the stroke at the end of the sixth division upon the vernier coincides with a stroke upon the limb: the reading, therefore, is 1·36 inches, or one inch three-tenths and six hundredths.

The limbs of the best sextants are now divided at every ten minutes, and 59 of these parts are made equal to 60 divisions of their verniers. In this case

$$\text{L} - \text{V} = \frac{\text{L}}{60} = \frac{10'}{60'} = 10'';$$

so that these instruments can be read off by the aid of their verniers to an accuracy of 10 seconds. The verniers occupy on the limbs spaces equal to 9° 50′.*

The limbs of circular protractors and of small theodolites are generally divided at every 30 minutes, and 29 of these

* That is, according to the graduation of the instrument; but, as the angles observed by a sextant are double the angles moved over by the index, the limb of the instrument is graduated, as though it were double the size, so that the verniers really occupy an arc of 4° 55′ only.

parts are made equal to 30 divisions of their verniers, which therefore enables us to read off to an accuracy of $\frac{30'}{30}$, or 1′.

In the mountain barometer the scale being divided into 20ths of an inch, 10 divisions of the vernier are made equal to 11 of these parts, which therefore enables us to read off to an accuracy of the 200th of an inch.

In plotting a survey due regard must be had to both accuracy and despatch, and we should aim to lay down the various points observed with an accuracy proportionate to the accuracy of the survey itself. To this end the principal points should be laid down by setting off with the beam compasses the computed sides of the triangles, the angles of which have been accurately observed with the large theodolite; and the direction of the meridian is to be laid down from an observation of the angle which it makes with a side of one of these triangles, by means of the computed chords,* which chord is also to be set off with the beam compasses.

THE CIRCULAR PROTRACTOR.

The principal points having thus been laid down, the boundaries may be put in by first laying down upon the paper a large circular protractor. This protractor may be pricked off by means of the circular metallic protractor represented in the accompanying figure, and the lines can then be transferred to any part of the paper by means of a large ruler and triangle, or by any parallel ruler. The circular protractor is a complete graduated circle, connected with its centre by four radial bars, *b b b b*. The centre is left open, and surrounded by a concentric ring, or collar, which carries two indices, I, I, with verniers working quite round the circumference of the protractor, and subdividing the primary divisions on the protractor to single minutes, and by estimation to 30 seconds. To the index bars are attached the branches, D, D′, carrying at their extremities fine steel prickers, whose points are kept above the surface of the paper by springs placed under their supports, which give way when the branches are pressed downwards,

* If a table of chords be not at hand, take out the sine of half the angle from a table of natural sines, and reckoning the first figure as integral, this will be the chord of the whole angle to radius 5, or, reckoning the first two figures integral, it will be the chord to radius 50.

and allow the points to make the necessary punctures in the paper. The branches, D, D′, are attached to the index bars, I, I, with a joint which admits of their being folded backwards over the instrument when not in use, and for packing in its case. The centre of the instrument is represented by the intersection of two lines drawn at right angles to each other on a piece of plate glass, which enables the person using it to place it so that the centre, or intersection of the cross lines, may coincide with any given point on the plan. If the instrument is in correct order, a line connecting the fine pricking points with each other would pass through the centre of the instrument, as denoted by the before-mentioned intersection

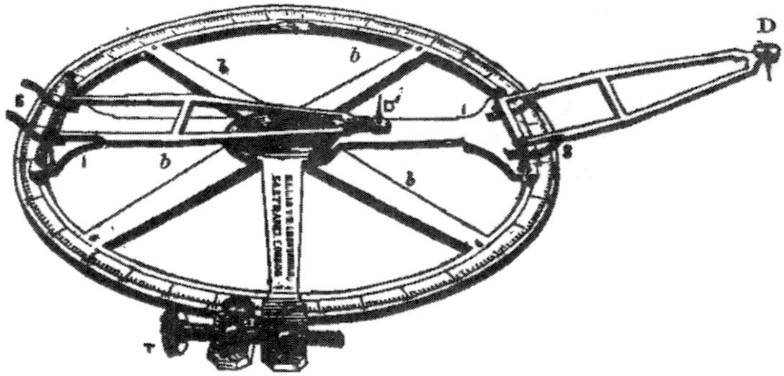

of the cross lines upon the glass, which, it may be observed, are drawn so nearly level with the under surface of the instrument as to do away with any serious amount of parallax, when setting the instrument over a point from which any angular lines are intended to be drawn. In using this instrument one of the verniers, called the forward-angle vernier, should first be set to zero (or the division marked 360°) on the divided limb, and the other, called the back-angle vernier, at the division marked 180°; and it should then be placed on the paper, so that the two fine steel points may be on the given line (from which the other angular lines are to be drawn), and the centre of the instrument coincide with the given angular point on such line. This done, press the protractor gently down, which will fix it in position by means of very fine points on the under side. It is now ready to lay off the given angle, or any number of angles that may be required, which is done by turning the indices till the forward vernier reads the required angle. Then press lightly downwards the branches D, D′, which will cause the points to make punctures in the paper at opposite sides of the circle; and the line joining these punctures will pass through the given angular point, if the instrument was first correctly set. In this manner, at one setting of the instrument, a great number of angles, or a complete circular protractor, may be laid off from the same point. The instrument is furnished with a clamp and tangent screw motion, T, by means of which the verniers are set with accuracy; and as there are two of them, the one forms a check upon the other, as the reading of the back-angle vernier should always be exactly 180° more than the reading of the first. In turning the indices to read any desired angle, the clamp must first be loosened, and the indices set to the division on the limb indicating the desired angle; then tighten the clamp, and perfect the reading by the tangent screw, T.

THE T-SQUARE AND SEMICIRCULAR PROTRACTOR.

A very valuable method, by which a traverse* can be

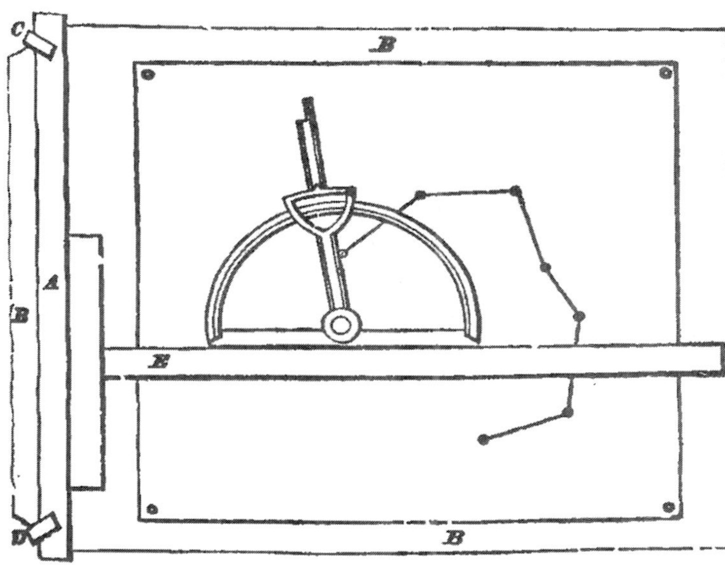

most expeditiously as well as accurately plotted with the T-

* The term "traverse" is applied to a survey made with a chain and theodolite.

square and semicircular protractor, is thus described by Mr Howlett, late chief draughtsman, Royal Ordnance Office, ir vol. i. of "Papers on Subjects connected with the Duties o the Royal Engineers:"—

"As, when away from home, it seldom happens that the surveyor can obtain a good drawing board, or even a table with a good straight edge, I fix a flat ruler, A, to the table B B B, by means of a pair of clamps, C, D, and against this ruler I work the pattern square, E, one side of which has the stock flush with the blade; or, if a strait-edged board be at hand, then the square may be turned over, and used against that edge instead of the ruler A. Here, then, is the most perfect kind of parallel ruler that art can produce, capable of carrying the protractor over the whole of a sheet of blotting paper of any size, and which may be used upon a table of any form. It is convenient to suppose the north on the left hand, and the upper edge of the blade to represent the meridian of the station.

"The protractor is held in the hand while the vernier is set, which is an immense comfort to the sight; and it will be seen that, as both sides of the arm are parallel with the zero and centre, the angle may be drawn on the paper against either side, as the light or other circumstances may render desirable.

"From this description and a mere glance at the engraving, it is clear that angles taken with the theodolite can be transferred to the plot as accurately as the protractor can be set, namely, to a single minute, and that, too, in a rapid and pleasant manner.

"By means of the notch at the end of the arm, this instrument may be used in the manner of a circular protractor, should a square not be at hand."

Metcalf's protractor is another semicircular protractor, a modification of Howlett's in which the radial arm is turned into a scale, so that distances may be set off at the same time that an angle is protracted. There is a scale on each edge of the radius, and the vernier is double, so as to apply to either. Diagonal scales corresponding to those on the arm of the instrument are engraved upon it. With either of these semicircular protractors there is no check as to the exact setting of the instrument over the centre of the station, such as is afforded by the circular protractor when both arms are used. It must also be borne in mind that, in plotting an extensive plan, the errors arising from the expansion and contraction of the paper can only be avoided by laying down in the first place a protractor and scales upon the paper itself, from which the angles and distances are afterwards to be measured.

THE STATION POINTER.

When the principal features of a country have been laid down by the methods already pointed out, the details may be put in with great rapidity by means of the instrument which we are now about to describe.

A A, B B, C C, are three arms movable about a common centre, O, and carrying three fine wires stretched quite tight, the prolongation of which would pass exactly through the centre, O. The arms B B and C C carry each a graduated arc, *b b* and *c c*; and the arm A A carries two verniers, *a a'*, adapted to the graduated arcs *b b* and *c c* respectively.

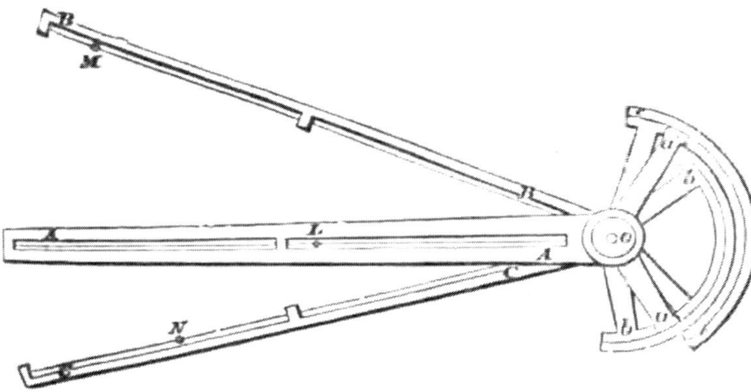

The angles L O M and L O N, subtended, at one of the points which we wish to put in, by L, M, and N, three of the principal points of the survey already laid down, having been observed by means of the prismatic compass, or pocket sextant, the arms of the station pointer are opened out till the verniers point out these angles upon the graduated limbs, *b b* and *c c*, respectively. The station pointer being then placed upon the paper, and moved about till the fine wires pass exactly over the stations L, M, N, as marked already upon the plot, the centre will be exactly over the point to be filled in, and its place is to be marked by passing a pricking point through a small opening which is made at the centre to serve this purpose.

The station pointer is peculiarly adapted for laying down the positions of the soundings in a marine survey. If the boat, when a sounding is taken, be in a line with two objects on shore, previously laid down on the chart, an angle is taken with the sextant between this line and a third object also previously laid down on the chart; either of the arms B B or C C is then set at this angle with the centre arm A A, and the latter being laid on the above line, and the former made to pass through the third object on the chart, the centre of the instrument will be over the required point to be pricked off. If the boat be not in a line with two objects previously laid down, the two angles between three objects previously laid down must be taken with the sextant, and the station point be set and used to fill in the required point in the manner already described.

CHAPTER V.

COPYING INSTRUMENTS.

TRIANGULAR COMPASSES.

COPIES of drawings are very frequently required either of the same dimensions as the original or upon some other scale. To assist in the production of copies of the same dimensions as the original, a convenient instrument is found in the triangular compass.

The most usual form of triangular compasses is constructed by jointing a third leg to the centre pin of the head-joint of

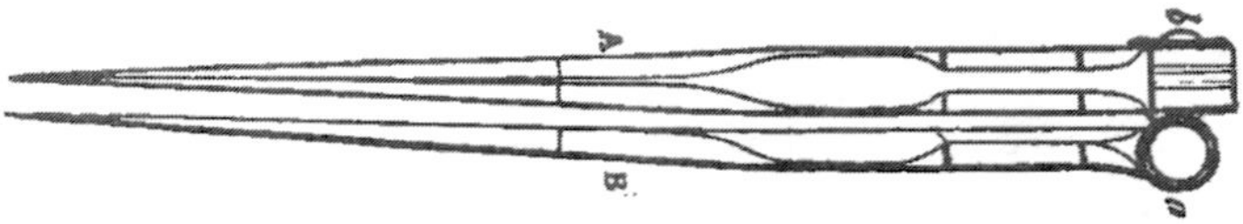

the ordinary dividers, so that this leg, turning both with the centre pin and upon its own joint, has a universal motion.

The mode of using this instrument is to set its three points to coincide, in the first instance, with any three points on the original, and then transfer them to the copy. After this first step the two points of the legs A must be set upon two points of the drawing already copied, and the point of the third leg B made to coincide with a new point of the drawing, that is, one not yet copied; then, by placing the two first points on the corresponding points in the copy, the third point of the compasses will transfer the new point to the copy.

WHOLES AND HALVES.

Drawings may be copied so that all lines and distances in them may be reduced to one-half, or enlarged to double their length, by means of the compasses called wholes and halves, because the longer legs being twice the length of the shorter, when the former are opened to any given length, the shorter ones will be opened to half that length. These compasses are also adapted for dividing lines by continual bisections—for which purpose, in fact, they are chiefly used. To effect this the compasses must first be opened until the distance between the points A, B, is equal to the entire length of the line to be divided, and the distance between the points D, E, set off from either end of the line, will then bisect it. The compasses being now closed until the distance between A and B is equal to one of the divisions just laid down, the distance between D and E will divide the line into four equal parts, and by repeating the operation, it will be divided into eight equal parts, then into sixteen, and so on.

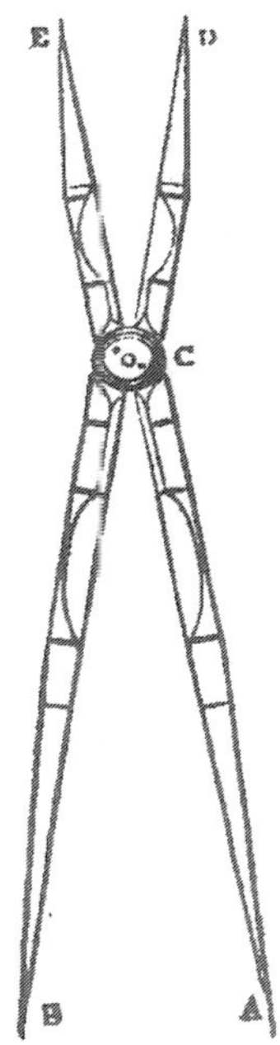

For purposes of copying, they are superseded by the proportional compasses.

PROPORTIONAL COMPASSES.

By mean of this ingenious instrument drawings may be reduced or enlarged, so that all the lines of the copy, or the area or solids represented by its several parts, shall bear certain proportions to the lines, areas, or solids of the original drawing. They will also serve to inscribe regular polygons in circles, and to take the square roots and cube roots of numbers. In the annexed figure the scale of lines is placed on the leg A E, on the left-hand side of the groove, and the scale of circles, on the same leg, on the right-hand side of the groove. The scale of plans and solids are on the other face of the instrument. To set the instrument, it must first be accurately closed, so that the two legs appear but as one; the nut C being then unscrewed, the slider may be moved until the line across it coinsides with any required division upon any one of the scales. Now tighten the screw, and the compasses are set.

To reduce or enlarge the Lines of a Drawing.—The line across the slider being set to one of the divisions, 2, 3, 4, &c., on the scale of lines, the points A, B, will open to double, triple, four times, &c., the distances of the points D,

E (Euc. vi. prop. 4). If, then, the points A and B be opened to the lengths of the lines upon a drawing, the points D and E will prick off a copy with the lines reduced in the proportion of ½ to 1, ⅓ to 1, ¼ to 1, &c.; but if the points D and E be opened to the lengths of the lines upon a drawing, the points A and B will prick off a copy with the lines enlarged in the proportions of 2 to 1, 3 to 1, 4 to 1, &c.

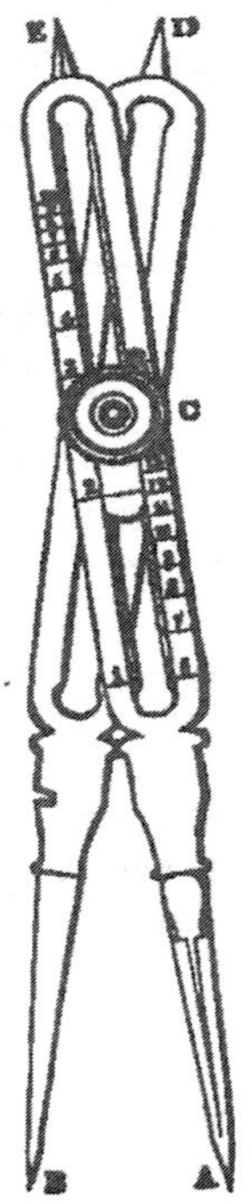

To inscribe in a Circle a regular Polygon of any number of Sides from 6 *to* 20.—The line across the slider being set to any number on the scale of circles, and the points A and B being opened to the length of any radius, the points D and E will prick off a polygon of that number of sides in the circle described with this radius: thus, if the line across the slider be set to the division marked 12 on the scale of circles, and a circle be described with the radius A B, D E will be the chord of a $\frac{1}{12}$th part of the circumference, and will prick off a regular polygon of 12 sides in it.

To reduce or enlarge the Area of a Drawing.—The numbers of the scale of plans are the squares of the ratios of the lengths of the opposite ends of the compasses, when the line across the slider is set to those numbers; and the distances between the points being in the same ratio as the lengths of the corresponding ends (Euc. vi. prop. 4), the areas of the drawings, and of the several parts of the drawings, pricked off by these points, will have to one another the ratio of 1 to the number upon the scale of plans to which the instrument is set (Euc. vi. props. 19, 20; and xii. prop. 2). Thus, if the line across the slider be set to 4 on the scale of plans, the distance between the points A and B will be twice as great as the distance between D and E; and if A and B be opened out to the lengths of the several lines of a drawing, D and E will prick off a copy occupying ¼th the area; if the line across the slides be set to 5 on the same scale, the distances between the points will be in the ratio of 1 to $\sqrt{5}$, and the area of the copy pricked off by the points D and E will be ⅕th of the area of the drawing, of which the lines are taken off by A and B: conversely, if the lines of the drawing be taken off by the points D and E, the points A and B will prick off a copy, of which the area will be 4 times or 5 times as great, according as the line across the slider is set to the division marked 4 or 5 on the scale.

To take the Square Root of a Number.—The line across the slider being set to the number upon the scale of plans, open the points A and B to take the number from any scale of equal parts (p. 22), then the points D and E applied to the same scale of equal parts will take the square root of the number. Thus, to take the square root of 3, set the line across the slider to 3, open out the compasses till A and B take off 3 from any scale of equal parts, and the points D and E will take off 1·73, which is the square root of 3, from the same scale of equal parts. A mean proportional between two numbers, being the square root of their product, may be found by multiplying the numbers together, and then taking the square root of the product in the manner explained above.

The numbers of the scale of solids are the cubes of the ratios of the lengths of the opposite ends of the compasses, when the line across the slider is set to those numbers; so that, when this line is set to the division marked 2 upon the scale of solids, the distance between the points A and B will give the side of a solid of double the content of that of which a like side is given by the distance of the points D and E; when the line is set to 3, the respective distances of the points will give the like sides of solids, the contents of which will be in the proportion of 3 to 1; and so on.

The Cube Root of a given number may be found by setting the line across the slider to the number upon the scale of solids, and, opening the points A, B to take off the square of number upon any scale of equal parts, the points D, E will then take off the required cube root from the same scale.

THE PANTAGRAPH.

The pantagraph consists of four rulers, A B, A C, D F, and E F, made of stout brass. The two longer rulers, A B and A C, are connected together by, and have a motion round, a centre at A. The two shorter rulers are connected in like manner with each other at F, and with the longer rulers at D and E; and, being equal in length to the portions A D and A E of the longer rulers, form with them an accurate parallelogram, A D F E, in every position of the instrument. Several ivory castors support the machine parallel to the paper, and allow it to move freely over it in all directions. The arms, A B and

D F, are graduated and marked $\frac{1}{2}$, $\frac{1}{3}$, &c., and have each a sliding index, which can be fixed at any of the divisions by a milled-headed clamping screw, seen in the engraving. The sliding indices have each of them a tube, adapted either to slide on a pin rising from a heavy circular weight called the fulcrum, or to receive a sliding holder with a pencil or pen, or a blunt tracing-point, as may be required.

When the instrument is correctly set, the tracing-point, pencil, and fulcrum will be in one straight line, as shown by the dotted line in the figure, and which may be proved by stretching a fine string over them. The motions of the tracing-point and pencil are then each compounded of two circular motions, one about the fulcrum, and the other about the joints at the ends of the rulers upon which they are respectively placed. The radii of these motions form sides about equal angles of two similar triangles, of which the straight line B C, passing through the tracing-point, pencil, and fulcrum, forms

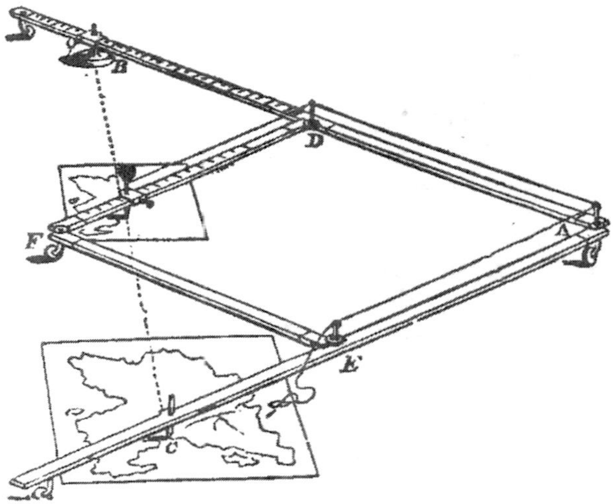

the third side. The distances passed over by the tracing-point and pencil, in consequence of either of these motions, have then the same ratio, and, therefore, the distances passed over in consequence of the combination of the two motions have also the same ratio, which is that indicated by the setting of the instrument.

Our engraving represents the pantagraph in the act of

reducing a plan to a scale of half the original. For this purpose the sliding indices are first clamped at the divisions upon the arms marked $\frac{1}{2}$; the tracing-point is then fixed in a socket at C, over the original drawing; the pencil is next placed in the tube of the sliding index upon the ruler D F, over the paper to receive the copy; and the fulcrum is fixed to that at B, upon the ruler A B. The machine being now ready for use, if the tracing-point at C be passed delicately and steadily over every line of the plan, a true copy, but of one-half the scale of the original, will be marked by the pencil on the paper beneath it. The fine thread represented as passing from the pencil quite round the instrument to the tracing-point at C, enables the draughtsman at the tracing-point to raise the pencil from the paper, whilst he passes the tracer from one part of the original to another, and thus to prevent false lines from being made on the copy. The pencil-holder is surmounted by a cup, into which sand or shot may be put, to press the pencil more heavily on the paper, when found necessary.

If the object were to enlarge the drawing to double its scale, then the tracer must be placed upon the arm D F, and the pencil at C; and if a copy were required of the same scale as the original, then, the sliding indices still remaining at the same divisions upon D F and A B, the fulcrum must take the middle station, and the pencil and tracing-point those on the exterior arms, A B and A C, of the instrument.

The successful use of the pantagraph in copying very minute and complicated drawings can only be attained by perseverance and experience, and we therefore proceed to mention some of the other means employed for the attainment of the same object. In fact, while the pantagraph affords the most rapid means of reducing a drawing, we cannot recommend its use for enlarging a copy, or even for copying upon the same scale.

To produce a Copy of the same Size as the Original. First Method.—Lay the original drawing upon the sheet of paper intended for the copy, and fix them together by means of weights or drawing-pins.* 2. With a fine needle † prick

* The drawing-pin consists of a brass head, with a steel point at right angles to its plane. A represents it as seen edgewise, and B as seen from above.

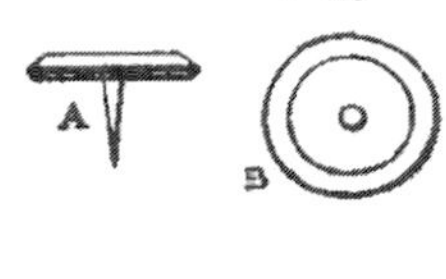

† See pricking point, page 4.

through all the angles and principal points, making corresponding punctures in the paper beneath. 3. Draw upon the copy such of the lines on the original as are straight, or nearly so, by joining the points thus marked upon the paper. 4. Set off such other points upon the copy, by means of the compasses, as may be desirable, and draw the curved lines upon tracing-paper placed over the drawing. 5. Fill in the lines indicated by the points set off by the compasses, and transfer the curved lines from the tracing-paper to the copy, by rubbing the back of the tracing-paper with powdered black-lead, placing it in its correct situation upon the copy, and passing a blunt tracing-point* over the lines drawn upon it.

Second Method.—A sheet of tracing-paper having the under side rubbed over with powdered black-lead may be placed upon the paper intended for the copy. The original being then placed over this, the tracing-point may be carefully and steadily passed over all the lines of the drawing with a pressure proportioned to the thickness of the paper; and the paper beneath will receive corresponding marks, forming an exact copy, which is afterwards to be inked in.

Third Method.—The drawing is placed upon a large sheet of plate glass called a copying-glass, and the paper to receive the copy placed over the drawing. The glass is then fixed in such a position as to have a strong light fall upon it from behind, and shine through it. By this means the original drawing becomes visible through the paper placed over it, and a copy can be made with precision and ease, without any risk of soiling or injuring the original.

To copy with nicety upon a Reduced or Enlarged Scale.—For this purpose we may have recourse to the method of squares, by which, with the aid of the proportional compasses, the most minute detail may be copied with great accuracy. This may, perhaps, be best shown by an example.

Let Fig. 1 in the following engraving represent a plan of an estate, which it is required to copy upon a reduced scale of one-half. The copy will therefore be half the length and half the breadth, and, consequently, will occupy but one-fourth of the space of the original. Our subject is a map of an estate, but the process would be precisely the same if it were an architectural, mechanical, or any other drawing.

* The eye end of the pricking needle, or the fine point of a porcupine's quill, may be used for this purpose.

1. *Draw* the lines F I and F G at right angles to each other. 2. From the point F, towards I and G, set off any number of equal parts, as F *a*, *a b*, *b c*, &c., on the line F I, and F *i*, *i k*, *k l*, &c., on the line F G. 3. From the points on the line F I

Fig. 1.

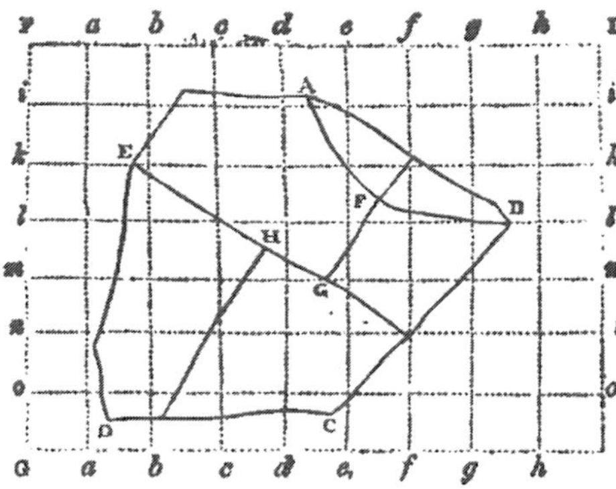

Fig. 2.

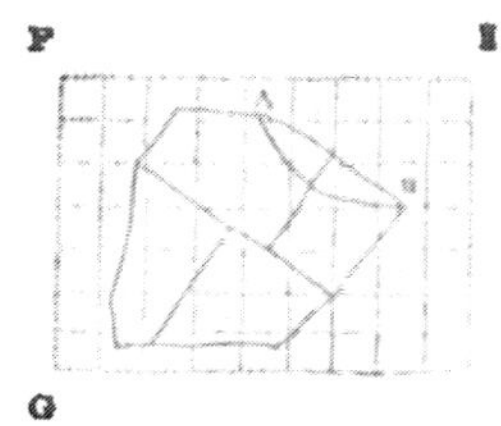

draw lines parallel to the other line F G, as *a a*, *b b*, *c c*, &c., and from the points on F G draw lines parallel to F I, as *i i*, *k k*, *l l*, &c., which being sufficiently extended towards G and I, the whole of the original drawing will be covered with a reticule of small but equally sized squares.

This done, draw upon the paper intended for the copy a similar set of squares, but having each side only one-half the length of the former, as is represented in Fig. 2. It will now be evident that, if the lines of the plan A B, B C, C D, &c., Fig. 1, be drawn in the corresponding squares of Fig. 2, a correct copy of the original will be produced, and of half the original scale. Commencing then at A, observe where, in the original, the angle A falls, which is towards the bottom of the square marked on the top *d e*. In the corresponding square, therefore, of the copy, and in the same proportion towards the left-hand side of it, which should be determined by the use of the proportional compasses described at p. 63, place the same point in the copy. From thence finding, by the proportional compasses, the point on the bottom line of that square where the curved line A F crosses, which is about two-fifths from the left-hand corner towards the right, cross it similarly in the copy. Again, as it crosses the right-hand bottom corner of the second square below *d e*, describe it so in the copy; and by means of the proportional compasses find the points where it crosses the lines *f f* and *g g*, above the line *l l*, by taking the distances of such crossings from the nearest corner of a square in the original, between the large points of the proportional compasses, and with the small points at their opposite end setting off the required crossing on the corresponding lines on the copy. Lastly, determine the place of the point B in the third square below *g h* on the top line; and a line drawn from A in the copy, through these several points to B, will be a correct reduced copy of the original line. Proceed in like manner with every other line on the plan, and its various details, and you will have the plot or drawing laid down to a small scale, yet bearing all the proportions in itself exactly as in the original.

It may appear almost superflous to remark, that the process of enlarging drawings by means of squares is a similar operation to the above, except that the points are to be determined in the smaller squares of the original, and transferred to the larger squares of the copy. The process of enlarging, under any circumstances, does not, however, admit of the same accuracy as that of reducing; and when an enlarged drawing is required, it is best to construct it anew from the original data.

THE EIDOGRAPH.

The pantagraph just described requires four supports upon the paper, and from this cause, and from its numerous joints, its action is apt to be unsteady. An instrument to avoid these defects was invented by Professor Wallace in 1821. This instrument, called the eidograph, is more regular in its action than the pantagraph, as will readily be understood from the following description of its construction, by which it will be seen that there is only one point of support upon which the entire instrument moves steadily and regularly; and the joints, if we may so term them, consist of fulcrums fitting in accurately ground bearings, the motion round these fulcrums being capable of adjustment for regularity as well as accuracy. It also possesses the further advantage over the pantagraph, that it may be set with equal facility to form a reduced copy bearing any proportion whatever to the original, while the pantagraph can only be set to vary the relations between the original and the copy in the few proportions which are specifically marked upon it.

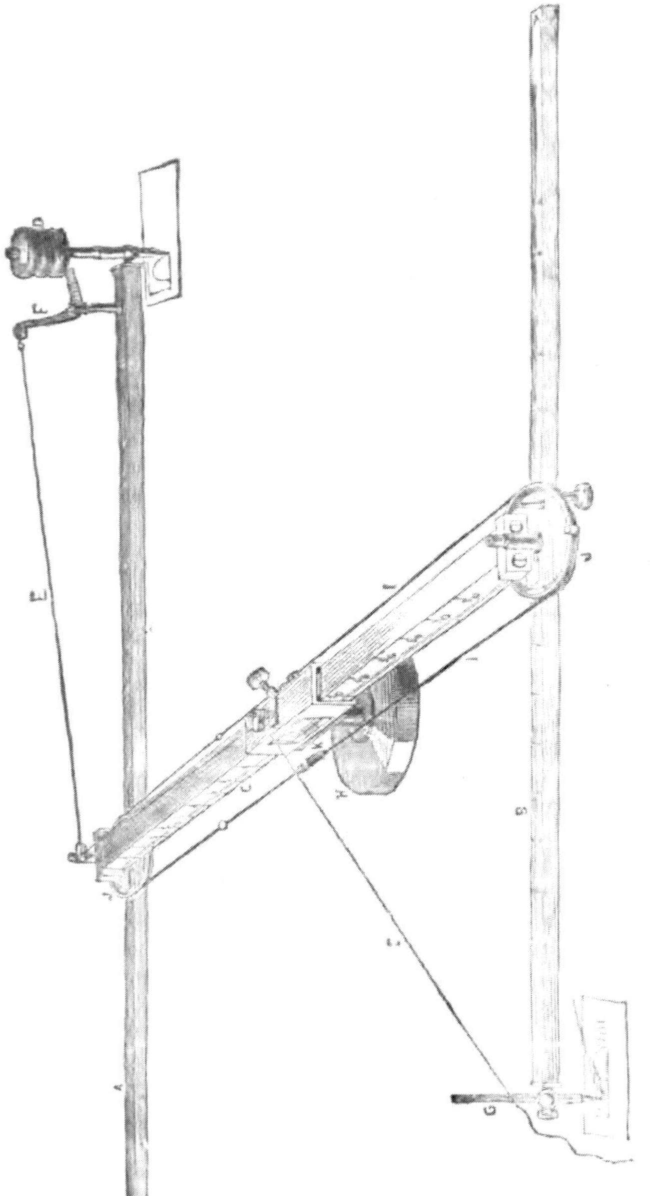

The point of support of the eidograph is a heavy weight, H, formed exteriorly of brass and loaded internally with lead, and having three or four small needle points to keep it steady on the paper. The pin, forming the fulcrum upon which the whole instrument moves, projects from the centre of this weight on its upper side, and fits into a socket attached to a sliding box, K. The fulcrums are ground to fit very accurately. The centre beam, C, of the instrument fits into and slides through, the box K, and may thus be adjusted to any desired position with respect to the fulcrum, and then fixed by a clamping screw attached to the box. Deep sockets are attached to each end of the centre beam, into which are accurately fitted the centre pins of two pulley wheels, J, J. These pulley wheels are made most exactly of the same diameter, and have two steel bands, I, I, attached to their circumferences, so that they can move only simultaneously, and to exactly the same amount. By means of screw adjustments these bands can have their lengths regulated so as to bring the arms of the instrument into exact parallelism, and, at the same time, to bring them to such a degree of tension as shall give to the motions of the arms the required steadiness, which forms one of the advantages of the instrument over the pantagraph. The arms, A and B, of the instrument pass through sliding boxes upon the under side of the pulley wheels, these boxes, like that for the centre beam, being fitted with clamping screws, by which the arms can be fixed in any desired position. At the end of one of the arms is fixed a socket with clamping screw, to carry a tracing-point, G, and at the end of the other is a socket for a loaded pencil, D, which may be raised when required by a lever, F F, attached to a cord which passes over the centre of the instrument to the tracing-point. The centre beam C, and the arms A, B, are made of square brass tubes, divided exactly alike into two hundred equal parts, and figured so as to read one hundred each way from their centres, and the boxes through which they slide have verniers, by means of which these divisions may be subdivided into ten, so that with their help the arms and beam may be set to any reading containing not more than three places of figures. A loose leaden weight is supplied with the instrument to fit on any part of the centre beam, and keep it in even balance when set with unequal lengths of the centre beam on each side of the fulcrum.

The pulleys, J J, being of exactly equal size, when the steel bands II are adjusted so as to bring the arms of the instrument into exact parallelism, they will remain parallel throughout all the movements of the pulleys in their sockets, and thus will always make equal angles with the centre beam. If, then, the two arms and the centre beam be all set so that the readings of their divisions are the same, a line drawn from the end of one arm across the fulcrum to the end of the other arm will form with the beam and arms two triangles, having their sides about equal angles proportionals, and being, therefore, similar; hence any motion communicated to the end of one arm will produce a similar motion at the end of the other, so that the tracing-point being moved over any figure whatever, an exactly similar figure will be described by the pencil.

To adjust the Eidograph, and examine its accuracy.—Set the indices of all three verniers to coincide with the zero divisions on the centre beam and arms, and make marks at the same time with the tracer and with the pencil; then move the pencil point round until it comes to the mark made by the tracer, and if the tracer at the same moment comes into coincidence with the mark made by the pencil, the arms are already parallel, and the instrument consequently in adjustment; but if not, make a second mark with the tracer in its present position, and bisecting the distance between this mark and the mark made by the pencil, bring the tracer exactly to this bisection by turning the adjusting screws on the bands. The instrument being now in adjustment, if the zero division be correctly placed on the arms and beam, the pencil point, tracer, and fulcrum will be in the same straight line, and they will still remain so when the instrument is set to give the same readings on the three scales, whatever those readings may be, if the dividing of the instrument be perfect.

The instrument being adjusted, we have next to set it so as to make the dimensions of a copy, traced by its means, bear the desired proportion to the original. It must be borne in mind that the divisions on the instrument are numbered each way from the centres of the beam and arms up to 100, and that the verniers enable us to read decimals or tenths of a division; so that if the indices of the verniers were a little beyond any divisions, as 26, and the third stroke of the verniers coincided with the divisions marked 29, the reading would be 26·3. Now suppose it were required to set the instrument so that the proportion of the copy to the original should be that of one number, a, to another number b. Suppose x to represent the reading to which the instrument should be set, then the centre beam and arms are each divided at their fulcrums into portions whose lengths are $100-x$ and $100+x$ respectively, and consequently $\frac{100-x}{100+x}=\frac{a}{b}$, from which we find that the required reading $x=\frac{100\,(b-a)}{b+a}$; thus, if the proportions are as 1 to 2, we have $x=\frac{100\,(2-1)}{2+1}=\frac{100}{3}=33{\cdot}3$, and the instrument must be set with the third divisions of the verniers beyond the indices on the third divisions of the instrument beyond the 33rd. We have, therefore, the following simple rule: Subtract the lesser term of the proportion from the greater, and multiply it by 100 for a dividend, add together the two terms of the proportion for a divisor, and the quotient will give the reading to which the instrument is to be set. The following readings are thus obtained :—

Proportions.	Readings.	Proportions.	Readings.
1 : 2	33·3	2 : 3	20
1 : 3	50	2 : 5	42·9
1 : 4	60	3 : 4	14·3
1 : 5	66·7	3 : 5	25
1 : 6	71·4	4 : 5	11·1

When the copy is to be reduced, the centre beam is to be set to the reading found, as above, on the side of the zero next to the arm carrying the pencil point, and this arm is

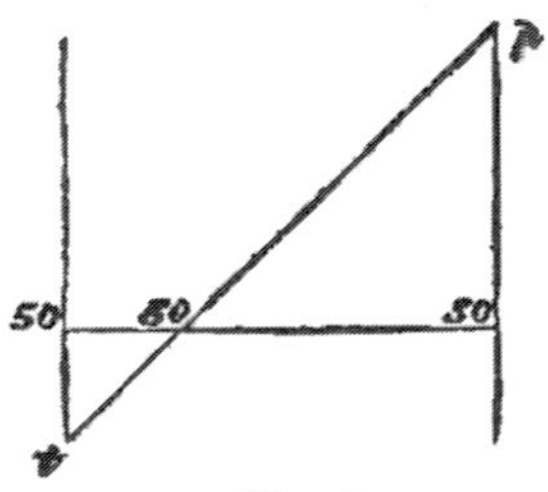

Fig. 1.

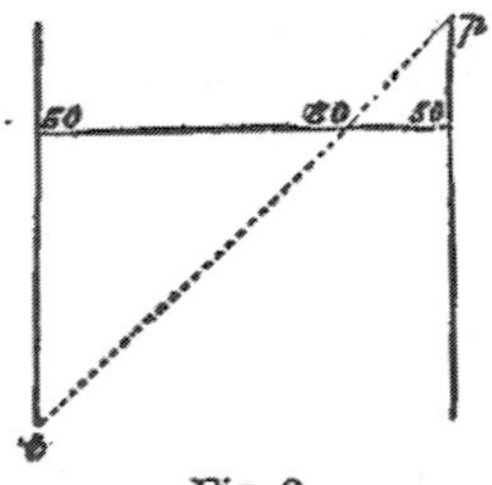

Fig. 2.

also to be set to the same reading on the side of its centre or zero nearest the pencil end, while the tracer arm is to be set with the reading furthest from the tracer. When the copy is to be enlarged, these arrangements must of course be reversed: thus, 50 being the reading for the proportion 1 : 3, Fig. 1 will represent the setting to make a copy having its linear dimensions three times those of the original; where *p* represents the position of the pencil point, *t* that of the tracer, and F the place of the fulcrum. Fig. 2 represents in the same way the setting to make the linear dimensions of the copy one-third of those of the original.

After all, the most perfect copying-machine is the photographic camera. In fact, photography has now fully taken its place as an auxiliary to engineering operations. By its use the mechanical engineer is enabled to send forth *correct* representations of his machinery; and works of civil engineering have their monthly progress recorded, and the results sent home, to give ocular proof of their truth to chiefs and directors. A practical knowledge of photographic processes has hence become almost a necessary, certainly a most desirable portion of every young engineer's education. Several special manuals on the art of photography, in English and French, are to be found. Of the apparatus used for the reduction of maps and plans a full description will be found in Part III. of this volume.

CHAPTER VI.

INSTRUMENTS FOR MEASURING LINES AND AREAS ON PLANS.

THE OPISOMETER.

This simple instrument measures the length of roads, rivers, fences, walls, &c., on any map or plan which is drawn to a scale, without requiring any arithmetical calculations.

The principle of the opisometer is that after having been applied to any line it retraces or measures backwards precisely the same length on the scale with which the line is to be compared. It consists of a milled wheel, with a steel screw for its axis, mounted on a convenient handle. To measure the length of a line, as the distance between two towns by the road traced upon a map, turn the milled wheel up to one end of the screw until it stops, and place the instrument on the map in an upright position, as represented in the draw-

ing, the wheel resting upon one extremity of the line to be measured; then run the wheel along the road, following every bend as closely as possible. Care must be taken to keep the wheel in contact with the paper, but the pressure need not be such as to injure the map. When the wheel has arrived at the other extremity of the line, lift the instrument carefully from the paper, and carrying it to the zero end of the scale, run the wheel backwards along the scale until it stops at the same end of the screw from which the measurement began; the division of the scale at which the wheel stops shows the length of the line measured on the map. Should the scale be shorter than the line measured, when the wheel arrives at the end, carry it to the zero mark again as often as may be necessary, counting the number of repetitions.

The difficulty of measuring lines of double curvature, such as a rhumb line, *i.e.*, the line of a ship's course on a globe, is entirely removed by the use of the opisometer.

The accuracy of the result given by the opisometer is unaffected by the dimensions of the instrument itself, and depends entirely on the care with which it is used. The chief point is to see that the handle of the instrument is perpendicular to the surface at the beginning and end of each step of the measurement.

The computation of the area represented by a plan, especially if the outline be very irregular, is both a tedious task and liable to many sources of error, which, after all, forbid us from looking at the result as any more than a careful approximation. In the first place, the irregularity of the outline is dealt with by drawing straight lines, which, in the estimation of the computer, shall go as far outside the figure on one part as they fall inside on another, so as to form a rectilinear figure estimated to be equal to the plan. This is called the give-and-take principle, and is, of course, a mere guess, depending for its degree of accuracy upon the skill and practice of the operator. Then the polygon thus formed has either to be divided into triangles, the area of which must be separately computed, or a triangle must be constructed, by the principles of geometry, equal to the whole polygon, and then the area of this triangle will be the estimated area of the surface represented by the plan. Any means of instrumentally getting at the computation with greater expedition, without producing greater inaccuracy than this mode of computation, must be hailed as a boon, and we have a very simple and inexpensive instrument, which is now much used for this purpose, in the

COMPUTING SCALE, OR COMPUTER.

The computer is designed to measure only the areas of figures constructed to one particular scale, and as many com-

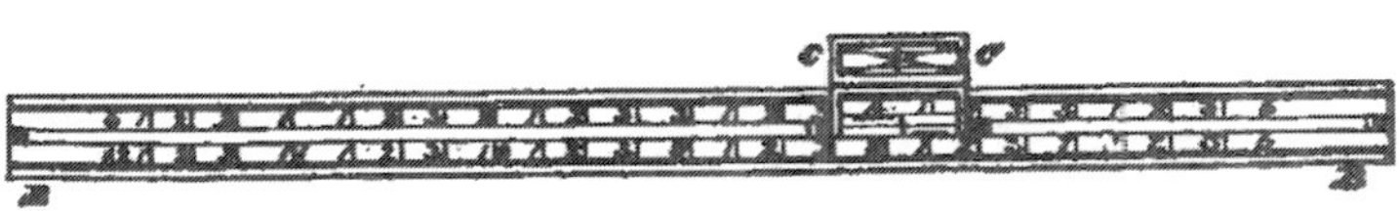

Fig. 1.

puters, therefore, would be required by the surveyor as he adopted scales for his plans. The simplicity, and consequently inexpensive nature, of the instrument prevent this from being felt as a serious disadvantage.

The computer (Fig. 1) consists of a rule, B B, of box-wood, graduated and figured to show the number of acres and roods in a slip of land of a given width. In that from which the annexed figure is drawn, the graduations on the rule itself show the number of acres and roods in a slip of land one chain wide, and it is adapted to a plan of 3 chains to an inch. The scale occupies 20 inches of the length of the rule, and therefore represents 60 chains in length, and an area of 60 square chains, or 6 acres, in a slip of 1 chain wide. These 20 inches are therefore divided into 6 equal parts for single acres, and each of these parts is divided into four equal subdivisions to represent roods. Along the centre of the rule is an undercut groove, and the rule is graduated on each side of this groove, so that it reads forward from left to right on the upper side of the groove, commencing at 0 and terminating at 6 acres, and then reads back again from right to left on the under side of the groove, commencing at 6 acres and terminating at 12 acres. Into the groove is fitted, so as to slide freely along it, a tongue or slip of brass, A A, represented in

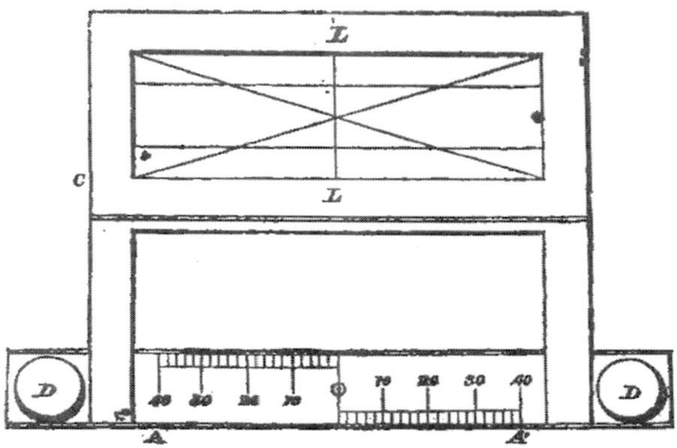

Fig. 2.

Fig. 2 on an enlarged scale, which has an index drawn across its centre, and on each side of this index a distance equal to one of the subdivisions on the rule is divided into 40 equal parts to represent perches. These divisions are placed, those on the left of the index to read with the divisions of the scale on the upper side of the groove, and those on the right of the index to read with divisions on the under side of the groove. To the tongue A A is attached a frame, C C, which passes over the side of the rule, and lies flat upon the paper on which the rule is placed when in use. The handles D, D, enable the tongue and attached frame to be moved with facility along the divided scale. The frame encloses a piece of thin horn, on which are ruled two lines parallel to the rule, at the distance apart which represents a chain, in the present instance ⅛ of an inch, and the centre of this enclosure being determined by the intersection of its diagonals, a line L L, called the index line, is drawn through this centre at right angles to the parallel lines, and in the same straight line with the index on the brass tongue.

The following is the mode of using the computer:—The plan, which, as has been already stated, has been constructed on the same scale as the computer to be applied to it, is ruled over with parallel lines at distances apart representing one chain, in this case ⅛ of an inch, and so dividing the whole plan into slips of one chain in width. The computer, having its index set to zero on the scale, is now placed on the paper parallel with these lines, in such a position that the parallel lines on the horn move along in coincidence with the first two lines, enclosing the first slip of land into which the plan has been divided, and that the index line, L L, lies across the boundary of one end of this slip; and the tongue being then moved along the scale until the index line crosses the boundary at the other end of the slip, the position of the index will indicate on the scale the content of the first slip of land. The instrument must now be moved without shifting the index, and placed with the index line across the boundary at the beginning of the next slip, and the tongue is again to be slid along until the index line crosses the boundary at the other end, when the position of the index will indicate the content of the two slips; and proceeding in this way from slip to slip until the whole have been passed over, the index will finally show the content of the entire surface represented by the plan. When this content is more than that represented by the entire length of the scale (which in the present case is 6 acres), as soon as the index reaches the 6 at the end of the scale the instrument should be turned round, care being taken to place the index line again in exactly the same position; and the operation can then be continued until the index comes back to its original position, and indicates 12 acres, when it must again be turned, and a note made that 12 acres have been computed. Proceeding in this way, the ultimate reading of the index, added to the number of 12 acre pieces already noted, will give the entire content in acres, roods, and perches.

At the beginning and end of each slip the operator must exercise his judgment as to the proper position of the index line, so that in crossing the boundary he shall add as much on the one side of the line as he cuts off on the other. Practice will also enable him to judge of the best direction for his parallel lines, so that the contents of the extreme slips may be best estimated.

Instead of ruling lines over the plan itself, if these be ruled on a transparent paper, such as is used for tracing, they can be shifted round upon the plan until they come into a position which will best suit the operator's purpose. This method has the additional advantage of avoiding any injury to the plan itself, either by ruling it over, or by making marks to assist in resetting the instrument each time the index arrives at one end of the scale.

THE POLAR PLANIMETER.

The method of taking the areas of figures by the computer just described, though more expeditious than that of dividing the figure into triangles, and computing their contents by geometrical rules, is still a troublesome operation, and the correction for the irregularity of outline upon the give-and-take principle is still but guesswork; consequently an instrument which will give the results with greater facility and accuracy has been a great desideratum.

There is a very elegant instrument, of French origin, called the totalisateur, which accomplishes this object; but its expense alone would prevent it from coming into common use. The instrument, however, which we proceed to describe leaves nothing to be desired either in point of simplicity of construction, and consequently moderate price, facility of use, or accuracy of the results obtained. The whole operation of taking an area with this instrument consists in passing a point round the outline of the figure, and the operator is astonished by the marvellous accuracy of the results obtained, no less than by the extreme simplicity of the instrument with which it is accomplished.

This instrument, called the polar planimeter, is the invention of Mr. J. Amsler, professor of mathematics at Schaffhausen, and is sold in England by Messrs. Elliot Brothers, of Charing Cross. It consists essentially of two arms jointed together, so as to move with perfect freedom in one plane, and a wheel which is attached to one of the arms, and turning on this arm as an axis, records by its revolutions the area of a figure traced out by a point on the arm to which it is attached, while a point on the other arm is made a fixed centre, about which the instrument revolves.

Thus, let A B, B C represent the two arms, of which B C carries a wheel at D, revolving freely in a plane at right angles

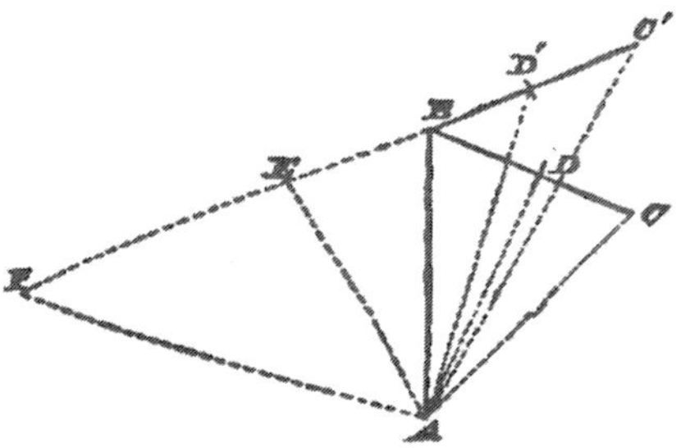

to B C, and let the arms be opened until the line A D drawn from a point A on the arm A B to the wheel is at right angles to B C; then if, without varying the angle between the arms, the instrument be moved round A as a centre, any point C on B C will describe a circle which will constitute a zero or datum line, the wheel at D then moving entirely perpendicular to its own plain, and consequently without turning upon its axis; again, let the arms be opened until B D C comes into the position B D′ C′, and from A draw A E perpendicular to C′ B, produced if necessary, and A F at right angles to A D′; then if, with this new position of the arms, the instrument is again moved round A as a centre, the point C′ will describe another circle, and while the point D′ moves through a distance represented by A F perpendicular to A D′, the wheel will slip through the distance F E at right angles to its own plane, and will run through the distance E A parallel to it, which latter distance alone it will therefore record. Now the triangles F E A and A E D′ being similar, E D′ has the same ratio to A D′ that A E has to A F, and, therefore, while the point D′ actually describes an arc of the circle whose radius is A D′, the circumference of the wheel rolls through a distance equal to the corresponding arc of a circle whose radius is D′ E. Now, from Euc. Bk. ii. props. 12 and 13, we have $A\,C^2 = A\,B^2 + B\,C^2 - 2\,B\,C, B\,D$, and $A\,C'^2 = A\,B^2 + B\,C'^2 + 2\,B\,C, B\,E$, and therefore $A\,C'^2 - A\,C^2 = 2\,B\,C\,(B\,E + B\,D') = 2\,B\,C \times D'\,E$: but the area of the ring contained between the circles described by the points C and C′ is equal to $\pi\,(A\,C'^2 - A\,C^2)$, and therefore to $2\,\pi\,D'E \times B\,C$, *i.e.*, to the *circumference* of the circle whose radius is D′ E, which is the distance rolled through by the circumference of the wheel, multiplied into the distance B C from the joint B to the tracing-point C. If the arms, instead of being opened further apart, had been brought closer together, so that C′ in its second position had described a circle within the datum or zero circle, the area of the ring between the two circles would still have been measured by the product of B C into the distance described by the circumference of the wheel. Now, when the point C is moved over any outline while the centre A remains fixed, we have two motions, one of the whole instrument round A, and the other of the arm B C round the joint B; but in going completely round any figure to the point of starting, the arms must at last come back into the same position as they held at first; and, hence, whatever amount the wheel may have turned one way in consequence of a movement about the joint B in one direction, it must have turned back an equal quantity the other way in consequence of a movement about the joint in the opposite direction, and hence the whole of the recorded motion is that due to the rotation of the instrument about the centre A, and measures the areas included between the outline and the datum circle; and since those areas are equal to the distance run by the circumference of the wheel multiplied by B C, the graduations on the wheel, whatever be their lengths, may be made to record the areas in any desired unit of measurement by making the length of B C contain as many of these units as the unit contains of the graduated lengths. The length of the arm A B, and the part of the arm B C upon which the wheel is placed, do not affect the scale of measurement, and the instrument may therefore be constructed without regard to these points, or to the size of the wheel, and regulated solely by the distance B C of the tracing-point C from the joint B. The forms given to the instrument of which the principle has been fully explained above are represented in Figs. 1 and 2. In Fig. 1 the arm A carrying the tracing-point passes through a slide-box J, so that the tracing-point may be set at different distances from the joint, and thus give the areas in different units of measure. In Fig. 2 the arm A is of invariable length, and consequently the areas to be measured can be read off from the instrument only in one kind of standard measure. Those manufactured of this pattern for this country are regulated to record the areas in square inches.

The essential points in the construction of the instrument are as follows:—1. The centre pin C C, the axis I, and the extremity of the tracing-point E must lie all in one plain; and hence great care must be taken not to bend the arm A or the tracer. 2. The roller D must play easily without coming into contact with the vernier. 3. The screw centres on which its axis, I, revolves must be adjusted so as to allow of perfect freedom of rotation, and so also must those on which the centre pin, C C, revolves; these centres have to be greased sometimes with a little sweet oil. 4. The roller D, which moves on the paper, will not bear the least spot of rust or slightest injury without impairing the accuracy of the instrument. 5. The needle point ought to project but very little from its socket. In the instrument represented in Fig. 2, the circumference of the roller measures nearly 2½ inches, and

Fig. 1.

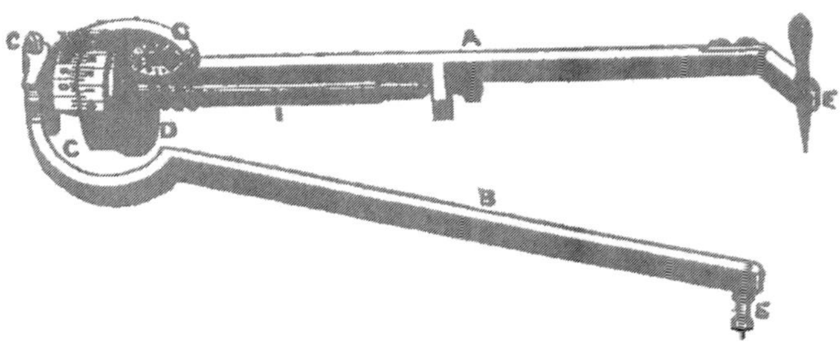

Fig. 2.

the tracing-point being set at such a distance (viz., differing a little from 4 inches) as will make the product of this distance into the circumference of the roller = 10 square inches, every complete revolution of the roller records an area of 10 square inches. The correct setting of the tracing-point is, of course, to be obtained by passing it over a test outline of a surface whose area is accurately known. The roller D, by means of an endless screw on its axis I, acting upon a rack, gives one revolution to the counting wheel, G, for every ten revolutions of its own, and this counting wheel being graduated into ten equal divisions, every one of these records an area of 10 square inches. The primary divisions on the roller D itself being ten in number, each of these records an area of 1 square inch; and, each of these being subdivided into ten equal parts, and these subdivisions again divided into ten by means of the vernier, the areas are read off to the 100th of a square inch.

In using the instrument, (1) place it upon the paper in such a manner that it rests upon a point of the circumference of the roller D, the tracing-point F, and the needle point E: press the latter point gently on the paper, but not enough to pierce it through, and place above the point E the small weight supplied with the instrument, to keep this point stationary during the operation; or the same result may be obtained by a light pressure of the thumb or finger. (2) Mark any point P (Figs. 3 and 4) on the outline of the figure; and, after setting the tracing-point F on this point, read off the counting wheel G, the roller D, and the vernier H: suppose, for instance, the readings indicated are 2 for the counting wheel, 91 for the roller, and 5 for the vernier; then the number 29·15 must be written down. (3) Carry the tracing-point F round, in the direction of the hands of a watch, with great exactness over the outline of the figure to be measured, until the starting-point is again reached: along straight lines the tracing-point may be guided by a rule: observe the readings again: suppose them to be 4 for the counting wheel, 76 for the roller, and 7 for the vernier; then the number 47·67 must be written down for the second reading. (4) The difference of the two readings gives the areas between the outline and the datum circle P R, and consequently we have two cases to consider: in Fig. 3, where the needle point, E, is outside the figure to

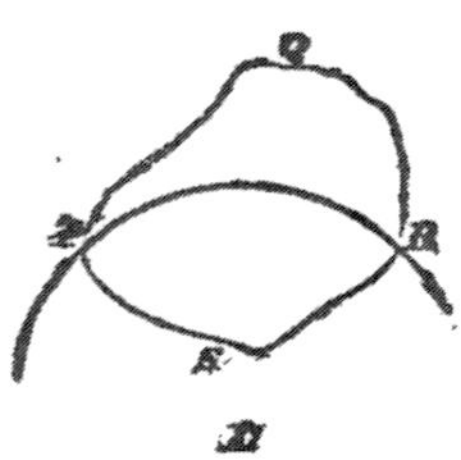

Fig. 3.

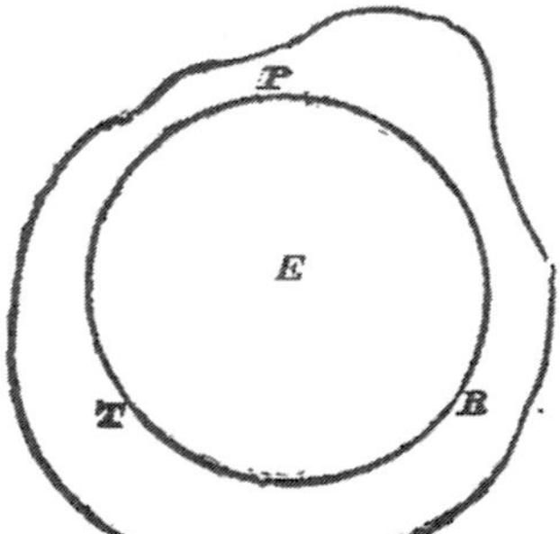

Fig. 4.

be measured, the difference of the two readings gives 18·52 square inches for the areas of the two portions, P Q R and P R S, and consequently of the whole figure P Q R S; in Fig. 4, where the needle point E is within the figure to be measured, the excess of the second reading over the first gives only the area beyond the datum circle P R T, the result being negative if the figure be less than the datum circle: consequently, the area of the datum circle must be added to the second reading before the first is subtracted from it, and the remainder will give the area of the figure required. The area of the datum circle may be found by first placing the needle point E outside a convenient figure, and taking the difference of the two readings: suppose it to be 171·05, this represents the area of the figure; again placing the needle point within the figure, let the excess of the second reading above the first be 10·64; this represents the excess of the area of the figure above that of the datum circle; consequently the area of this circle is 171·05 — 10·64 = 160·41; this quantity 160·41 is engraved either on the bar A, or upon the small weight supplied with the instrument. Suppose now, in going over a figure with the needle point E within it, the first reading to be 20·95, and the second to be 17·34, then the area would be found as follows:—

Second reading	17·34
Add datum circle	160·41
	177·75
Deduct first reading . . .	20·95
Remainder . . .	156·80

The measured area of the figure on the paper is, therefore, 156·8 square inches.

If, in measuring an area, the 0 or zero on the counting wheel passes the index point while moving forwards, 100·00 must be added to the second reading, and if it passes while moving backwards, 100·00 must be added to the first reading.

If the figure is so large that the tracing-point cannot conveniently be passed round it at one operation, it can be subdivided by drawing straight lines through it, and the contents of the several parts measured separately can be added together to give the total area.

If a large amount of work had to be done in measuring a series of maps or plans, all constructed to one scale, the instrument represented in Fig. 1 could be adapted to measure the areas to this scale by constructing on the same scale a rectangular figure of a given area, and sliding the arm A to such a part of the tube as to give the known area correctly.

Roods...	0	1	2	3
Perches.	..	·250	·500	·750
1	·006	·256	·506	·756
2	·012	·262	·512	·762
3	·019	·269	·519	·769
4	·025	·275	·525	·775
5	·031	·281	·531	·781
6	·037	·287	·537	·787
7	·044	·294	·544	·794
8	·050	·300	·550	·800
9	·056	·306	·556	·806
10	·062	·312	·562	·812
11	·069	·319	·569	·819
12	·075	·325	·575	·825
13	·081	·331	·581	·831
14	·087	·337	·587	·837
15	·094	·344	·594	·844
16	·100	·350	·600	·850
17	·106	·356	·606	·856
18	·112	·362	·612	·862
19	·119	·369	·619	·869
20	·125	·375	·625	·875
21	·131	·381	·631	·881
22	·137	·387	·637	·887
23	·144	·394	·644	·894
24	·150	·400	·650	·900
25	·156	·406	·656	·906
26	·162	·412	·662	·912
27	·169	·419	·669	·919
28	·175	·425	·675	·925
29	·181	·431	·681	·931
30	·187	·437	·687	·937
31	·194	·444	·694	·944
32	·200	·450	·700	·950
33	·206	·456	·706	·956
34	·212	·462	·712	·962
35	·219	·469	·719	·969
36	·225	·475	·725	·975
37	·231	·481	·731	·981
38	·237	·487	·737	·987
39	·244	·494	·744	·994

From the areas, however, of the figures on the paper, as given in square inches by the instrument in Fig. 2, the areas of the figures represented according to the scales of so many chains to an inch ordinarily made use of, can readily be obtained by removing the decimal point one place to the left, and multiplying the result by the square of the number of chains to the inch, since one square chain is a tenth of an acre; thus, if the measured area be 51·65, and the scale be one chain to an inch, the area represented will be 5·165 acres; if the scale were three chains to an inch, the area represented would be 9 × 5·165 = 46·485 acres. The decimal of an acre contained in the area may be readily converted into roods and perches by multiplication, or these may be taken from the foregoing table, where the roods corresponding to any decimal will be found at the top of the column in which it stands, and the perches will be found on the same horizontal line with the decimal in the column on the extreme left: thus, if the decimal be ·587, at the top of the column in which ·587 is found stands the number 2, and opposite ·587 in the left-hand column is 14; and 2 roods 14 perches is the value of ·587 of an acre.

PART II.

ON RULES AND GAUGES.

The instruments described in this part are of two classes: viz., (1) instruments for the performance of arithmetical computation; and (2) instruments designed for accuracy of measurement.

Under the first head will be included descriptions of

Gunter's Lines,
Gunter's Scale,
Slide-rules.

Under the second,

Callipers and Gauges for the measurement of timber,
Spherometer,
Extensometer,
German-silver Wire-gauge.

SECTION I.

INSTRUMENTS FOR ARITHMETICAL COMPUTATION.

GUNTER'S LINES.

A series of numbers increasing by equal differences is said to be in arithmetical progression; thus, each of the series

0	1	2	3	4	5, &c.,
	1	3	5	7	9, &c.,

is in arithmetical progression, the common difference being 1 in the first series and 2 in the second.

A series of numbers increasing by equal ratios is said to be in geometrical progression; thus, each of the series

1	2	4	8	16	32, &c.,
	2	8	32	128	512, &c.,

is in geometrical progression, the common ratio being 2 in the first series and 4 in the second.

In the two examples of arithmetical series given above, it is to be noted that 0 forms a term of the first, but not of the second; and in the two examples of geometrical series 1 forms a term of the first, but not of the second.

Now if the terms of a series of numbers in arithmetical progression, of which 0 forms a term, be written in order over the terms of a series of numbers in geometrical progression, of which 1 forms a term, taking care to place the 0 of the arithmetical series over the 1 of the geometrical series, the terms of the arithmetical series are called logarithms of the corresponding terms of the geometrical series: thus, if we select the first of the series of each class given above,

0	1	2	3	4	5, &c.,
1	2	4	8	16	32, &c.,

the numbers 0, 1, 2, 3, 4, 5, &c., are respectively logarithms of the numbers 1, 2, 4, 8, 16, 32, &c., over which they stand: so also of the series

0	3	6	9	12	15, &c.,
1	5	25	125	625	3125,

the terms of the first are logarithms of the terms, over which they stand, of the second.

Hence it appears that the variety of systems of logarithms is infinite. Only one, however, is made use of for the purposes of arithmetical computation, viz., that in which the arithmetical series whose common difference is 1 is made to correspond with the geometrical series whose common ratio is 10: thus,

Logarithms	0	1	2	3	4	5, &c.,
Numbers	1	10	100	1000	10000	100000, &c.

The logarithms of the numbers between 1 and 10, 10 and 100, &c., are computed by methods which it is not our present business to investigate, and, being registered in due order, form what is termed a table of logarithms.

On examining any system of logarithms, the following remarkable properties will be observed :—

If any two of the logarithms are added together, their *sum* will be the logarithm of the *product* of the numbers to which they belong: thus, in the system,

Logarithms	0	1	2	3	4	5, &c.,
Numbers	1	2	4	8	16	32, &c.,

2 is the logarithm of 4, 3 is the logarithm of 8, and 5, the sum of 2 and 3, is the logarithm of 32, the product of 4 and 8.

If any one of the logarithms be substracted from another, their *difference* will be the logarithm of the *quotient* resulting from dividing the second of the corresponding numbers by the first: thus, in the system,

Logarithms	0	3	6	9	12	15, &c.,
Numbers	1	5	25	125	625	3125, &c.,

9 is the logarithm of 125, 15 is the logarithm of 3125, and 6, the defect of 9 from 15, is the logarithm of 25, the quotient of 3125 divided by 125.

If any one of the logarithms be multiplied by a number, the product will be the logarithm of that power of the corresponding number which is denoted by the multiplier; and if any one of the logarithms be divided by a number, the quotient will be the logarithm of that root of the corresponding number which is denoted by the divisor: thus, in the system to which the table of logarithms is computed,

Logarithms	0	1	2	3	4	5, &c.,
Numbers	1	10	100	1000	10000	100000, &c.,

2 is the logarithm of 100, and 2 multiplied by 2 is 4, the logarithm of 10000, which is the square or second power of 100. Again, 6 is the logarithm of 1000000, and 6 divided by 3 is 2, which is the logarithm of 100, the cube root of 1000000.

Thus, then, the addition of logarithms answers to multiplication of the corresponding numbers; subtraction, to division; multiplication, to involution; and division, to evolution.

Logarithms were invented by John Napier, Baron of Merchiston, and were first published in 1614. In 1620 Edmund Gunter, then Gresham Professor, inventor of the Sector, laid them down on a scale, by the application to which of a pair of compasses arithmetical questions might be solved instrumentally.

The difficulty that formerly existed of laying down the

logarithmic divisions correctly considerably marred their utility; but in the present day rules are engine-divided, and most useful instruments for computations of all kinds are produced.

Besides lines graduated to form a scale of logarithms of numbers, others are so graduated as to represent the logarithms of the sines, tangents, &c., of the degrees and minutes in a quadrant; and to these are added, for the use of navigators, lines of the logarithms of the sine-rhumbs and tangent-rhumbs. All these are called GUNTER'S LINES.

They may be constructed as follows :—

1. *To construct the Line of Logarithmic Numbers marked N.* —Having fixed upon a convenient length for the entire scale, bisect it, and figure it 1 at the commencement on the left hand, 1 again in the middle, and 10 at the end. The half line, then, is taken for unity, or the logarithm of 10, and, consequently, the whole line represents 2, or the logarithm of 100. The half line being now divided into 1,000 equal parts, the lengths corresponding to the three first figures of the logarithms of 2, 3, &c., up to 9, as found in the common table of logarithms, may be taken, and set off from the 1 at the commencement of the line for the logarithms of 2, 3, &c., to 9, and again from the 1 at the middle of the line for the logarithms of 20, 30, &c., to 90. The divisions thus formed are to be subdivided by setting off, in the same manner, the three first figures of the logarithms of 1·1, 1·2, 1·3, &c., to 1·9; of 2·1, 2·2, 2·3, &c., to 2·9, and so on, each of the primary divisions being thus subdivided into ten; and these again are to be subdivided each into ten, or five, or two, as the length of the secondary divisions may admit, by setting off the logarithms of 1·11, 1·12, 1·13, &c.; or of 1·12, 1·14, &c.; or of 1·15, 1·25, &c.; and the scale is completed.

2. *To construct the Line of Logarithmic Sines marked S.*— The whole length of the scale is taken as the logarithm of the radius, and, since this extent upon the line of numbers

represents 2, or the logarithm of 100, it follows that the line of sines is to be a scale of the logarithms of the sines, to radius 100, of which the logarithm is 2; whereas the logarithmic table of sines is set down to a radius, of which the logarithm is 10. By taking 8, then, from each of the tabulated values of the logarithmic sines, we should obtain the logarithmic sines to radius 100, and the three first figures of these reduced values are to be set off from the left hand towards the right; 1st, for every 10 degrees, then for every degree, and then for every half degree, every 10 minutes, and every 5 minutes, as far as the length of the several primary divisions will admit. The line is then numbered 1, 2, 3, &c., at every degree to 10; and afterwards 20, 30, 40, &c., at every ten degrees to 90, which stands at the extreme right, since sine 90° equals radius.

The tabulated logarithmic sine of 34′ 23″, being 8·0000669, will coincide, or nearly so, with the zero point upon our scale, and consequently angles smaller than this cannot be taken off from the sines. This remark applies equally to the line of tangents, the tabulated logarithmic tangent of 34′ 23″ being 8·0000886.

By taking the extents backwards from right to left, and reckoning them as forward distances, the line of sines becomes a line of cosecants,* giving us, in fact, the excesses of the logarithmic cosecants above the logarithmic radius; and, by taking the complements of the required angles, the line of sines becomes a line of cosines when measured forwards from left to right, and a line of secants when measured backwards from right to left.

3. *To construct the Line of Logarithmic Tangents marked T.*—8 being taken from each of the tabulated values of the logarithmic tangents up to 45°, the extents corresponding to these values are to be set off upon the scale, and numbered from left to right, in a similar manner to that in which the logarithmic sines were set off and numbered upon the line of logarithmic sines. The logarithmic tangent of 45°, extends

* Cosecant $= \frac{r^2}{\text{sin.}}$, and sec. $= \frac{r^2}{\text{cos.}}$;

and, therefore, log. cosecant = 2 log. rad. — log. sine;
or, log. cosecant — log. rad. = log. rad. — log. sine;
and log. secant = 2 log. rad. — log. cos.;
or, log. secant — log. rad. = log. rad. — log. cos.

to the extreme right of the scale, coinciding in extent with the sine of 90°, since tangent 45° equals radius; and the logarithmic tangents of the angles from 45° to 90° are measured backwards from the extreme right to the complement of the angle required, these extents giving us, in fact, the excesses of the logarithmic tangents sought above the logarithmic radius.* When, then, the angle is greater than 45°, the distance from the radius to the angle, though measured backwards upon the scale, must be reckoned a forward distance, and *vice versâ*.

The lines of logarithmic sine-rhumbs, marked S.R., and tangent-rhumbs, marked T.R., are formed in the same way as the lines of logarithmic sines and tangents, but are set off for the angles corresponding to the points and quarter points of the compass, instead of for degrees and minutes.

We shall now proceed to explain the uses of Gunter's lines.

1. *The Line of Logarithmic Numbers.*—The primary divisions upon this line, as explained in its construction, represent the logarithms of all the integers from 1 to 100, while the extents to the first subdivisions will indicate tenths of an unit from the beginning of the scale to 1 in the middle, and units from 1 in the middle to 10 at the end, where the figures 2, 3, &c., stand for 20, 30, &c., as has been explained in the construction. If any of the subdivisions be further subdivided into ten parts, each of these last divisions will indicate hundredths of an unit from 1 at the beginning to 1 in the middle, and tenths of an unit from 1 in the middle to 10 at the end. By this line the multiplication and division of numbers of any denomination either whole or fractional may be readily accomplished, questions in proportion solved, and all operations approximatively performed with great rapidity, which can be performed by the common table of logarithms; but the numbers sought must always be supposed to be divided or multiplied by 10 as many times as will reduce them to the numbers, the logarithms of which are actually set off upon the line of numbers, and these tens must be mentally accounted for in the result.

* $\text{Tan.} = \frac{r^2}{\text{cotan.}} = \frac{r^2}{\text{tan. of compt.}}$;

$\therefore$ log. tan. = 2 log. rad. — log. tan. of compt.;
or, log. tan. — log. rad. = log. rad — log. tan. of compt.

Multiplication is performed by extending from 1 on the left to the multiplier; and this extent will reach forwards from the multiplicand to the product. Thus, if 125 were given to be multiplied by 250, extend the compasses from 1 at the left hand to midway between the second and third subdivision, in the first primary division from 1 to 2, for the 125. This extent is really the logarithm of 1·25. Set off this extent towards the right from the fifth subdivision after the primary division marked 2, which is taken to represent the logarithm of 250, but is really the logarithm of 2·5, and the compasses will reach to a quarter of the next subdivision beyond the first subdivision after the primary division marked 3. The extent to this point is really the logarithm of 3·125 ; but in this case it represents the number 31250, because two powers of ten have been cast out from both the multiplier and multiplicand, and therefore the product must be multiplied by the product of four tens, or ten thousand; or, in other words, the first figure of the product must be reckoned as so many tens of thousands.

Division, being the reverse of multiplication, is performed by extending from 1 on the left to the divisor; and this extent will reach backwards from the dividend to the quotient. Thus, if 31250 were to be divided by 250, extend the compasses from 1 on the left to 2·5, and this extent will reach backwards from 3·125 to 1·25. Then, since the divisor contained 2 powers of ten and the dividend 4, the quotient must contain 2, and therefore the result is 125.

Proportion being performed by multiplication and division, extend the compasses from the first term to the second, and this extent will reach from the third to the fourth, taking care to measure in the same direction, so that, if the first be greater than the second, the third may be greater than the fourth, and *vice versâ*. *Example.*—If the diameter of a circle be 7 inches, and the circumference 22, what is the circumference of another circle, the diameter of which is ten? Extend the compasses from 7 to 10, and this extent will reach from 22 to 31·4, or nearly 31½ inches, the circumference required.

The same thing may also be performed by extending from the first term to the third, and this extent will reach from the third term to the fourth (Euc. v., prop. 16). Thus, the extent from 7 to 22 will reach from 10 to 31·4, as before.

To measure a Superficies, extend from 1 to either the

breadth or length, both being reduced to the same denomination, and this extent will reach forwards from the length or breadth to the superficial content. *Example.*—Required the superficial content of a plank 27 feet long by 15 inches broad. Extend from 1 to 1·25, for 15 inches equals 1·25 feet, and this extent will reach from 27 feet to 33·75 feet, the superficial content required.

Second Method.—Extend from 12 to the number of inches in the breadth, and this extent will reach in the same direction from the number of feet in the length to the number of square feet in the superficial content. Thus the extent forwards from 12 to 15 will reach forwards from 27 to 33·75, as before; while the extent backwards from 12 to 9 will reach backwards from 27 to 20·25 or 20¼, showing the superficial content of a plank 27 feet long by 9 inches broad to be 20·25, or 20¼ feet.

To measure a Solid Content.—The breadth, depth, and length being all reduced to the same denomination, extend from 1 to either the breadth or depth, and this extent will reach from the depth or breadth forwards to a fourth number, which will represent the superficial content of the section at the place measured: then, if the breadth and depth be the same throughout the entire length, the extent from 1 to the superficial content thus found will reach forwards from the length to the solid content. *Example.*—What is the solid content of a pillar 1 foot 3 inches square, and 21 feet 9 inches long? The extent from 1 to 1·25 reaches forward from 1·25 to 1·56, the superficial content of a section of the pillar; and the extent from 1 to 1·56 reaches from 21·75 to 34, or more accurately to 33·93, the solid content in feet.*

2. *The Lines of Logarithmic Sines and Tangents.*—These lines are generally used, in connection with the line of numbers, for solving all proportions in which any of the terms are functions of angles, as sines, tangents, &c., and, in fact, all questions in which such quantities appear as factors or divisors. We will exemplify their use by giving the solution, by their aid, of the several cases of right-angled trigonometry.

* The uses of the line of logarithmic numbers will be more fully appreciated after studying the slide-rule, to be presently described but the student will, we hope, from what he sees here, perceive that it can be readily applied to every case of mensuration, and, in short, to almost every arithmetical operation. Additions and subtractions, however, cannot be performed by it.

Case 1*. The hypothenuse and angles being given, to find the perpendicular and base.

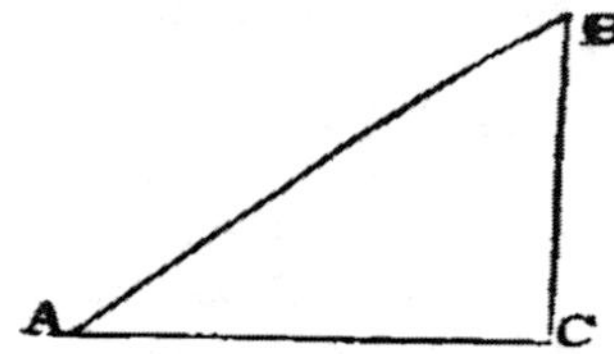

NOTE.—One acute angle of a right-angled triangle being the complement of the other, or the sum of the two acute angles being equal to 90°, when one of the acute angles is given, the other is also given.

Solution.—Extend the compass from 90°, or radius, on the line of sines to the number of degrees in either of the acute angles, and that extent will reach backwards, on the line of numbers, from the hypothenuse to the side opposite this angle. *Example.*—Given the hypothenuse AB = 250, and the angle A = 35° 30′

	90° 0′
Extend from 90° to 35° 30′ on the line of sines, and this extent will reach from 250 to 145 on the line of numbers	BAC = 35 30
	ABC = 54 30
	∴ BC = 145
Extend from 90° to 54° 30′ on the line of sines, and this extent will reach from 250 to 203·5 on the line of numbers	∴ AC = 203·5

Case 2. The angles and one side being given, to find the hypothenuse and the other side.

Solution.—Extend from the angle opposite the given side to 90°, or radius, on the line of sines, and this extent will reach forwards from the given side to the hypothenuse on the line of numbers. Again, extend from the angle opposite the given side to the angle opposite the required side, and this extent will reach in the same direction on the line of numbers, from

* These cases are, in fact, the solutions, by the aid of Gunter's lines, of the following proportions, which will be obvious to the student upon inspection of the accompanying figure.

Rad. : sin. A : : AB : BC	Case 1.
Rad. : sin. B : : AB : AC	Case 1.
Sin. B : rad. : : AC : AB	Case 2.
Sin. B : sin. A : : AC : BC	Case 2.
Rad. : tan. A : : AC : BC	Case 2.
AB : AC : : rad. : sin. B	Case 3.
Rad. : sin. A : : AB : BC	Case 3.
AC : BC : : rad. : tan. A	Case 4.
BC : AC : : rad. : tan. B	Case 4.
Sin. A : rad. : : BC : AB	Case 4.

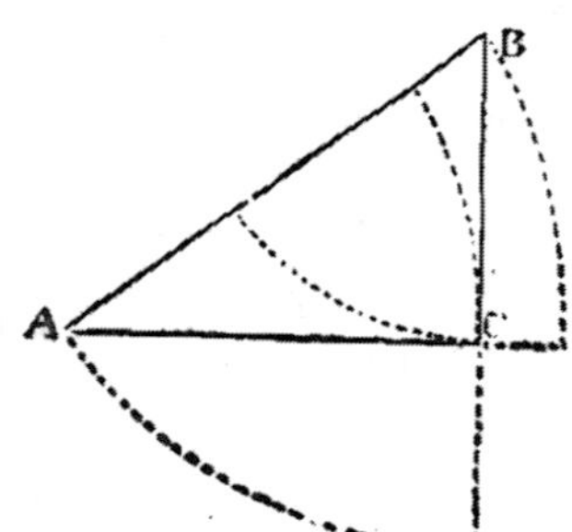

the given side to the required side. Or, extend from radius, or 45°, on the line of tangents, to the angle opposite the required side, and the extent will reach, in the same direction on the line of numbers, from the given side to the required side; recollecting that, when the angle is greater than 45°, the extent is to be taken on the scale backwards from rad. or 45° to the complement of the angle, but is to be reckoned a forward distance, the logarithmic tangents of angles greater than 45° exceeding the logarithmic tangents of 45°, or radius, by as much as the logarithmic tangents of their complements fall short of it. *Example.*—Given the angle A = 35° 30′ and side A C = 203·5.

	90° 0′
Extend from 54° 30′ to 90°, or rad., upon the line of sines, and this extent will reach forwards from 203·5 to 250 on the line of numbers	B A C = 35 30 A B C = 54 30 ∴ A B = 250
Again, extend from 54° 30′ backwards to 35° 30′, on the line of sines, and this extent will reach backwards from 203·5 to 145 on the line of numbers .	∴ B C = 145
Or extend backwards from 45°, rad., to 35° 30′ on the line of tangents, and this extent will reach backwards from 203·5 to 145 on the line of numbers, as before.*	

Case 3. The hypothenuse and one side being given, to find the angles and the other side.

Solution.—Extend from the hypothenuse to the given side on the line of numbers, and this extent will reach from 90 or rad. to the angle opposite the given side upon the line of sines. The other angle is the complement of this. Extend upon the line of sines from the rad. to the angle last found, which is opposite the required side, and this extent will reach from the hypothenuse to the required side. *Example.*—Given the hypothenuse A B = 250, and the side A C = 203·5.

Extend backwards from 250 to 203·5 on the line of numbers, and this extent will reach from 90° to 54° 30′ on the line of sines	90° 0′ ∴ A B C = 54 30
Extend from 90° to 35° 30′ on the line of sines, and this extent will reach backwards from 250 to 145 on the line of numbers	B A C = 35 30 ∴ B C = 145

* The property that tan. : rad. : : sine : cosine, may be made a test of the accuracy of the scale, since the distance from 45 to any angle upon the line of tangents ought to be the same as the distance from the angle to its complement upon the line of sines.

Case 4. The two sides being given, to find the angles and the hypothenuse.

Solution.—Extend from one side to the other upon the line of numbers, and this extent will reach backwards upon the line of tangents from rad. to the least angle, and to the same point, considered as a forward distance, representing the greatest angle, which is the complement of the least. Again, extend on the line of sines from one of the angles just found to rad., and this extent will reach from the side opposite the angle taken to the hypothenuse. *Example.*—Given AC = 203·5 and BC = 145.

Extend backwards upon the line of numbers from 203·5 to 145, and this extent will reach backwards from 45° to 35° 30′ on the line of tangents, which is the angle opposite the side 145	90° 0′ ∴ BAC = 35 30
If we measure forwards from 145 to 203·5, then from rad. to 35° 30′ is to be considered a forward distance, and the angle to be taken as the complement of 35° 30′, that is, 54° 30′, which is the angle opposite the side 203·5	∴ ABC = 54 30
Again, extend from 35° 30′ to 90° on the line of sines, and this extent will reach from 145 to 250 upon the line of numbers	∴ AB = 250

GUNTER'S SCALE.

Gunter's scale, for the use of navigators, which still goes by the name of its inventor, the Gresham Professor of 1619, consists of a rule, generally 2 feet in length, and about 1¾ inches broad, having one edge bevelled to answer the purpose of a straight-edge. On one side of it are placed lines of logarithmic numbers, from 1 to 100, marked N U M, of logarithmic sines, versed sines and tangents, marked respectively S I N, V S, and T A N, and of sine rhumbs and tangent rhumbs, marked S R and T R. Upon the same side are also placed a scale of meridional parts and a corresponding scale of equal parts. These latter scales are used in the construction of a Mercator's chart. On the other face of the instrument are placed, at the top edge, a scale of inches, forming an ordinary two-feet rule, a combined inch and half-inch diagonal scale, scales of rhumbs, chords, sines, tangents, and semitangents, and corresponding scales of rhumbs, chords, and longitudes. Several scales of equal parts are sometimes added, also a scale of 100 parts to a foot, ranging throughout the entire length of the rule, and corresponding scales of roots and cubes, the numbers on the cube line being the cubes of those on the root line under which they stand.

The construction and use of these scales have been described, with the exception of the scale of meridional parts.

In a Mercator's chart the earth is conceived to be projected on a plane surface in such a manner, that the bearings between places, and the proportions between the latitude, longitude, and nautical distance, shall be the same as those on the globe. Now, in proceeding from the equator to the poles, the lengths of the degrees of longitude diminish in the proportion of radius to cosine of the latitude, or of secant of the latitude to the radius; and as in Mercator's projection, in which the meridians are represented by parallel straight lines, the lengths of the lines representing equal differences of longitude are everywhere equal, the proportion between the true lengths of the degrees of latitude and longitude must be restored by increasing the lengths of minute portions of the meridians, in the continually increasing ratio of radius to secant of the latitude. The meridional parts for any given latitude are therefore found by adding together the products of successive minute portions of that latitude by the natural secants of the latitude of each portion; and the smaller those are taken, the more correct will be the table so formed. Constructing, then, a scale of equal parts to represent the actual lengths of degrees of latitude, a second scale having the divisions increased, as explained above, will represent the individual parts, from which the distances between the parallels of latitude on the chart may be taken off, while the distances between the meridians are taken from the corresponding scale of equal parts.

The sliding Gunter's scale has a slide working in its centre throughout its entire length, having on one of its faces lines of logarithmic numbers, and on the other lines of logarithmic sines and tangents. The advantages of this arrangement will be understood after reading the following articles upon the slide-rule; but both faces of the scale being now occupied with logarithmic lines, the numerous scales of equal parts and protracting scales, placed on the ordinary Gunter's scale, are omitted.

THE SLIDE-RULE.*

The slide-rule is an instrument containing the logarithmic lines, which have been already described, so arranged, as by means of the motion of a slide, to enable the sums and

* See also: "*The Slide-Rule, and How to Use It*," by Chas. Hoare, C.E. Price 3s. WEALE SERIES. With Slide-Rule in text.

differences of the logarithmic lengths to be at once seen without the assistance of the compass. We shall find, as we proceed, that the lines can be so arranged as to give special facilities for particular classes of questions, and hence several slide-rules are now in common use, adapted to the wants of persons engaged in different pursuits. All these rules, however, contain on one of their faces the four lines, A, B, C, D, (Plate III., Figs. 1, 2, 3). In most rules all the three lines, A, B, C, are called double lines, because the figures from 1 to 10 are contained twice in the length of the rule; and in the fourth line, D, the same figures are contained but once, and consequently the lengths upon this line, being twice the lengths denoted by the same numbers on the others, denote the logarithms of the squares of the numbers placed against the divisions.* The lines B and C are graduated along the two edges of a stick which works in a groove between the lines A and D. The lines B and C, being identical, are on some rules denoted by the single letter B or C; but both edges of the slide are still graduated, one to work with the A line, and the other with the D line. The B and C lines are mere copies of the A line.

The method of notation on the rule, and the manner of estimating any number upon it, are the same as have already been fully explained, when treating of the line of logarithmic numbers (p. 93).

When the slide is laid evenly in the groove, the divisions on A and B coincide, and the numbers on the coincident divisions are the *same;* and when the slide is in any other position, the numbers on the divisions of A are *proportional* to the numbers on the coincident divisions of B.

Also, when the slide is laid evenly in the groove, the numbers of the divisions on C are the squares of the numbers on the coincident divisions of D; and when the slide is in any other position, the numbers of the divisions on C are *proportional* to the squares of the numbers on the coincident divisions of D.

In using the A and B lines, it is immaterial whether a number is chosen upon the first or second divisions of those lines; but in ascertaining squares and square roots by means

* No advantage is obtained however by making the A line double, provided the B and C lines be so. But if the A line be made single, and equal to the whole length of the rule, the D line must be double the length of the rule, and must therefore be laid down half above and half below a groove.—See Coulson's Slide-Rule, page 144.

of the lines C and D, if the number of figures representing the square be odd, it must be selected from the first division of C; if even, from the second division. In extracting the square root of a number containing decimals, the number of decimals, if odd, must first be made even by the addition of a zero, and the preceding rule be then applied in the same manner as though the figures represented whole numbers.

In Coulson's rule the A line is only a single line, and the D line is laid half above and half below the slide: by this means the rule is made of only half the length for the same sized length of the graduations. This convenient arrangement presents no difficulties in the working; but the numbers on C must be considered as belonging to the first division when used with the first half of D above the slide, and to the second division when used with the second half of D below the slide.

ON THE USES OF THE A AND B LINES.

In whatever position these lines may be placed with respect to each other, the numbers in the coincident divisions represent a series of proportionals; in fact, we have here an instrument exhibiting at once to the eye the solutions of all problems in the rule of three. In its application, then, three terms of a proportion are always given to find a fourth. In the cases of multiplication and division, one of these terms is unity; thus, in multiplication, 1 : either of the factors : : the remaining factor : the product; and in division, divisor : 1 : : dividend : quotient, or divisor : dividend : : 1 : quotient.

We proceed to give a few examples of the application of the lines A and B.

Problem 1. *To multiply two Numbers together.*—Set 1 on A to the multiplier on B; and against the multiplicand on A will be found the required product on B. *Example.*—To multiply 33 by 23. Set 1 on A to 2·3 on B, and against 3·3 on A will be found 7·59 B; 759 is therefore the product required.

B 23	759
A 1	33

NOTE.—It is to be remarked that, in the solution of questions by the slide rule, the figures representing the result are alone given by the rule, and the power of ten involved in the result must be mentally supplied by the operator: thus, in the above example, each of the factors containing one power of ten, the product contains two powers of ten or hundreds; had the factors been 2·3 and 330, both the operation and the result would have been the same; but had the factors been 2·3 and 3·3, or 230 and 300, the operations would have been the same, but the results would have been 7·59 and 75900, respectively. It is also to be noted that, provided the tens in the result be duly supplied, either ends of the lines A or B may be taken for the 1, or the division marked 1 in the middle may be used, when the lines are double, or numbered twice, from 1 to 10.

The subject of numeration upon the rule is therefore one of considerable importance.

NUMERATION ON THE A AND B LINES.

In some treatises upon the slide-rule that we have seen, great and unnecessary difficulties have been introduced into the subject of numeration on the rule, and it appears to us that this has in a great measure arisen from treating as necessary that which is, in fact, superfluous, and, inasmuch as it has led to perplexity, injurious, viz., the laying down the lines A, B, and C twice over, from 1 to 10. It is, in fact, only desirable that either the line A or the line on the slide should be so laid down, and in this case the rule of numeration would be extremely simple. In fact, as already stated, it is only the making the A line, as well as the lines on the slides, double, that has created any difficulty in the matter. When they are so, the same numbers are always coincident on both sections of the lines; but we should advise the learner, although supplied with a rule having the A line laid down twice, to use at first one part of it only, until he thoroughly understands the use of the rule. Making use, then, of only one section of the A line, the rule for numeration will be as follows:—let the first term of the proportion be taken on A, so that the fourth or sought term is on B, then, if the two numbers on the B line coinciding with two on the A line be both on one section of the B line, the difference between the numbers of integral places of the numbers on the B line must be the same as that for the numbers on the A line; but when the numbers coinciding with two numbers on the A line are on different sections of the B line, the difference between the numbers of integral places on the B line being first considered to be the same as those on the A line, as in the former case, the number of integral places in the fourth or sought term must be increased by one, if it be on the second, or right-hand, section of B, or diminished by one if it be on the first, or left-hand, section. Thus, in the example to Problem 1, the 23 on the line B being set against 1 on A, the

33 on A, taken on the same section as the 1 (the other section on this line being altogether neglected), comes against 759 on the first section of B, and since the number of integral places in 33 is one more than the number in 1, the number in the 759 must be one more than the number in 23, so that the answer is 759. Had the second number on A been 3·3, the integral places in 759 must have been the same as in the multiplier 23, so that if this had been ·23, 2·3, or 23, the number sought would have been respectively ·759, 7·59, or 75·9. Had the second number on A been 330, having two more integral places than 1, the number sought would have had two more integral places than the multiplier, so that if this had been 230, the product would have been 75900. If the problem had been to multiply 43 by 33, the 43 on the first, or left-hand, section of B being set against the 1 on A, the 33 on A will come against the 1419 on the second, or right-hand, section of B, and the number of integral places in 33 being one more than the number in 1, the number in 1419 must, on this account, be one more than the number 43, and again another one more because it comes on the right-hand section of B, so that the answer would be 1419.

In the following examples one section only of the A line is made use of.

Problem 2. *To divide one Number by another.*—Against the divisor on A set the 1 on B, and against the dividend on A will be found the required quotient on B. *Example.*—To divide 216 by 24. Against 24 on A set 1 on B, and against 216 on A will be found 9 on B, which is the quotient required.

B	9	1
A	216	24

In this example the number of integral places in 216 being one more than the number in 24, the number in the result, if it came on the same section of the B line as the given term 1, would be one more than the number in this term, so that the result would in that case contain two integral figures; but, the 1 being taken on the right-hand section of the B line, the result, coming on the left-hand section, must on this account have its number of integral places diminished by one, so that it will contain only the single integral figure 9.

Second Method.—Against the divisor on A set the dividend on B, and against the 1 on A will be found the required quotient on B. *Example.*—To divide 551 by 29. Against

29 on A set 551 on B, and against 1 on A will be found **19**, the required quotient on B.

B 19	551
A 1	29

Here the numbers on the B line being both on the same section of it, and the number of integral figures in the 1 on the A line being one less than the number in the 29 on the same line, the number in the result 19 on the line B is to be one less than the number in the 551 on the same line.

Problem 3. *To find a Fourth Proportional to three given Numbers.*—Set the third term on B to the first term on A, and against the second term on A will be found the required fourth term on B. Or, against the first term on A set the second term on B, and against the third term on A will be found the required fourth term on B. *Example.*—To find a fourth proportional to the three numbers, 3½, 11, and 14. Against 3½ or 3·5 on A set 11 on B, and against 14 on A will be found, on B, 44, the fourth proportional required.

B 44	11
A 14	3·5

Or against 3·5 on A set 14 on B, and against 11 on A will be found 44 on B, as before.

B 44	14
A 11	3·5

The number of integral figures in 14 or 11 on the A line is one more than the number in 3·5 on the same line, and consequently the number in the result would be one more than the number in 11 or 14, if it came on the same section of the line B as this term; but the numbers on the B line being on different sections of it, the number of integral places in the result which comes on the left-hand section must on this account be reduced by one, so that instead of three integral figures there must be two only.

Problem 4. *To find a Third Proportional to two given Numbers.*—This is the same problem as the preceding, repeating the second number for the third term of the proportion. *Example.*—To find a third proportional to the two numbers 3½ and 11. This is to find a fourth proportional to the three numbers, 3½, 11, and 11. Set, therefore, 3½ or 3·5 on A to

11 on B, and against 11 on A will be found, on B, 34·6, the third proportional required.

B 34·6	11
A 11	3·5

Problem 5. *To square a given Number.*—Against 1 on A set the given number on B, and against the given number on A will be found its square upon B. *Example.*—Required the square of 23. Against 1 on A to 23 on B, and against 23 on A will be found its square, 529, on B.

A 1	23
B 23	529

The numbers on the B line being here both on the same section of it, and the number of integral figures in the 23 on the A line being one more than the number in the 1 on the same line, the number in the result on the B line must also be one more than the number in the 23 on this line.

Problem 6. *To find the Area of a Board or Plank. First Method.*—Set 12 on A to the mean breadth in inches on B, and against the length in feet on A will be found upon B the required area in feet and decimals of a foot. If the plank taper regularly, the mean breadth is half the sum of the extreme breadths; but, if the plank be irregular, several breadths should be measured at equal distances from each other, and their sum divided by their number may be taken as the mean breadth. In the latter case, however, greater accuracy would be obtained by finding separately the areas of portions of the plank, and adding them together for the whole area, or by the following. *Second Method.*—Take the measure in inches of several breadths at equal distances from each other, and add together half the two extreme breadths, and the sum of all the intermediate breadths. Set 12 on A to the sum thus found upon B, and against the distance in feet at which the breadths have been measured upon A will be found upon B the required area in feet and decimals of a foot. *Example* 1.—A board, 15 feet long, being 14 inches broad at one end, and 8 inches broad at the other, required its area. The mean breadth is 11 inches, half the sum of 8 and 14. Set, then, 12 on A against 11 on B, and against 15 on A will be found upon B 13·75 or 13¾ feet, the area required.

B 11	13·75
A 12	15

Example 2.—An irregular board, 18 feet long, being 7 inches broad at one extremity, 11 inches broad at the other, and the intermediate breadths at each 3 feet of the length being 13 inches, 25 inches, 23 inches, 32 inches, and 22 inches, required its area. By the first method, the sum of the seven breadths divided by 7, gives 19 inches for the mean breadth; and, setting 12 on A against 19 on B, against 18 on A will be found upon B 28·5 or 28½ feet, the area required.

B 19	28·5
A 12	18

By the second method, half of the two extreme breadths added to the intermediate breadths, gives the sum, 124 inches; and setting 12 on A against 124 upon B, against 3 on A will be found upon B 31, the area required, a more accurate result than the preceding.

B 124	31
A 12	3

In these examples the numbers on the B line are in each case both on the same section of it, and consequently the difference in the numbers of integral figures in the terms on the B line must be the same as that for the terms on the A line.

The following example of reduction will afford a further illustration of the method of numeration we have explained.

What is the value of ·27 of a ton in cwts., qrs., lbs., and ounces?

Since 20 cwts. make 1 ton, set 20 on B to 1 on A; and then against ·27 on A is 5·4 cwts. on B.

B 20	5·4
A 1	·27

Next, since 4 qrs. make 1 cwt., set 4 on B to 1 on A; and against ·4 on A is 1·6 qrs. on B.

B 4	1·6
A 1	·4

Again, as 28 lbs. make 1 qr., set 28 on B to 1 on A; and against ·6 on A is 16·8 lbs. on B.

B 28	16·8
A 1	·6

Lastly, as 16 ozs. make 1 lb., set 16 on B to 1 on A; and against ·8 on A is 12·8 ozs. on B.

B 16	12·8
A 1	·8

Hence, ·27 of a ton is equivalent to 5 cwts. 1 qr. 16 lbs. 12$\frac{8}{10}$ ozs.

In the first step of the above reduction, the two numbers on B occur on the same section of this line, and the second number on A having one integral figure less than the first number on the same line, the result, 5·4, which is the second number on B, must also have one integral figure less than 20, the first number on this line.

In the remaining steps of the reduction the two numbers on the B line are on different sections; and the results, coming on the right-hand section, must on this account have their number of integral figures increased by one; so that, instead of being one less than in the first numbers on B, they are to be the same.

It remains to consider the case of pure decimals, *i.e.*, of proper fractions decimally expressed, which consequently have no integral part. It is evident that here no integral figures can be taken away; but the insertion of any number of zeros after the decimal point must be considered as equivalent to a diminution, and the removal of any number of them as equivalent to an increase of as many integral places: thus, instead of ·21, to put ·021, ·0021, ·00021, &c., is equivalent to depriving it of one, two, three, &c., integral places; and conversely, instead of ·0021, to put ·021 or ·21 is equivalent to an increment of one or two integral places respectively.

In fact, whether a number be whole or fractional, or partly whole and partly fractional, the removal of the decimal point any number of places to the right is equivalent to increasing the number of integral figures by that number, and the removal of the decimal point any number of places to the left is equivalent to diminishing the number of integral figures by that number, the requisite number of places, when deficient, being in both cases supplied by zeros: thus, to increase by five the number of integral places in ·021, we must write 2100, the decimal point when it comes to the extreme right of a number not being expressed; and to diminish by 4 the number of integral places in 21 we must write ·0021.

Example 1.—To divide 6·3 by ·023. Against 23 on A set 1 on B, and against 6·3 on A will be found 274 on B.

B 1	274
A ·023	6·3

The numbers on B being both on the same section, and 6·3 being considered to have two more integral figures than ·023, the 274 has two more integral figures than the 1. Had the divisor been ·23, the quotient would have been 27·4, and had the divisor been 2·3, the quotient would have been 2·74.

Example. 2.—Find a fourth proportional to the three numbers 5100, 19, and 11. Against 5100 on A set 19 on B, and against 11 on A will be found ·041 on B.

B ·041	19
A 11	5100

Since 11 has two integral figures less than 5100, the result must on this account have two integral figures less than 19, *i.e.*, it must have no integral figure; but, the two numbers on B being on different sections of the line, the number of integral places in the result, which comes on the left-hand section, must be still farther reduced by one, and, as there are already none left, an equivalent operation must be performed by inserting a zero after the decimal point, making the result ·041.

Problem 7. *To find the Reciprocal of any Number.*—Against the given number on A set the 1 on the second section of B, and against the 1 on A will be found the required reciprocal on B; or set the given number taken on the first section of B against the 1 on A, and against the 1 on the second section of B will be found the required reciprocal on A. The two numbers are, in fact, mutually the reciprocals one of the other. *Example.*—To find the reciprocal of 55. Against 55 on A set 1 on B, and against 1 on A will be found the reciprocal ·0182 on B.

B ·0182	1
A 1	55

The proper numeration of the result will be given by the method already explained, or by the following simple rule. Of the two numbers mutually reciprocal, one has always an integral part, and the other is a pure decimal, and the number of zeros after the decimal point in the pure decimal, is always one less than the number of integral places in the other number: thus, if the given number contain only one integral figure, the reciprocal will have no zero after the decimal point; if the given number contain two integral figures, the reciprocal will have one zero after the decimal point; and if it contain 3, 4, &c., integral figures, the reciprocal will have 2, 3, &c., zeros after the decimal point. On the other hand, if the given number be a pure decimal, and have no zero after the decimal point, the reciprocal will contain one integral figure, and if it have 1, 2, 3, &c., zeros after the decimal point, the reciprocal will contain 2, 3, 4, &c., integers.

In the rules that have been given for the use of the A and B lines, only one division of the A line has been required, and consequently these methods are applicable to every form in which the rule is manufactured, whether the A line upon it be double or single; but in explaining the uses of the C and D lines, we have to consider three different modes of construction:—1. The most usual form, in which the A line is double, or has the numbers from 1 to 10 laid down twice over on the length of the rule, while on the D line these numbers are only laid down once in the same length. 2. The form, adopted in the timber-contenting rule, and in some other rules, in which the A line is double, as in the preceding construction, and the D line is laid down once in the length of the rule, but instead of commencing at 1 and terminating at 10, it is made to commence at 4 and terminate at 40. 3. The form adopted in Coulson's rule, in which the A line is laid down only once in the length of the rule, and consequently the D line, which must occupy twice the length of the A line, is laid down, one half of it above the groove in which the slide works, and the other half below it. In all three constructions, and in all positions of the slide, the numbers on the divisions of C are *proportional* to the squares of the numbers on the coincident divisions of D.

Problem 8. *To square a given Number by the Lines* C *and* D.—In the first of the instructions just explained, the D line is numbered from 1 to 10, and, when the slide is laid evenly in the groove, the numbers on the divisions of C are the squares of the numbers on the coincident divisions of D, so that to square any number, we have merely to set the slide so that the 1 on the beginning of C coincides accurately with the 1 upon D, and then, looking for the number on D, we shall find the square against it on C. With respect to the numeration of the result,—1. If the number to be squared, taken on D, contain any number of integral figures, its square, found upon C, will contain twice that number of integral figures if it comes on the second section of C, and one less than twice the number if it comes on the first section. 2. If the number to be squared, taken on D, be a pure decimal, whatever number of zeros it may contain after the decimal point, the square, found upon C, will contain twice that number of zeros after the decimal point if it come on the second section of C, and one more than twice that number if it come on the first section. Hence it appears that when the slide is laid evenly in the rule, the squares taken on its first section must contain either an odd number of integral figures, or an odd number of zeros after the decimal point, and the squares taken from the second section of the slide contain either an even number of integral figures, or either none or an even number of zeros after the decimal point.

Example.—To find the squares of 1·6125, 16·125, 161·25, ·16125, and ·016125.

Against 16125 on D is 26 on the *first* section of C, and therefore the square of 1·6125, 16·125, and 161·25, containing one, two, and three integral figures respectively, will be 2·6, 260, and 26000, containing one, three, and five integral figures respectively, *i.e.*, one less than double the number in their respective roots; while the squares of ·16125, and ·016125 will be ·026 and ·00026, containing after the decimal point one more than double the number of zeros in their respective roots.

C 1	26	26
D 1	16125	51

Example 2.—To find the squares of 5·1, 51, ·51, and ·051. Against 51 on D will be found 26 on the *second* section of C, and therefore the squares of 5·1 and 51, containing one and two integral figures respectively, will be 26 and 2600, containing two and four integral figures respectively, *i.e.*, double the number in their respective roots; while the squares of ·51 and ·051 will be ·26 and ·0026, containing after the decimal point double the number of zeros in their respective roots.

In the second construction of slide rule explained above, in which the D line is numbered from 4 to 40, the following are the modes of proceeding :—

First Method.—If the number to be squared lie between 1 and 4, or 10 and 40, or 100 and 400, &c., so that its first significant digit is less than 4; set the 1 on C to 10 on D, and against the digits on D, expressing the given number, will be found on C the digits expressing the required square. Then, the square of 1 being 1, of 10, 100, of 100, 10,000, &c., and of ·1 being ·01, of ·01 being ·0001, &c., the digits upon C must be estimated at the actual values represented by them as numbered upon the scale, viz., 1, 2, &c., to 16, or at 100 times their values, from 100 up to 1600, or at 10,000 times their values from 10,000 up to 160,000, &c., or, again, at the $\frac{1}{100}$th part of these values from ·01 up to ·16, or at $\frac{1}{10,000}$th part of these values from ·0001, up to ·0016, &c., according as the highest denomination in the number to be squared is units, or tens, or hundreds, &c., or again, tenths, hundredths, &c. *Example.*—Required the square of 23. The 10 on D being set against the 1 on C, against 23 on D will be found 5·29 on C, and, the highest denominations in 23 being tens, the square required is 529. Also the squares of 2·3, 230, 2300, ·23, ·023, would be 5·29, 52,900, 5,290,000, ·0529, ·000529, respectively, the highest denominations in the proposed numbers being respectively units, hundreds, thousands, tenths, and hundredths. *Second Method.*—If the number to be squared lie between 4 and 10, or 40 and 100, or 400 and 1000, &c., so that its first significant digit is not less than 4; set the 100 on C against the 10 on D, and against the digits on D, expressing the given number, will be found on C the digits expressing the required square. *Example.*—Required the square of 51. The 100 on C being set against the 10 on D, against 5·1 on D will be found 26 on C, and, the highest denomination in 51 being tens, the square required is 2600*.

In the third construction of slide rules, in which the D line is laid down, one-half above the groove in which the slide containing the C line works, and the other half below it, when the numbers on the slide are used with the numbers on the D line above the groove, they must be considered to belong to the first section of the C line, and to its second section

* The accurate square is 2601, but the fourth figure cannot be estimated upon a foot rule. The solution, in fact, may be considered as obtained to within the 1000th part of the whole; but, if greater accuracy is required, arithmetical methods must be resorted to.

when used with the numbers below the groove. With this proviso, the methods for finding squares are the same as in the first construction of slide-rule: thus the 1 upon the c line being made to coincide accurately with the 1 upon the D line, 26 upon the c line will be found to coincide with 16125 on the D line above the groove, and with 51 below the groove. When, therefore, the decimal point is so placed, that the number of integral places, or the number of zeros after the decimal point, in the number represented by the figures 26, is odd, the number represented by these significant figures is the square of a number represented by the figures 16125; and when the number of integral places, or the number of zeros after the decimal point, in the number represented by 26, is even, it is the square of a number represented by the figures 51.

D	16125
C	26
D	51

Also, when the number to be squared lies above the groove, the number of integral places in the square upon the line c is one less than double the number in the root upon D, or the number of zeros after the decimal point is one more than double the number in the root; and when the number to be squared lies below the groove, the number of integral places in the square upon c, or the number of zeros after the decimal point, is double the number in the root upon D.

Problem 9. *To extract the Square Root of a given Number.*—This problem being the converse of the preceding, set the rule in the same manner. Then in the first construction of rule, if the given number have an odd number of integral figures or of zeros after the decimal point, take it upon the first section of c, and the figures of the root will be found against it on D, and adding one to the number of integral places, or subtracting one from the number of zeros after the decimal point, half the sum, or remainder will give the number of integral places, or of zeros after the decimal point in the root; but if the given number have an even number of integral figures or of zeros after the decimal point, take it upon the second section of c, and half the number of these integral places or zeros will give the number of integral places or of zeros in the root found against it on D.

Example 1.—To find the square roots of 260, and ·00026.

Take 26 on the first section of C, and against it on D will be found 16125; then adding one to the integral places in 260, and subtracting one from the zeros in ·00026, we have four integral places and two zeros respectively, and, taking the halves of these, we have two integral places and one zero respectively for the roots, which are, therefore, 16·125 and ·016125.

Example 2.—Find the square roots of 26 and ·000026. Take 26 on the second section of C, and against it in D will be found 51, and taking half the number of integral places in ·26, and of zeros in ·000026, we have one integral place and two zeros respectively for the roots, which are, therefore, 5·1 and ·0051.

In the second construction of the slide-rule, set the 1 on C against the 10 on D, if the given square be between 1 and 16, or 100 and 1600, or 10,000 and 160,000, &c., or again between ·01 and ·16, or ·0001 and ·0016, &c., and with the 100 on C against the 10 on D, if the given square be between 16 and 100, or 1600 and 10,000, &c., or again between ·16 and 1, or ·0016 and ·01, &c.; and then against the given number on C will stand its square root on D.

Example 1.—Required the square root of 529. The given number being between 100 and 1600, set the 1 on C against the 10 on D, and against 5·29 on C will be found 23 on D, the square root required.

Example 2.—Required the square root 2601. The given number being between 1600 and 10,000, set the 100 on C against the 10 on D, and against 26 on C will be found 5·1 on D, and 51 is therefore the root sought.

In the third construction of the slide-rule, having set the 1 upon C in exact coincidence with the 1 upon D, find the given number upon C, and if it contain an odd number of integral places or of zeros after the decimal point, the figures in the root will be found above it on the first half of D; and if it contain an even number of integral places or of zeros after the decimal point, they will be found below it on the second half of D. The numeration of the result will then be the same as for the first construction of the rule, explained above; the roots found on the first half of D, above the groove, containing half the number of integral places of those in the given number increased by one, or half the number of zeros after the decimal point of those in the given number decreased by one; and the roots found on the second half of D, below the groove, containing half the number of integral places or of zeros after the decimal point of those in the given number.

The particular cases of numeration just considered, are included in the general rules for numeration upon the C and D lines, which we proceed to explain, and which will lead us to an examination of the relative merits of the three forms of construction.

In every position of the slide, and in all these constructions, the numbers on the divisions of C are proportional to the squares of the numbers on the coincident divisions of D, and the numbers on the divisions of D are proportional to the square roots of the numbers on the coincident divisions of C: hence every problem solved upon these lines must resolve itself into finding the fourth term of a proportion in which the two first of the three given terms are either (1) the squares or (2) the square roots of the given numbers. In the former case the two first terms are taken upon D, and the third term, coincident with first term, upon C, upon which also the result is found, coincident with the second; and in the latter the two first terms are taken upon C, and the third, coincident with the first, upon D, upon which also the result is found, coincident with the second term.

In the first mode of construction, in which the D line is numbered from 1 to 10, upon the whole length of the rule, (1) if the first two given terms are squares, and therefore taken on D, and the coincident divisions upon C are upon the same section of this line, then the difference between the numbers of integral places in the two terms upon C must be made double the difference between the numbers in the two terms upon D, and if any of the terms are pure decimals, the addition of zeros after the decimal point are to be considered equivalent to the subtraction of as many integral places, and the removal of zeros after the decimal point, to the addition of as many integral places. (2.) If the first two terms are squares, and therefore taken on D, and the coincident divisions upon C are upon different sections of this line, then the difference between the numbers of integral places in the two terms upon C, being first made double the difference between the numbers in those upon D, the number of integral places in the result must be increased by one if it come on the right-hand section of C, and diminished by one if it comes on the left-hand section. (3.) If the first two terms are the square roots of two numbers, and are therefore to be taken on C, then if the difference between their numbers of integral places be even, the same proviso as before being made with respect to zeros after a decimal point, these two terms must be taken on the same section of the c line, and the difference between the numbers of integral places in the two terms upon the coincident divisions of the D line must be made equal to half the difference between the numbers of those taken upon C. (4.) If the first two terms are the square roots of two numbers, and therefore to be taken on C, and the difference between their numbers of integral places be odd, these two terms must be taken on different sections of the c line; then to obtain the numeration of the result, the difference between their numbers of integral places must be diminished or increased by one, according as the term having the greater number of integral places is on the right-hand or on the left-hand section of C, and the half of the difference, thus diminished or increased, must be taken for the difference between the numbers of integral places in the coincident terms on D.

In the second construction of the rule, on which the D line is numbered from 4 to 40, throughout the entire length of the rule, when the two terms on D have their first significant digits either both less than 4, or both not less than 4, the rules for numeration are the same as in the first mode of construction; but, when one of the terms has its first digit less than 4, and the other has its first digit greater than 4, then, if the two terms coincident with those on c come both on the same section of this line, the difference between their numbers of integral places is two less or two more than twice the difference of the numbers of integral places in the two terms upon D, according as the term on the right-hand or that on the left has the greater number of integral places; but if the two terms on C come on different sections of the line, the difference between their number of integral places is one less or one more than twice the difference of the numbers of integral places in the two terms upon D, according as the term on the right hand or that on the left has the greater number of integral places.

With reference to both modes of construction just considered, it is to be remarked that, except in the case in which the slide is laid evenly in the rule, there are numbers on the D line which lie beyond one of the extremities of the C line, and *vice versâ*, and consequently that in a vast number of cases the rule is inadequate, as far as the C and D lines are concerned, to serve the purpose for which it is intended. Those questions which could not be solved on one of the rules could most frequently be solved upon the other, but some questions could be solved upon neither of these constructions of rule.

In the first mode of construction all questions in which 1 is one of the terms to be taken upon D, and the other term upon D lies beyond the extremity of the slide, the difficulty can be got over by using the 10 at the end of the D line for 1, and obtaining the numeration of the result by making the difference between the numbers of integral places in the terms upon the C line two less or two more than twice the difference between the numbers of those on D line, according as the term on the right hand or that on the left has the greater number of integral places.

In the measurement of timber, where the dimensions of a section are given in inches and the length in feet, to find the contents in cubic feet, the failure of the first construction of rule is most conspicuous; thus, if the length were 9 feet, and the section a square, the result would be given by the proportion 12^2 : breadth^2 : : 9 : the cubical content, and the 9 on C being placed against the 12 on D, the extremity of the slide would only reach to 4 on D, and no solution would be given if the breadth were between 4 inches and 10 inches. In like manner, if the section were a cylinder, no solution would be given for diameters between $4\frac{1}{2}$ inches and 10 inches. The second form of construction has therefore been adopted in rules specially intended for this kind of work, the 12 upon the D line being in this case not far from the centre of the line, and the cases of failure to solve this particular class of problems being thereby greatly diminished.

The failures of these rules could be remedied by the addition of a third section to the C line; and in the third mode of construction, in which the D line is placed half above and half below the groove, we have, in effect, such a third section, the first half of the C line, playing the part of a first section for the numbers on D above the groove, and that of a second section for the numbers below the groove, while the second half plays the part of a second section for the numbers above the groove, and that of a third section for the number below the groove.

In making use of this third construction of the rule, as long as the two terms upon C occur on the same section, or on two adjacent sections of this line, the rules for numeration are the same as those already explained for the first construction of the rule. It remains to consider the cases in which the two terms upon C occur, one on the first, and the other on the third section of this line. If, then, the two first terms are the squares of given numbers, and therefore taken on D, the third term, to be made coincident with the first term on D, is to be taken on the first or third section of C, according as this term is above or below the groove, and the fourth term sought will be found, coincident with the second term on D, on the third or first section of C. The difference between the numbers of integral places in the two terms on C being, then, first made equal to twice the difference between the numbers of those on D, the number of integral places thus found must be increased by two if the result come on the third section of C, and diminished by two if it come on the first section. Secondly, if the two first terms are square roots, and taken on C, and if the differences between their number of integral places be even, their two terms must be taken either on the same section of C, or one on the first and one on the third section. Putting then the first term on C in coincidence with the third term on D, if the second term taken on the same section of C as the first lies beyond the rule, these terms must be sought on a first and third section, and then, the difference between the numbers of integral places in the two terms being increased or diminished by two, according as the second term is on the third section or on the first, the half of the difference thus increased or diminished will give the difference between the numbers of integral places in the two terms upon D. When the second term taken on the same section as the first is coincident with a term upon D, the problem might be solved on a rule of the first mode of construction, and the rule for numeration is the same as that described for that mode of construction.

It is to be borne in mind that in all cases of terms in the form of pure decimals, the removal of zeros after the decimal point is equivalent to the addition of as many integral places, and the addition of zeros after the decimal point to a diminution of as many integral places.

It is also to be noted that throughout the whole of the rules here explained, the differences are to be understood to be taken in the same order, that is to say, the two terms on C and D having the greater numbers of integral places are to be coincident terms, and also the two having the lesser numbers of integral places.

Lastly, an equality in two numbers of integral places is to be treated as an even difference.

We now proceed to further examples of the uses of the C and D lines.

Problem 10. *To find a Mean proportional between two given Numbers.*—If x be a mean proportional between a and b, $x^2 = a\,b$, and $\therefore 1 : b : a : x^2$, and consequently $a : b :: a^2 : x^2$. Set, then, one of the numbers upon C to the same number on D, and against the other number on C will be found upon D the mean proportional required. *Example.*—Required a mean proportional between 4 and 49. The difference between the numbers of integral places in 4 and 49 being odd, their numbers must be taken upon different sections of the C line; then, in a rule of the first construction, the 4 upon the second section of C being set against the 4 on D, against the 49 on the first section of C will be found 14 upon D.

C	49	4
D	14	4

Again, in a rule of the second construction, the 4 upon the first section of C being set against the 4 upon D, against the 49 upon the second section of C will be found 14 upon D.

D	4	14
C	4	49

And again, in a rule of the third construction, the 4 upon C being set against the 4 on D, which is below the groove, against the 49 on C will be found above the groove 14 on D. The 4 is thus considered to belong to the second, or right-hand section of the C line, and the 49 to the first, or the left-hand section.

D		14
C	4	49
D	4	

To find the correct numeration of the result, we have to increase by one the difference between the numbers of integral places in the two terms upon the C line, and taking the half of this increased number gives us one for the difference between the numbers of integral places in the two terms upon the D line.

Problem 11. *To cube any given Number.*—If a represent the given number, we have the proportion $1^2 : a^2 :: a : a^3$. Set, therefore, the given number or root on C to 1 on D; and against the root on D is the required cube on C. *Example.*—Required the cube of 17. Set 17 on C to 1 on D; and the 17 on D will reach more than a tenth beyond 49 on C, giving 4913 for the cube required.

D	1	17
C	17	4913

The setting of the instrument in this example is the same for all these rules, and the two terms upon C coming on the same section of this line, the difference between their numbers of integral places must be made double the difference between the numbers of the integral places in the two terms upon D.

Problem 12. *To multiply any Number by the Square of another.*—Here we have the proportion 1^2 : (multiplier)^2 : : multiplicand : product. Set, therefore, the multiplicand on C against 1 on D, and against the multiplicand on D will be found the product on C. *Example.*—Multiply 5 by the square of 7. In this example the setting is the same for both the first and second form of the rules, the 10 on D being used for 1, and the two terms upon C coming upon the same section of this line. Set, then, 5 on C against 10 on D; and against 7 on D will be found 245 on C.

D	7	10
C	245	5

In a rule of the third form of construction set 5 on C against 1 on D; and against 7 on D will be found 245 on the third section of C.

D 1	
C 5	245
D	7

The rules given for the numeration in these cases show that the difference between the number of integral places in the two terms upon C must be made two more than twice the difference between the numbers in the 1 and 7 upon D, and since this difference is nothing, the result 245 must have two more integral places than the 5.

Problem 13. *To multiply a Number by the Square Root of another.*—Here we have the proportion $\sqrt{1}$: $\sqrt{\text{(multiplier)}}$: : multiplicand : product. Set, therefore, 1 on C to the multiplicand on D, and against the multiplier on C will be found the product on D. *Example.*—Multiply 12 by the square root of 5. Set 1 on C against 12 on D; and against 5 on the same section of C is 26·83 on D.

D 12	26·83
C 1	5

There being no difference between the numbers of integral in the two terms taken on C, they are both taken on the same section of this line, and the setting of the instrument is the same for all three forms of the rule.

Also, there will be no difference between the numbers of integral places in the numbers on D.

Problem 14. *To divide any Number by the Square Root of another.*—Here we have the proportion $\sqrt{}$ (divisor) : $\sqrt{}$ 1 : : dividend : quotient. Set, therefore, the divisor on C against the dividend on D; and against 1 on C will be found the quotient on D. *Example.*—Divide 99 by the square root of 9. There being no difference between the numbers of integral places in the two terms taken upon C, they are to be taken on the same section of C in rules of the first and third modes of construction. Set, then, 9 on C against 99 on D; and against 1 on the same section of C will be found 33, the required quotient, on D.

C 1	9
D 33	99

In a rule of the second form, 9 on the first section of C being set against the 99 on D, the 10 at the end of the second section of C will come against the 33 on D.

D 99	33
C 9	10

Problem 15. *To divide any Number by the Square of another.* Here we have the proportion $(\text{divisor})^2 : 1^2$: : dividend : quotient. Set, therefore, the dividend on C against the divisor on D, and against 1 on D will be found the quotient on C. *Example.*—Divide 84 by the square of 4. Set 84 on the second section of C against 4 on D; and against 1 on D will be found 5·25, the required quotient, on the first section of C. We represent the setting on all three constructions of the rule.

1st construction :—

C 5·25	84
D 1	4

2nd construction :—

D 10	40
C 5·25	84

3rd construction :—

D 1	
C 5·25	84
D	4

The two terms upon C coming on different sections of this line, the numeration of the result, which comes on the first, or left-hand, section, is given by first making the difference between the numbers of integral places in the terms on C double that of the numbers in the terms on D, which is nothing, and then taking one integral place away.

Problem 16. *To find the Fourth Term of a Proportion of which the two First Terms are the Squares of given Numbers.*—Set the third term on C against the first term on D; and against the second term on D will be found the result on C.

Example 1.—Find the value of $\frac{8 \times 16^2}{5^2}$, or a fourth proportional to 5^2, 16^2, and 8. In the first and third constructions of the rule, set 8 on the second section of C against 5 on D, and against 16 on D will be found 81·9, the result on the first section of C. In the second form of the rule, set 8 on the first section of C against 5 on D; and against 16 on D will be found 81·9 on the second section of C.

1st construction :—

C 81·9	8
D 16	5

2nd construction :—

D 5	16
C 8	81·9

3rd construction :—

D		16
C	8	81·9
D	5	

Example 2.—Find the value of $\frac{32^2 \times 41}{45^2}$, or a fourth proportional to 45^2, 32^2, and 41. This example cannot be solved on a rule of the second form.

In the first and third constructions, set the 41 upon one of the sections of C against the 45 upon D; and against the 32 upon D will be found upon the same section of C, the result, 20·7.

C 20·7	41
D 32	45

Example 3.—Find the value of $\frac{6{\cdot}5^2 \times 5{\cdot}2}{12^2}$, or a fourth proportional to 12^2, $6{\cdot}5^2$, and 5·2. This example cannot be solved on a rule of the first form.

In the second construction, against 12 on D set 5·2 on either section of C; and against 6·5 on D will be found the result, 1·53, on the same section of C.

D 6·5	12
C 1·53	5·2

In the third construction, against 12 on D set 5·2 on the first section of C, and against 6·5 on D will be found 1·53 on the third section of C.

D 12	
C 5·2	1·53
D	6·5

Example 4.—Find the value of $\frac{96 \times 38{\cdot}5^2}{11^2}$, or a fourth proportional to 11^2, $38{\cdot}5^2$, and 96. This example can be solved on neither of the two first constructions of the rule.

In the third construction, against 11 upon D set 96 on the first section of C; and against 38·5 upon D will be found the result, 1176, on the *third* section of C.

D 11	
C 96	1176
D	38·5

Problem 17. *To find the Fourth Term of a Proportion of which the two First Terms are the Square Roots of given Numbers.*—Set the first term on C against the third term on D; and against the second term on C will be found the fourth term

on D. *Example.*—Find the value of $\frac{4 \times \sqrt{27}}{\sqrt{12}}$, or a fourth proportional to $\sqrt{12}$, $\sqrt{27}$, and 4. Set 12 on either section of C against 4 on D; and against 27 on the same section of C will be found 6 the result, on D. The setting for this example is the same for all the three forms of rule.

D 4	6
C 12	27

When a problem arises in which the result involves the sums or difference of different quantities, the several portions of the result must be obtained separately from the rule, and then added together. *Example.*—Find the value of $\frac{40\,(24^2 + 32^2 + 59{\cdot}5^2)}{46^2}$.

$$\frac{40\ C}{46\ D}\left\{\begin{array}{rl} 24\ D = & 10{\cdot}9\ \ C \\ 32\ D = & 19{\cdot}36\ C \\ 59{\cdot}5\ D = & 66{\cdot}9\ \ C \\ \hline & 97{\cdot}16 \end{array}\right.$$

Against 46 on D set 40 on C, and against 24, 32, and 59·5 respectively on D will be found on C the numbers 10·9, 19·36, and 66·9; and 97·16, the sum of these, will be the value required.

USE OF THE A, B, C, AND D LINES CONJOINTLY.

In the first and third forms of construction of the rule, explained above, in every position of the slide the same numbers on C and B coincide with numbers upon D and the squares of those numbers upon A. In the third construction the A and B lines will be found on the opposite face of the instrument to the C and D lines. On either of these constructions, then, the A, B, C, and D lines may be used conjointly.

Problem 18. *To divide by a Number two Numbers multiplied together, one of which is squared.*—Set the unsquared factor on B against the divisor on A, and against the squared factor on D will be found the result on C. *Example.*—Find the value of $\frac{5^2 \times 32}{6}$. Set 32 on B against 6 on A; and against 5 on D will be found 133, the required result, on C.

Solution on *First* form of Rule :—

A 6	
B 32	133
C 32	133
D	5

Solution on *Third* form of Rule :—

On one face set	B / A	32 / 6
On the other face find	C 133 / D 5	32

With regard to the numeration of the result in these conjoint solutions, it is only necessary to consider the result as always taken on B, and the number on D, if it occur on the second half of this line, as a number with twice as many integral places taken on A, and, if it occur on the first half, as a number on A with twice as many integral places less one, the rule to be applied will, then, be the same as that for solution on the A and B lines (p. 102).

Problem 19. *To divide the Product of two Numbers by the Square of a Third Number.*—Set the third or squared number on D against one of the factors on C, and against the other factor on A will be found the result on B. *Example.*—Find the value of $\frac{8 \times 16}{2{\cdot}5^2}$. Against 2·5 on D set 8 on C; and against 16 on A will be found 20·5 the result on B.

Solution on *First* form of Rule :—

D 2·5	
C 8	
B 8	20·5
A	16

Solution on *Third* form of Rule :—

On one face set	D / C	2·5 / 8
On the other face find	B 20·5 / A 16	8

THE INVERTED SLIDE.

It has been seen that, when the slide is inserted in the groove in the ordinary manner, the numbers on the line A are always proportional to the corresponding numbers on the line B, and the numbers on the line C proportional to the squares of the numbers on the line D. Now, if we invert the slide, the numbers on A and the squares of the numbers on D become reciprocally proportional to the numbers on the slide, and, as a matter of course, division becomes multiplication, and multiplication division: also the products of the corresponding numbers on A and on the slide are everywhere the same, and the products of the numbers on the slide and the squares of the numbers on D are everywhere the same.

D 2	5	10
ꓭ 09	8	ᘔ
Ↄ 09	8	ᘔ
A 4	25	100

The rules for numeration with the inverted slide are deduced from those for the direct slide, by making the differences inverse, so that the term with the greatest number of integral places on the slide coincides with the term with the lesser number of integral places on the rule, and *vice versâ*, also observing that the second section of the slide is now on the left hand, and the first section on the right hand.

EXAMPLES OF THE USE OF THE INVERTED SLIDE.

Problem 20. *To find the Reciprocal of a given Number.*—Set the inverted slide evenly in the groove, so that 10 on C coincides with 1 on A, and 1 on C with 10 on A, and any two coincident numbers will be reciprocals the one of the other, and *vice versâ*, the numeration being so arranged that the number of zeros after the decimal point in one number shall be one less than the number of integral places in the other, thus—

A 1	314	3184	10
Ↄ I	ᔭ8I8	ᔭI8	I

314 and 3184 being mutually coincident on the inverted slide and the rule, ·3184 and 3·14 are mutually reciprocal; *i.e.*,

$\frac{1}{3\cdot14} = \cdot3184$ and $\frac{1}{\cdot3184} = 3\cdot14$; so also 31·84 and ·0314, 314 and ·003184, &c.

Problem 21. *To multiply one Number by another.*—Set one of the numbers on the slide against the other on A, and against 1 on A will be found the product on the slide. *Example.*—Find the product of 55 by 24. Set the slide so that 24 and 55 on it coincide with 55 and 24 on A, and 1 on A will coincide with the required product, 1320, on the slide.*

C 1320	1	55	24	1320
A 1	1320	24	55	10

Problem 22. *To divide one Number by another.*—Set the slide so that the dividend and 1 are coincident, and the divisor and quotient will also be coincident. *Example.*—Divide 58 by 16. Set the slide so that 58 and 1 on A and C coincide, and 16 will coincide with the required quotient, 3·625.

C 58	3·625	16	1	58
A 1	16	3·625	58	10

Problem 23. *To find the Product of two Numbers, one of which is squared.*—Set the unsquared number on the slide against the squared number on D, and against 1 upon D will be found the product on the slide. *Example.*—Find the value of $17^2 \times 23$. Set 23 on the slide against 17 on D, and against 1 on D will be found 6647, the required product, on B.

D 1	17
B 6647	23

Problem 24. *To cube a given Number.*—Set the given number on the slide against the same number on D; and against 1 on D is the cube required. *Example.*—Find the cube of 8. Set 8 on the slide against 8 on D, and against 1 on D is 512, the required cube, on B.

D 8	10
B 8	512

Problem 25. *To find the Cube Root of a given Number.*—Set the given number on the slide against 1 on D; and look for a number on the slide standing against the same number on D, and this number will be the root required. *Example.*—Find the cube root of 7. Set 7 on the slide against 1 on D; and

* This double coincidence enables us both to set the rule and to read off the result with greater accuracy.

1·914, the required cube root, will be found coincident on the slide and on D.

D 1	1·914
B 7	1·914

Problem 26. *To square a given Number.*—Set the given number on the slide against the same number on A; and against 1 on A will be found on the slide the square required. *Example.*—Find the square of 31·5. Set 31·5 on the slide against 31·5 on A; and against 1 on A will be 992, the required square, on C.

C 992	31·5
A 1	31·5

Problem 27. *To find the Square Root of a given Number.*—Set the given number on the slide against 1 on A; and look for a number on the slide standing against the same number on A, and this number will be the root required. *Example.*—The square root of 8. Set 8 an the slide against 1 on A; and 2·828, the required square root, will be found coincident on the slide and on A.

C 8	2·828
A 1	2·828

Problem 28. *To find a Mean Proportional between two given Numbers.* Set one of the numbers on the slide against the other on A; and look for a number on the slide standing against the same number on A, and this number will be the mean proportional required. *Example.*—Find a mean proportional between 4 and 49. Set 4 on the slide against 49 on A; and 14, the required mean proportional, will be found coincident on the slide and on A.

C 14	4
A 14	49

Problem 29. *To find the Fourth Term of an Inverse Proportion.*—Against the first term on A, set the third term on the slide; and against the second term on A will be found the fourth term on the slide. *Example.*—Find a number which shall have to 25 the inverse proportion of 42 to 29. Against 42 on A set 25 on C; and against 29 on A will be found 36·2, the result, on C.

C 36·2	25
A 29	42

Problem 30. *To find the Fourth Term of an Inverse Propor-*

tion the First Two Terms of which are the Squares of given Numbers.—Against the first term on D set the third term on the slide; and against the second term on D will be found the fourth term on the slide. *Example.*—Find a number which shall have to 20 the inverse proportion of 27^2 to 35^2. Against 27 on D set 20 on B; and against 35 on D will be found 11·9, the result, on B.

20 B	11·9
D 27	35

Problem 31. *To find the Fourth Term of an Inverse Proportion, the First Two Terms of which are the Square Roots of given Numbers.*—Set the first term on the slide against the third term on D; and against the second term on the slide will be found the fourth term on D. *Example.*—Find a number which shall have to 20 the inverse proportion of $\sqrt{27}$ to $\sqrt{35}$. Set 27 on the slide against 20 on D; and against 35 on the slide will be found 17·57, the result, on D.

35 B	27
D 17·57	20

THE INVERTED LINE.

As far as we have proceeded hitherto, we have been able at one operation to obtain the *product* of two numbers only, one of which may or may not be squared, divided either by unity or by a third number either squared or not; but as solids have the three dimensions of length, breadth, and thickness, and their contents are to be found by multiplying together these dimensions, and dividing by some number dependent upon the peculiar form of the solid whose content is sought, it must manifestly be a very great advantage, if, by any arrangement of the logarithmic lines, the entire result can be obtained at one operation. Now suppose we replace the D line by an inverted line, which we will call the E, or A I, line;* and, at first, let it be laid down so that its 1, 10, and 100 come against the 100, 10, and 1 of the direct line A; then, whatever be the position of the slide, laid in in the direct manner, the products of the corresponding numbers on the lines C and E are everywhere the same; and, hence, under the 100 of E, will be found on the slide the figures denoting the product of any of the two coincident numbers on the C and E lines, the

* On the timber and cubing rules the inverted lines are denoted by E, and on Coulson's Rule they are denoted by A I.

real product being these figures multiplied by 100. Now these figures repeated on the B line of the slide are coincident with the 1 on the A line, and hence, to multiply by another factor, we have only to look for this factor on the A line, and over it, on the B line, will be found the figures representing the continued product of the three factors; for we have, in fact, found the result to be obtained by multiplying together the three factors and dividing by 100. Thus, setting 5 on C against 60 on E, against 100 on E we have 3 on C and B, and 1 on A, and over 3 on A we have 9 on B, which is the product of $\frac{5 \times 60 \times 3}{100}$.

E	100	60	
C	3	5	
B	3		9
A	1		3

Again, let the E, or A I, line* be laid down so that any other number comes over the commencement of the A line; then, setting against any number on E, the inverted line, a second number on the slide, and taking a third number on A, we shall find on the slide, against this last number, a number which will be the product of the three numbers divided by the number on the inverted line, which falls opposite the 1 upon the A line.

EXAMPLES.

Problem 32. *To construct a Rule to give the Content in Cubic Feet of a piece of Timber or other material, when the Width and Thickness are given in Inches, and the Length in Feet.*—The width and thickness, being given in inches, must each be divided by 12 to bring them to feet, and consequently the product must be divided by 144 to bring the result to cubic feet. Let the E line, therefore, be so laid down that 144 on it may be over the 1 of the A line, and the object desired will be accomplished. *Example.*—Find the content of a piece of timber 17 feet long, 11 inches wide, and 8 inches thick. Set 8 and 11 on the slide against 11 and 8 on E, and against 17 on A will be found 10·39 cubic feet, the required content.

E	11		8
C	8		11
B		10·39	
A		17	

* See page 128.

Problem 33. *To construct a Rule to give the Content in Cubic Yards of an Excavation, when the Length, Breadth, and Depth are each given in Feet.*—Each of the dimensions must be divided by 3, to bring them to yards, and consequently the product must be divided by $3 \times 3 \times 3 = 27$, to bring it to cubic yards. Let the A I line, therefore, be so laid down that 27 on it may be over the 1 on the A line, and the object desired will be accomplished. *Example.*—Find the number of cubic yards in an excavation 9 feet deep, 20 feet broad, and 63 feet long. Set 9 on the slide against 20 on E, and against 63 on A will be found 420 cubic yards on B, the required content.

A I	20	
C	9	
B		420
A		63

Problem 34. *To construct a Rule to give the Weight, in Pounds, of a rectangular piece of Cast Iron, when the Length, Breadth, and Thickness are all given in Inches.*—A cubic inch of cast iron weighs $\frac{1}{3.836}$ of a pound; and hence the content in cubic inches, divided by 3·836, gives the weight in pounds. Let the A I line, therefore, be so laid down that 3·836 on it may be over the 1 on the A line, and the object desired will be accomplished. *Example.*—Find the weight of a cast-iron plate, half an inch thick, 8 inches broad, and 48 inches long. Set 0·5 on the slide against 8 on E, and against 48 on A will be found 50 lbs. on the slide, the required weight.

A I	8	
C	·5	
B		50
A		48

NOTE.—It is immaterial on which of the lines, A, C, and E or A I, the several dimensions are taken; but it is to be observed that the result is always to be looked for on B, opposite to a dimension which has been taken on A. To prevent confusion, however, the student is advised to practise an uniform method; for example, to make the *breadth* and *thickness* coincide on C and E, and against the *length* on A will be found the content on B.

In the preceding description we have supposed the A and E lines to be laid down twice in the length of the Rule as they are on the Timber Contenting Rule (p. 149). In this rule also, as well as in the Cubing Rule (p. 147), the C and B lines are both required because they are differently divided and numbered, although coincident points upon them really represent the same numbers: thus in the Timber Rule, the C line, on which inches are taken, is subdivided into fourths, while the B line is divided decimally; and in the Cubing Rule, instead of the numbers representing inches, on C are placed their equivalents in feet and inches. When, however, the lines are all divided decimally, the best arrangement is to have the A line and the inverted line laid down once only in the length of the rule, and to have only one line, B or C, on the slide, as in Coulson's Rule (p. 144). Using, then, only one section of the A line and the corresponding portion of the inverted slide, which consists of portions of two sections, all the divisions to the right of 1 or 10 belong to the first section, and the portion on the left to the second section; while on the slide the left-hand section is the first, and the right-hand section is the second. The rules for numeration are then extremely simple, viz.: add together the number of integral places in the three given terms, and subtract from the sum thus found the number of integral places in the constant divisor on the inverted line over the 1 of the A line, and the remainder will be the number of integral places in the result, if all the numbers come on first sections; but if a given term come on a second section, this number of integral placed must be diminished by one for each such term, and if the result come on the second section of B, the number must be increased by one. Hence if one of the given numbers and the result both come on a second section, the number of integral places is unaltered, *i.e.*, the same as when all the numbers are on first sections. If two of the given terms come on second sections, the integral places must be reduced by two on this account; so that if the result also come on a second section, the resulting number of integral places will still be one less than the sum of the numbers in the three terms diminished by the number in the constant divisor.

We proceed to apply these rules to the numeration of the results in the three preceding examples.

In the example to Problem 32 the sum of the numbers of integral places in the given numbers, 11, 8, and 17, is five, from which subtract three, for the number in the constant divisor 144, and there remain two, and since one of the given numbers is on a second section, for which one integral place is to be added, and the result is also on a second section for which one is to be added, two is the number of integral places in the result.

In the example to Problem 33 the three given numbers, 9, 20, and 63 contain together five integral places, from which must be subtracted two for the number in the divisor 27, and there remain three, which, since one of the given numbers comes on a second section, and so also does the result, must be the number of integral places in the result.

In the example to Problem 34, the number of integral places in the given numbers, 0·5, 8, and 48 is three, from which subtracting one for the number in the divisor 3·836, we have two remaining, and the given numbers and result being all on first sections, two must be the number of integral places in the result.

THE TRIGONOMETRICAL SLIDE.

The trigonometrical slide contains upon it, instead of the B and C lines, lines of the logarithmic tangents and sines (p. 95) of all angles from 0° to 90°; the former of which works to the line A, and the latter to the line D; so that when the slide is laid evenly in, the numbers on A are the natural tangents, and the numbers on D the natural sines, of the degrees marked opposite to them, on the slide; and when the slide is in any position, the numbers on the A and D lines are proportional to the natural tangents and sines of those degrees. If the trigonometric lines are laid down to a radius equal to the whole length of the line of numbers, it is manifest that, since, in any position of the slide, the beginning and end of the trigonometric lines come against the same figures upon the line of numbers, supposed to be repeated, either the beginning or end of the line may be taken for radius, as may be most convenient. Sometimes the line of tangents is laid down to a radius equal to half the length of the slide only, and then either the beginning, middle, or end of the line may be taken for radius; but the middle, marked 45°, will generally be found most convenient.

In the sliding Gunter's scale the lines of logarithmic sines and tangents are laid down upon one face of the slide, and two lines of numbers on the other face, whilst the rule has upon one face a line of sines above the groove, and a line of tangents below it; and, on the other face, a line of numbers above the groove, and a line of sine-rhumbs below it.

There only remains one more line to be described, viz., the line of cubes, E. This line is laid down on a slide three times in the same length in which A is laid down twice, and D once; and, consequently, when this slide is laid evenly in, between the lines A and D, the numbers on D will be coincident with their cubes on the slide, and their squares upon A; and when the slide is in any other position, the numbers on E will be proportional to the cubes of the corresponding numbers on D, and the square of the numbers on E will be proportional to the cubes of the numbers on A; or the numbers on D, the square roots of the numbers on A, and the cube roots of the numbers on E, at coincident divisions, will be proportionals.

RATIOS.

When similar figures vary, their linear dimensions vary in simple proportion, but their areas vary as the squares of their linear dimensions. Thus, if the diameter of one circle be 5 times the diameter of another, the circumference of the former will also be 5 times the circumference of the latter; but its area will be 25 times the area of the latter. If then two numbers be found having the same ratio to one another as two lines in some figure, and if these two numbers be made coincident on the lines A and B, any other coincident numbers on these lines will represent the lengths of the like lines in a similar figure: thus, the ratio of the diameter to the circumference of a circle being 113 to 355, if 355 on B be set against 113 on A, then, if any diameter be taken on the line A, the corresponding circumference will be found against it on the line B, and *vice versâ*.

Again, if two numbers be found, one of which has to the square of the other the same ratio as the area of a figure to the square of one of its linear dimensions, and if the number representing the area on C be made coincident with the number representing the linear dimension on D, then any other coincident numbers will represent the linear dimension on D and the corresponding area on C of a similar figure: thus, 43 represents the area of a circle, the diameter of which is 7·4; and if 43 on C be set against 7·4 on D, and any diameter be taken on D, the corresponding area will be found on C, and *vice versâ*.

The advantage of having the ratios in whole numbers, consisting of only a few figures, is obvious, as they can be set on the rule with ease and exactness.

TABLES OF RATIOS.

TABLE I.—*Ratios relating to the Circle.*

On the Line A.*	On the Line B.*
113† to diameter.	355† to circumference.
44 ,, ,,	39 ,, side of equal square.
99 ,, ,,	70 ,, ,, inscribed square.
355 ,, circumference.	113 ,, diameter.
39 ,, ,,	11 ,, side of equal square.
40 ,, ,,	9 ,, ,, inscribed square.
70 ,, side of square.	79 ,, diameter of equal circle.
24 ,, ,,	85 ,, circumference ,,
50 ,, ,,	76 ,, side of equal equilateral triangle.
11 ,, area of circle.	7 ,, area of inscribed square.
452 ,, ,, square.	355 ,, ,, circle.
35 ,, ,,	29 ,, ,, octagon.
57·3 ,, No. of degrees	radius ,, length of arc.

On the Line D.	On the Line C.
7·4 to diameter.	43 to area.
17 ,, circumference.	23 ,, ,,

TABLE II.—*Polygons within and without Circles.*

No. of Sides.	Inside Circle on Line A.*	Outside Polygon on Line B.*	Outside Circle on Line A.*	Inside Polygon on Line B.*
3	15 to diameter.	26 to side.	15 to diameter.	13 to side.
4	1 ,,	1 ,,	99 ,,	70 ,,
5	11 ,,	8 ,,	80 ,,	47 ,,
6	26 ,,	15 ,,	2 ,,	1 ,,
7	56 ,,	27 ,,	30 ,,	13 ,,
8	29 ,,	12 ,,	34 ,,	13 ,,
9	11 ,,	4 ,,	41 ,,	14 ,,
10	40 ,,	13 ,,	320 ,,	99 ,,
11	17 ,,	5 ,,	32 ,,	9 ,,
12	56 ,,	15 ,,	58 ,,	15 ,,

* The terms taken on A and B may be interchanged so as always to bring the result on B, which is desirable, in order to apply the rules for numeration we have laid down.

† That is, distance from 113 to diameter on line A is equal to distance from 355 to corresponding circumference on B; so that if the 113 on A be set against 355 on B, any diameter on A will coincide with corresponding circumference on B.

TABLE III.—*Areas of Polygons.*

No. of Sides.	On Line D.	On Line C.
3	30 to side.	390 to area.
4	1 ,,	1 ,,
5	5 ,,	43 ,,
6	5 ,,	65 ,,
7	5 ,,	91 ,,
8	6 ,,	174 ,,
9	4 ,,	99 ,,
10	6 ,,	277 ,,
11	4 ,,	150 ,,
12	4 ,,	179 ,,

TABLE IV —*Miscellaneous Ratios.*

On Line D.	On Line C.
7·4 to diameter of sphere.	172 to surface of sphere.
17 ,, circumference of sphere.	92 ,, ,,
Side of square to its diagonal.	1 ,, 2.
Side of square to diameter of circumscribing circle.	1 ,, 2.
Side of cube to its diagonal.	1 ,, 3.
Side of cube to diameter of circumscribing sphere.	1 ,, 3.
6 seconds to time.	579 ft. / 193 yds. } to distance a body falls *in vacuo.*
80 seconds to vibrations per minute.	22 to length of pendulum in inches on inverted slide C.

EXAMPLES ON THE USE OF RATIOS.

The diameter of a circle 7 inches; what is the circumference?

A 113 ——— 7
B 355 ——— 22 inches. *Ans.*

The diameter of a circle is 9 inches; what is the side of an equal square?

A 44 ——— 9
B 39 ——— 7·97 inches. *Ans.*

What is the side of a square inscribed in the preceding circle?

A 9 ——— 99
B 6·36 inches. *Ans.* ——— 70

Supposing the circumference of the earth to be 25,000 miles, what is its diameter?

A 355 ——— 25000
B 113 ——— 7960 *Ans.*

The circumference of a circle is 8 inches; what is the side of an equal square?

A 39 ——— 8
B 11 ——— 2·25 inches. *Ans.*

What is the side of a square inserted in the preceding circle?

A 40 ——— 8
B 9 ——— 1·8 inches. *Ans.*

The side of a square is 4 inches; what is the diameter of an equal circle?

A 4 ——— 70
B 4·51 inches. *Ans.* ——— 79

What is the circumference of the preceding circle?

A 24 ——— 4
B 85 ——— 14·2 inches. *Ans.*

The side of a square is 10 inches; what is the side of an equal equilateral triangle?

A 50 ——— 10
B 76 ——— 15·2 inches. *Ans.*

The area of a circle is 18; what is the area of the inscribed square?

A 11 ——— 18
B 7 ——— 11·45 *Ans.*

The area of a square is 18 square inches; what is the area of the inscribed circle?

A 452 ——— 18
B 355 ——— 14 sq. in. *Ans.*

What is the area of a regular octagon inscribed in the same square?

A 18 ——— 35
B 14·9 sq. in. *Ans.* ——— 29

The radius of a circle is 9 inches; what is the length of an arc of it containing 32 degrees?

A 32	57·3
B 5 inches. *Ans.*	9

The diameter of a circle is 5 inches; what is its area?

C 19·63 sq. in. *Ans.*	43
D 5	7·4

The circumference of a circle is 12 inches; what is its area?

C 11·46 sq. in. *Ans.*	23
D 12	17

The diameter of a circle is 12 inches; what is the side of the equilateral triangle described about it?

A 12	15
B 20·8 in. *Ans.*	26

What is the side of a regular pentagon inscribed within the same circle?

A 12	80
B 7·05 in. *Ans.*	47

The side of an equilateral triangle is 1; what is its area?

C ·433 *Ans.*	390
D 1	30

The side of an octagon is 5 feet; what is its area?

C 121 sq. ft. *Ans.*	174
D 5	6

The circumference of a sphere is 12 inches; what is its surface?

C 45·8 sq. in. *Ans.*	92
D 12	17

The side of a cube is 7 inches; what is the length of its diagonal?

C 3	1
D 12·12 in. *Ans.*	7

What is the length of a pendulum making 60 vibrations per minute?

39·2 B	22	
D 60	80	39·2 *Ans.*

DIVISORS AND GAUGE POINTS.

We have already seen that any person, whose occupation requires him to make calculations in which the same number constantly appears as a divisor, may have a rule constructed with an inverted line, to perform with facility this particular class of computations. For general purposes, however, recourse must still be had to the A, B, C, D lines; and as the formulæ for many surfaces, and almost all solids, require the use of the D line, it is far more convenient to compute the square roots of the divisor, and take these roots on the D line, than to take the divisors on the A line. The points so taken on the D line are called gauge points, and it is manifest that, when the D line commences with unity, the gauge points on D are exactly opposite the divisors on A, so that in this case the slide is at the same time set to a gauge point and the corresponding divisor, and, therefore, except as a matter of convenience, it would be immaterial whether gauge points or divisors were made use of. The most useful gauge points are usually marked upon the rule with brass pins.

It is not necessary to register a table of divisors, since setting the C line to a gauge point on D also sets the B line to the divisor on A.*

Denoting by □ G and ○ G the square and circular gauge points for areas and surfaces, the following table exhibits their values:—

Dimensions in	Surfaces in	□ G.	○ G.
Inches	Square inches	1	1·128
„	„ feet........	12	13·54
„	„ yards	36	40·62
Feet..........	„ „	3	3·385
„	„ rods	16·5	18·62
Links	„ „	25	28·2
Yards	„ „	5·5	6·206
„	„ acres	69·57	78·5
Chains........	„ „	3·162	3·567

The following diagrams represent computations of areas by means of the above gauge points:—

* On rules with the D line laid down half above and half below the groove, the rule must be turned over, and the B line on the opposite face is set to the divisor.

Figure	Scale	Setting	Reading
Square	C	1	area
	D	□ G	side
Parallelogram	A		height
	B		area
	C	base'	
	D	□ G	
Triangle	A		height
	B		area
	C	½ base	
	D	□ G	
Parabola	A		height
	B		area
	C	⅔ base	
	D	□ G	
Circle	C	1	area
	D	○ G	diameter
Ellipse	A		other axis
	B		area
	C	one axis	
	D	○ G	
Cycloid	C	3	area
	D	○ G	diameter of generating circle
Prism or cylinder	A		height*
	B		surface
	C	perimeter	
	D	□ G	
Pyramid or cone	A		slant height
	B		surface†
	C	½ perimeter	
	D	□ G	
Sphere	C	4	surface
	D	○ G	diameter

---

* This will not include the areas of the ends, the rule for finding which will vary with their form. The diagram in the text will give the whole surface for a square prism if the height be increased by the eighth part of the perimeter, and for a cylinder if the height be increased by the radius.

† This does not include the area of the base. It will be included if the slant height be increased by the eighth part of the perimeter for a square pyramid, and by the radius for a cone.

Spherical zone

A	diameter
B	area
C 4 height	
D ○ G	

EXAMPLES.

The side of a square is 34 inches; what is its area in square feet?

C 1	8·03 sq. ft. *Ans.*
D 12	34 in.

A piece of land measures 95 links by 70; how many square rods does it contain?

A	95 links.
B	10·61 sq. rods. *Ans.*
C 70 links	
D 25	

The diameter of a circle is 17 inches; how many square feet does it contain?

C 1	1·576 sq. ft. *Ans.*
D 13·54	17 in.

TABLE OF GAUGE POINTS FOR SOLIDS.

Dimensions all taken in inches.

Contents in	□ G.	○ G.	Contents in lbs.	Specific Gravity.	□ G.	○ G.
Cubic inches	1	1·128	Platinum	21·45	1·135	1·28
" feet	41·57	46·9	Standard gold	17·61	1·252	1·412
Pints	5·88	6·64	Mercury	13·58	1·427	1·61
Gallons	16·65	18·79	Lead	11·35	1·56	1·76
Bushels	47·09	53·14	Standard silver	10·2	1·646	1·857
Pounds of tallow	5·5	6·2	Copper	8·33	1·822	2·055
" hot soap	5·29	5·97	Brass, cast	8·3	1·825	2·059
" cold "	5·21	5·88	Cast iron	7·21	1·958	2·21
" flint glass	2·94	3·316				
" plate "	3·36	3·79				

The following formulæ and diagrams describe the computation of contents by means of the above gauge points:—

Square prism $= \dfrac{l\,s^2}{\square\text{ G}}$

C length	content
D □ G	side

Square pyramid $= \dfrac{\frac{1}{3}\,l\,s^2}{\square\text{ G}}$

C ⅓ length	content
D □ G	side

Cylinder $= \dfrac{l\,d^2}{\bigcirc\text{ G}}$

C length	content
D ○ G	diameter

Cone $= \dfrac{\frac{1}{3}\, l\, d^2}{\bigcirc\ G}$ $\qquad \dfrac{\text{C } \frac{1}{3} \text{ length}}{\text{D } \bigcirc \text{ G}} \qquad \dfrac{\text{content}}{\text{diameter}}$

Sphere $= \dfrac{\frac{2}{3}\, d\, d^2}{\bigcirc\ G}$ $\qquad \dfrac{\text{C } \frac{2}{3} \text{ diameter}}{\text{D } \bigcirc \text{ G}} \qquad \dfrac{\text{content}}{\text{diameter}}$

Spheroid $= \dfrac{\frac{2}{3}\, f\, v^2}{\bigcirc\ G}$ $\qquad \dfrac{\text{C } \frac{2}{3} \text{ fixed axis}}{\text{D } \bigcirc \text{ G}} \qquad \dfrac{\text{content}}{\text{revolving axis}}$

Rectangular prism $= \dfrac{l\, b\, d}{\square\ G}$ $\left\{ \begin{array}{l} \dfrac{\text{A}}{\text{B}} \qquad \dfrac{\text{depth}}{\text{area}^*} \\ \dfrac{\text{C breadth}}{\text{D } \square \text{ G}} \\ \dfrac{\text{A 1}}{\text{B area}} \qquad \dfrac{\text{length}}{\text{content}} \end{array} \right.$

EXAMPLES.

What is the content, in gallons, of a vessel in the shape of a square prism, 4 inches deep and 14½ inches along each of the sides?

$\dfrac{\text{C 3·03 gallons}}{\text{D 14·5}} \qquad \dfrac{4}{16·65}$ *Ans.* 3·03 gals.

Each side of a square pyramid is 46 inches, its perpendicular depth is 90 inches; what is its content in gallons?

$\dfrac{\text{C 30}}{\text{D 16·65}} \qquad \dfrac{\text{229 gallons.}}{46}$ *Ans.*

The depth of a cylinder is 20 inches, its diameter is 21½ inches; how many bushels will it contain?

$\dfrac{\text{C 3·275 bushels}}{\text{D 21·5}} \qquad \dfrac{20}{53·14}$ *Ans.* 3·275 bush.

The depth of a cone is 24 inches, its diameter 17; how many pounds of tallow will it hold?

$\dfrac{\text{C 60 lbs.}}{\text{D 17}} \qquad \dfrac{8}{6·2}$ *Ans.* 60 lbs.

What is the weight of a hollow globe of brass, the exterior diameter being 9 inches, and the interior 6 inches?

---

* The content of a solid whose length is unity is called an area, and finding such content is called averaging areas.

C 6		115	115
,, 4	34·		34
D 2·059	6	9	*Ans.* 81 lbs.

The fixed diameter of an oblate spheroid is 33 inches, its revolving diameter is 54 inches; how many bushels will it contain?

C 22	22·7 bushels. *Ans.*
D 53·14	54

A cistern is the shape of a rectangular prism, is 112·5 inches long, 41 inches broad, and 18 inches deep; how many gallons will it contain?

A 18	
B 2·66	
C	41
D	16·65
A 1	112·5
B 2·6	300 gallons. *Ans.*

Almost all the slide-rules now manufactured agree in having the four lines, A, B, C, D, upon one of their faces, the only exception being rules constructed solely for the purpose of one particular class of operations, an instance of which is afforded in the timber-cubing rule (p. 147). We proceed to notice a few of the best slide-rules.

The Routledge Slide-Rule has a joint in the middle, and is 24 inches long when opened out. One of the faces is graduated to inches, and also to a drawing scale; so that when looking at this face it has the appearance, and answers every purpose, of a common two-feet rule. One of the edges is divided to decimals of a foot, and the other to inches subdivided into tenths and twelfths. On closing the rule and turning it over we find one half of it forming a slide-rule 1 foot in length, with the lines A, B, C, D on it, the D line being numbered from 1 to 10; and on the other half are tables of gauge points, for square, cylinder, and globe, for regular polygons, for proportions of the circle, square, and triangle, and also one for pumping-engines. The old Coggleshall's rule was very similar to this, the only difference consisting in other tables holding the place of the table of gauge points, and the line D ranging from 4 to 40 instead of from 1 to 10.

Bayley's Slide-Rule is a very neat instrument, manufactured by Messrs. Elliott. It is 1 foot in length, having two grooves, one on the front, and one on the back; and as the slides for these grooves are each graduated on both sides, the instrument is equivalent to a rule with four slides. On one face are the A line above the groove and the D line below it, while the slide has on one face the B and C lines, as usual, and on the other the E line, on which the logarithmic line is repeated thrice, and the numbers on the divisions are the cubes of the numbers on the corresponding divisions of the D line, with which it works. On this slide is also a scale of equal parts marked Num, the whole length being divided into 1,000 equal parts, every two of which are graduated.

This line is used in conjunction with the D line; when placed evenly in the groove, the number of equal parts or thousandths of the entire length of the D line, reaching to any division of this line, represents the first three decimal places of the fractional part of the logarithm of the number marked against this division. The line is divided first into ten principal divisions, numbered 1, 2, 3, &c., indicating so many tenths; these are subdivided, each into ten secondary divisions, denoting hundredths; and each of these is again subdivided into five tertiary divisions, so that the half of one of these is a thousandth part of the whole line which represents unity. Laying the slide evenly in the groove, so that the first and last divisions on the line N coincide with the first and last divisions on the D line, the division of this line marked 2 will be found midway between the primary division on the N line, marked 3, and the first tertiary division beyond it; thus showing that the logarithm of 2 is ·301; and, in like manner, the other numbers on the D line will coincide with divisions above them, denoting the fractional portions of their logarithms. The integral part of any one of the logarithms is known from the consideration that it is one less than the number of integral places in the number of which it is the logarithm. These two lines, therefore, form a table of natural numbers on the D line, and of their logarithms on the line N, correct to the extent of three decimal places. A few logarithms to six places of decimals are recorded on one of the edges of the rule; viz., those that are required in the computation of compound interest, at ordinary rates, and of annuities and leases.

The other slide contains on both faces lines of sines and

tangents, those on one face being laid down to radius considered as 100, and equal to the whole length of the line, as already described in the construction of Gunter's lines (p. 114), and on the other face the line of sines being constructed to a radius equal the whole length considered as 10, and the line of tangents to radius equal half the length of the line considered as 10. On the edges of the instrument, and on the faces of the grooves, are recorded a number of useful ratios, gauge points, logarithms, comparisons of measures, &c.

The Kentish Slide-Rule has a square section, thus offering four faces of similar dimensions, and its length is 1 foot. On three of these faces the lines A and D are laid down, the A line at the top, and the D line at the bottom, the D line in each case being numbered from 1 to 10, and the A line twice from 1 to 10 or from 1 to 100. The grooves between the A and D lines contain the following slides:—(1) the B and C slide, the B line being over the A line, and the C line under the D line; (2) the E slide, on which the logarithmic line is repeated thrice, so that the numbers range from 1 to 1000, and are the cubes of numbers on the corresponding divisions of the D line, while the numbers on the A line are the squares of the same numbers; (3) a trigonometrical slide graduated with logarithmic lines of sines and tangents, the former of which works with the line D, and the latter with the line A. These lines are each laid down to radius 10, and not to radius 100, as described in the construction of Gunter's lines, and as they are laid down on the Gunter's scale, whether sliding or not. On the fourth face are placed comprehensive tables of ratios and gauge points.*

Coulson's Slide-Rule (Plate III., fig. 3) is a portable flat instrument, 9½ inches long, with two slides the same thickness as the rule, divided and numbered on both sides, so that in effect it forms a rule with four slides.

The A lines on this rule are only numbered once from 1 to 10, so that the divisions on them are equal to the divisions on a rule 18 inches long.

The two slides are numbered on from one to the other, so as to be used in conjunction in any of the grooves, and

* The above description has been written from the inspection of one of these instruments forwarded to me by the makers, Messrs. Dring and Fage: it is manifest that they might be constructed without repeating the lines A, D, the three slides working to the same lines, and being held in spare grooves when not in use.

Plate III.

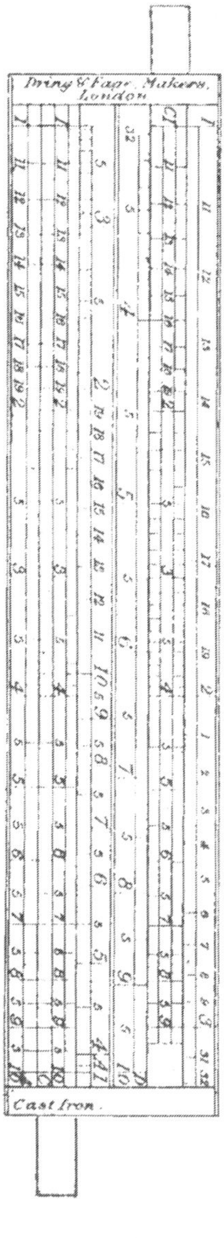

Fig. 3.

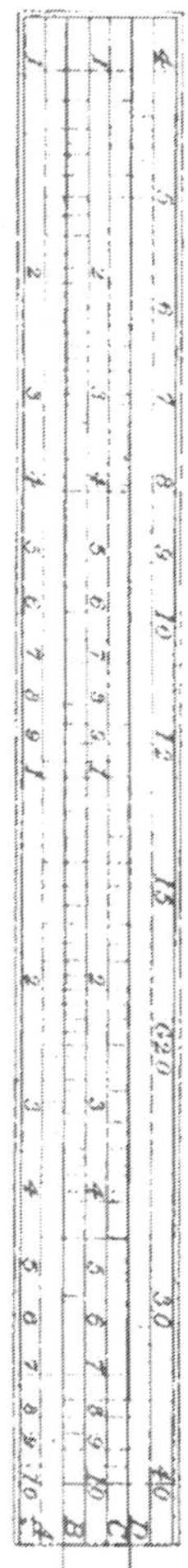

Fig. 2.

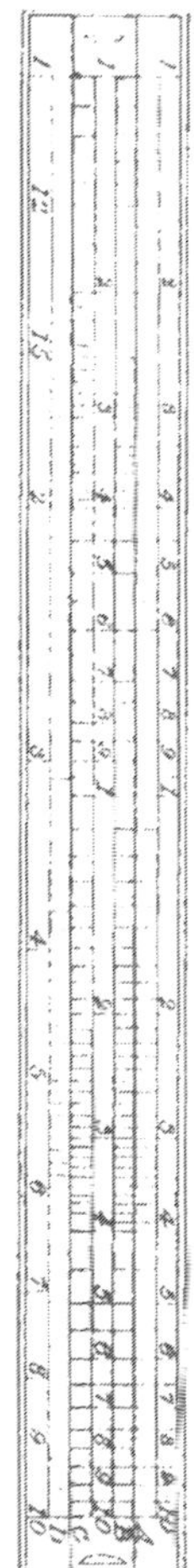

Fig. 1.

thus form a double line: by this means all the advantages of a rule of double the length, with double lines, are obtained.*

The line D, which is double the length of the rule, begins on the upper edge of the rule, and is numbered from left to right from 1 to 3·2, which is on the right end of the rule; the line is then continued below the groove, from 3·2 on the left to 10 on the right. This arrangement requires the numbers on the C line of the slide to be considered to belong to the first section of this line when used with the divisions above the slide, and to the second section when used with the divisions below the slide.

On the line D several gauge points are marked. At 1·128 is a brass pin, marked C C. In., signifying the cylindrical gauge point for cubic inches: this gauge point is also used for finding the content of a cylinder in cubic feet or cubic yards, when the length and diameter are given in feet or yards. In fact, in whatever dimensions the linear measurements may be given, this gauge point serves to determine the content in the same dimensions cubed. The gauge point and brass pin at 13·54, marked C C. Ft., signifies cylindrical gauge point for cubic feet: this gauge point is used for finding the content of a cylinder in cubic feet, when the length is given in feet, and the diameter in inches. The pin and gauge point at 42·53, marked C. Ft., is used for finding the content of a cylinder in cubic feet, when the length is given in feet, and the girth or circumference in inches. The gauge point and pin at 16·65, marked Im. G., is used for finding the content of a square vessel in imperial gallons. The gauge point and pin at 18·79, marked Im. G., is used for finding the content of a cylindrical vessel in imperial gallons. The pin and gauge point at 2·126, marked R. B. I., is used for finding the weight of a round bar of malleable iron in pounds, when the length and diameter are given in inches. The pin and gauge point at 2·21, marked R.C.I., is used for finding the weight of cylindrical cast iron in pounds, when the length and diameter are given in inches.

The gauge point and pin at 27, marked OX, is used for finding the neat dead weight of cattle, when the length and

* This is all that Mr. Coulson claims for it; but the simplification of the numeration on the rule, and the gain of a third section on the slide, when used with the D line, are advantages of far greater importance obtained by this form of construction.

girth are given in inches; and that at 27·6, marked Kyloes, is the gauge point for finding the neat dead weight of small Kyloes. The pin and gauge point at 47·1, marked M. B., is used for finding the content of a square vessel in malt bushels.

The lines marked "Cast Iron," on the same side of the rule, are used for finding the weight of that metal; the upper line, marked A I, signifies the A line *inverted*, which is done in such a manner that 3·836, the square divisor for cast iron, is directly opposite to 1 or 10 upon A; whence any three dimensions are multiplied together, and divided by the square divisor at one operation. The construction and use of these lines have been already explained (p. 128).

It has also been remarked, that the same result is obtained if the dimensions are used indiscriminately on the lines, only observing that the content is always found on C, opposite some one of the dimensions on A.

On the opposite side of the rule are lines marked "Cubic Yards," for finding the content of excavations, &c., in cubic yards, which are used in a similar manner to those for cast iron, the upper line A I being so arranged, that 27, the number of cubic feet in a cubic yard, is opposite to 1 or 10 upon A; the dimensions of course must be taken in feet and decimal parts.

For the convenience of ironfounders and others, to whom the lines marked "Cubic Yards" may not be useful, rules are also made with lines marked "Bar Iron," which are used for finding the weight of malleable iron bars, in a similar manner to those for cast iron.

On this side of the rule are also lines marked "Cubic Feet," which are used for finding the content of timber or other materials in cubic feet, the inverted line, marked A I, being so disposed that the square divisor, 144, is directly opposite to 1 or 10 upon A; the thickness and breadth, must be given in inches, and the length in feet, and these dimensions may, like those for cast iron, or cubic yards, be taken indiscriminately on any of the lines, always observing that the content must invariably be found on B, opposite one of the dimensions on A; to prevent confusion, however, it will be better for the learner to practise an invariable method, and I have, therefore, in the examples given on these lines, set the thickness *in inches* on B, to the breadth in inches in A I;* then, against the length *in*

* In setting the thickness on B to the breadth on A I, the *breadth* on B is also set to the *thickness* on A I at the same time.

feet on A, is the content in cubic feet on B. It must be carefully observed, that the figures and divisions *on all* the inverted lines, marked A I, *must be read from right to left.*

There is a brass pin and gauge point on the line A, at 2218·192, marked Im. B., for finding the content of a square or rectangular vessel in bushels at 1 inch deep; also a pin and gauge point at 277·274, marked Im. G., for finding the content of a square or rectangular vessel in imperial gallons at 1 inch deep; and a gauge point and pin at 353, marked C. Im. G., for finding the content of a circular or elliptical vessel, in imperial gallons at 1 inch deep. A gauge point and brass pin is placed at 3·1416 for finding the circumference of a circle, or the pitch of a wheel.

One of the edges of the rule is divided into inches and tenths, and numbered from the left to right with 1, 2, 3, 4, 5, 6, 7, 8, 9. Under this are two lines, the first marked "Spheroid," and the third line "Second Variety," and both are numbered, from the left to the right, with the figures 1, 2, 3, 4, 5, &c., and the spaces between each of these figures are divided into ten equal parts. These three lines are used for finding a mean diameter between the head and bung diameters of casks of the first and second varieties, which is performed in the following manner:—

Find the difference between the bung and head diameters on the first line, or line of inches, then against such difference, for each variety, is a number, which, being added to the head diameter, will give the mean diameter sought: hence the cask is reduced to a cylinder.

On the other edge of the rule are two A line radii, so disposed as to suit the finding of a mean diameter of a cask of the third variety, which is performed in the following manner:—

Find the difference between the bung and head diameters on the second or undermost line, then against that number is another number on the top line, which, if added to the head, will give a mean diameter; hence the cask will be, as in the preceding case, reduced to a cylinder.

For timber-measuring two slide-rules are used, one called the cubing rule, and the other the timber or timber-contenting rule.

The Cubing Rule.—This rule is three feet in length, and contains four lines, marked A, B, C, E. The E line is above, and the A line below, the groove; the lines B and C are on the slide, the B line above the A line, with which it works, and the C line below the E line, with which it works. The A, C, and E lines are divided duodecimally, that is, into feet and inches; and the last, E, is inverted, so that the numbers taken on the three lines are multiplied together; and as the product of the three commencing numbers is placed against them on the commencement of the B line, the same is everywhere the case, and the rule gives the content at once in cubic feet.

On one side of the rule the E line commences at 5 inches, and runs back to twelve feet. The 1, 2, 3, and 4 inches are supplied by the 10, 20, 30, and 40 feet of the B line, which numbers have the 0 separated by a star, thus—1*0, 2*0, 3*0, 4*0, to signify that the 0 is to be rejected, and that they are to count only as 1, 2, 3, and 4 inches. In reality, however, though they are marked on the B line, they are the numbers on the C line; for the 20 of the B line cuts across the 1 foot 8 inches, that is the 20 inches, of the C line, on which, as it works to the E line, the two inches ought to be taken. The content in this case must, consequently, be reduced to one-tenth: for instance, if the content is found marked on the B line as 70, it is to be taken as 7; if 80, as 8; and so on.

The second side of the rule is a continuation of the first, the one being for small numbers, the other for large.

When none of the dimensions are under 5 inches, they are taken upon the lines in the order denoted by the word E C A B; when one is under 5 inches, and therefore marked with a star, by the word E B A B. Consequently, when taken in the order E C A B, the figures on the rule show the correct content; when in the order E B A B, one-tenth of the content shown by the rule is to be taken.

The B line is divided decimally, giving the solidity in cubic feet and tenths.

EXAMPLES.

A pile of deals is 20 feet 7 inches wide, 4 feet 7 inches high, and 14 feet long: required its content.

20 ft. 7 in. E : 4 ft. 7 in. C : 14 ft. A : 1320 cub. ft. B.

E 20 ft. 7 in.	
C 4 ft. 7 in.	
B	1320 cub. ft.
A	14 ft.

A deal is 14 feet long, 2 feet 3 inches wide, and 3 inches thick: required the content.

2 ft. 3 in. E : 3 in. B : 14 ft. A : 7·88 cub. ft. B.
E 2 ft. 3 in.
C 2 ft. 6 in.
B 3*0 78·8 cub. ft.
A 14

The Timber-contenting Rule.—This is usually 2 feet long. One side contains four lines, A, B, C, and E, the last of which is inverted, and so graduated that 144 falls opposite the commencement of the slide; consequently it divides the product by 144 (p. 129), and reduces square inches to square feet. Therefore, by taking the length of a piece of timber in feet, and the thickness and width in inches, it gives the content in cubic feet. The dimensions are taken, as in the cubing rule, in the order of the letters E C A B, the last or B line exhibiting the content. The A, C, and E lines are divided into fourths, the B line decimally.

The other side of the rule contains the usual A, B, C, and D lines; the last of which, termed the girt line, or more properly the quarter-girt line, is made to commence at 4 and run on to 40. On this face of the rule the following points are marked with brass pins:—the division on the A line at 12, for reducing linear inches to feet; the square-gauge point on the D line at 12; the circular, hexdecagonal, and octagonal quarter-girt gauge points between the 10 and 11, marked O, 16, 8; and the octagonal, hexdecagonal, and circular diameter gauge points between the 13 and 14, marked 8, 16, O. The O denotes the gauge points for round timber; the 16, for hexdecagonal or sixteen-sided timber; and the 8, for octagonal or eight-sided timber.

The words girt and diameter explain themselves; the former denoting the measure round, the latter across. As applied to octagonal and hexdecagonal timber, by diameter is to be understood the perpendicular distance between opposite sides, in contradistinction to diagonal, which is the distance between opposite angles; or as applied to polygons in general, the term diameter is applied to the diameter of the inscribed circle, the term diagonal to the diameter of the circumscribed circle. It is to be borne in mind that with the diameter-gauge points the *whole* of the diameter is taken; with the girt or more properly quarter-girt points only *one-fourth* of the girt is taken. We proceed to explain fully the meaning and use of these gauge points.

In a square, the diameter, side, and quarter-girt are equal, and, if they each measure 1 inch, 144 such squares make a foot. Consequently, the square divisor is 144, to be taken on A, and 12, the square root of 144, is the square-gauge point to be taken on D, to reduce to square feet the area of a square, the side of which is given in inches.

If the diameter of a regular octagon be 1 inch, then 173·82 such octagons make a square foot. If, then, the diameter of a regular octagon be taken in inches, 173·82 is the divisor, to be taken on A, and its square root, 13·18, is the diameter-gauge point, to be taken on D, to reduce the area of the octagon to square feet.

If the perimeter of a regular octagon be 4 inches, so that the quarter-girt is 1 inch, 119·29 such octagons make a square foot: consequently 119·29 is the quarter-girt divisor, and its square root, 10·92, the quarter-girt gauge point, for this figure.

If the diameter of a regular hexdecagon be 1 inch, then 180·98 such hexdecagons make a square foot; and if the perimeter of a regular hexdecagon be 4 inches, then 114·57 such hexdecagons make a square foot. Consequently, 180·98 and 114·57 are the diameter and quarter-girt divisors; and their square roots, 13·45 and 10·7, are the gauge points for this figure.

If the diameter of a circle be 1 inch, then 183·35 such circles make a square foot, and if the circumference be 4 inches, then 113·1 such circles make a square foot. Consequently, 183·35 and 113·1 are the diameter and quarter-girt divisors, and their square roots, 13·54 and 10·63, are the gauge points for the circular area to reduce it to square feet.

The squares of the diameters or quarter-girts in inches, and the divisors being taken on A, their quotients on B will give the sectional areas of a piece of timber or other material in square feet; and the same result will be obtained if the diameters or quarter-girts in inches, and the gauge points be taken on D, and their quotients on C.

If, then, the diameters or quarter-girts be taken in inches, and the lengths in feet, the products of the lengths and sectional areas will give the contents in cubic feet.

EXAMPLES WITH TIMBER RULE.

A plank is 36 feet long, $9\frac{3}{4}$ inches broad, and $4\frac{1}{2}$ inches thick: required its content.

E $9\frac{3}{4}$
C $4\frac{1}{2}$
B 10·97 cub. ft. *Ans.*
A 36

A piece of timber is $63\frac{1}{2}$ feet long, and $17\frac{1}{2}$ inches square: required its content.

C $63\frac{1}{2}$ 135 cub. ft. *Ans.*
D 12 $17\frac{1}{2}$

An octagonal mast is $72\frac{1}{4}$ feet long, and $21\frac{3}{4}$ mean diameter: required its content.

C $72\frac{1}{4}$ 196 cub. ft. *Ans.*
D 8 diameter $21\frac{3}{4}$

A hexdecagonal mast is $72\frac{1}{2}$ feet long, and 69 inches round in the middle: required the content.

C $72\frac{1}{2}$ 188 cub. ft. *Ans.*
D 16 girt $17\frac{1}{4}$

A round mast is 85 feet long, and $23\frac{3}{4}$ inches mean diameter: required the content.

C 85 261 cub. ft. *Ans.*
D O diameter $23\frac{3}{4}$

MEASURING INSTRUMENTS—TIMBER MEASURERS.

Having now discussed rather fully those instruments which are used for computations based upon given dimensions, we proceed to describe some of the best instruments for taking dimensions—first describing those used by timber-measurers for taking the dimensions required for the computations to be performed with the cubing and timber rules just described. These are the hull callipers, bow callipers, graduated steel bar, plank rule, girt tape, timber rod, and scribing-iron.

The *Hull Callipers* are used for taking the thickness of square timber, and consist of a steel stock, graduated upon both sides and the edge in inches and quarters, and with the end bent at right angles, so as to form a fixed arm and claw. To ascertain the thickness of a log of timber, open the arms of the callipers, and place the fixed claw against one side, then slide the other arm towards it until stopped by contact with the log; the required dimensions, in inches and quarters, upon the steel stock are now shown against the lower edge of the sliding arm. Care must be taken that neither is placed in any hole or crack in the timber, or a considerable error in the estimated content will be the result.

The callipers should be tested occasionally, to see that the indications of the graduated stock agree with the distance between the claws. They are usually made in sizes from 18 to 48 inches; the 18-inch callipers commence at one inch, the larger sizes are not graduated below 4 inches.

The Bow Callipers are used for taking the diameters of large, round, and irregular timber. They are to be placed so that the points touch at the required parts; the points are then to be transferred to the graduated steel bar which will give the distance in inches and quarters.

The Steel Bar is used in conjunction with the bow callipers, and is graduated on each side with inches and quarters, from 3 to 33 inches, so as to indicate the dimensions taken. It is usually carried in the left hand; and, being shaped like a crowbar, is useful for shifting the logs, so as to place them in the best position for measuring.

The Plank Rule is a sliding boxwood calliper, furnished with steel tips, and is used for ascertaining the width and thickness of planks. The graduations upon it are in inches and quarters, from ½ inch to 24 inches.

The Girt Tape is a broad strong tape, marked on one side in plain inches, after the usual manner, so as to show the perimeter, or girt, of a piece of timber; on the other side every inch counts but as a quarter, so that when the tape is passed round the log under measurement, the quarter-girt is shown at once where the brass ring intersects the tape. This is a far more accurate and expeditious mode than the old plan of taking the girt with a piece of string, and then measuring the latter with a rule.

The Timber Rods are painted deal rods, with a hook at one end, for ascertaining the length of timber. They are made in different lengths, from 5 to 20 feet, and are marked with feet and quarters, or, for more accurate measurement, with feet and inches. When they are of a greater length than 10 feet, it is advisable to have them jointed.

The Scribing Iron is for marking the content: it is made of two sizes—large for the hand, and small for the pocket.

Instruments similar to those just described are manifestly adapted to the measurement in any direction of bodies of a great variety of forms, where an approximate estimate of their dimensions is all that is required. We now proceed to describe a few instruments which are adapted to measure distances and dimensions with the greatest amount of accuracy that is practically available.

THE SPHEROMETER.

This is a delicate instrument for measuring the diameter of bullets and spherical bodies. It consists of a

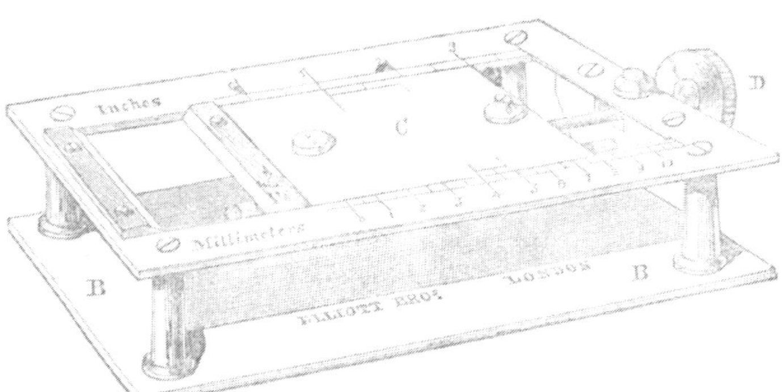

Fig. 1.

frame supported on pillars rising from a stand, B B. In the frame works an accurately fitting slide, C, to which a slow motion is given by the screw D. On the inner edges of the frame are two scales—that on one side being divided into $\frac{1}{10}$ths of an inch, and that on the other into millimètres; while on the edges of the slide are two verniers—that on one side sub dividing the divisions into 20, so as to read to $\frac{1}{1000}$th of an inch, and that on the other subdividing the millimètres into 10. In using it, the bullet is passed between the bevelled edges shown at the ends of the frame and of the slide.

THE EXTENSOMETER.

This is an instrument for measuring the extension of bars or rods, produced by given strains. It consists of two clamps, A and B, to one of which, A, is attached a frame, C C, and to the other a slide, D D, working acccurately in that frame. The clamps, being placed upon the rod or bar to be experimented upon, are held in their place by turning the milled-headed screws S S. The rod being then subjected to a strain, which is

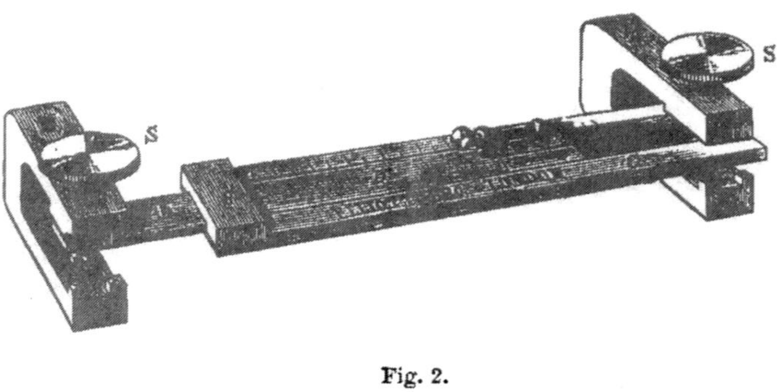

Fig. 2.

accurately measured by a balance composed of a system of levers, the points of the screws retain the positions to which they have been turned down upon the bar, and the distance to which the slide is drawn out shows the amount of extension of the rod in a length equal to the distance between the original points of the screws. The measurements are read from a scale on the frame divided to $\frac{1}{50}$th of an inch, by means of a vernier subdividing each division of the scale into 20, and thus giving the reading to the $\frac{1}{1000}$th of an inch.

GERMAN-SILVER WIRE GAUGE.

In this beautiful little instrument, accurate minute measurements are arrived at by a process differing from those just described. It consists of a claw, A B C, having its arms A and

C at right angles to its stock, B. The arms A and B terminate in cylindrical heads, pierced by female screws having their axes in the same straight line, and through these pass two male screws having their extremities accurately ground, so that when brought into contact no light can penetrate between them. The cylindrical head is prolonged and graduated on its surface to $\frac{1}{50}$th of an inch, which is the distance between the threads of the screw, and the screw working through this cylinder is attached to a milled head; and to this head is also attached a hollow cylinder, F, concentric with the screw and cylinder in which it works, and just working over the surface of this cylinder. The extremity of the cylinder F is bevelled, and graduated into 20 equal parts, so that the passage of one of those parts past an index on the cylinder E indicates a $\frac{1}{20}$th part of a turn of the screw, and therefore a motion of the extremity of this screw through a distance of $\frac{1}{1000}$th of an inch. The screw in the cylinder D is adjusted by the hand so that the graduations on the cylinders E and F shall both read zero, when the two screws are brought into contact; and, consequently, when a wire or any other substance, the thickness of which it is desired to measure, is placed between the screws, and the milled head turned until it is held firmly between them, without being compressed, the readings in the cylinders E and F give the required dimension to the $\frac{1}{1000}$th part of an inch, those on E recording the number of turns, and those on F the parts of a turn, in the distance between the extremities of the screws.

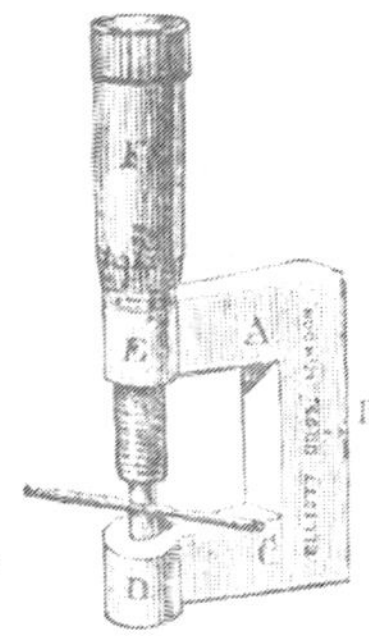

Fig. 3.

INDEX TO VOL. I.

A and B lines, on slide rule, uses of, 101; numeration on, 102; problems on, 103.
A, B, C, and D lines, uses of conjointly, 123.
Angles, methods of setting off and measuring, 28, 48.
Arcs of large circles, mode of describing, 19.
Area, of a drawing reduced or enlarged, 64; of a plan measured, 77, 80; of a board or plank computed, 105.

Bar, steel, for timber measuring, 152.
Bayley's slide-rule, 143.
Beam compasses, 17.
Bow, compasses, 15; line, 6.

C and D lines, on slide-rule, different construction of, 109; comparative merits of these constructions, 115; numeration on, 114; squaring by, 109; extraction of square root by, 112; problems on, 118.
Callipers, hull, 152; bow, 152.
Centres, horn, 16.
Centrolinead, 8.
Chords, construction of line of, 26; uses of line of, 28, 48.
Compasses, 13; beam, 17; bow, 15; with movable points, 14; proportional, 63; triangular, 62; tubular, 16.
Computer, 77.
Content, of a drawing reduced or enlarged, 68; of a solid computed, 95; of timber, construction of rule for finding, 129, 147, 149; of excavation, construction of rule for finding, 130.
Copying drawings, various modes of, 62, 63, 64, 65, 67, 68, 73.
Coulson's slide-rule, 144.
Course, method of setting off a ship's, 29.
Cube of a number found, 118, 126; root extracted, 65, 126.
Cubing rule, 147.

Development of a portion of a sphere, 34.
Diagonal scale, 24.
Dial, horizontal, 36; erect south, 37; east and west, 37.
Division of one number by another, 94, 103, 126; of one number by the square of another, 120; of one number by the square-root of another, 120; of a line into equal parts, 45, 54.
Divisors, 138.
Drawing, a, copied, 62, 67, 68; its linear dimensions reduced or enlarged; 47, 63, 67, 68, 73; its area reduced or enlarged, 64; its solid content reduced or enlarged, 65.
Drawing boards, requisites of, 2; construction of, 3.
Drawing compasses, 13.
Drawing paper, mounting of, 1; dimensions of, 2.
Drawing pen, 4.

Edge, straight, 5.
Eidograph, 70.
Ellipse, construction of, 20.
Equal parts, division of a line into, 45; scales of, 22, 47.
Excavation, construction of rule for finding content of, 130.
Extensometer, 154.

Flamstead's projection of the sphere, 31

Gauge, German-silver wire, 154; points 138, 140.
Geometrical construction, general rules for, 15.
Girt tape, 152.
Gnomon, construction of, 37.
Gnomonic projection of the sphere, 31.
Gunter's lines, 88; scale, 98.

Horn centres, 16.
Hours and latitudes, scales of, 27, 28; uses of, 36.
Hyperbola, construction of, 21.

Inverted line, 129; slide, 128.

Kentish slide-rule, 144.

Latitudes, hours and, scales of, 27, 28; uses of, 36.
Line, straight, rules for drawing, 5; divided into equal parts, 45, 54; of lines, uses of, 45; of logarithmic numbers constructed, 91, 93; of logarithmic sines and tangents constructed, 91; used, 95.
Linear dimensions of a drawing, reduced or enlarged, 47, 63, 67, 68, 73; measured, 75
Logarithms, 89; lines of, their construction, 90; their use, 93.
Longitudes, scale of, constructed, 27; used, 29.

Marquois's scales, 52; parallels and perpendiculars drawn by, 53; lines divided into equal parts by, 54.
Mean proportional found, 118, 127.

Method of squares, 68.
Military protractor, 39.
Mounting paper, 1.
Movable points, compasses with, 14.
Multiplication performed by Gunter's rule, 94; by slide-rule, 101; of one number by the square of another, 119, 126; of one number by the square-root of another, 119.

Numbers, lines of logarithmic, constructed, 91; their use, 93; reciprocals of, found, 108, 125.
Numeration, on the A and B lines, 102; of decimals on the A and B lines, 107; on the C and D lines, 114; with inverted line, 129.

Opisometer, 75.
Orthographic projection of the sphere, 30.

Pantagraph, 65.
Parabola, construction of, 22.
Parallel, rulers, 7; lines drawn at equal distances, 53.
Pen, drawing, 4.
Perpendicular, lines drawn, 53.
Plain scale, 38.
Plan, lines on, measured, 75; area of, measured, 77, 80.
Plank, area of, computed, 105.
Plotting scales, 23.
Polar planimeter, 80.
Polygon, in a circle or on a line, drawn, 49; tables of ratios for, 134, 135.
Pricking point, 4.
Projection of the sphere, Flamstead's, 35; gnomonic, 31; orthographic, 30; stereographic, 32.
Proportion, 45, 94, 104; inverse, 127; involving squares and square roots, 121, 127.
Proportional compasses, 63; mean, found, 118, 127.
Protracting scales, 26.
Protractor, upon plain scale, 38; circular, 57; military, 39; semi-circular, 59.

Radius, found from the length of the sine, tangent, or secant of an arc, 51.
Ratios, 133; tables of, 134, 135.
Reciprocals of numbers found, 108.
Rhumbs, scales of, their construction, 27; their use, 29.
Root, square, extraction of, 65, 112, 127; cube, extraction of, 65, 126.
Routledge's slide-rule, 142.

Scales, diagonal, 24; Marquois's, 52; plotting, 23; protracting, 26; of equal parts, 22, 47; of feet and inches, 23, 47.
Secant, scales of, their construction, 26; their use, 30; to any desired radius found, 50.
Sector, 42.
Semi-tangents, scales of, their construction, 27; their use, 30.
Sines, scales of, their construction, 26; their use, 30; to any desired radius found, 50; line of logarithmic, 91; sine rhumbs, line of logarithmic, 93.
Slide-rule, 99; Bayley's, 143; Coulson's, 144; Kentish, 144; Routledge's, 142.
Slide, inverted, 125; trigonometrical, 132.
Solid contents computed, 95, 140, 147, 149.
Sphere, development of portion of, 34; projection of, Flamstead's, 35, gnomonic, 31; orthographic, 30; stereographic, 32.
Spherometer, 153.
Square, T, 7; set, 8; of a number found, 105, 109, 127; root of a number extracted, 112, 127.
Station pointer, 60.
Stencilled letters, 5.
Stereographic projection of the sphere, 32.
Straight edge, 5.
Superficies, or areas, computed, 94, 105, 138.

T-square, 7.
Tables, of roods and perches in decimals of an acre, 86; of ratios, 134; of gauge points for surfaces, 138; of gauge points for solids, 140.
Tangents, scales of, construction of, 26; use of, 30; to any radius found, 50.
Templets, 20.
Timber contenting rule, 149; measuring instruments, 151.
Triangle, construction of isosceles having each of the angles at the base double of the third angle, 50; solution of right-angled, 96.
Triangular compasses, 62.
Tubular compasses, 19.

Vernier, 54.
Versed sines, scale of, 26.

Weight of cast iron, construction of rule for finding, 130.
Wholes and halves, 62.
Wire, German-silver, gauge, 154.

THE END.

PRINTED BY J. S. VIRTUE AND CO., LIMITED, CITY ROAD, LONDON.

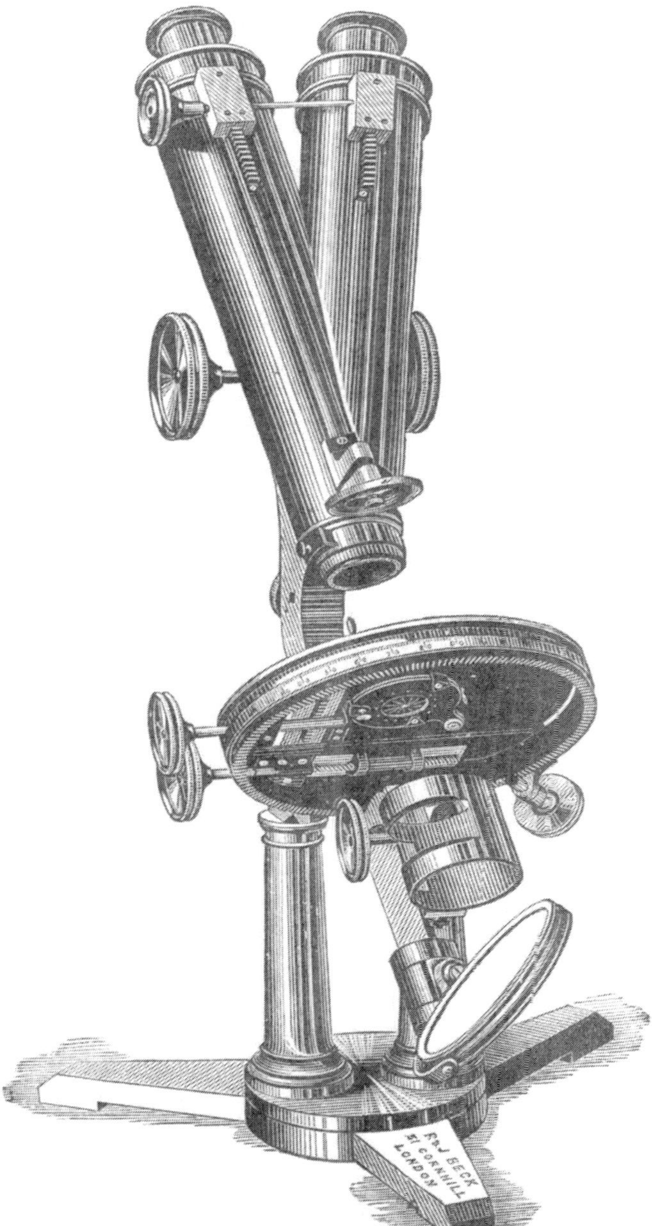

Beck's First-class Microscope.

MATHEMATICAL INSTRUMENTS

VOL. II.

OPTICAL INSTRUMENTS

INCLUDING

(MORE ESPECIALLY) TELESCOPES, MICROSCOPES, AND APPARATUS FOR PRODUCING COPIES OF MAPS AND PLANS BY PHOTOGRAPHY

PREFACE TO VOL. II.

In the original edition of this work, published in a single volume at one shilling, so small a space could be afforded for the description of Optical Instruments, that it was necessarily very imperfect. In this new edition, in which a small volume is entirely devoted to the subject, much pains have been taken to produce a work, that, without pretending to rival the merits of more elaborate and expensive treatises, shall describe clearly the construction and use of the most important Optical Instruments, and the means of testing their merits.

The various mechanical modes of making copies of plans and drawings upon any desired scale, have been fully described in vol. i.; and the subject is completed in this volume by the introduction of a full description of the method employed in producing reduced copies of maps and plans by the assistance of photography, and in fixing the pictures produced by the camera, derived from the instructions of, and for the information of, engineer officers employed in the survey of India, which have been kindly forwarded to the author by Captain R. S. Palmer, R.E.

Valuable assistance has been received from several eminent opticians, and more especially from Messrs. Parkes and Son, of Birmingham, who supplied the author with many capital instruments for examination, and from Messrs. R. and J. Beck, who have furnished several of the illustrations, including that of the admirable first class microscope of their construction, which forms the frontispiece to this volume.

Watford,
1871.

CONTENTS OF VOL. II.

PART III.

OPTICAL INSTRUMENTS.

CHAPTER I.

PLANE AND CURVILINEAR REFLECTORS.

PAGE

Definitions of Terms 1
Reflection from a Plane Surface 2
Reflections from two Parallel Plane Surfaces 2
Reflection from two Inclined Plane Surfaces 4
Repeated Reflections from Inclined Mirrors 5
Construction of the Kaleidoscope 6
Spherical, Parabolic, and Elliptical Reflectors 9

CHAPTER II.

PRISMS.

Definitions of Terms 12
Deviation and Decomposition of Light produced by Prism . . 12
Refracting power of Medium determined 14
Total Reflection produced by Prism 15
Mounting of Prism 15
The Camera Lucida 15

CHAPTER III.

LENSES AND SPECTACLES.

The term Lens defined 18
Different forms of Lenses 19
Conditions of Distinct Vision 20
Spherical Aberration and Confusion 20
Chromatic Dispersion 21
Focal Lengths of Lenses 21
Real and Virtual Images formed by Lenses 22
Image formed on the Retina 24
Causes of Weak or Long Sight, and Strong or Short Sight . . 24
Weak Sight and Short corrected by Lenses 25
Requisites of good Spectacles 25
The Visometer 27
Magnifying power of Lenses 27

CHAPTER IV.

BINOCULAR VISION.

	PAGE
The Stereoscope defined	31
Parallax	32
Phenomena of Double Vision	33
Wheatstone's Reflecting Stereoscope	34
Brewster's Lenticular Stereoscope	35
Photographic Pictures for Stereoscopes	38

CHAPTER V.

MICROSCOPES.

Simple Microscope	38
Coddington Lens	39
Doublets	40
Darwin's Dissecting Microscope	41
Quekett's ditto	43
Beck's Improved ditto	43
Beck's Binocular Prisms	45
Nachet's ditto	47
Compound Microscope	47
Magnifying Apparatus of Compound Microscope	48
Magnifying Power and Dimensions of Compound Microscope	49
Modern Improvements in Compound Microscopes	50
Object-pieces	51
Eye-pieces	52
Use of Varying Powers	54
Draw Tube and Erector	55
Binocular Body, Nachet's	56
———————— Wenham's	57
————————- Nachet's Improved	58
Powell and Ireland's Binocular Prism	59
Illumination of Objects	60
Stage, Mirror and Condenser, and Diaphragm	61
Lieberkuhn	62
Stage with delicate Movements	62
White's Lever Stage	63
Stands for Compound Microscopes	65
Beck's Popular Microscope	66
Chevalier's Universal Microscope	69
First class English Microscopes	72
Accessories of the Microscope	73
Brown's Iris Diaphragm	74
The Spot Lens	74
Wenham's Parabolic Reflector	75
Beck's Parabolic Illuminator	76
Beck's Patent Illuminator	77
Sorby's Spectroscope Eye-piece	78
Polarising Apparatus	78
Application of Camera Lucida to Microscope	79

PAGE
The Stage Micrometer 80
Eye-piece Micrometer 80
Jackson's Eye-piece Micrometer 81
Stage Forceps 82
Live-box and glass Trough 83
Management of Light and Care of Eyes 83
Management of Opaque Objects 84
Tests for Microscopes 85
Care of the Instrument 86

CHAPTER VI.

TELESCOPES.

Astronomical Telescope 86
Eye-pieces 87
Galilean and Binocular Telescopes 89
Magnifying Power suitable for Terrestrial Telescopes . . 90
Reflecting Telescopes 91
Pouillet's Method of determining the Magnifying Power . . 94
Adjustment and Testing of Refracting Telescopes . . . 94
——————— Reflecting Telescopes . . . 97
Mounting Telescopes 98

CHAPTER VII.

INSTRUMENTS FOR EXHIBITING MAGNIFIED PICTURES OF OBJECTS TO AN ASSEMBLAGE OF SPECTATORS.

The Magic Lantern 100
Phantasmagoria 101
Dissolving Views 102
Powerful sources of Light for the Lanterns 103
Solar Microscope 104
Oxy-hydrogen and Photo-electric Microscopes 107

CHAPTER VIII.

THE AMERA OBSCURA.

The Camera Obscura 108
The Photographic Camera 109
Apparatus for Reduction of Plans 110
Photographic Process for the Production of Reduced Copies . 122

CHAPTER IX.

INSTRUMENTS FOR MEASURING THE ANGLES OF CRYSTALS.

The Common Goniometer 130
Wollaston's ditto 132
Charles's ditto 137

MATHEMATICAL INSTRUMENTS.

PART III.

OPTICAL INSTRUMENTS.

UNDER this head our principal object will be to consider the construction and principles of action of such instruments as are indispensable to assist the vision in making observations upon objects too minute to be examined by the unassisted eye, and upon distant objects, whether upon terrestrial objects for the purposes of the surveyor, or upon celestial objects for the purposes of astronomy and navigation. A full description is also given of the method employed in producing reduced copies of maps and plans by the assistance of photography, and in fixing the pictures produced by the camera, derived from the instructions of, and for the information of, engineer officers employed on the survey in India, which have been kindly forwarded to me by Captain R. S. Palmer, R.E. Such other optical instruments as by their utility, or by the frequency with which they are brought before us, appear to demand attention, will also meet with due consideration.

CHAPTER I.

PLANE AND CURVILINEAR REFLECTORS. THE KALEIDOSCOPE.

PLANE REFLECTORS.

A COLLECTION of straight lines, either conical or cylindrical representing rays of light, is called a *pencil* of light, and the axis of the cylinder or cone is called the *axis* of the pencil.

The term *medium* is used in optics to signify any *transparent* substance, that is, any substance through which a portion of the light falling upon it can pass.

When a ray of light is reflected at a plane surface, the reflection takes place in a plane perpendicular to the reflecting surface, and the incident and reflected rays make equal angles with this surface. Thus, if Q A represent a ray of light incident upon a plane reflector at the point A, and the plane of the paper represent the plane which contains Q A, and is perpendicular to the reflecting surface, intersecting it in the line R A R′, then, making the angle Q′ A R′ equal to the angle Q A R, and in the same plane with it, A Q′ will represent the course of the reflected ray.

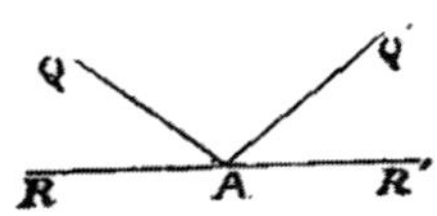

The effect of a plane reflector upon the pencils of light which fall upon it is to change the direction of all the rays forming each pencil, without altering the angles at which the several rays of the pencil are inclined to one another, so that the divergency or convergency of the pencils remains the same after reflection as before, and the objects from which they proceed appear to be at the same distances behind the mirror as they really are in front of it. Thus, a pencil of light diverging from a point of an object at P, after reflection at the point R of a plane mirror, appears to proceed from the point P′ on the line P M P′, perpendicular to the mirror, at the distance M P′ behind the mirror equal to the distance M P. The point P′, from which, after reflection, the pencil appears to have diverged, is called a virtual focus; and the apparent image of the object behind the mirror is called a virtual image.

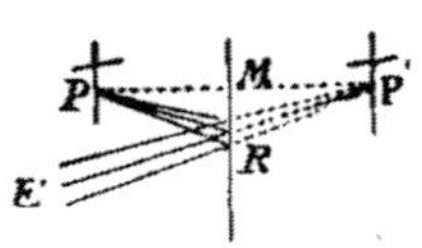

Reflection from two Parallel Plane Surfaces.—When an object is placed between two plane reflectors, the rays proceeding from the object and falling upon one of these reflecting surfaces, are turned back, and fall upon the second surface: from this they are reflected back again to the first, and so on alternately, forming an endless series of images behind the mirrors. In like manner, the rays falling first upon the second mirror form a second endless series of images. These images, however, grow fainter and fainter, and thus gradually die away, until they are no longer bright enough to render themselves visible. By colouring an object red on one side and blue on the other, the series of images, which results from rays reflected first from one of the mirrors, can readily be distinguished from that which results from a first reflection at the other. Plane reflectors are usually made of glass plates, well polished, on both faces and coated with metal, or, as it is called, silvered, on one of these faces. In this manner a highly polished plane metallic surface is obtained, the brilliancy of which is effectively preserved by the glass covering. The rays of light which pass through the medium of the atmosphere and fall obliquely on the surface of the glass, are bent, or refracted, on entering the denser medium of the glass, are next reflected at the metallic surface at the back of the glass, and are again refracted upon emerging from the glass into the

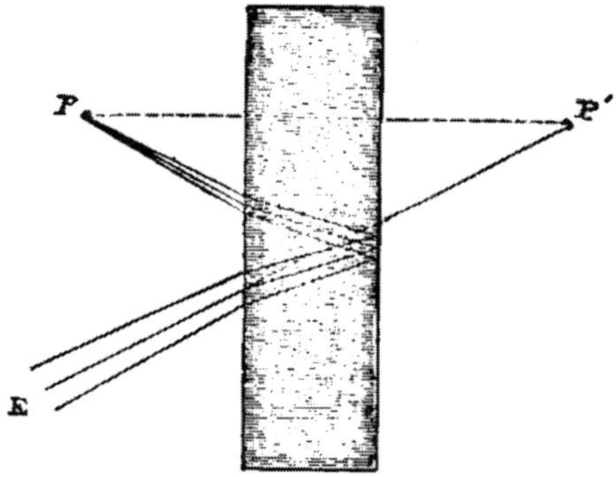

atmosphere. The distance behind the mirror of the apparent image, formed by the combination of the refractions with the reflection at the metallic surface, varies with the position of the eye, being greatest when the eye is as nearly as possible in the same line with the object and its image, without the latter being hidden by the former. This greatest distance of the apparent image behind the mirror is equal to the distance of the object in front of the mirror together with four-thirds the thickness of the glass. If then the object be placed in immediate contact with the glass, the distance between the object and its image, when viewed as directly as possible, is four-thirds the thickness of the glass; and, conversely, *the thickness of the glass is three-fourths the distance between the object and its apparent image.*

The light, however, proceeding from the object does not all pass through the glass to the metallic surface at its back; a portion of it is reflected from the first surface without entering the glass at all: neither do the rays reflected from the metallic surface all pass out again after recrossing the thickness of the glass; a portion are again reflected internally back to the metallic surface, again reflected there, and again in part emerge, and in part are again reflected, thus forming a series of apparent images, growing fainter and fainter, and, in fact, so far exceeded in brightness by the apparent image formed at their first meeting with the metallic surface, as in general to be entirely eclipsed by it. By placing, however, a lighted candle before an unsilvered glass in front of a dark background, the series of images may be easily seen. It follows, therefore, that when a pencil light falls on a glass surface, it does not all pass through it; but a portion is reflected back from each surface of the glass, and a portion lost for all practical purposes by a series of internal reflections.

The plane reflector, formed of polished glass with silvered back, is universally employed as an article of domestic furniture, under the name of a *looking-glass*.

The test of a good looking-glass is the non-distortion of distant objects looked at very obliquely.

The formation of a true plane surface is by no means an easy problem, and in the plane reflector we must have not only one, but two, true planes, and they must, moreover, be strictly parallel.

The more highly polished the surfaces are, the brighter will be the images formed; and the mirrors which look the darkest are generally the best.

REFLECTION FROM TWO INCLINED PLANE SURFACES.

When a ray of light, proceeding in a plane at right angles to each of two plane mirrors, which are inclined to each other at any angle whatever, is successively reflected at the plane surfaces of each of the mirrors, the total deviation of the ray is double the angle of inclination of the mirrors. For let I *i* and H *h* represent sections of the two mirrors made by the plane of incidence at right angles to each of them, and let S I represent the course of the incident ray: then the ray S I is reflected at I into the direction I H, making with I *i* the angle H I A, equal to the angle S I *i*, and is again reflected at H into the direction H E, making, with H *h*, the angle E H A equal to the angle I H *h*. Now the angle A H V, being equal to the exterior angle I H *h*, is also equal to the two interior angles H I A and H A I; and because the vertical angles A V H and I V E are equal, and that the three angles of every triangle are equal to two right angles, therefore the two angles V I E and S E H are, together, equal to the two angles A H V and H A I, and therefore to the angle H I A and twice the angle H A I (since A H V has been proved equal to H I A and H A I); but V I E, being equal to the vertical angle S I *i*, is also equal to the angle H I A: therefore, taking away these equals, the remainder, the angle S E H, is equal to the remainder, twice the angle H A I. *Q.E.D.*

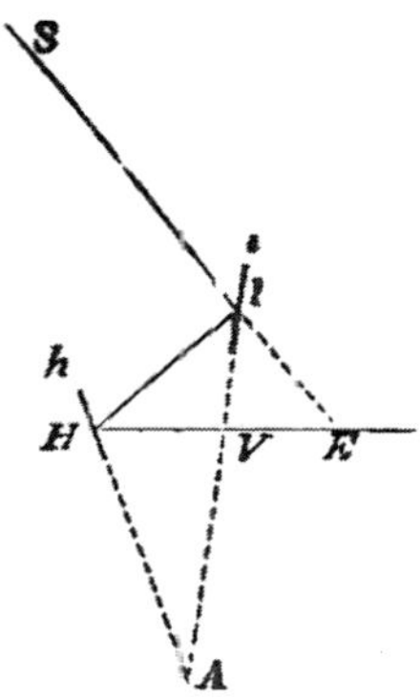

This property of two plane reflectors enables us by their aid to measure the angle subtended at the eye by any two objects whatever, and is the foundation of the construction of Hadley's Quadrant, and the improvements upon it, in Hadley's Sextant, hereafter to be described.

THE KALEIDOSCOPE.

Repeated Reflection at Inclined Mirrors.—When an object is placed between two plane mirrors which are inclined to one another at an angle, a limited number of images are formed, arranged in a circular form round the intersection of the two mirrors; and, by properly choosing the angle of inclination, very interesting and beautiful effects are produced. Let us take, for instance, two mirrors, A B and A C at right angles to each other, and let O represent an object placed between them; then the rays of light diverging from O, which fall upon the mirror A B, will, after reflection from this mirror, diverge from the point O′, at the same distance behind A B, that O is before it; again, the rays from O that fall on A C will, after reflection, diverge from O″, at the same distance behind A C that O is before it; and, lastly, rays which, after reflection at A B, fall upon A C, will then diverge from O‴, at the same distance behind A C that O′ is before it; and, in the same manner, rays proceeding from O, first to A C, and then reflected to A B, will again be reflected from this mirror, and appear to proceed from O‴. The point O‴ being behind both mirrors, the rays diverging from it can be no more reflected, and hence there will be three images, o′, o″, o‴, of the object o, and no more, formed by reflection at the two mirrors, and they manifestly lie on the circumference of the circle described from A, at the distance A o. Now, a ray o*a*, proceeding from o, is reflected in the direction *a* E, as though proceeding from o′, a ray o*b* is reflected in the direction *b* E, as though proceeding from o″, and a ray o*c* is reflected first into the direction *cd*, and then into the direction *d* E, as though proceeding from o‴, and thus an eye placed opposite to the point E, and looking along the intersection of the two mirrors, will see at the same time the object o and its three images, o′, o″, and o‴.

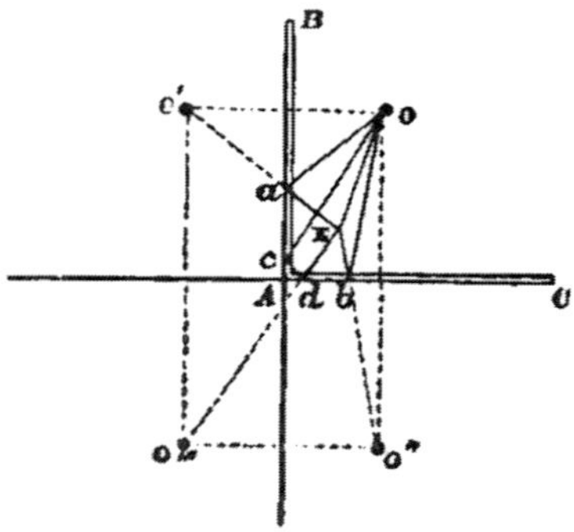

Symmetry.—Since the image of each point of an object is at the same distance behind a plane mirror as the point itself is before it, it follows that, however irregular the form of an object may be, this object, together with its image, produce a figure perfectly symmetrical with respect to the reflecting surface. Again, we have seen in the case of two mirrors, at right angles, that the images o‴, formed by reflection first at one mirror and then at the other, coincide; and this coincidence of the last images formed will always occur, whenever the number of degrees in the angle of inclination of the two mirrors is an even divisor of 360°; consequently we shall have then a number of pairs of images, forming a series of symmetrical figures symmetrically arranged round the intersection of the two mirrors.

Construction of the Kaleidoscope.—This property of the symmetrical arrangement of the figures produced by the reflections, is taken advantage of in the construction of the instrument invented by Sir D. Brewster, and named the *kaleidoscope.* Two plane mirrors, A B, A C, formed of slips of glass silvered at the back, are placed in a tube, A B C, blackened on the inside, and set at an angle of 60°; at one end of the tube are then fixed two discs of glass, between which any such objects as pieces of coloured glass, petals of flowers, beads, &c., can be placed; and by shaking or turning round the tube, fresh combinations of some of these objects are continually brought between the two reflecting surfaces, and thus present to an eye looking through a hole in an opaque disc at the other end of the tube, an endless variety of beautifully symmetrical figures. Thus, an object with the irregular outline *a b c d* would present the three similar symmetrical forms, of which *d c b a a′ b′ c′ d′* is one, represented in Fig. 1, the whole series being symmetrically

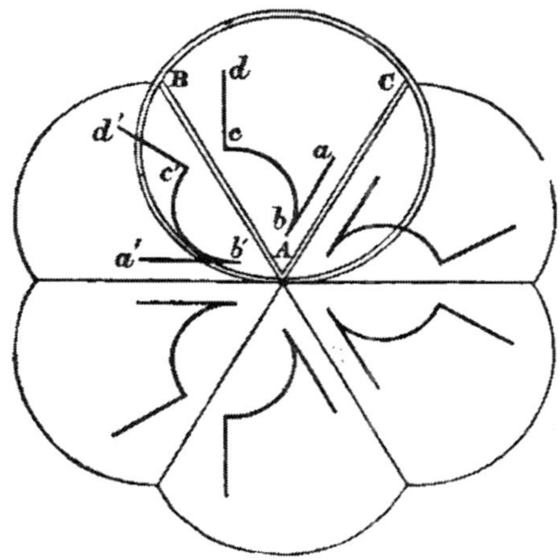

Fig. 1.

arranged witn respect to each of the three diameters of the hexagonal diagram.

Fig. 2 shows the effect that would be produced by the same object, if the two mirrors were at right angles: the symmetry would still be perfect; but from the paucity in the number of the symmetrical figures produced, the effect would be inferior to that produced by an inclination of 60°. With an inclination of 45° four symmetrical figures would be produced; with one of 36°, five, and with any smaller inclination, which should be an *even* divisor of 360°, still more symmetrical combinations would appear; but the diminished light at each successive reflection detracts from the effect produced; so that the inclination of 60° has been found to give the best effect.

When the number of degrees in the angle of inclination is

an *odd* divisor of 360, the symmetry is no longer perfect, and the images of different portions of the object may ultimately

Fig. 2.

overlie each other. Figs. 3 and 4 represent the effects produced by inclinations of 120° and 72° respectively.

If the number of degrees in the angle were not an exact divisor at all of 360°, still greater confusion would be produced.

Fig 3.

By placing a third reflector in the tube, so as to form with the two others a hollow equilateral prism, an additional number of symmetrical combinations will be produced, and,

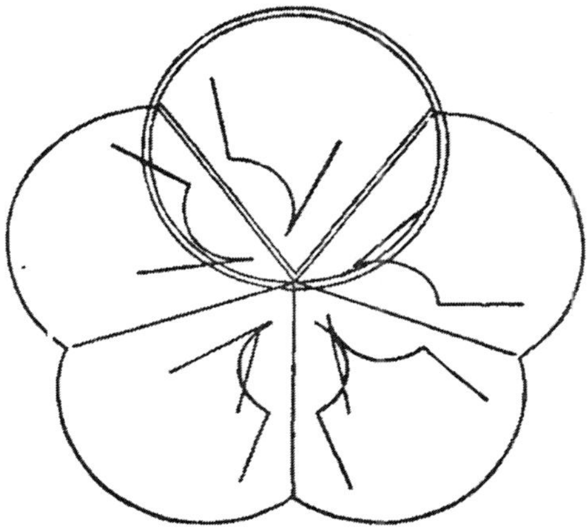

Fig. 4.

in fact, will be repeated, like the images in parallel mirrors, becoming fainter and fainter, until lost in obscurity.

ON CURVILINEAR REFLECTORS.

Spherical reflectors, or specula, as they are called, produce upon pencils of rays results precisely similar, with one exception, to those produced by lenses (p. 19). Thus, a *concave reflector* makes the rays of the pencils incident upon it more convergent, and corresponds in its uses with a convex lens; while a *convex reflector* makes the rays of the incident pencils more divergent after reflection, and corresponds in its uses with a concave lens. The exception to the similarity of the results produced by lenses and reflectors is, that with the latter there is no chromatic dispersion (p. 21), and the only sources of error are the aberration and spherical confusion, which are common to both spherical reflectors and lenses. For astronomical observations, however, in which case the rays incident upon the object-speculum are parallel, these sources of error are removed by making this speculum of a parabolic form, and another speculum, if it be used, of the form of the vertex of a prolate spheroid. There is great difficulty in procuring flint glass in pieces of large size without flaws, and we are consequently limited as to the size of the lenses of good quality that can be formed with such glass; and, without its use, we have not hitherto been able to form available achromatic object-glasses. Recourse is, therefore, had to parabolic or spherical specula in the formation of telescopes of large power for examining the heavens.* These specula are formed of metal, and the chief objection to them is the impossibility of producing an accurate surface. Even supposing its general form to be correct, there are always minute inequalities arising from the nature of the substance, which cause a waste or dispersion of light. Great pains are, consequently, taken in their construction to obtain the form and surface of the best possible quality.†

* Sir William Herschel's largest telescope was 40 feet long, and the mirror 4 feet wide. Lord Rosse's largest telescope is 56 feet long, and the mirror 6 feet wide.

† The following description of the methods employed in forming and polishing parabolic reflectors is extracted in an abridged form from an account of Skerryvore Lighthouse, by Alan Stevenson, LL.B., F.R.S.E., M.I.C.E., Engineer of the Northern Light Board:—

"The reflector plate is formed of Virgin silver and the purest copper, from the ingot, in the proportion of 6 oz. silver to 16 oz. of copper. The two metals are formed into pieces of the form of rectangular parallelopipeds about 3 inches in length, and the same in breadth, and are then tied together with wire, placed in the furnace, and united with a flux of burnt borax and nitre, mixed to the consistence of cream. The metal thus united is repeatedly passed through the rolling mill, and annealed in the furnace after each time of passing through, until it comes out a plate 28 inches square. It is then cut into a circular disc ready for hammering; and great care must be taken to keep the metal clean during the process of hammering and polishing now to be performed.

"The hammering is commenced by placing the plate with the copper side upon a block slightly concave, and beating it on the inner or silver side with a box-wood mallet, rounded at each end, *c* and *d* (Fig. 1). The beating is commenced on the edge and continued round and round till the centre is gradually reached. After the disc has been raised sufficiently by this means, it is taken to a machine called the horse, and beaten with a wooden mallet upon the copper side, its concave face being turned about upon a bright steel head *a* (Fig. 2), until it has nearly reached the proper height for the reflector, which is ascertained by a mould *m* (Fig. 3).

"After each course of raising with the wooden mallet the reflector is annealed, as follows: first damped with clean water, and dusted over with powder, composed of one pint of powdered charcoal to one ounce of saltpetre; then put on a clear charcoal fire, till the powder flies off and shows when it is duly heated. It is next plunged into a pickle, composed of one quart of vitriol, in five or six gallons of water; and, lastly, washed with clean water and scoured with Calais sand.

"The next step is to put the reflector into an iron stool, and, having drilled a small hole in its vertex, to describe a circle from this point with beam compasses, and cut the paraboloid to the proposed size.

"The reflector is now *hard-hammered* with a planishing hammer, or planished, as it is called, on the bright steel head *a*; and then *smoothed* with a lighter hammer muffled with parchment. Then comes

Fig. 1. Fig. 2. Fig. 3.

the finishing, called also filling up to the mould, which is thus performed. It is constantly tried with the mould *m*, and those portions which do not meet it are marked with fine slate pencil, and then gone over with the muffled hammer, till every point touches the mould. Great care must be taken in this process that no part of the surface be raised above the gauge, or the reflector would have to be reformed with the wooden mallet, and the whole process repeated. The reflector is then tried with a lamp brought to its focus, and, if the whole surface is luminous, it is fit for polishing; but, if not, it must be again tested by the mould, and carefully filled up with the muffled hammer, till the result of the lamp test is perfectly satisfactory.

"The edge of the reflector is next turned over to stiffen it, and the bizzle *w* (Fig. 1), and back belt *g* (Fig. 2), having been soldered on, the final process of polishing may be proceeded with. This process is commenced by scouring, first with a piece of pure charcoal of hard wood, and next with a mixture of Florence oil and finely-washed

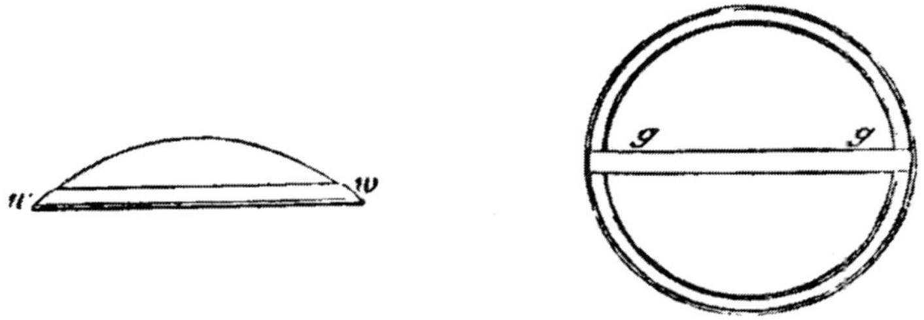

Fig. 1. Fig. 2.

rotten stone, applied by means of a large ball of superfine cloth. The reflector is then cleansed with a fine flannel dipped in Florence oil, and afterwards dusted over with powder of well-washed whiting, and wiped out with a soft cotton cloth. Lastly, it is carefully rubbed by the naked hand with finely washed rouge and clean water, and wiped with a smooth chamois skin. In all the polishing and cleansing processes some skill is required in the manipulation, as the hand must be moved in successive circles parallel to the lips of the reflector, and having their centres on the axis of the generating curve."

The speculum of Lord Rosse's great telescope is composed of 1264 parts of copper and 589 of tin, fused together and cast in a mould, the bottom of which is formed of hoop iron bound closely together with the edges uppermost. By this means the heat is conducted away through the bottom so as to cool the metal towards the top, while the interstices between the hoops, though small enough to prevent the metal from running out, are sufficiently open to allow the air to escape. After casting, the speculum is annealed in a brick oven, which is heated almost to a red heat, and shut up with the speculum in it, and allowed to cool gradually. The speculum is then placed with its face upwards upon a turning apparatus, and the grinding and polishing performed entirely by the aid of mechanical contrivances, so that the proper parabolic form is accurately given to it. To test the work, the dial plate of a watch is placed upon the top of a mast at 90 feet distance from the speculum, and the image of this dial-plate formed by the speculum, being viewed through an eye-glass properly placed, the distinctness of this image denotes the accuracy of the speculum.

CHAPTER II.

PRISMS. THE CAMERA LUCIDA.

THE PRISM.

THE term *prism* in optics is used to signify a portion of any medium bounded by plane surfaces which are inclined to one another. The bounding surfaces are called the *faces* of the prism; the line in which the faces intersect is called the *edge* of the prism; and the angle formed by any two adjoining faces through which the pencil of light passes is called the *refracting angle.*

The prism is to be placed so that the axis of the pencil, by which an object is seen through it, may be in a plane perpendicular to the edge of the prism, and during and after its passage through the prism still remain in this plane.

One effect of a prism of denser material than the surrounding medium is to bend every ray of light passing through it, and, consequently, the whole pencil, further from the refracting edge of the prism.

Another effect of such a prism is to decompose each single ray of white light into several rays of different colours, which rays are bent at different angles, so as to form a lengthened image of different colours, of the point from which the ray proceeds. This image is called the spectrum, and these colours the colours of the spectrum.

When, then, any object is viewed through a prism, the two following effects are produced. 1stly. The apparent position of the object is changed, so that, if the prism be held with its edge downwards, as in the accompanying figure, the

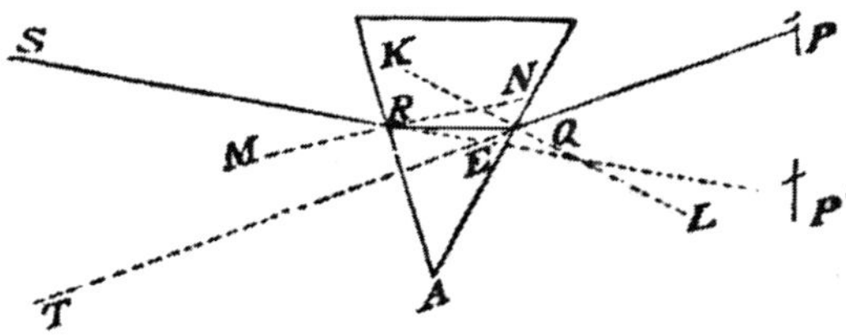

object appears lower than it really is, while, if the prism were held with its edge upwards, the object would appear in a position higher than its actual position. 2ndly. The boundaries of the object are indistinctly defined, and fringed with colours.

The figure represents the section of the prism made by the plane of incidence, that is, by the plane which is perpendicular to the edge of the prism, and contains the incident ray of light P Q, forming the axis of the pencil under consideration, which proceeds from one point of an object P. A Q and A R are sections of the faces of the prism; A is a point in its edge; and the angle Q A R is its refracting angle. Now the ray of light P Q, proceeding from the object at P through the medium of the atmosphere, is bent, upon entering the denser medium of the prism, from the direction Q T into the direction Q R, nearer to L Q K, the perpendicular, to the face A Q of the prism, at the point of incidence Q; and, upon emerging from the prism into the rarer medium of the atmosphere, is again bent from the direction Q R into the direction R S, further from M R N, the perpendicular to the face A R at the point of emergence R. The eye, being placed at S, sees the point P, therefore, by means of a pencil of light of which S R is the axis, and P consequently appears at P′ on the prolongation of the line S R. A similar effect being produced upon every other point, the entire object is apparently depressed, as represented in the figure.

The angle T E S, or P E P′, between E R S, the direction of emergence, and P E T, the direction of incidence, is called the angle of deviation.

The amount of refraction, when a ray of light passes from one medium into another, varies with the angle of incidence, so that the sine of the angle of refraction bears a constant ratio to the sine of the angle of incidence. This ratio varies for each different medium, and is called the refracting power of the medium. The deviation of a ray in passing through a prism varies also with the angle of incidence, and has a minimum value when the angles of incidence and emergence are equal. The refracting power can be determined by finding practically this minimum deviation, as follows:—Place the prism with its edge downwards, so as to receive a small beam of solar light, admitted into a dark room through a hole in a shutter, and let the beam of light, after refraction, be received upon a screen behind the prism. The prism must then be turned round an axis parallel to its edge, so as to vary the angle of incidence, and, consequently, the position of the bright spot upon the screen; and, in one particular position, we shall find the bright spot to remain stationary for an instant, though the motion of the prism is continued. The deviation will then be a minimum, and will be equal to the sum of the sun's altitude and the inclination of the emergent beam to the horizon. Let S represent this minimum deviation, and A the refracting angle of the prism, and let the refracting power be represented by μ: then $\mu = \dfrac{\sin. \frac{A + S}{2}}{\sin. \frac{A}{2}}$

Since a ray of light emerging from a prism into the rarer medium of the atmosphere is bent farther from the perpendicular to the emergent face of the prism, it follows that the angle of incidence on this face may be increased until the angle of emergence becomes a right angle, and the emergent ray is consequently parallel to the face of the prism. If then the angle of incidence be still further increased, the ray cannot emerge at all at this face, but is totally reflected within the prism. The angle of total reflection for glass is about 41°, and hence, when a ray passing into a glass prism falls on the emergent face at an angle with the perpendicular to this face, exceeding 41°, it is totally reflected.

The consideration of the properties of the prism is of great importance, as exhibiting in the simplest manner the principles of the refraction and dispersion of light.

For the purpose of examining these properties, the prism is usually mounted on a stand, similar to that represented in the annexed figure. This stand consists of a hollow pillar, P, in which slides a tube, T. This tube is fixed at any desired height by a tightening screw, S; and to it the frame carrying the prism is attached by a knuckle-joint, C. The prism is attached to the piece *a a*, which has an axis passing through the upright portion of the frame, and can be turned round by the milled head *b*. It will thus be seen that the apparatus can be adjusted so as to place the prism in any desired position whatever with respect to the incident light.

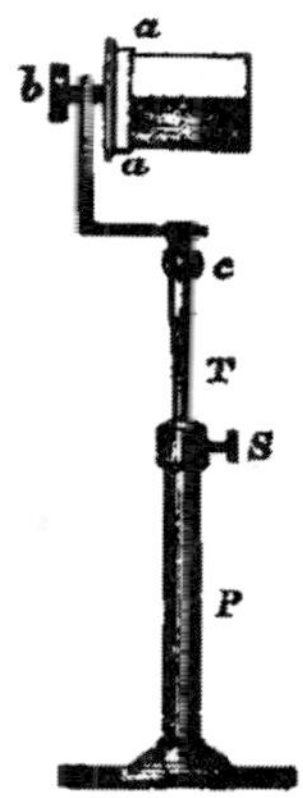

Plane reflectors were formerly used in mathematical instruments to change the directions of pencils of light, whenever such a change was necessary or desirable; but in all cases in which the deviation, or amount of the change, is constant, prisms have now superseded the use of the plane reflector. The prisms are made of such a form that the immergent and emergent faces are perpendicular respectively to the incident and emergent rays, and thus the loss of light from reflection at these surfaces is rendered as small as possible while the faces upon which rays fall within the prism at angles less than the angle of total reflection (p. 14) form the most perfect reflectors that can be obtained.

THE CAMERA LUCIDA.

This ingenious instrument, the invention of Dr. Wollaston, consists of a quadrilateral prism, of which A B C D represents a section made by a plane at right angles to each of its edges: A B is equal to B C, and A D to D C, and the angles of the prism are a right angle at B, an angle of 135° at D, and angles each 67° 30′ at A and C. Over the face B A, and projecting beyond A, is a plate of metal having in it a narrow longitudinal aperture, which is just bisected by the edge A of the prism.

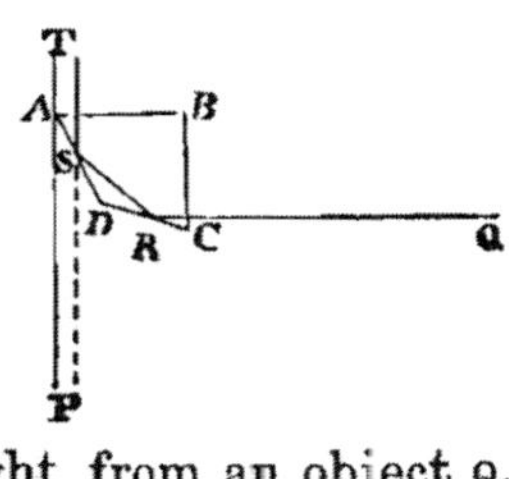

The axis, Q R, of a small pencil of light from an object Q, directly in front of the face B C, passes straight through this face, and falls upon the face D C, making with it an angle of

22° 30′. It is there reflected into the direction R S, and falling upon the face D A, at the same angle, is again reflected into the direction S T, perpendicular to the face A B, and consequently passes straight through this face without refraction. Looking down through the aperture in the metal plate, an image of the object Q is seen at P, at a distance from A B equal to the distance of the object itself from B C; and if A B be placed at a distance from a table not less than the distance of distinct vision, and a sheet of paper be laid upon the table at P, the image of the object, as seen through the prism, will be thrown upon the paper, seen through the projecting part of the aperture, and an accurate drawing of the object may be traced upon the paper.

It is necessary, for distinct vision of both the object and the drawing at the same time, that the pencils of rays proceeding from them should have nearly the same degree of divergence. This is accomplished by placing the prism, as above directed, at a distance from the table not less than the distance of distinct vision, when such an arrangement is possible; but when such an arrangement is impossible, or inconvenient, a lens may be placed between the table and the prism. To adapt the instrument for use without fatiguing the eye, coloured glasses should be used, to give nearly the same brightness to the two images.

To effect the arrangements described above, the prism is mounted upon an axle, X, parallel to its edges. This axle is attached to the end of a rod sliding in a tube which has at the other end a clamp, C, for fixing it to the edge of the table. The tube is connected with the clamp by a knuckle-joint, and the setting piece *t*, is also jointed to the clamp at one end, and at the other to a collar, *k*, through which the tube can slide, and be fixed at any desired inclination by the screw *s*; and, by sliding the rod that holds the prism in or out of the tube, the distance of the prism from the table may be shortened or lengthened at pleasure.

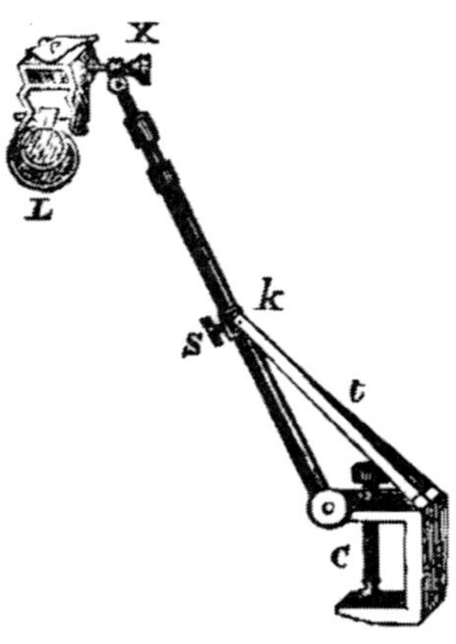

Lenses and coloured glasses are mounted at L upon an axis below the prism.

The purpose accomplished by the two reflections, first at the face C D, and then at the face D A of the prism, is to render the image of the object erect, otherwise an inverted image,

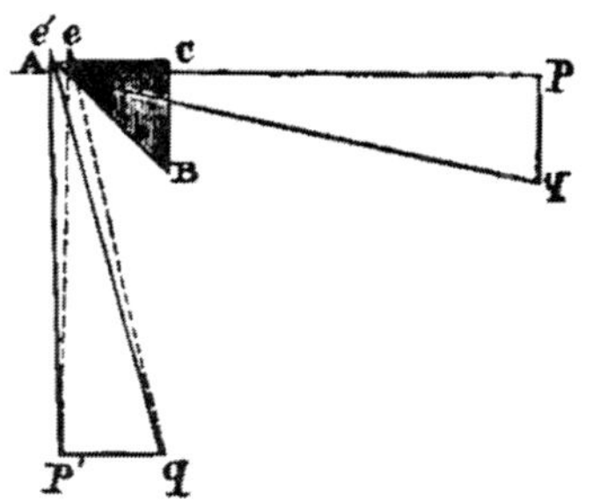

$q' p'$, of the object $p\,q$, might be formed by a single reflection at the face A B of a rectangular prism, A C B.

Whatever be the form of the prism, however, the eye must be kept steadily in such a position that the pupil is bisected, or very nearly so, by the edge A of the prism. For let $p\,q$ represent an object in face of the prism, and $p'\,q'$ its image in apparent coincidence with the paper on the table: then, drawing the lines p' A e and q' A e', from the points p' and q', through the edge of the prism, to meet the pupil of the eye in the points e and e', it is manifest that no ray from the point p of the object can be reflected to the left of the point e, nor can any ray from the point q' of the paper reach the eye to the right of the point e'. If, therefore, the eye be so far advanced over the prism that the point e' falls below the pupil, a portion of the paper at the bottom of the image will be invisible, and if the eye be drawn back until the point e fall above the pupil, a portion of the image at its upper part, p' will be invisible. Hence, to obtain distinct vision at the same time of the entire image, and of the paper on which it is projected, the eye must be so placed that the limiting reflected ray $p'\,e$ from the object shall pass below the upper edge of the pupil, and the limiting direct ray $q'\,e'$ from the paper in coincidence with the image shall pass above the lower edge of the pupil. To make this possible, the distance between the points e and e' must be less than the diameter of the pupil. Again, the entire diameter of the pupil being usually about one-fifth of an inch, it follows that a very slight motion of the head will serve to put out of view a portion either of the image or of the paper in coincidence with it. A considerable amount, then, of practice and patience will be needed in order to become expert in the use of this instrument.

There is another form of the camera lucida, in which an image of the hand and paper is seen by reflection, and the object is looked at through a small hole about the eighth of an inch in diameter. To render the image erect, the rays may first be reflected at the interior surface of a prism, and then at the surface of the plane reflector M M: thus, *a b c* E represents the course of a ray proceeding from the paper in the direction *a b*, and reflected first from the point *b* of the prism to the point *c* of the mirror, and thence from the mirror to the eye at E; while the rays from the object pass to the eye directly through the hole in the mirror M M. This is the form of the instrument used by Amici with the microscope; and, as the image of the hand and paper is certain to be reflected to the eye from some portion of the reflector, it is only necessary to direct the sight through the aperture so as to obtain a full view of the object; and thus the difficulty experienced in adjusting the position of the eye, when using the Wollaston form of the instrument, is in this form obviated.

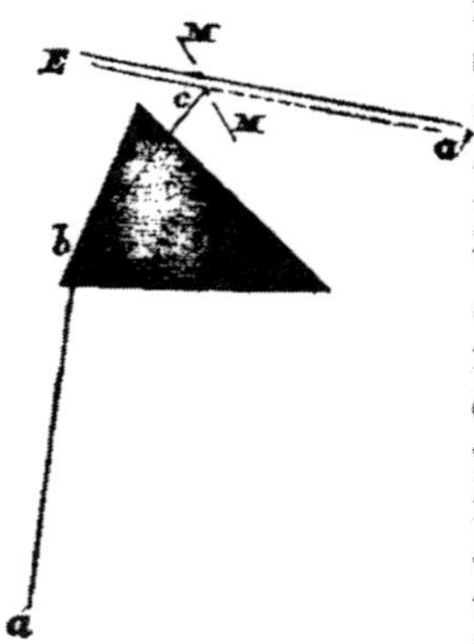

CHAPTER III.

LENSES AND SPECTACLES.

LENSES.

A PORTION of any medium bounded by two spherical surfaces having a common axis, or by a spherical surface and a plane one, is called a *lens*.

The effects produced by lenses upon pencils of light depend both upon the form of the lens itself, and upon the direction in which the pencil is proceeding with respect to the lens. Lenses consequently receive distinguishing names, to mark either different forms, or different positions with respect to the light falling upon them. These distinguishing names are the following, and the forms and positions of the corresponding lenses are represented in the accompanying diagram, the light being considered to be proceeding from left to right:—

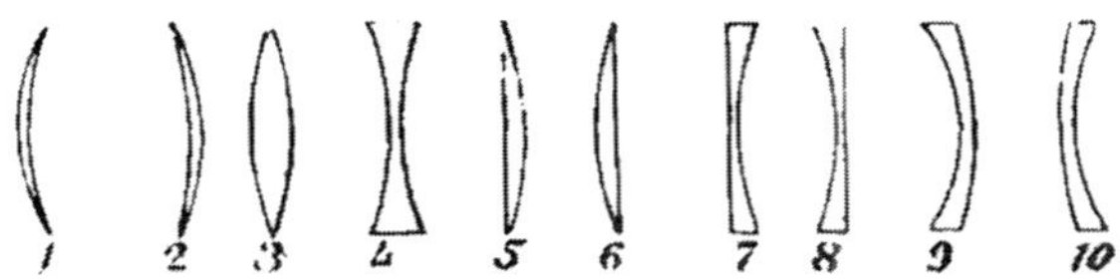

1. Convex meniscus. 2. Concave meniscus. 3. Double convex. 4. Double concave. 5. Plano-convex. 6. Convexo-plane. 7. Plano-concave. 8. Concavo-plane. 9. Concavo-convex. 10. Convexo-concave.

The rays forming any pencil of light must either be divergent, parallel, or convergent; and when a pencil of light passes through an essentially convex lens, that is, one which is thicker in the middle than at the edges, as 1, 2, 3, 5, 6, the rays are made more convergent, so that a pencil of converging rays becomes still more convergent, a pencil of parallel rays becomes convergent, and a pencil of diverging rays becomes either less divergent, parallel, or convergent: but when a pencil of light passes through an essentially concave lens, that is, one which is thinner in the middle than at the edges, as 4, 7, 8, 9, 10, the rays are made more divergent, so that a pencil of converging rays becomes either less convergent, parallel, or divergent, a pencil of parallel rays becomes divergent, and a pencil of diverging rays becomes still more divergent.

The sensation of vision is produced by pencils of rays proceeding from every point of the visible object, and entering the pupil of the eye; and in order to produce distinct vision the rays of each such pencil must either be parallel or slightly divergent. Thus the sun, moon, and planets are seen distinctly, although so distant, by parallel rays; and the least distance from the eye at which an object can be seen distinctly varies in different persons, according to the power of the natural lenses which, in fact, form the eye. 1. When any object is brought nearer to the eye than this without the intervention of a lens, the vision becomes confused. If, however, the rays of light proceeding from the object were, by the interposition of a convex lens, rendered less divergent or parallel, the vision would be again distinct. 2. It is further necessary for distinct vision that the intensity of the light be not less than a certain intensity, as may easily be exemplified by gradually closing the shutters of a room, and thus diminishing the intensity of the light proceeding from the objects in the room, when they will grow more and more indistinct. 3. Lastly, it is also necessary, for distinct vision of any object, that the axes of the pencils proceeding from its extreme parts enter the eye under an angle not less than a certain angle, so that, however strong a light be thrown upon an object, if this object be very minute, or removed to a distance very great with respect to its magnitude, it will not be seen by the naked eye. If, however, by the assistance of a lens, or combination of lenses, a sufficient number of rays to produce the required intensity of light can be collected from each point of an object, and passed through the pupil of the eye, and if at the same time the axes of the extreme pencils are bent by this lens, or combination of lenses, so as to enter the eye under a sufficiently large angle, while the rays of each pencil are made parallel, or but slightly divergent, then vision will ensue, no matter how minute, how distant, or how imperfectly illuminated the object may be.

When a pencil of rays proceeding from a point of an object passes through a lens, the rays which pass through at different distances from its centre will diverge from or converge to different points, so that the whole pencil will not any longer diverge from, or converge to, a single point; and from this cause the image of one point will overlap the image of another, and an indistinctness of the object will arise. This source of indistinctness is called the spherical *aberration*. A combination of lenses may, however, be formed, so that the aberration of one shall be corrected by the aberration of the others. Such a combination is said to be *aplanatic*.

Since rays of light proceed in every direction from the points of visible objects, the pencils of light intercepted by a lens for the first time are all *centrical*, that is, their axes all pass through the centre of the lens; but the pencils, upon emerging from this first lens are already determinate both in extent and direction, and consequently will fall, some of them at least, *eccentrically* upon a second lens, that is, their axes will meet this lens at different distances from its centre. The axes of the most eccentrical pencils will then, after emergence, cross the axis of vision nearer to the lens than will those of the more nearly central pencils; and thus, while the centre of the object is seen distinctly, the parts at a distance from the centre will be distorted, or *vice versâ*. This source of error is called the *spherical confusion*. The spherical confusion is diminished by dividing the desired deviations, or bendings of the axes, between two or more lenses; and it is found by opticians to be a good rule to divide the deviations equally amongst the lenses employed.

The most important source of indistinctness, however, is the dispersion of each ray into rays of different colours refracted at different angles (p. 12), which is called the *chromatic dispersion*. The effect of this dispersion upon a centrical pencil is partly analogous to the spherical aberration, causing the images of neighbouring points to be of finite extent and overlap one another; and it, moreover, fringes the image with colour. An eccentrical pencil is separated by this dispersion into pencils of rays of different colours, the axes of which are bent at different angles; and the imperfection arising from this cause is far greater than that from both the spherical aberration and spherical confusion.

Before stating the manner in which the imperfections arising from the chromatic dispersion are remedied, it will be expedient to explain what is understood by the focal length of a lens. Now a pencil of parallel rays, after passing through a lens, becomes either convergent or divergent, as the lens is convex or concave, and the distance from the surface of the lens to the point to which the pencil converges, or from which it diverges, is called the *focal length* of the lens.

To find practically the Focal Length of a Convex Lens.—Place a lighted candle at one extremity of a scale of inches and parts, with which the lens has been connected in such a manner as to slide along, and always have its axis parallel to the scale. A flat piece of card is also to be made to slide along, so as to be always in a line with the light and the lens, the lens being between the light and the card. The lens and card are then to be moved along, backwards and forwards, till the least distance between the card and light is discovered at which a clear image of the light is formed upon the card; and this distance is four times the focal length.

The imperfection arising from the chromatic dispersion is

remedied, for the centrical pencil, by making a compound lens of two or more lenses of different substances, as flint glass and crown glass, which are fitted close together, and are of such focal lengths that the chromatic dispersion of one is counteracted by the chromatic dispersion of the other. The effect of the chromatic dispersion upon an eccentrical pencil is remedied by setting two or more lenses at proper distances depending upon their focal lengths: such a combination of lenses, as applied to telescopes and compound microscopes, is called an achromatic eye-piece.

When an object, O O, is placed at a distance from a convex lens greater than its focal length, the divergent pencils of

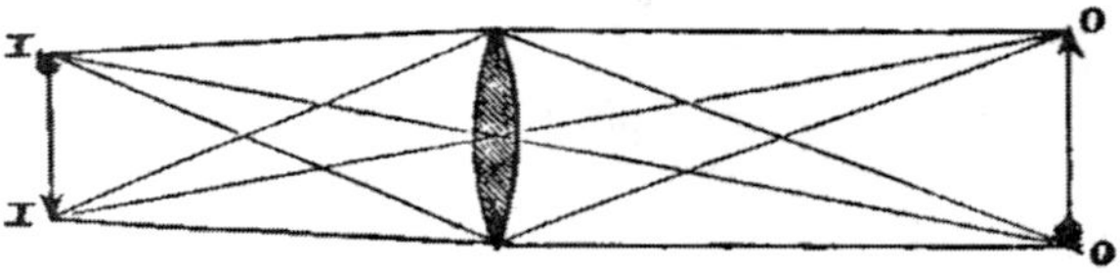

rays, proceeding from every point of the object, become, after passing through the lens, convergent, and at a certain distance from the lens,* having converged nearly to points again, form there an inverted image, I I, of the object. The essential difference between any point in this image, and the corresponding point in the object itself, is, that the latter emits light in all directions, while the light from the former is limited to the pencil which has been transmitted through the lens, and is consequently determinate both in magnitude and direction. If, however, a screen be placed at the required* distance from the lens, a picture of the object in an inverted position will be formed upon this screen, and from each point of this picture light will be emitted in all directions in the same manner as from the points of the object itself. The single pencil of light from any point of the

* If u be the distance from the lens at which the object is placed, f the focal length of the lens, then v, the distance from the lens at which the image is formed, is determined from the equation $\frac{1}{v} = \frac{1}{u} - \frac{1}{f}$; and, hence, when u is greater than f, v is negative, so that the image is on a different side of the lens from the object. When u is less than f, v is positive, and the image is imaginary, and on the same side of the lens as the object. The linear magnitude of the image is to that of the object as v to u.

object, transmitted through the lens, supplies, however, in this case, the light for all the pencils emitted from the corresponding point of the image; and a very strong light must therefore be thrown upon the object to give a moderate brightness to the picture; more especially if the picture be of larger dimensions than the object. A portion of the light is also absorbed by the lens itself.

When an object, O O, is placed at a distance from a convex lens less than its focal length, the divergent pencils of rays, proceeding from every point of the object, become, after

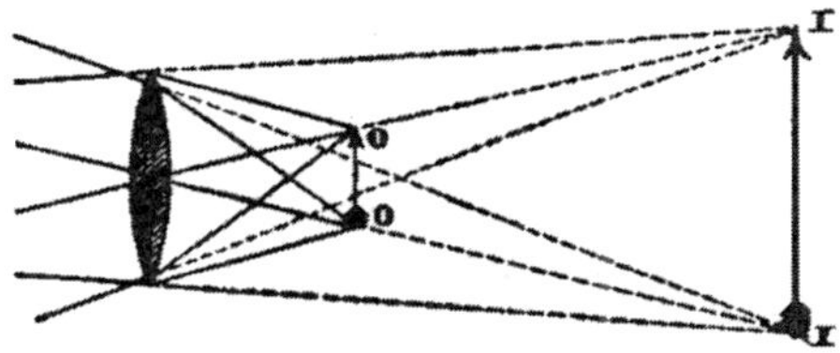

passing through the lens, less divergent, as though proceeding from an enlarged image, I I, at a greater distance from the lens than the object, O O. No image could in this case be formed on any screen; but as the rays come to the eye, as though proceeding from an image, I I, this is called an imaginary image.

When then an object is placed before a convex lens, if its distance from the lens be greater than the focal length, the image is on the opposite side of the lens, and is real, and may be either greater or less than the object; but if the distance from the lens is less than its focal length, the image is at a greater distance from the lens on the same side of it, and is erect, imaginary, and greater than the object.

When an object, O O, is placed before a concave lens, the

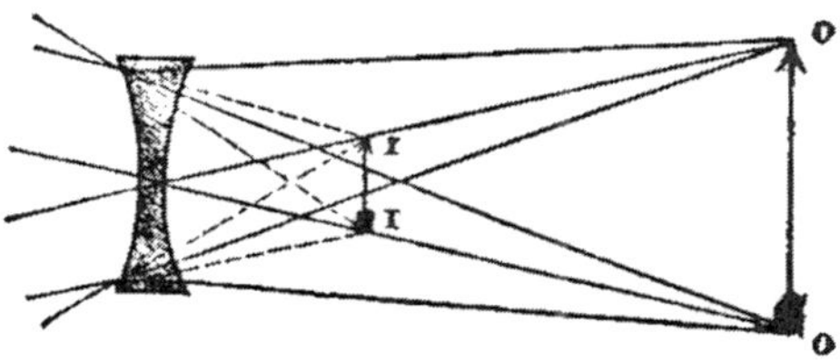

divergent rays proceeding from every point of the object will, after passing through the lens, become more divergent, as though proceeding from points between the object and the lens, and will thus form an image, I I, at a diminished distance from the lens, erect, imaginary, and less than the object.

SPECTACLES.

The defects of vision capable of being remedied by spectacles are those called weak sight and short sight.

The sensation of vision is conveyed to the brain by the intervention of an image of the object on the retina, or back portion of the interior of the eye. This image is formed by the rays of light, which enter the pupil, there meeting with, and passing through, a natural double-convex lens, called the crystalline lens. By this lens the pencils of parallel rays from the different points of distant objects, or of divergent rays from objects at a distance of not less than ten inches from the eye, should be made to converge, so as again to be brought back to points on the retina of the eye. By some arrangement, the nature of which has hitherto escaped detection, from the minuteness of the change necessary to produce the effect, the eye is enabled to vary its convergent power within the limits necessary to form distinct images on the retina of objects at various distances from it. When there is no defect in the vision, this distance may be any, greater than ten inches.

The defect called weak sight arises from the convergent power of the eye being insufficient to bring to a focus upon the retina the divergent pencils of rays proceeding from near objects. The focus of the unassisted eye, in this case being behind the retina, must be brought forward by the interposition of convex spectacles. The nearer the object looked at, the greater should be the converging power of the glasses; and hence it comes that very weak-sighted persons require to be provided with more than one pair of spectacles, those adapted to more distant objects being less convex, and those adapted to nearer objects more so.

The defect called short sight arises from the convergent power of the eye being too great. When its defect is not very great, a book or other object can without inconvenience be placed sufficiently near to the eye to throw the focus back upon the retina; but the pencils from the more distant objects presented to the eye, when walking abroad, or in large rooms, converge to a focus before the retina, and to produce distinct vision of such objects the focus must be thrown back by concave spectacles.

When weak-sighted persons can see distant objects distinctly, but fail to see objects clearly unless they are placed at too great a distance from the eye to be convenient for working or reading, glasses of ten inches focal length will cause the pencils of rays from objects at ten inches distance from the eye to enter the pupil exactly as they would if they proceeded directly from objects at several feet distance, and will therefore bring them to a focus on the retina. When even distant objects are not seen distinctly, glasses still more convergent must be used.

When short-sighted persons can without effort see objects distinctly at a short distance from the eye, concave glasses having a focus equal to this distance will give to parallel rays from distant objects the same degree of divergence as pencils would naturally have proceeding from objects whose distance is equal to the focal length of the glasses; consequently, according to the supposition, the eye will bring such rays to a focus on the retina, and the vision of distant objects will therefore be rendered distinct by such glasses. With eyes so near sighted as to see distinctly objects only when placed at such short distances from them as to be altogether inconvenient, one pair of spectacles will be required for distant objects, and another of less divergent power to see with facility objects at about ten inches distance.

The requisites of good spectacles are twofold: firstly, that the lenses shall be perfectly adapted to perform their office; and secondly, that they shall be so mounted as to enable them to perform their office without fatiguing or injuring the eyes.

With respect to the lenses, in consequence of their slight magnifying power, it is not easy to detect either imperfections in the material of which they are composed, or errors in the form of their curves; nevertheless it is most important that the material should be of equal density throughout, and that the lenses should be ground into a correct optical figure.

I have received from Messrs. Parkes and Son, of Birmingham, some specimens of spectacles, formed from a glass for which they have a patent, and of which they have also forwarded specimens in the unworked state. This material they denominate patent crystal, and the spectacle lenses ground by them from it appear to me to merit every commendation. I cannot do better than transcribe a portion of their letter accompanying these specimens:—

"You are of course aware that many persons who wear spectacles have a decided preference for Brazilian pebble, on account of its being a better conductor of heat than glass. But you are also as fully aware as myself that it is a *doubly refracting crystal;* and, as it is always cut by opticians, for the sake of economy, in the worst possible direction for optical purposes, viz., in the direction of the axis, instead of at right angles to it, it follows that *optically* this material is much less worth than common glass, on account of the confusion of image produced by it. It is for this reason that we have sought out an improved optical glass, termed 'optical crystal,' which, although not possessed to the same extent of the hard and cool properties of pebble, yet more than compensates for this by its *optical* qualities."

The lenses being good, the next consideration is their mounting, which evidently ought to be such as (1) to render their axes parallel, and (2) to bring their centres exactly opposite the centres of the pupils, when the eyes look straight forward.

To ensure the first of these conditions, great care ought to be taken that, in the original construction of spectacles, the axes of the lenses should be truly parallel, and the mounting ought to be sufficiently strong to prevent any after-derangement of the parallelism of the axes.

To ensure the second condition, it is necessary that the distance between the centres of the lenses should be precisely equal to the distance between the centres of the pupils; for the clearest vision being obtained by looking through the centre of the lenses, and the eyes consequently always endeavouring to look through those centres, their axes, if the distance between the centre of the lenses be too great, will have a tendency to diverge from parallelism, as in the case of an outsquint; while if the distance between the centres of the lenses be too small, the axes of the eyes will have a tendency to converge, as in the case of an insquint.

Persons using spectacles should therefore have the distance between the centres of their eyes exactly measured, and select mountings for their spectacles corresponding with this distance.

To enable spectacles to be selected, whose focal length

is best suited to distinct vision for reading or working, opticians have an instrument fitted up termed a *visometer*, which contains an object to be looked at, placed at about ten inches distance from its front. In this are two holes, at a distance apart equal to the ordinary distance of the centres of the eyes, and, by turning a handle, while looking through these holes, a variety of lenses are brought in succession before the eyes, which are thus enabled to select the focal length that affords distinct vision with the greatest ease and comfort to them.

As the most perfect vision with spectacles is produced when the eye looks in the direction of the axis of the lens, and imperfection always attends oblique vision through them, which imperfection is greater the greater the obliquity, persons using spectacles are obliged to turn their heads, when those whose sight does not require their assistance merely turn the eye. To diminish this inconvenience the late Dr. Wollaston suggested the use of menisci or concavo-convex lenses, instead of the double concave or double convex lenses ordinarily used. The effect of these lenses, as compared with the double concave or convex, is that objects seen obliquely through them are less distorted, and that, consequently, there is a greater freedom of vision by turning the eye without turning the head, from which property they were named *periscopic spectacles.*

MAGNIFYING POWER OF LENSES.

The apparent magnitude of an object, seen through a lens, depends upon the angle at which the axes of the pencils from its extreme points enter the pupil of the eye: the further the eye is placed from the lens, the more eccentric are these pencils, the more their axes are inclined, and the greater is the apparent magnitude of the object. When, however, the eye is placed close to the lens, which we shall suppose it to be in estimating the magnifying power, the axes, *a a′*, and, *b b′*, of the pencils from the points *a* and *b*, after passing through the centre *c* of the lens, diverge at an angle, *a′ c b′*, equal to the angle, *a c b*, that the points subtend at this centre. Suppose the object A B, when placed before a lens at a distance less than its focal length, to be seen distinctly by

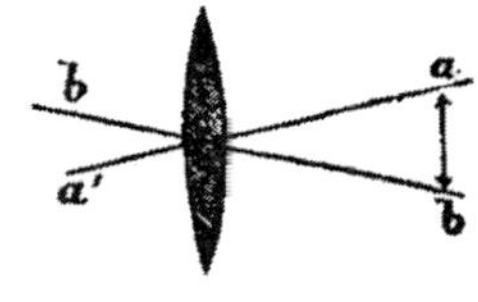

the eye placed immediately behind the lens. The most distinct view of the object will evidently be obtained when the object is placed at such a distance, that the virtual image A′ B′, from which the rays diverge after passing through the lens, is at that distance from the lens at which the eye has the power of distinct vision. This varies in different indivi-

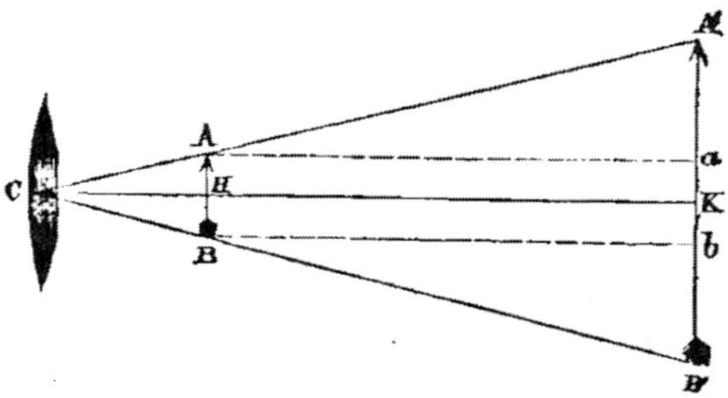

duals, and consequently the magnification produced by the use of the lens depends not only on the focal length of the lens, but also on the range of sight of the individual.

The exact distance, C H, from the centre of the lens at which the object should be placed is readily computed as follows:—Let x represent this distance, f represent the focal length of the lens, and d represent the distance, C K, of distinct vision in the case of the observer: then it is evident that the pencils of rays proceeding from the object, A B, at the distance x, must, after passing through the lens, have the same divergence as though they proceed from a virtual image, A′ B′, at the distance d, or, in other words, their virtual foci must be at this distance. Hence we have the condition expressed by the equation,

$$\frac{1}{d} = \frac{1}{x} - \frac{1}{f}; \text{ and, therefore, } x = \frac{d f}{d + f}.$$

Now, without the intervention of the lens, the object, in order to be distinctly seen, would have to occupy the position a b, under the angle a C b, whereas, by the use of the lens, it is seen under the angle A C B, and being mentally referred to the distance of distinct vision, appears of the same magnitude as though it really occupied the space A′ B′ of its virtual image. The magnification of the object is therefore correctly represented by the ratio of A′ B′ to a b, which is the same with that of A B to $a\ b$, or of C K to C H, that is of d to x, and is therefore represented by the value of the fraction $\frac{d}{x}$. Now, we have seen that

$$x = \frac{d\,f}{d + f};\ \text{and, therefore,}\ \frac{d}{x} = \frac{d + f}{f};$$

and this expression represents the magnification of the object for the individual whose range of sight or distance of most distinct vision is d.

It appears, then, that the magnification is greater for long-sighted than for short-sighted persons. Assuming, however, the magnifying power to be that number which represents the magnification for a person of the average range of vision, say ten inches, this magnifying power would then be properly represented by the fraction $\frac{10 + f}{f}$.

There is no objection to this mode of estimating magnifying power, and it is only necessary to consider that for long-sighted persons the actual magnification is somewhat greater than this, while for short-sighted persons it is somewhat less.

For lenses of very short focal length, the expression $\frac{10 + f}{f}$ becomes nearly equal to $\frac{10}{f}$, and hence the magnifying power of such lenses has sometimes been estimated by this ratio. Unfortunately, this has led to some writers unthinkingly giving the ratio $\frac{10}{f}$ as an approximate estimate of magnifying power for all lenses. The absurdity of such a blunder will be manifest, when we apply such a formula to estimate the magnifying power of a convex lens of ten inches focal length; for it would make it appear to produce no magnification at all, while convex lenses of more than ten inches focal length would by the same rule actually produce a diminution instead of an enlargement, of the apparent size of an object.

The only unambiguous mode, then, of designating the power of lenses is by stating their focal length, and any person, by observing his own range of natural vision, can compute the magnification which they will produce for himself from the expression $\frac{d + f}{f}$; thus, *add the focal length to the range of sight or distance of most distinct vision, and divide the sum by the focal length, and the quotient will represent the magnification, or apparent enlargement, of the object.*

The following table exhibits the magnifications produced for persons with three different ranges of sight:—

Focal Length of Lens.	Magnification $\frac{d+f}{f}$		
	Short Sight Range, 5 inches.	Ordinary Sight Range, 10 inches.	Long Sight Range, 20 inches.
$\frac{1}{2}$ inch	11	21	41
1 ,, ...	6	11	21
5 inches ..	2	3	5
10 ,, ..	$1\frac{1}{2}$	2	3
20 ,, ..	$1\frac{1}{4}$	$1\frac{1}{2}$	2

From the examination of these numbers, it will be seen that the long-sighted person derives far more advantage from the use of a lens, especially from a lens of short focal length, than the short-sighted person, the advantage being estimated by a comparison of the apparent magnitude of the object seen with the lens, with that seen by the unassisted eye: thus, a person whose range of most distinct vision is five inches, sees an object through a lens of half-inch focal length magnified to eleven times the size in linear dimensions, at which it appears to his unassisted vision, while to a person who, with the eye alone, sees objects most distinctly at a distance of twenty inches, they would appear through the same lens to be increased forty-one times in linear magnitude.

It must, however, be borne in mind that the terms long sight and short sight are, in fact, synonymous with the terms weak sight and strong sight, so that the short-sighted person may be said to have a microscopic eye. The actual magnitude of the pictures formed upon the retina, which may be considered as a measure of the power of vision, are directly proportional to the angles subtended by the object at the pupil of the eye, and are therefore very nearly in the inverse proportion of the distance of the object from the eye.

The natural vision, then, of a person whose distance of distinct vision is five inches may be considered four times as powerful as that of a person who has to remove objects to a distance of twenty inches in order to view them most clearly. Minute objects, which the former could examine with the naked eye, could only be examined with equal facility by the latter through a lens of six inches and two-thirds focal length, enabling him also to bring the object within five inches of his eye, and thus producing for him a fourfold magnification of the apparent magnitude. With a lens of a focal length of half an inch, the place of the object to be viewed most distinctly would be $\frac{5}{11}$ inch from the lens for a person whose natural distance of distinct vision is five inches, and $\frac{20}{41}$ for one whose natural distance of distinct vision is twenty inches. Hence it appears that with such a lens the object would have very nearly the same angular magnitude to both individuals; and, consequently, that with lenses of high powers all eyes are enabled to see minute objects with nearly equal facility and distinctness.

CHAPTER IV.

BINOCULAR VISION.

THE STEREOSCOPE.

The stereoscope is an instrument which, by an exaggeration of the effects of binocular parallax, creates such strong impressions of perspective and relief as raise emotions of delightful surprise and astonishment. Pictures delineated upon a flat surface, being placed in the instrument and viewed by its means, give to the mind ideas of the solidity of objects, not only stronger than can otherwise be represented by the best executed drawings or paintings, but actually stronger, in the case of large objects, than could be produced by direct observation of the objects themselves. In fact, it presents to us not merely views on a reduced scale of the pictures presented to the eye by nature, but gives to these views all the effects of solidity, which would be perceived in models of the objects executed on a sufficiently small scale to be looked at from a very short distance.

To comprehend the effects produced by the stereoscope, it will be necessary to revert to the principles upon which the impressions derived from vision are based. The perception of relative position, form, and magnitude, is originally due to impressions acquired by the sense of touch; and it is only by a comparison with these previously acquired impressions, of their apparent outlines and varieties of light and shade, that the mind can make use of the eye in forming its judgment; but the continual repetition of this act of comparison, every time that the sense of vision is exercised, causes it to be performed with such facility and rapidity, that the mind, unconscious of any mental effort, imagines magnitude, form, and position to be the immediate subjects of visual perception, instead of their merely being judged of by an experience that, reckoning from the time of unconscious infancy, has been imperceptibly acquired.

This unconscious experience, among its other effects, not only causes us to regard as derivations from a single object the pictures presented by them to each eye, but, moreover, when there is a sensible difference in these two pictures, conveys to us the impression not of the form and position of the object as presented to either eye, but of a modification of the two.

This case does not occur in looking at distant objects, that is, so distant that the distance between the two eyes is as nothing in comparison; and in this case the effects of perspective and relief are produced solely by varieties of light and shade, and of distinctness; and if these varieties be transferred by the painter to his canvas, the impressions of distance and relief conveyed by his picture will approach those received from the immediate view of the objects themselves with a closeness limited only by his skill in transferring them with fidelity and precision.

Parallax is a term applied to the cause which produces an apparent difference in the appearance of objects, arising not from any actual changes in their real positions, but to a change in the position from which they are viewed. It is, in fact, the angle subtended at the place of the object by two positions, from which it is viewed, that is termed parallax. Now it is manifest that, in looking at objects first with the right eye and then with the left, a change in the point of view takes place, and the apparent difference in their aspects from this cause is due to binocular parallax, or the angles subtended by the centres of the two eyes at the places of the objects. In the case of objects whose distance bears such a large proportion to the distance between the eyes that there is no sensible difference between the picture formed by them on the retina of one eye, and that formed on the retina of the other, the phenomena due to binocular parallax are not developed, and perceptions of perspective and relief are due solely to varieties of light and shade. But in objects whose distances from the observer are not many times the distance between the two eyes, the binocular parallax becomes considerable, and the phenomena developed by it very remarkable. For instance, at the distance of ten inches from the eye, which is about that at which small objects are usually viewed with most satisfaction, the binocular parallax is as much as 13 or 14 degrees.

In looking at such objects, then, and to a considerable degree in looking at objects but a few feet distant from us, the following phenomena occur :—(1) parts of such objects are seen by either eye which are not seen by the other; (2) the parts which are seen at the same time by both eyes have not the same apparent relative positions; so that if the observer were to make an exact drawing of the object as viewed first with the left eye closed, and then with the right eye closed, not only would one show a part of the object on the extreme right which would not be shown in the other, and the latter a portion on the extreme left not included in the former, but the more remote parts of the objects concealed by those more prominent would be different in the two pictures, and in point of fact the two pictures would in no part of them be identical. Now in looking at such objects with both eyes open, it is manifest that two such *different* pictures as above described are simultaneously impressed upon the vision, and the mind regarding them as, what from experience it conceives them to be, the production of single objects, takes them as indications of the forms and positions of these objects. Hence it follows that the perception of relief, form, and relative position is much stronger when assisted by the phenomena of binocular parallax than in the case of more distant objects, where this parallax disappears, and the mind can form its judgment only from the outline and varieties of light and shade.

Now in different persons the distance between the centres of the eyes varies to a small extent, and the distance from the eyes best adapted for distinct vision also varies according as they are longer or shorter sighted, and hence the perception of relief when looking at near solid objects must be stronger in some individuals than in others.

If, however, a skilful artist were to make pictures of some object, sufficiently near to him to produce considerable binocular parallax, as viewed first with one eye closed, and then the other, and if by some optically contrived instrument these pictures could be so presented, each to the eye with which it was drawn, and shut off from the other, that the optic axes directed to them should converge at the same angle as when directed to a single object, each eye obtaining the same view as if the object itself were presented simultaneously to both, the visual perception will be the same as that derived from the view of a single object, and the judgment will be cheated into the conception that it sees but one picture, although astonished at an appearance of relief and solidity equal to, or even exceeding, that derived from looking with both eyes at the object itself. It is manifest, however, that in looking at the two pictures in the manner just described, every one, whether long-sighted or short-sighted, would receive the same view with either eye, and would therefore receive the same impression of relief; and if the draughtsman were to vary his position slightly, so as to make the distance between the two points of view exceed the distance between the centres of his eyes, the appearance of perspective and relief would be exaggerated beyond any that could be produced by a direct view of the object delineated.

An optical contrivance for enabling an observer to obtain such a view of two pictures as above described is termed a *stereoscope*, the name being derived from the Greek words—*stereon*, "a solid object," and *scopeo*, "I look at," because by its means the two really flat pictures produce the same effect upon the vision as though a solid object were presented to it.

The discovery of the remarkable phenomena that were due to binocular parallax was made by Professor Wheatstone, and in the form of instrument used by him two plane reflectors, placed at right angles to each other, were enclosed in a box having two eye-holes in its front, through which one eye was directed upon one reflector, and the other eye upon the other. The two pictures being placed in the box so that one was seen by reflection from one mirror with the right eye, and the other with the left eye by reflection from the other mirror, the position being so adjusted that the rays from the points of the two pictures, which are the representations of a single point in the original object, should enter the eye as though they had diverged from a common point, the same impression was made upon the vision as though the point of divergence had been a point in a single object from which they had proceeded.

Professor Wheatstone carried his researches further, into the effects which might be produced by presenting to the eyes views of objects in such aspect as they could not possibly be seen simultaneously, except by artificial means; such as presenting to the left eye that view which would naturally be seen with the right, and *vice versâ;* and a variety of curious and astonishing appearances were the result—objects really hollow being made to appear as standing out in bold relief, and objects in strong relief being made to appear hollow. As a philosophical instrument for the investigation of all the remarkable phenomena due to binocular parallax, the *reflecting form of the stereoscope*, given to it by its inventor, cannot probably be surpassed; but as an artistic instrument for heightening the effects to be produced by the graphic productions of photography, in a manner impossible to be imitated by any skill of the painter, the *lenticular form of the stereoscope*, given to it by Sir David Brewster, has been preferred everywhere by the general public. In this arrangement the visual rays from the two pictures are deflected, and made to proceed to the eye as though diverging from a single object by means of two eccentric double-convex lenses.

These are formed by first cutting a double-convex lens, A B C D, into two semi-lenses, B A D and B C D, and then cutting the two eccentric lenses out of these, their diameters, A E, E C, being semi-diameters of the original lens: thus, if A E C F represent a section of the original lens made by a plane passing through A E C at right angles to its surface, A E F and C E F will represent sections of the two eccentric lenses, which will therefore still have their optical centres at their thickest part, E, while their geometrical centres, O and O′, will lie midway between E and the points A and C.

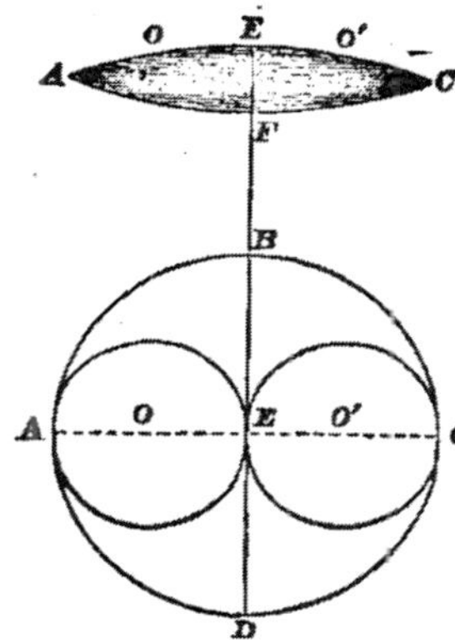

These two lenses are set in two tubes, which are fitted to a box of a convenient form for receiving the pictures to be viewed by them, the distance between the tubes being made equal to the distance between the eyes. The lenses are so set that their edges, A, and C, are turned towards each other, and their thickest parts, E and E′, are furthest from each other, and that their distance from the back of the box, where the pictures are placed, is equal to their common focal length.

Now, suppose P and P′ to be points on two pictures so placed, representing the same point of an object, as viewed from the right eye, and left eye, respectively; the pencils of rays proceeding from P and P′ will, in passing through the lenses, be so refracted as to form pencils of parallel rays, each refracted ray being parallel to the lines P E or P′ E′, joining P and P′ with the optical centres, E and E′, of the lenses; and, conse-

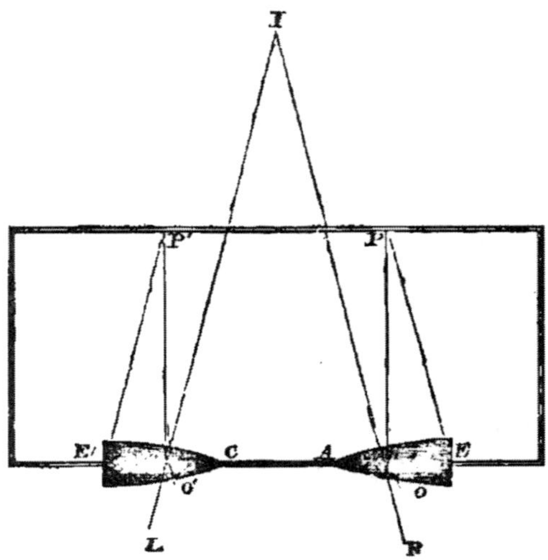

quently, the points P and P′ will be seen in the directions O I and O′ I, parallel to E P and E′ P′, as the single point I, in which O I and O′ I meet. In like manner every pair of points in the pictures, representing the same point of the object, will be seen as a single point of a solid, viewed by both eyes. The effect produced, then, by the combined pictures is identical with that of a solid object, placed at the same distance from the eyes as I is from the lenses. The pleasing effects produced by this instrument are heightened by the magnifying power of the lenses.

The annexed figure shows the form in which the lenticular stereoscope is usually constructed. The tubes, T, T′, containing the lenses, draw in and out, so as to adapt the foci to different eyes, while they are prevented from turning round by pins and slots, so as to keep the lenses in their proper relative positions. The pictures, if opaque, are illuminated through the opening A B C D, and the inside of the lid C D E F, may be formed into a mirror, to increase the illumination. Transparent pictures are illuminated through the back of the

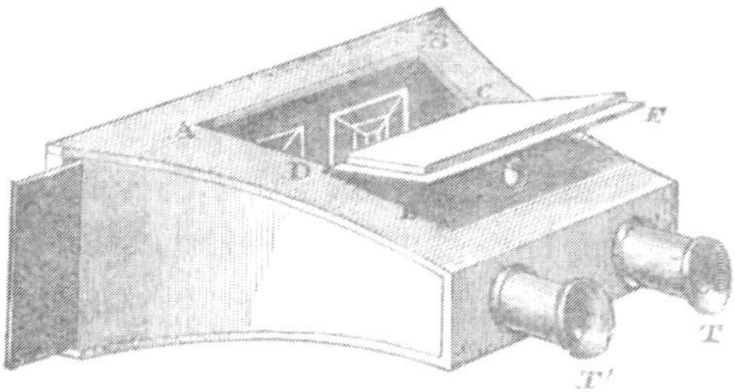

instrument which is formed of ground glass, the lid in this case being closed.

Views of geometrical figures, as they would appear from the different positions of the two eyes, may, by rule and compass, be made with great precision, and in presenting such views to the instrument, some of the most interesting and instructive effects of the stereoscope are exhibited; but the impracticability must be apparent of producing two pictures by any process of hand-drawing, which should represent a portrait, or any artistic object, as viewed by each eye in succession, with the precision indispensable for the production of the stereoscopic effect. The contemporaneous discovery of the photographic art has lent its aid to the stereoscope, and produced with facility, not only pictures representing the effects of binocular parallax upon small objects at short distances from the eye, but also pictures imitating the effects of binocular parallax on objects so distant that they would really have no sensible binocular parallax at all.

Thus, let two lines be drawn from the centre of an object at an angle equal to that at which the lines of vision are made to converge by the lenses of the stereoscope. Let two photographic cameras be placed on these lines, at equal distances from the centre of the object, and having their optic axes directed to that centre. The pictures thus produced on a small scale are such as might have been taken from near points of view of a model of the building, constructed on an equally small scale; in which case the distance between the points of view with the same amount of parallax, would not have been greater than the distance between the eyes. When these pictures are viewed in the stereoscope, the perception of perspective and relief, owing to the exaggeration of binocular parallax, is much stronger than any that could be produced by the view of the building itself; and is, in fact, the same as would be derived from a near view of a reduced model of the building, the distance between any two points in the model bearing the same relation to the distance between their originals in the building, as the distance between the two eyes bears to the distance apart at which the two cameras were placed, to take the pictures.

CHAPTER V.

MICROSCOPES.

THE SIMPLE MICROSCOPE.

The microscope is an instrument for magnifying minute but accessible objects. A convex lens is a microscope, but the imperfections of such an instrument have been already explained (p. 20), and the greater the power of the lens the greater will be these imperfections. For small magnifying powers, then, convex lenses may be used, as they are for spectacles; but for obtaining good images with high magnifying powers a combination of lenses must be used.

Single lenses, then, of five or six inches focal length, constitute the microscope in its feeblest form, and are very useful for many purposes; such as the reading of very small print, the examination of photographs, the inspection of plants by the naturalist, and of anatomical subjects by the medical profession.

Simple microscopes of the above form are called magnifiers, or magnifying glasses. Fig. 1 represents a mode of mounting a glass of this kind for the library table, and Fig. 2 a convenient form of mounting for a magnifier to be carried in the pocket. In Fig. 3 two such magnifiers are mounted in one frame with a diaphragm. The lenses are of different powers, and can be used separately or in combination. In the latter case, the diaphragm is placed between them, and diminishes the effects of spherical aberration.

Glasses of short focal lengths are used by miniature

painters and engravers. They are usually mounted in frames of tortoiseshell or horn. Watchmakers, jewellers, gem sculptors, &c., make use of glasses of about one inch focal

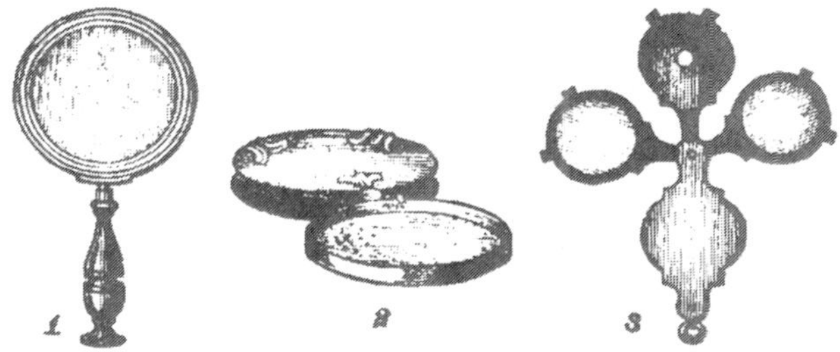

1 2 3

length. They are usually mounted in a norn cell, in the shape of a truncated cone. The larger end of the cell is made of such a size as to be conveniently held by the contraction of the muscles surrounding the eye, and the lens being fixed at the smaller end of the cell, the object to be examined or manipulated is brought to a distance of about an inch from the lens, and seen by pencils of parallel, or nearly parallel, rays, emerging from the lens and entering the eye. Sometimes, to avoid the inconvenience of holding the magnifier either in the hand or in the manner just described, a stand is made use of, consisting of a jointed arm attached to a ring, which slides upon a vertical rod, and can be fixed at any desired height by a tightening screw : the arm carries a socket at its extremity, into which the cell that holds the lens can be dropped, and the object, placed on the table beneath it, can be viewed without inconvenience or fatigue.

Small glass spheres are used as microscopes of high powers; but a thin lens composed of any more highly refractive substance is preferable; because, the focal length of the sphere measured from its centre being but three semi-radii, the distance of the object from the surface is only one semi-radius, which prevents its being used in the examination of delicate objects. The refracting sphere is much improved as a microscope by cutting a groove round it in a diametrical plane, and filling it up with some black opaque substance By this contrivance the aperture is diminished, without con-

tracting the field of view, and all the pencils are necessarily centrical. A lens formed in this manner from a sphere of glass is called a Coddington lens, from its supposed invention by the eminent mathematician of that name. It appears, however, to have been one of the contributions of Sir David Brewster to optical science.

The annexed figure represents an elegant and convenient form of mounting the Coddington lens, as a pocket microscope.

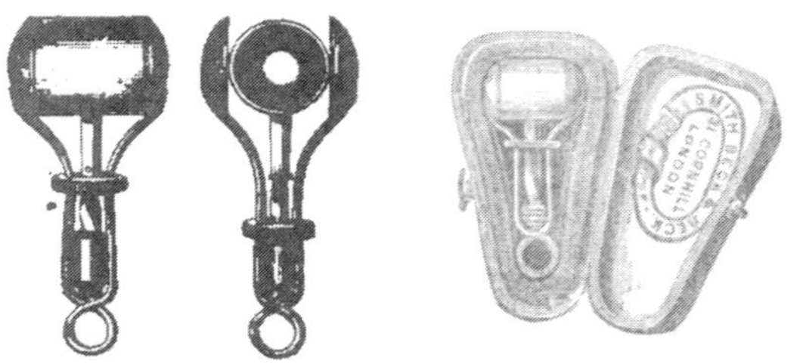

Microscopes have been made of diamond and sapphire, and the aberration is much less than with glass. Sir David Brewster employed, as a microscope, a drop of Canada balsam or turpentine varnish upon a thin plate of glass, of which the surfaces were exactly parallel. This is a very ready way of forming a plano-convex lens, and if kept free from dust will last some time.

Such glasses as have been just described may be conveniently used, for the inspection of transparent objects, when high powers are not required; but when higher powers are necessary, or the objects are opaque, combinations of two or three plano-convex lenses, with their convex sides towards the eye, may be used. These are called doublets or triplets. The triplet is the most powerful of these combinations, and in it the two lenses nearest to the object are in contact, and play the part of the single lens, which is sufficient for lower powers, the construction being in all other respects the same as for a doublet. Between this lens, D D, and the posterior lens, E E, a diaphragm, S S, is interposed, consisting of a circular disc of metal, with a hole in its centre, to cut off scattered rays not necessary for the formation of the imaginary image.

The lenses are usually set in two thimbles, one of which

screws into the other, in order that their mutual distance may be adjusted so as to produce the best effect. When unscrewed, the lenses can be used singly for very low magnifying powers; and, again, the same posterior glass, E E, can be used either with a single anterior lens, D D, to form a doublet, or a double lens to form a triplet, according to the magnifying power desired.

The advantages of the doublet form are, firstly, that the injurious effect of spherical aberration is much diminished; and, secondly, that the object can be placed at a much greater distance from the anterior lens D D, and can thus be more conveniently manipulated; while it can also be illuminated by a light thrown on it from that side which is presented to the glass.

Such lenses, or combinations of lenses, as those above described, in which the object is viewed directly, are called *simple* microscopes. In their most useful form they are mounted in a framework, either to be held in the hand, or placed on a table, having a contrivance for holding the object, so that it may be kept steadily in the axis of the lens or lenses, and moved to or from them with a smooth and easy motion.

Hand-frames are made to fold into a small compass and fit into a case to be carried in the pocket; but, when portability is not a chief desideratum, a more stable mode of mounting should be adopted. The following is the arrangement of *Darwin's Dissecting Single Microscope.*

The base A of the instrument is formed of a solid block of oak or mahogany. Into this is firmly screwed the pillar B, into which the triangular bar C C slides, and is raised or lowered by a rack and pinion with a slow and easy motion, by turning either of the milled heads L, L. On the top of the bar revolves a square head, through which passes a horizontal arm, D E, which is advanced or retired by turning either of the milled heads H, H, and carries at one extremity a socket, E, for holding a lens that forms the microscope. At the end D of the arm is a slot, to receive the stem of a ring that holds the large doublet S, this stem being fixed by the screw G. The stage I, is a ring, three inches in diameter, revolving with friction upon a fixed piece firmly attached to the top of the pillar B. Into this ring fit the following pieces:—1. Either of the large saucers, T and U, with glass bottoms for holding objects in water; a piece of cork, loaded underneath with lead, to sink it to the bottom, being placed in the saucer, when illumination from beneath is not required. 2. A piece, O, of boxwood, covered with cork, on which objects may be pinned. 3. A brass plate, R, with central aperture one inch and a half in diameter. Into the aperture of the plate R fit smaller saucers and various pieces for holding objects, and also a holder, P, for glass slips. K is

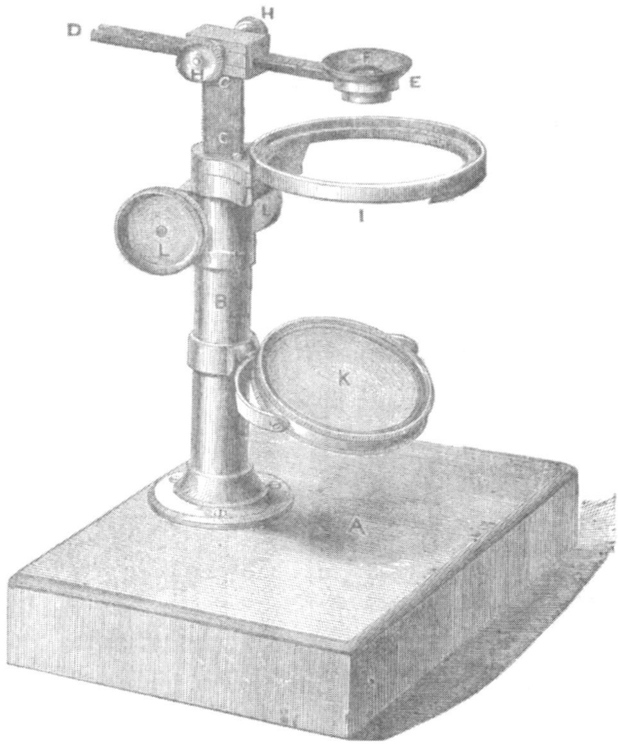

a concave mirror, having a cap with small flat mirror, for the illumination of transparent objects: it can be set at any desired angle, and the collar to which it is attached slides stiffly on the pillar B, so as to be set at any desired distance below the stage. The light from the mirror can be also regulated by the diaphragm M, fitting below the stage; or it can be entirely excluded by turning the stop, N, over the aperture of the diaphragm, when opaque objects are under observation, such objects being illuminated from above by an independent condensing lens.

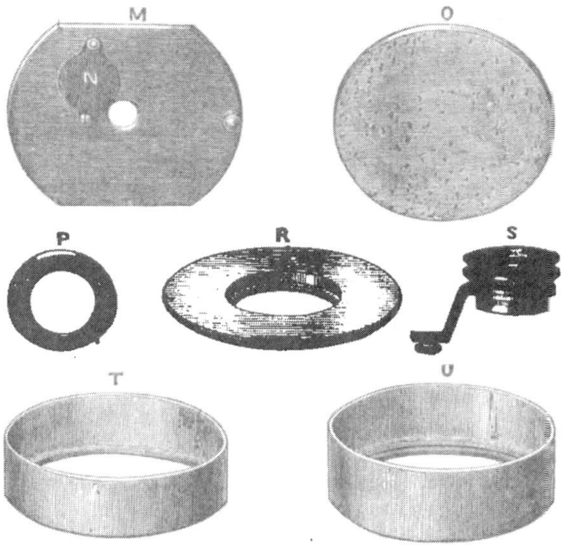

The object is brought into focus by raising or lowering the arm D E, by means of the rack and pinion movement of the bar C C; and by revolving the arm D E, and moving it backwards and forwards, by turning one of the mill heads E, the lens or doublet can be carried over every part of the field of view.

Queckett's Single Dissecting Microscope is very simple and effective in its construction, and very moderate in price.

Beck's Improved Dissecting Single Microscope.—The base A, of this instrument is of solid mahogany, and from it rise four pillars, B, supporting, at the height of four inches, a stage six and a half inches square, rounded at the corners. This stage carries a circular plate of six and a half inches diameter, which is revolved by turning the milled head E, beneath the stage. Above this is a smaller circular plate, with sliding motion, having a central aperture of one inch and a half diameter, for receiving various pieces to hold the objects to be examined. The horizontal arm H, with socket at its extremity for carrying the lenses, has a limited revolving motion on the top of the bar I, so that it can be turned away from the stage to the left hand, but, when turned to the right, is stopped with the centre of the socket over the centre of the stage; the bar I, which rises from the more remote left-hand pillar, is raised or lowered with a slow and easy motion by turning the milled head K; and as the amount of this motion extends to the whole length of the pillar B, objects of considerable thickness can be placed between the lens and the stage. The mirror L, for the illumination of transparent objects, rises from a recess in the base A, and above the stage is a condensing lens, M, for the illumination of opaque objects.

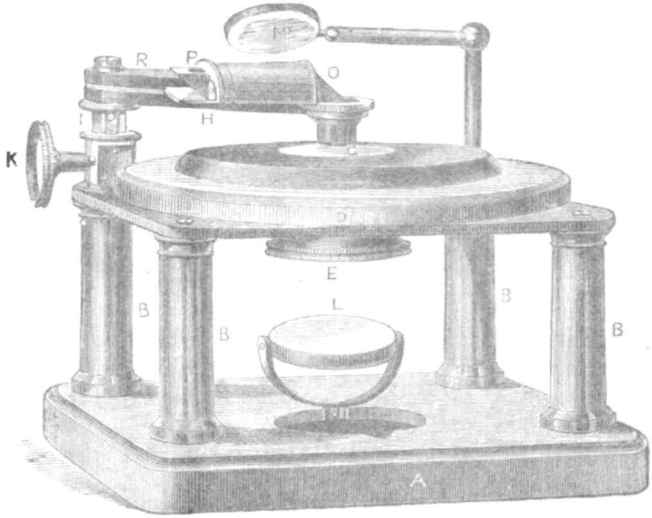

In this microscope the lens remains stationary over the object, and the latter is moved by the revolving and sliding plates, so as to bring every portion of it into the most favourable position for examination, while the axes of the lens and mirror maintain their coincidence: the sliding plate has a motion of three quarters of an inch in every direction, so as to bring every part of a circular field of an inch and a half diameter under the axis of the magnifier. The magnifiers most in use are Coddington lenses, from one inch to one-eighth of an inch focus.

Single Microscope, with Binocular Vision.—Not only does the employment of a single eye in lengthened microscopic examination fatigue and distress it, but the continued use of a monocular instrument will produce an unequal range of vision in the two eyes; moreover, the stereoscopic effects due to binocular parallax (p. 32) are entirely lost. With such low powers as are employed for ordinary spectacles, each eye is supplied with a separate lens, and all the comfort and perfection of binocular vision are retained; but when magnifiers of less than two inches in focal length are made use of, the employment of two lenses is impracticable, and it becomes an important problem to find a mode of realising the effects of binocular and stereoscopic vision, when an object is to be viewed by means of pencils proceeding from a single lens. Now the distance from centre to centre of the two eyes being about two inches, and the ordinary range of most distinct vision about ten inches, it is manifest that if the latter distance were reduced to one inch, there would be the same amount of binocular parallax with two imaginary eyes having a distance from centre to centre of one-fifth of an inch; if the distance of the object were reduced to half an inch, the distance between the imaginary eyes need only be one-tenth of an inch, and still the amount of binocular parallax would be the same as with the natural vision, and consequently the same stereoscopic effects would be produced. Hence, with a lens of short focal length, the two halves of the lens itself may be made to perform the office of the two imaginary eyes, and by any contrivance for directing the rays from the right half of the lens to the right eye, and those from the left half of the lens to the left eye, not only the ease and comfort of binocular vision, but also the stereoscopic effects of binocular parallax may be restored.

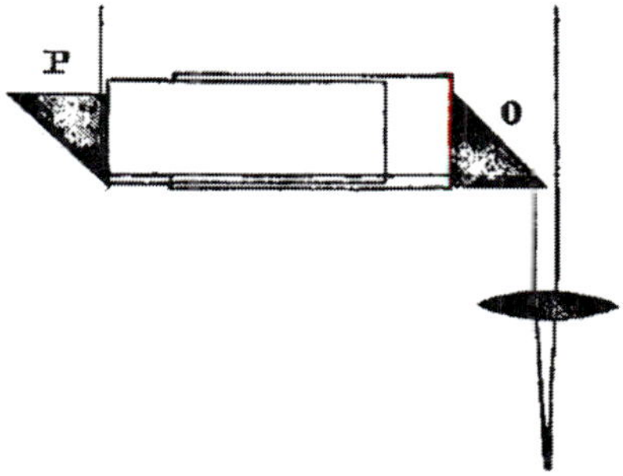

Beck's Binocular Prisms.—Our figure (p. 44) of Beck's improved single dissecting microscope is represented with these prisms applied. Since rays diverging from an object very nearly in the focus of a lens become almost parallel after their passage through the lens, the eye can be placed nearer to the lens, or removed further from it, to a considerable extent, without imparting indistinctness to the view of the object, and thus it can be viewed by both eyes at the same time, though one may be, in effect, farther removed from it than the other. The prisms O and P are placed at the ends of two semi-circular tubes, one of which, carrying the prism O, is fixed to the arm R, and the other, carrying the prism P, slides in the first, so that the prisms can be set at different distances, to suit the varying distances between the eyes of different persons. The arm R, being placed on a pivot above the arm H, carrying the lens, is turned to the right, and meets with a stop when the prism O covers the left-hand half of the lens. The tube carrying the prism P being, then, pulled out, until the distance between the prisms is equal to the distance between the eyes of the observer, the object is

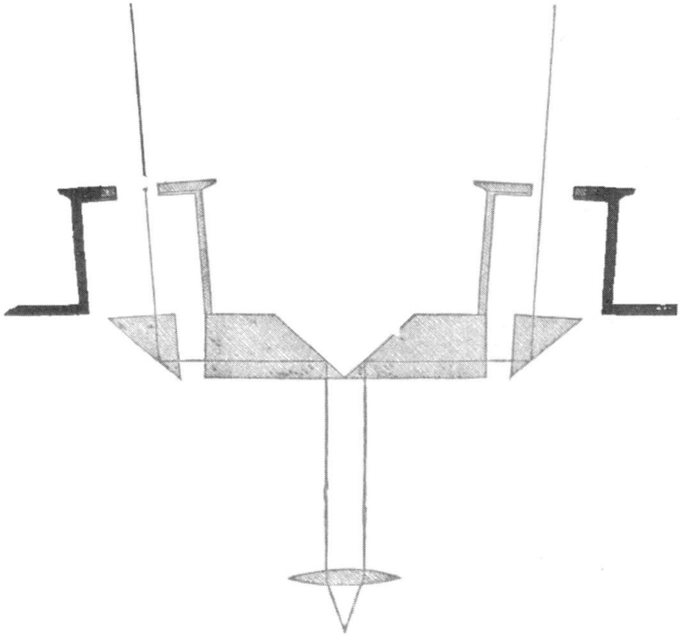

viewed directly by the right eye through the right half of the lens, while the rays passing through the left half are reflected by the prism O along the tube to the prism P, and thence to the left eye, which thus obtains the view of the object presented to the left half of the lens. By this arrangement the field of view presented to the left eye is smaller than that presented to the right, and the centre part only, that is common to both eyes, is stereoscopic; nevertheless the addition of these prisms to the microscope presents a very great improvement in the delineation of the objects.

Nachet's Prisms.—The form of these prisms and their arrangement, as adapted to the single microscope, is represented in the accompanying figure. The rays from either half of the lens are reflected through paths of equal length, those from the right half to the right eye, and those from the left half to the left eye; and the prisms G, G, are so inclined as to make the pencils of light to either eye, from the same point of the object, diverge, as though proceeding from a single point at the distance of most distinct vision, as they do in the stereoscope (p. 35). The object is, by this arrangement, presented to both eyes with exactly the same degree of magnification; and, the extents of the fields of view being also equal, the delineation is admirable, and the stereoscopic effect complete.

Beck's Improved Dissecting Single Microscope, with Nachet's prism, may be considered, therefore, an instrument of high perfection.

THE COMPOUND MICROSCOPE.

Microscopes in which an enlarged optical image of the object is first formed, and this image then viewed by a combination of lenses, as an object would be by a simple microscope, are called *compound* microscopes.

To avoid the indistinctness arising from chromatic aberration, microscopes used to be constructed in which the enlarged optical image was produced by a concave reflector; but the very great improvements which have, in the present generation, been made in the construction of object-glasses, has rendered them so superior to reflectors for the formation of the enlarged image, that the reflecting microscope has fallen entirely into disuse, and we shall, therefore, confine our attention entirely to the refracting microscope.

The necessary conditions of distinct vision with the compound microscope, like those for vision with the naked eye, are —1. That the pencils of light which enter the eye from every point of the object under examination, should have the same degree of divergence as if proceeding from the distance at which the most distinct vision is obtained with the naked eye; 2. That the axis of the pencils proceeding from the extreme parts of the object, enter the eye under a sufficient angle, or, in other words, that the apparent or virtual image subtend an angle of sufficient magnitude; 3. That there may be a sufficient quantity or intensity of light entering the eye from each point of the apparent or virtual image, or, in other words, that every point of this image be sufficiently illuminated—a condition which necessitates an illumination of the object in proportion to the magnitude of the angle under which its image is presented to the eye. The two first conditions must be satisfied by the apparatus for the magnification of the object, and the third by that for its illumination.

The magnifying apparatus of the compound microscope is constructed upon the two following principles:—

1. The formation of an enlarged optical image of the object to be examined.

2. The magnification of this enlarged image, as if it were the object itself viewed by a simple microscope.

In the rudest and most imperfect state of the instrument, as constructed by its first inventors in the beginning of the seventeenth century, these principles were carried out by the means of two lenses, called respectively the object-glass and eye-glass, fixed at the opposite ends of a tube, or rather of two

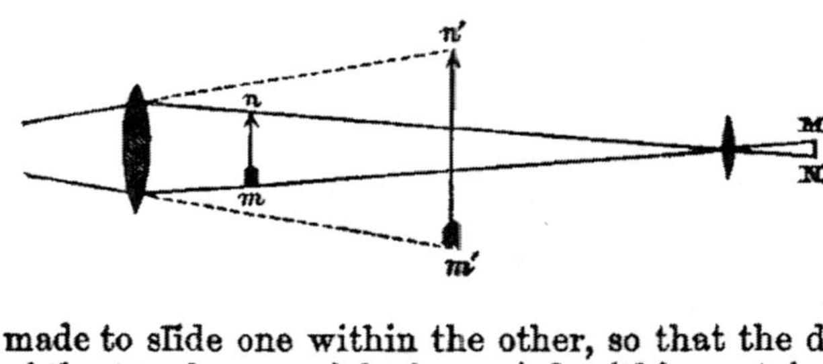

tubes made to slide one within the other, so that the distance apart of the two lenses might be varied within certain limits. The object-glass O, was a lens of short focal length, and the object M N, being placed at a very little distance beyond its focus, an inverted and enlarged image, *m n*, of the object was formed at a much greater distance from the object-glass, than that of the object, M N.

The image *m n* was then viewed through the eye-glass E, which performed the office of a magnifying-glass, and was adjusted at such a distance from the image *m n*, that the pencils of rays, after emergence, had the same divergence as though they proceeded directly from an object, *m′ n′*, at the distance from the eye best adapted to distinct vision. The same impression was then produced as though the object were really of the magnitude *m′ n′*, and viewed by the naked eye.

The ultimate magnification produced by the instrument was evidently the product of the magnifications produced by each of the lenses. The magnification produced by the object-glass depended upon the focal length of this lens, and the distance from it at which the object was placed: this distance, however, must exceed the focal length. The magnification produced by the eye-glass depended upon its focal length, and upon the distance of most distinct vision for the naked eye of the observer, and was almost proportional to this distance; so that for an observer seeing most distinctly with the naked eye at twenty inches distance, it was nearly four times as great as for one seeing most distinctly at five inches distance. To the two observers the magnified image would, in fact, have very nearly the same angular magnitude, whereas, the object, viewed with the naked eye, to the long-sighted observer would appear under an angular magnitude one-fourth only of that under which it would appear to him with the stronger vision.

From the mere consideration of the fundamental principles of the instrument, the dimensions to be given to it, as well as the effects produced by it, may be easily computed. Suppose, for example, that the focal length of the object-glass be one-fifth of an inch, and that of the eye-piece three-fourths of an inch: if then the object be placed at $\frac{1}{250}$ of an inch beyond the principal focus, or $\frac{51}{250}$ of an inch from the centre of the lens, a real inverted image will be formed at the distance of ten and one-fifth inches from the lens, on the other side of it, and the magnification of the object-glass will be fifty, so that for an object $\frac{1}{250}$ of an inch in diameter, the diameter of the image would be one-fifth of an inch. Then, for an observer whose distance of most distinct vision is ten inches, the eye-piece must be placed at a distance of $\frac{30}{43}$ of an inch behind the image, and the magnification produced by the eye-piece will be $14\frac{1}{3}$; so that the entire magnification will be $50 \times 14\frac{1}{3} = 716\frac{2}{3}$. On these hypotheses the distance between the eye-glass and object-glass, or the length of the instrument, would be $10\frac{1}{3} + \frac{20}{33} = 10\frac{31}{33}$ inches, or very nearly 11 inches.

The angular magnitude of the image, as viewed by the above observer, would be 16°, while the angular magnitude of the object itself, viewed by him with the naked eye, would be only 1′ 22½″.

A short-sighted person, whose length of sight was five inches only, would see the image with an angular magnitude of 17° 3′; and a weak-sighted person, whose length of sight was twenty inches, would see it with an angular magnitude of 15° 27′; while the angular magnitude of the object itself viewed by them with the naked eye, would be 2′ 45″ for the strong sight, and only 41¼″ for the weak sight.

With the same object-glass and eye-glass other degrees of magnification could be obtained by placing the object at a greater or less distance from the object-glass; but it would be necessary at the same time, to shorten or lengthen the distance between the two glasses, since the place of the image would approach towards the object-glass or recede from it, as the object was moved farther off or brought nearer. It appears, then, that by placing the eye-glass and object-glass in separate tubes, sliding one within the other, great variations can be made in the magnifying power, without making any change in the lenses. It will appear hereafter that a contrivance of this kind not only enables us to arrange the magnification to an exact extent, to meet special purposes, but also allows the introduction of additional apparatus for reinverting the image, and supplies the means of setting a binocular instrument to suit varying distance between the eyes of different observers.

The description of the instrument in its original crude form given above, has served to illustrate its fundamental principles; but the imperfections arising from the spherical and chromatic aberrations and distortions, being magnified with the magnification of the object, rendered it almost useless. The researches of modern microscopists, however, have led to the substitution for each of the simple lenses, originally forming the object-glass and eye-glass, of compound pieces, in which the imperfections produced by the several lenses, are so balanced as to counteract one another to a most marvellous extent; and a degree of perfection in the construction of microscopes has thus been arrived at which leaves little to be desired.

The Object-piece, or Objective.—Microscopes, then, are now constructed with compound object-glasses, or, more properly, object-pieces, in which the chromatic aberration is over-corrected, so as to produce what may be called negative chromatic aberration. Were the object-piece itself made perfectly achromatic, it is manifest that the chromatic aberrations thus effaced would be more or less reproduced by the eye-piece, unless this also were made perfectly achromatic; but the desired result is obtained more simply by constructing over-corrected object-pieces, to produce such amounts of aberrations and distortions, as will be effaced by aberrations and distortions of an opposite character in the eye-piece; and this expedient has been so far successfully carried out, as to give a result, practically speaking, free from all optical imperfection.

An object-glass consisting of a single lens, having a very high magnifying power, would have so short a focal length, and such great curvature, as to produce great spherical aberration, besides other inconveniences. Several compound achromatic lenses, or rather compound lenses in which the achromatic aberration is over-corrected, are therefore combined to form an object-piece, and by this means the spherical aberration is to a great extent eliminated, and a much larger aperture obtained. The compound lenses of which object-pieces consist, are generally composed of plano-concave and double-convex lenses, forming together plano-convex lenses; and the plane sides are presented towards the object. Were the convex side turned towards the object, the rays diverging from this object would fall upon the convex surface with greater and greater obliquity, as they approached its borders, and would, in fact, be in great part reflected without passing through the lens at all, so that a very small aperture only would be available. This defect is remedied to a great extent by presenting the plane side of the plano-convex lens to the object; but such improvements have been made in object-pieces as to allow of greatly increased apertures, and for this purpose the meniscus form has been given to the object-glasses, their concave sides being presented to the object. In the most powerful microscopes, three compound lenses, mounted so that their axes shall be precisely in the same straight line, constitute the object-piece, or, as it is usually but inappropriately called, the object-glass; and with the great magnifying powers now attained, the slightest imperfection would be so magnified as to be highly injurious. It is therefore manifest that the greatest precision in the balance of the aberrations must be arrived at.

Now it is necessary for the preservation of microscopic objects to cover them with thin films of glass, and the aberration produced by these films, if only the 100th of an inch in thickness, will, under a power of 1,500 or 2,000, impair the distinctness of the image. In microscopes, therefore, of the highest powers, the triple lenses which compose the object-piece, are so mounted that their mutual distances can be varied slightly by a delicate screw-action, and by this means a sufficient effect can be produced on the aberrations to neutralize the small amount of positive aberration produced by the glass film which covers the object.

The Eye-piece.—For eye-pieces of low power single lenses may be used. As higher powers are used, with the increased magnification of the image the imperfections would be magnified also, and compound eye-pieces are used, in which a second lens, called the field-glass, is placed in front of the eye-glass. The field-glass intercepts the rays from the object-piece, before they arrive at the position for the formation of an image, and by giving them increased convergence, causes them to produce a smaller image, the whole of which is seen with the eye-glass, while without the intervention of the field-glass the central portion only of the image would be so seen. It is called the field-glass from this property of enlarging the field of view. But not only does it do this, but by concentrating on the reduced image the light, which would otherwise be diffused over the larger, the brightness is greatly increased. The image formed by the object-piece is not flat, but forms a portion of a spherical surface, concave towards the object-piece, and, consequently, convex to the eye; the effect of a single lens as an eye-piece would be to increase this convexity, and it is therefore desirable that the field-glass should have the further effect of altering the form of the image into one convex to the object-piece, just to such an extent as to be corrected by the like tendency in the eye-glass to produce an alteration of curvature, and thus to render the imaginary image seen through the eye-glass flat, or, as it is termed, to produce a flat field.

The Huyghenian Eye-piece.—The preceding requisites of a good eye-piece are all fulfilled by the negative eye-piece, invented by Huyghens for the telescope, in order to produce a flat field, but without recognising its power to correct the chromatic aberration. It consists of two convexo-plane lenses, with their plane sides consequently turned towards the eye, their focal lengths and the interval between them being as 3, 1, and 2. The lens, *f*, of greatest focal length forms the field-glass; and when the microscope is arranged for distinct vision of an object, this lens stands between the object-piece and the focus at which it would produce an image of the object, at a distance from the latter equal to half its own focal length. Thus, the pencils of rays from the object-piece, tending to form an image at a distance from the field-lens equal to three-fourths of the interval between the two lenses of the eye-piece, are intercepted by the field-glass, and brought sooner to a focus, so as to form the image, *i*, half way between the two lenses, and consequently in the focus of the eye-glass. A diaphragm is placed in the tube between the lenses, at the point where the image is formed, to shut off all extraneous light, and another diaphragm is placed in front of the eye-glass. In this eye-piece the refractions of the axes of the pencils are equally divided between the two lenses, by which means the spherical confusion is much diminished; the forms of the lenses are also such as to produce but a small amount of spherical aberration, and the relation between the focal lengths of the lenses and the interval between them is such, as to produce a combination, whose aberrations shall so balance those of the over-corrected object-piece, as entirely to eliminate the spherical aberration, and reduce the chromatic aberration to a minimum, leaving only a very pale green tint from the combination of the blue and yellow rays, which is in no way objectionable. This eye-piece, called a *negative eye-piece*, is always to be preferred when we are only seeking to obtain the best defined and most distinct view of an object; but for some purposes it has been necessary that an image should be formed by the object-piece in front of the eye-piece.

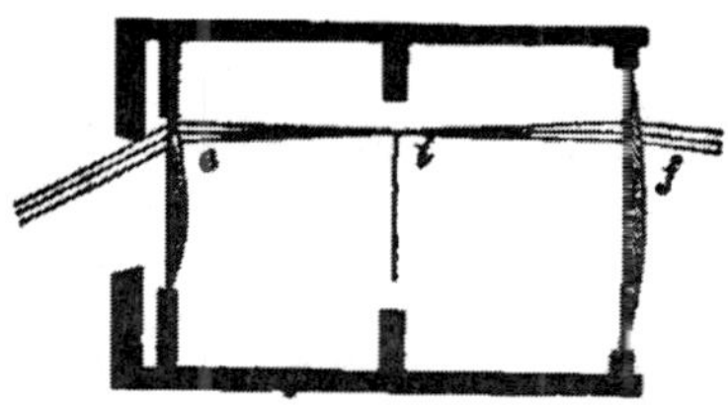

Ramsden's Eye-piece.—We have then recourse to the eye-piece, originally constructed for astronomical telescopes by Ramsden, called a *Positive Eye-piece.* This consists of two lenses of equal focal lengths, one plano-convex and the other convexo-plane, so that the convex sides are turned towards one another, the interval between them being equal to two-thirds of the focal length of either. When the microscope is arranged for distinct vision of an object, the field-lens *f*, is placed at a distance behind the image formed by the object-piece, very nearly equal to one-fourth of its own focal length. The focus of the entire eye-piece is, in fact, situate in front of the field-lens, at a distance equal to one-fourth the focal length of this lens and is therefore a very little

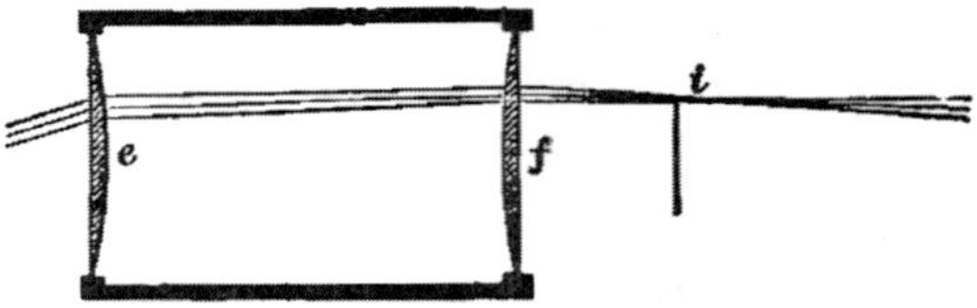

distance beyond the image, and the rays proceeding from the image at *i* emerge from the eye-lens, *e*, parallel, or in the condition best adapted for distinct vision. This eye-piece is not achromatic, but it has very little spherical aberration. Its principal defect is that it does not produce a flat field; so that while the delineation in the centre of the field is good, it becomes indistinct towards the edges.

Kellner's Eye-piece.—This eye-piece will give a field of fourteen inches diameter, and is consequently well adapted for a general view of an object; but the delineation is not sufficiently distinct for the examination of minute details. The field-glass is a double convex lens placed in the focus of the eye-glass, and the latter is an achromatic combination of a plano-convex flint lens with a double convex crown lens. The aberrations, however, are under-corrected, so as to balance the aberrations of the object-piece. The field-glass in this eye-piece must be frequently removed, and wiped with a soft leather or piece of silk, as the least speck of dirt upon it, being in the focus of the eye-glass, is seen with magnified dimensions.

Use of Varying Powers.—In the prosecution of microscopic researches, a great variety of magnifying powers are required. Not only are high powers required for very minute objects, and lower powers for larger objects, but, for the complete examination of the same object, a low power is first required, to view its form and character as a whole, and then, powers successively higher and higher must be used, to disclose more accurately its several parts, and the more and more minute features which characterize it. The best microscopes are therefore supplied with several object-pieces and eye-pieces, each eye-piece being adapted for use with every one of the object-pieces; and by this means a number of powers, varying from ten or twelve upwards, is obtained. For powers below thirty or forty, however, simple microscopes may be used.

With the largest and best instruments the makers supply four or five eye-pieces and eight or nine object-pieces, thus producing from thirty-two to forty-five varieties of power; and intermediate powers are produced to a large extent, by changing the distance between the object-piece and the eye-piece. For this purpose the object-piece is fitted into one tube and the eye-piece into another, moving within the former with a fine motion, produced by a rack and pinion. The most masterly skill is required to construct these tubes, so that their axes shall be always in exactly the same right line, and to set the several lenses in these tubes, so that their axes shall also range in the same straight line with the utmost precision. The least deviation in this respect would be fatal to the efficiency of the instrument.

Mechanical Portion, or Body of the Microscope.—We thus come to the consideration of the mechanical portion of the instrument. This consists then of a tube called the *body*, into one end of which the object-piece is screwed, and into the other the eye-piece slides. If it be desired to produce varying powers by changing the distance between the object-piece and eye-piece, the eye-piece is fitted into a tube called a draw-tube. This tube can be pulled out to the extent of six inches, and is graduated at every tenth of an inch, so as to be pulled out with exactness to any desired length. By drawing out the tube the balance of the spherical aberration is slightly disturbed; but this balance is again restored by an arrangement for varying the distance between the lenses which form the object-piece, in a mode to be explained hereafter.

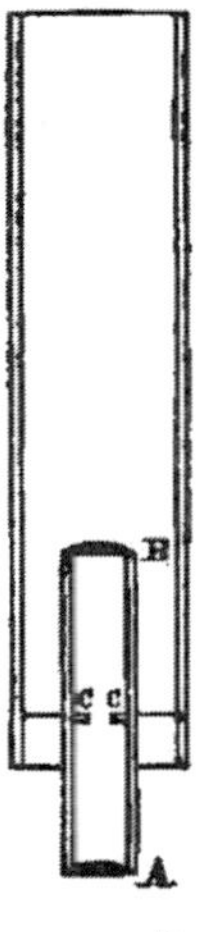

Erector.—Microscopes supplied with a draw-tube can have adapted to them the erector, or erecting eye-piece. It consists of a brass tube, three inches long and five-eighths of an inch in diameter, into the opposite ends of which are screwed two plano-convex lenses, A and

B, and between them is placed a diaphragm or stop, C. It is screwed into the lower end of the draw-tube, with the convex surfaces of the lenses turned towards the eye-piece, which is fitted into the upper end of the draw-tube. This appendage to the microscope fulfils the same office as the two additional lenses introduced into the terrestrial eye-piece of the telescope (p. 88); that is, it reinverts the inverted image of the object formed by the object-piece, and thus gives an erect view of the object with all its parts represented in their relative positions, so that the hand can follow the eye in dissecting.

Binocular Body.—The stereoscopic effects of binocular vision are to be obtained with the compound microscope, as in the simple microscope, by dividing the pencils of light, and removing the rays from one half of the object-piece to one eye, and from the other half to the other. In the first attempts, however, to effect this, the results produced were not stereoscopic, but pseudoscopic; for since the object-glass forms an *inverted* image of the object, it is necessary, in order to produce *stereoscopic* effects, to invert also the two imaginary eyes formed by the two halves of the object-piece, and remit to the *right* eye the rays from the *left* half of the object-piece, and to the *left* eye those from the *right* half.

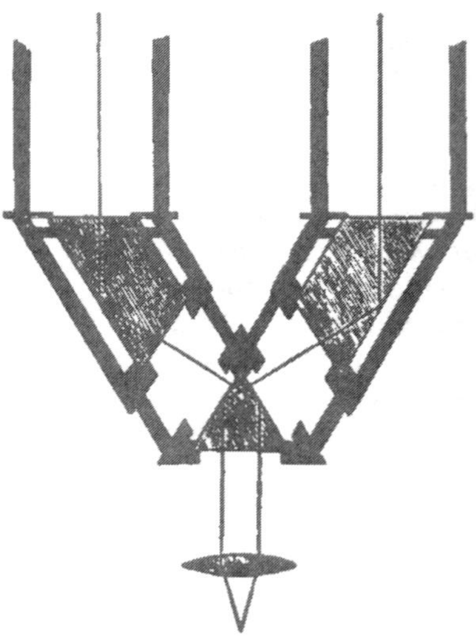

Nachet's Binocular Body.—This instrument consists of two parallel tubes, the distance between them being regulated by screws; these tubes carry at one end the eye-pieces, and at the other end prisms, which reflect up the tubes the rays reflected to them from the central prism A B C. This prism receives on the face A B the rays proceeding from the left half of the object-piece, and reflects them to the right, and on the face A C the rays from the right half of the object-piece, and reflects them to the left. The distance between the tubes being so adjusted, then, as to correspond with the distance between the eyes of an observer, upon looking into both tubes, his right eye sees the picture presented to the left half of the object-piece, with the relative positions of all its parts inverted; and, in like manner, the left eye sees the picture presented to the right half of the object-piece. The position of the eyes, as it were, being thus inverted, at the same time with the inversion of the picture, true stereoscopic effects are produced.

The defects incident to this arrangement are the loss of light from the reflections at the four surfaces, through which the whole of the light has to pass, and the fatigue to the eyes from their axes being directed in parallel directions instead of converging to a point, as is natural to them when looking at a near object.

Wenham's Binocular Body.—The arrangement invented by Mr. Wenham is now universally adopted by the microscope makers of this country. It possesses the singular properties that it can be used either as a binocular or monocular instrument, and that it can be adapted to any instrument already constructed as a monocular microscope. It consists of a small prism mounted in a brass box, which slides into an opening immediately above the object-piece, and reflects one half of the rays that pass through this object-piece into an additional tube, attached at an inclination to the ordinary body. The other half of the rays take their usual course entirely unobstructed, and the diverted rays, reflected twice within a single prism, pass through only two reflecting surfaces, and thus suffer only half the loss of light, to which both halves of the rays are subjected in Nachet's Binocular Body, described above. The eye-pieces are placed in draw-tubes, and by racking them out or in, the distances between the axes of the eye-pieces are increased or diminished, to correspond with the distance between the eyes of the observer. This adjustment is very important, and is to be tested by closing each eye alternately, without moving the head; if each eye has then a full and clear view of the object, the adjustment is correct, otherwise the draw-tubes must be racked out or in, until this result is obtained. If the prism is drawn back until stopped by a small milled head, the whole of the rays from the object-

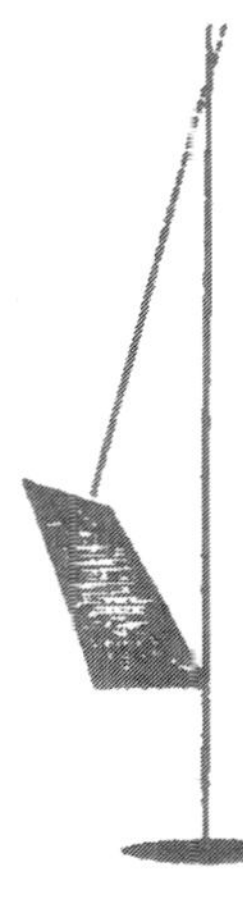

piece pass into the main body, neither the prism nor the additional body interfering in any way with the use of this body only as a monocular microscope. This is a point of considerable importance, as with the highest powers now attained, the division of the light between the two bodies cannot be afforded.

Nachet's Improved Binocular Body.—M. Nachet has adopted Mr. Wenham's plan of diverting only one half of the rays into an additional inclined body, thereby removing the defects noticed above in his original plan; and has also, by a very ingenious arrangement, obtained the means of instantaneously changing

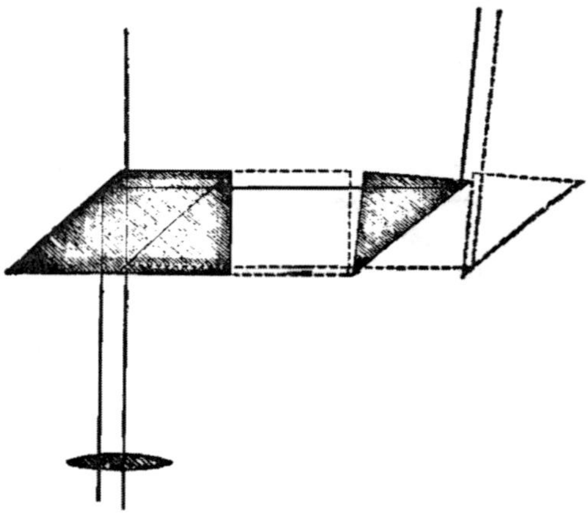

the pictures presented to the two eyes, so as to produce in one case a stereoscopic, and in the other a pseudoscopic, view of the object. Half the rays transmitted through the object-piece are reflected, first from the surface of the prism A to the surface of the prism, B, and then, from this surface into the additional body. When the prisms are in the position represented by the figure, the effects produced are stereoscopic; but when they are in the position represented by the dotted lines, the effects produced are pseudoscopic.

Though so great an improvement when using object-pieces of low powers, the *stereoscopic* binocular should not be employed with the highest powers for the following reasons:—1. The large apertures (p. 51) of these object-pieces produce too great an exaggeration of the stereoscopic effects; 2. Their want of penetrating power (p. 51) renders stereoscopic effects of little value; 3. The division of the rays of light into two halves is equivalent to a diminution of aperture to an equal amount, and consequently impairs greatly the resolving power in which consists the value of these object-pieces. The use of a single eye, however, for a lengthened observation is very fatiguing; and, moreover, the constant use of a monocular microscope will in time impair the sight by making the range of distinct vision of the two eyes unequal.

To remedy these inconveniences Messrs Powell and Ireland have constructed a prism which, being introduced into the binocular body in place of the Wenham prism, may be used with the very highest power to produce a non-stereoscopic binocular view of the object. The whole of the rays proceeding from the object-piece fall first upon the surface of a glass plate with parallel surfaces; the greater portion of these rays pass through the plate, and pursue their usual course to form an image with hardly any deterioration, while that portion of the light which is reflected from the anterior face of the plate passes into a prism at its side, and is reflected into a second body, where it forms an image for the second eye somewhat darker than the principal image. The mind, however, takes cognizance of the brighter picture in the principal body, while the comforts of binocular vision are fully realised.

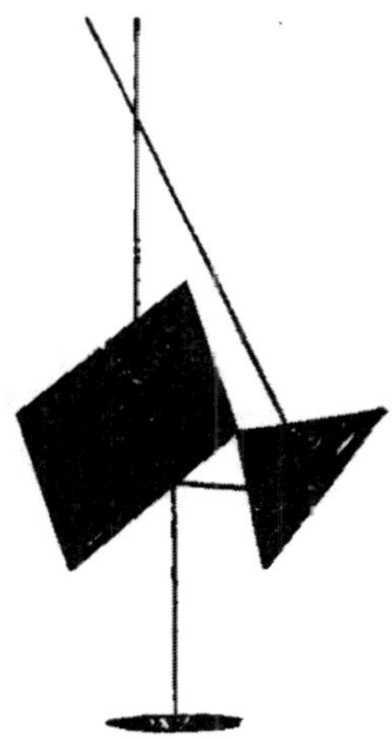

Having now discussed both the optical portions of the compound microscope, and the mechanical arrangements of the body in which they are placed, there remain to be considered the accessories for illuminating, and holding the object, and the construction of the stand for supporting both them and the body of the microscope.

Illumination.—The illumination of the object is intended to produce such an effect that the enlarged virtual or apparent image shall be viewed as though it were a real object, seen under the most favourable quantity, quality, and direction of light.

Quantity or Degree of Illumination.—The quantity of light which, proceeding from the object, passes through the instrument, and ultimately enters the eye of the observer, must be sufficient to make every point of the apparent image over which it is diffused, as bright as would be desirable for a natural object observed with the naked eye, and must therefore be intensified in exact proportion to the magnification of the object; and a much greater amount even than this must enter the object-piece, to allow for the quantity lost by reflection at each of the refracting surfaces of the several lenses, and cut off by the diaphragms to correct aberrations and distortions.

Angular Aperture.—The quantity of light which, after proceeding from the object, enters the instrument, and is transmitted through it, depends, *cæteris paribus*, upon the angular aperture; that is, the angle between the extreme rays, which, diverging from the object to the object-piece, pass through it, and arrive at the eye-piece. Object-pieces of two-inch focal length—the lowest power generally used—are constructed with apertures of about 14°, while object-pieces of a quarter-inch focal length are constructed with apertures of 100°, and even of 140°; and with higher powers, apertures of 170° are reached. To comprehend the effect produced by such apertures, it must be borne in mind that the angular aperture of the eye, or angular magnitude of the pencils of rays that enter the pupil from an object at the ordinary distance of distinct vision, is not more than half a degree.

Illumination of Translucent Objects.—The great majority of microscopic objects are translucent in consequence of their tenuity, and to render their translucence more perfect, they are placed on slips of glass, and saturated, with a drop of pure water, if intended merely for temporary inspection, or with Canada balsam, if intended to be preserved as permanent objects. They are next covered each with a small piece of very thin glass, made expressly for the purpose, the thickness in some cases not exceeding the hundredth of an inch, and the

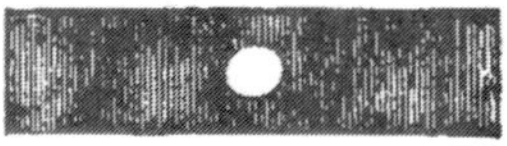

object is ready for temporary inspection. Permanent objects, however, are fixed in their place by gumming paper round the slip of glass, so as completely to envelop it, and cutting out a small circular piece on both sides of the slip, in the centre of it, where the object is placed. The slips of glass, with the objects thus mounted on them, are called *sliders*. They are usually about two inches long and half an inch wide.

The slider containing the object to be examined, is laid upon a stage, s s, the surface of which is blackened, as it is desirable to exclude as much as possible all extraneous light from entering the object-piece, in order to enhance the effect of that passing through it from the object. In the centre of the stage is a circular hole, and the slide is so placed that the object may be over the centre of the hole, which is so arranged as to be in a line with the axis of the microscope.

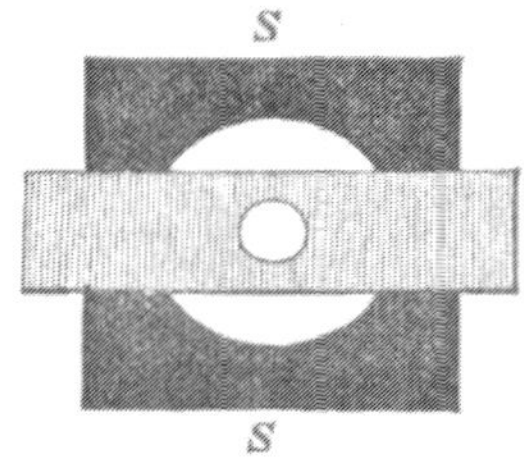

The concave mirror M M, collects the light from the sky (that from a white cloud being preferred), or from a lamp, and concentrates it upon the object, a compound achromatic lens, L L, called the condenser, being interposed, if a greater degree of condensation be required.

To limit the diameter of the pencil of light thus transmitted to the object to agree with the extent of the object, an opaque diaphragm, D D, with holes in it of different sizes, is attached to the underneath part of the stage by a pivot, round which it can be turned, until the condensed light passes to the object through the opening best suited to it. Again a plate of ground glass may be turned round beneath the diaphragm, so as to moderate the intensity of the pencil of light, when it is received on the mirror directly from the sun, or from a very brilliant lamp.

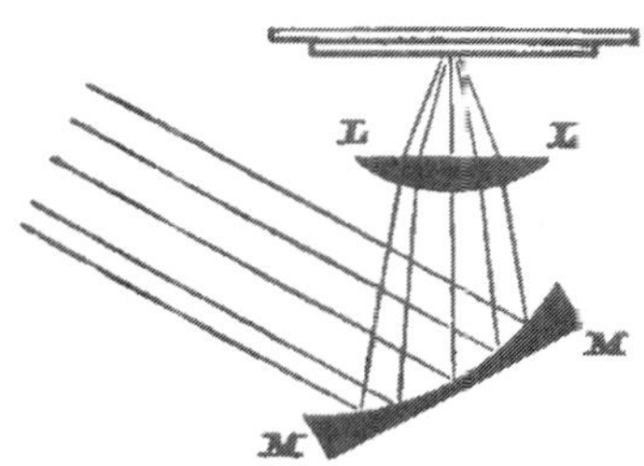

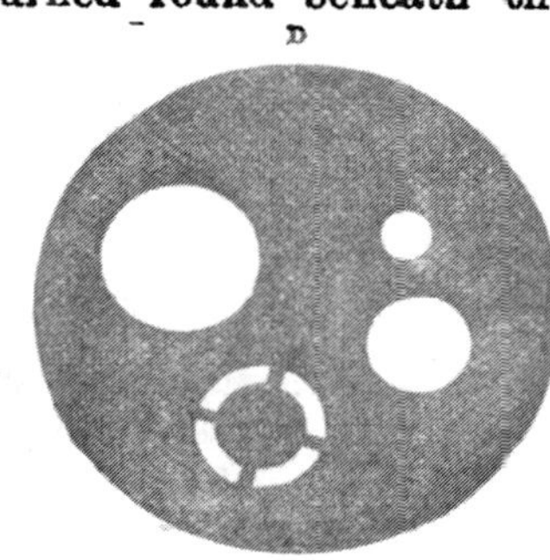

Illumination of Opaque Objects.—Opaque objects may generally be placed on a blackened plate of sufficient size to cover the opening in the centre of the stage, and then illuminated obliquely from above, by means either of a condensing lens, D D, or of a concave mirror, N N.

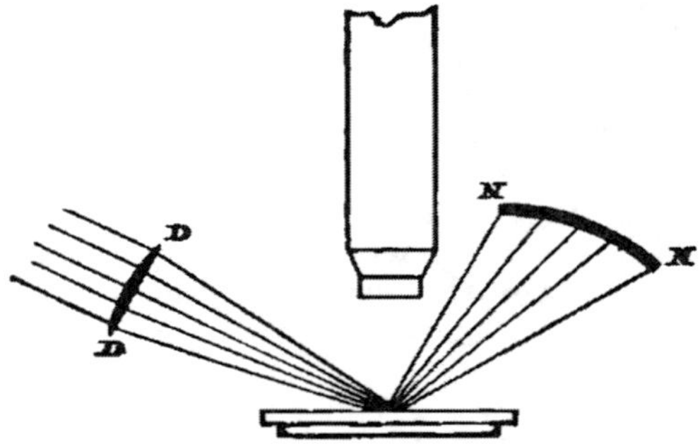

The Lieberkuhn.—With object-pieces of the highest powers, however, these methods are inapplicable, since the object is then so near the object-piece that the light from the lens or mirror would be either wholly, or in great part, intercepted by it. The object is in this case supported in a dark cell, which fits into the opening in the centre of the stage, and intercepts the central rays proceeding from the mirror

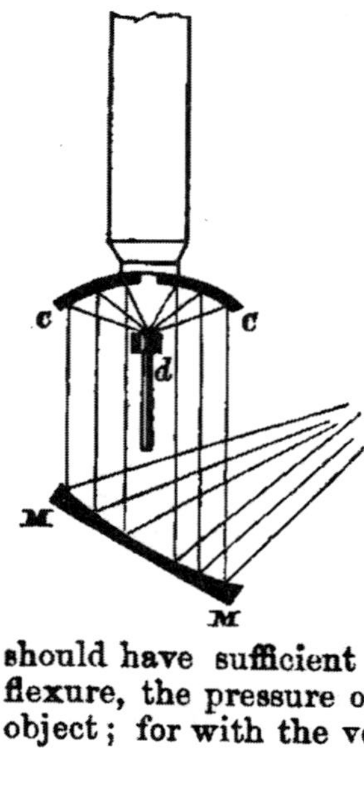

M M, while it allows the remaining rays of the pencil to pass to a silver cup, C C, called a Lieberkuhn, from the name of its inventor, with a hole in its centre, through which the object end of the microscope passes. By this cup, or concave reflector, the light received from the mirror M M, is condensed upon the object, and strongly illuminates it.

Support and Movement of the Object with Microscopes of High Powers.—In making use of high powers, it is necessary that the stage supporting the object should be perfectly free from all tremor, the slightest motion being so magnified as to prevent a good view from being obtained. It is also necessary that it should have sufficient strength and solidity to bear, without flexure, the pressure of the hands in the manipulation of the object; for with the very high powers now attained, the focal adjustment must be so exact, that a displacement produced by the pressure of a finger, upon a stage having the least elasticity, or not fixed to the instrument with perfect firmness, would throw the object out of focus. At the same time it is necessary that it possess facilities for imparting great delicacy of motion to the object, so as to bring the minutest part of it into view with ease and certainty. For this purpose the stage is made of two or more pieces moving one upon another, so that the object can either be turned completely round, without passing out of the field of view, or can be moved in two directions at right angles to each other by means of micrometric, or fine-motion screws, *b*, and *c*.

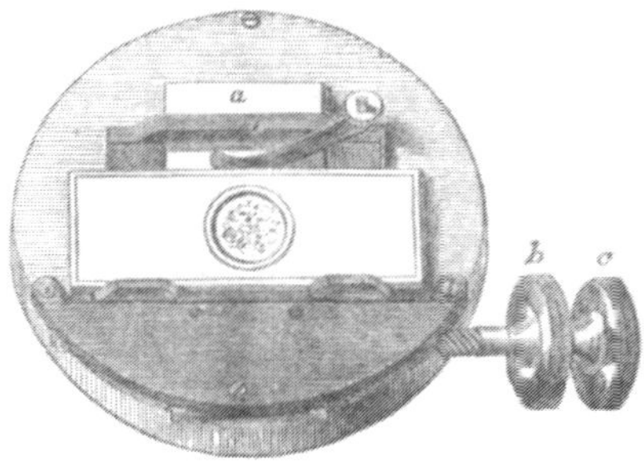

White's Lever Movement.—Instead of the two slow motions given by the two micrometer screws, White's lever, by a single motion of the hand, moves the object at once in *any* desired direction. It consists of a lever, D E (Fig. 1), to the shorter arm of which the ball B, is firmly screwed, and moves in a socket, formed by the upper plate of the stage A, and the cap

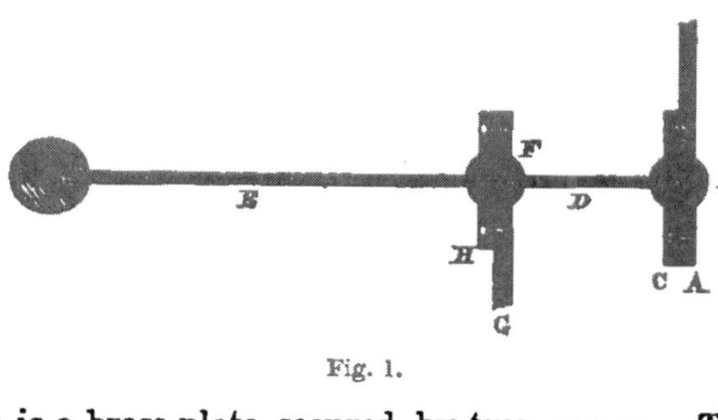

Fig. 1.

C, which is a brass plate secured by two screws. This lever passes through the perforated ball F, moving in a socket formed by the arm G, which is attached to some immovable part of the microscope-stand, and the cap H, similar, and similarly secured, to the cap C, as represented in Fig. 1; the longer arm E, of the lever is nearly three times the length of the shroter arm D; but, when the lever is in use, this proportion is not constant, as will be seen by Fig. 2; for the longer arm of the lever having been elevated, and the shorter one depressed, and the ball B having travelled in a straight line to *b*, instead of in an arc of a circle, it will be found that the shorter arm has increased by a quantity equal to the space between 1 *f* and 2 *f*; and for this reason the ball F is perforated, to enable that part of the lever to move easily, though without shake. As the end of the lever to which the hand is applied moves in all cases in an opposite direction to the ball B; and, as the compound microscope inverts the

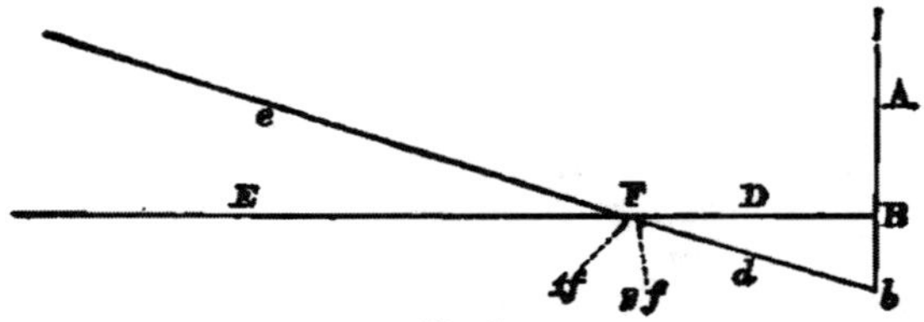

Fig. 2.

image of the object viewed, it follows the object will appear to move in the direction of the hand: moreover, as for each position of the ball B, there is but one in which the other end of the lever can be placed, if a certain portion of an obiect is seen at any time, and it is required to bring the same part into the field again, this will be done by placing the hand in the original position.

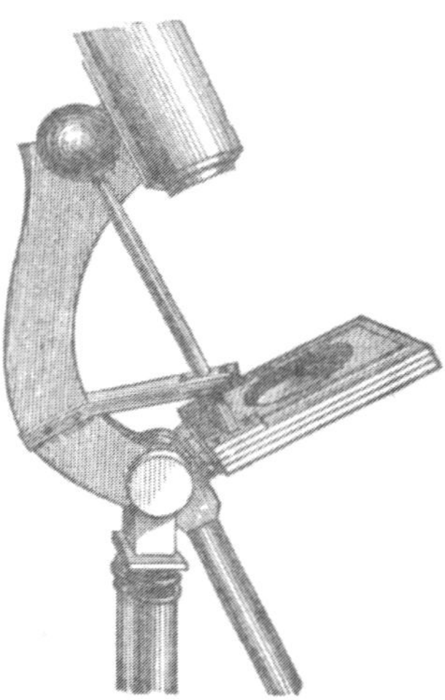

Fig. 3.

This lever may be applied to a stage, whatever be the construction of the guides for its motion. When, however, it is applied to a stage having plates moving horizontally and vertically by means of dovetail slides, it is better to have the upper plate move vertically, in which case the longer arm of the lever and its knob have to balance that plate only; but should the middle or second plate move vertically, then the lever must be sufficiently heavy to support the two plates, when the stage is in the position represented by Fig. 1. Fig. 3 represents White's lever applied to a three-plate stage.

Stands for Mounting Microscopes.—The most eminent makers of microscopes, both British and foreign, have devoted their attention to make the mounting of the instrument rival the perfection of its optical construction. Simplicity and portability have been combined with the necessary solidity and firmness, and perfect smoothness and precision in the action of the moving parts.

The amount of labour and skill bestowed on these beautiful instruments necessarily enhances their cost to a degree corresponding with their perfection; but very useful instruments are also manufactured at a very moderate price.

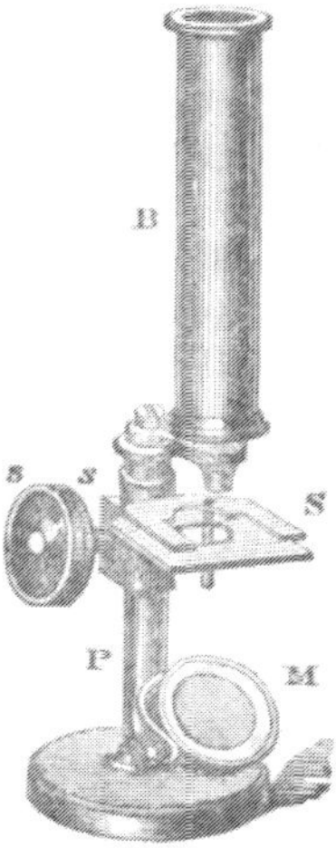

Fig. 1.

The annexed figure (Fig. 1) represents one of the simplest modes of mounting them. P is a vertical pillar rising from a stand of sufficient weight to give the instrument the required steadiness. On the top of this pillar is pivoted a socket, into which the body, B, of the microscope fits. The stage S, is attached to a tube which slides on the pillar, and is fixed in the position desired by a screw *s*, which clamps it to the pillar. The mirror M, at the bottom of the instrument, has a reflector on each side of it, one plane and the other concave, either of which can be turned to face the stage, and placed at any desired angle.

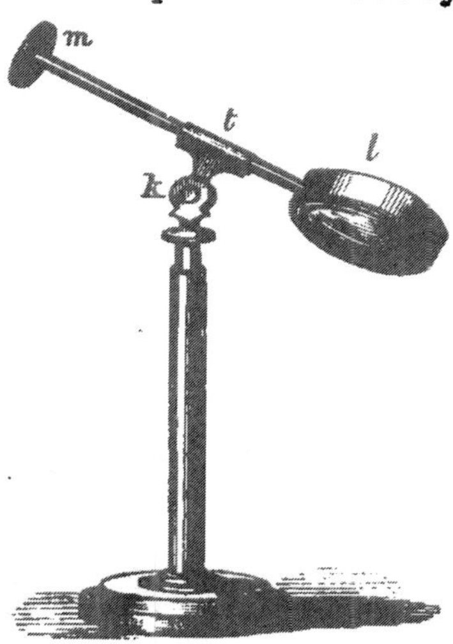

Fig. 2.

For the illumination of opaque objects, a condensing lens, *l*, is made use of: it slides through a tube, *t*, with friction, and can be turned round in it by means of the milled head *m*; and the tube is attached to its supporting pillar by a knuckle-joint, *k*. By these arrangements it is evident that the lens can be brought into any desired position.

In Fig. 3 the instrument is supported on a cradle C C, to which the stage and mirror are also attached, and it can thus be inclined at any angle that may be most convenient.

Fig. 4 is a cheap form of mounting well adapted for a dissecting microscope. It has a fixed stage firmly supported on three legs, and the condensing lens for illumination of opaque objects is attached to the stage itself. One of the legs is hollow, and the arm which carries the body of the instrument fits into this leg, and is moved up and down by a rack and pinion by turning the milled head *h*, by which means the focus is adjusted for distinct vision of the object.

Fig. 3.

R. and J. Beck's Popular Microscope.—This is undoubtedly one of the best *second-class* microscopes: its construction will be readily understood from the accompanying figure (p. 68). The body, A, which is Wenham's Binocular (p. 57), is carried by a strong arm, B, attached to a square bar, C, that may be moved up or down by a rack and pinion in the tube H, forming the lower part of the stand, to which the stage D, and the mirror E, are attached.

The base F is triangular, and connected with the tube H by a broad stay, G, moving on centres at the top and bottom, so as to allow this tube, and with it the body of the instrument, to be placed either in a vertical or in a horizontal position, or in a position more or less inclined. In every one

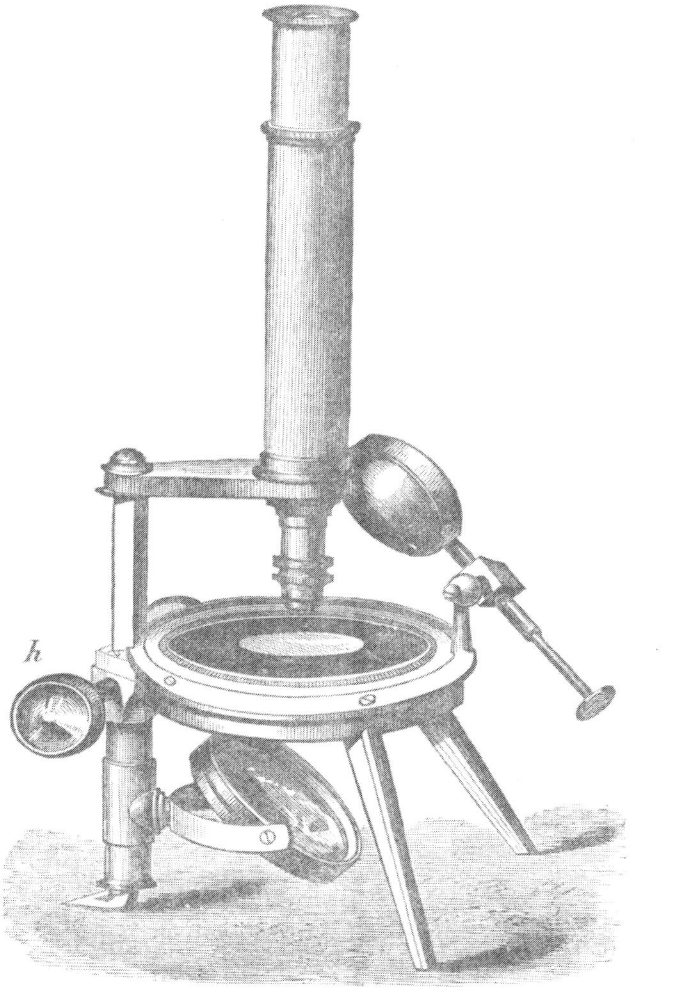

Fig. 4.

of these positions the instrument can be fixed firmly and steadily by fitting the projecting pin at the end of the tube H into one of six holes (one is seen near the stout pin L), of which four are ranged along the medial line of the base, and the sixth is in the stud K, at the extremity of the base. For a

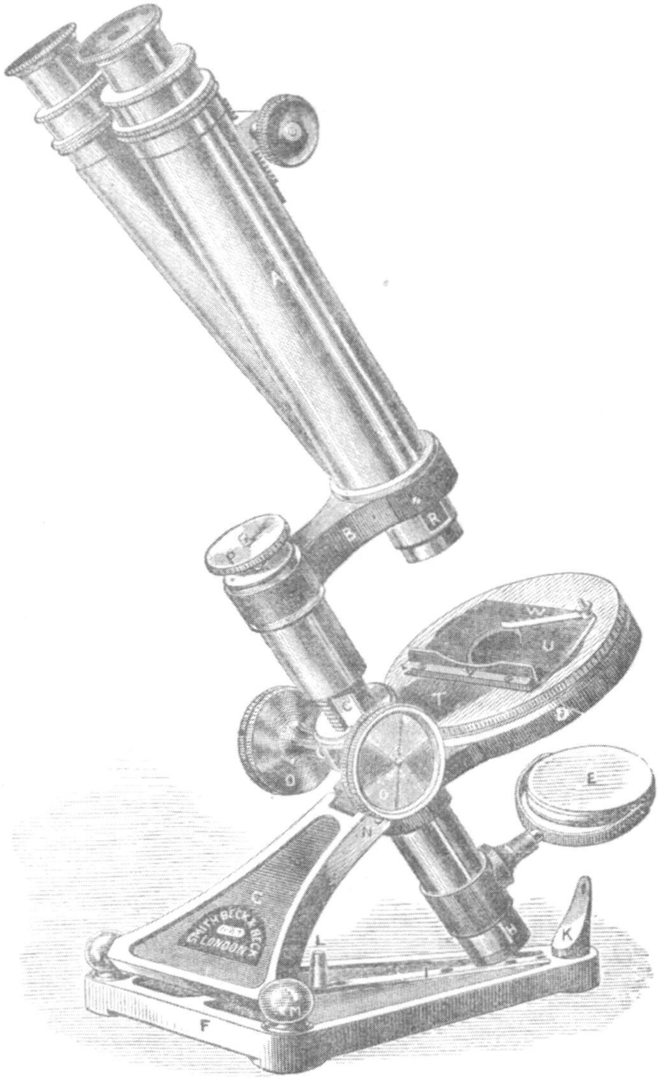

vertical position the pin is placed in the hole near L, for an inclined position it is placed in one of the holes towards the extremity of the base, and for a horizontal position it is placed in the hole in the stud K, the inner surface of the stay G then resting on the stout pin L. This is a light form of stand, and yet firm and strong; the instrument cannot shift in the least degree after being placed in any of its positions, and the weight is always well balanced nearly over the centre of the base.

The stage is circular, and consists of a bottom stage-plate, upon which is fitted the stage proper, which consists either of a plain stage, T, or a stage consisting of two plates, with slow-motion adjustment by means of micrometric screws. In either case the stage can be revolved upon the stage-plate, which is concentric with the body, so that the part of the object under examination will always remain in the field of view during the rotation. The object-holder U is provided with a ledge, V, and a light spring, W: it is held to the stage by a spring, so that it can be moved about easily by one or both hands. The small spring W is fastened to the object-holder by a milled head, which will unscrew, so that the position of the spring may be altered, to give more or less pressure upon the edge of the object.

When only a flat surface is required, the object-holder may be removed from the plain stage T, by unscrewing from its under side two small milled heads, which connect the holder to a circular spring; or, by removing the plain stage T, altogether, a simple flat plate may be substituted.

Beneath the stage plate is a cylindrical fitting for the reception of a diaphragm, or any additional apparatus that may be required there.

The mirror E, besides swinging in a rotating semicircle, will slide up and down on the tube H, and it will turn on either side for oblique illumination.

The object is focused by turning the milled head O, which gives a quick movement to the body of the instrument; and then the milled head P acts upon a lever which gives a very delicate motion to the body, and perfects the adjustment with great nicety.

Charles Chevalier's Universal Microscope.—This instrument, intended for all kinds of microscopic observation, is represented in the accompanying figure on a scale of one-fourth its natural size. The objective is at *b*, the eye-piece at *e*. The pencils of light by which the object is seen, rise at first vertically; but, being totally reflected at the hypothenuse of the prism *r*, they are turned back horizontally towards the eye-piece, an arrangement which enables the observer to take a convenient position, either for prolonging his observations

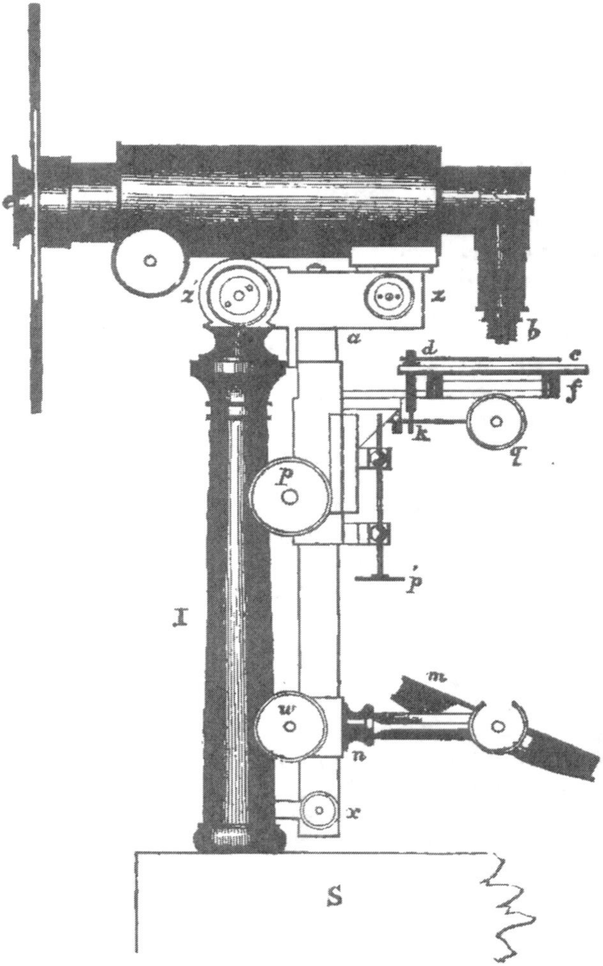

or for making drawings of the magnified images presented to his view (p. 79).

The objective is composed of two or three achromatic lenses whose focal lengths are from eight to ten millimeters;* they

* That is, from about ·355 in. to about ·394 in.

are marked with the numbers 1, 2, 3. No. 1 can be used alone, or No. 1 and No. 2 can be used together, taking care to screw No. 1 first to the body of the instrument, and then No. 2 upon No. 1; or No. 1, No. 2, and No. 3 may be all used together, taking care to screw No. 1 first to the body, then No. 2 on to No. 1, and, lastly, No. 3 on to No. 2. In the first case we have the least magnifying power, and the object is placed at the greatest distance from the objective; in the second, the magnifying power is higher and the object nearer; and in the third, the magnifying power is higher still, and the object is brought very close to the objective.

With each combination of the object-lenses there can be used any one of six eye-pieces, numbered 1, 2, 3, 4, 5, and 6. The four first are Huyghenian eye-pieces (p. 52), and 5 and 6 are simple magnifying glasses of very short focal length.

The instrument is supported on a pillar I, firmly screwed into the stand S, which also forms the case for packing away the instrument when not in use; the body of the instrument is attached to this pillar by the intervention of the arm *z z′*, represented in the figure in a horizontal position, and two knuckle-joints at *z* and *z′*.

At right angles to the arm *z z′* is attached a bar *a x*; and, when the instrument is in the position represented in the figure, the lower end of the bar fits into a piece projecting from the pillar, and is fixed by the pin at *x*. The bar *a x* has a rack on the face next the pillar, and on the bar are two brass boxes *n* and *t*, to one of which is attached the mirror, and to the other the stage.

The mirror is moved up or down by means of the pinion *w*: it has a concave reflector on one side, and a plane reflector on the other, and can be set with either side upwards, and inclined at any desired angle, since it moves on two pivots at the extremity of a diameter.

The stage *o* has a course adjustment by means of the pinion *p* working in the rack on the bar *a x*, and a fine adjustment by the slow motion screw *p′*, to set the object on the stage in exact focus. The piece *d*, which moves up and down with friction, serves to hold the objects on the stage, and to press them when necessary. Underneath the stage is the diaphragm *f*, and below this again a plate of ground glass, which can be turned to intercept and moderate the light when received directly from the sun, or from a very strong lamp.

The upper plate of the stage is pushed forward or drawn

back by the slow-motion screw *k*, and moved to the right or left by the slow-motion screw *q*. By these two motions the whole extent of the object can be passed before the view without losing sight of the image.

These screws, *k* and *q*, move at each complete turn through a very small and very accurately-determined distance, and have, moreover, their heads divided on the circumference, so as to note the number of turns, and fractions of a turn, that are required to pass an object from one side to the other of each of two micrometric threads, fixed, at right angles to each other, to the diaphragm of the eye-piece, which is at the exact place where the real image of the object is formed. The actual size of the object is thus determined.

To place the microscope in a vertical position, unscrew the prism, place the objective in the prolongation of the body, and turn this round the knuckle-joint *s*. Again, by drawing out the pin at *x*, and turning the instrument round the second knuckle-joint *s′*, the limb *s s′*, can be inclined to any extent, or rendered vertical, and the body of the instrument be thus inclined at any angle between the horizontal and vertical, the stage at the same time moving, and taking a position always at right angles to the tube. Lastly, to arrange the instrument for the examination of chemical objects, turn round the piece *b r*, which is connected with the body by a bayonet-joint, till it points vertically upwards. A second stage and mirror for illumination is to be adapted to it in this position. A watch-glass placed on the stage holds the solution to be examined, and the evaporation does not injure the objective.

There are three patterns of *first-class* microscopes manufactured in this country — the Ross microscope, Messrs. Powell and Ireland's microscope, and Messrs. R. and J. Beck's microscope. In the first and third, the instrument is supported by two uprights rising from a triangular base; in the second it is supported on a tripod; and in all three it turns upon two centres, about which it is so nicely balanced as to rest with equal steadiness at whatever inclination it may be placed.

In the Ross pattern, and in that of Messrs. Powell and Ireland, the piece to which the centres are attached consists of a square box, terminating at the bottom in a cylindrical tube for carrying the mirror, and the body is inserted in a ring at the end of an arm, which fits over a pivot on the head of a square bar that slides in the square box, and is moved in or out by a rack and pinion acted on by a milled head. This arrangement allows the arm to be turned on the pivot away from the stage, by which the latter and also the bottom of the object-piece can be got at and cleaned, or the object-piece itself can be changed with the greatest convenience. The arm also furnishes a lengthened space for a lever, by which the fine adjustment for focusing is made.

In R. and J. Beck's pattern, the part to which the centres are attached is carried up in one solid piece to a bed nearly the whole length of the body, upon which the latter rests, and along which it is moved by a rack and pinion, acted on by a milled head.

This arrangement does not allow the body to be moved from over the stage, nor for the delicate fine adjustment does it furnish a space equal to that in the patterns of Ross and Powell and Lealand; but these advantages are fully, if not more than fully, compensated by the steadiness of the support given to the body of the instrument, and consequent removal of all tremor from it, of which the slightest amount is injurious when its effects are magnified by the highest powers. Another advantage is the easy and even motion of its quick adjustment, while in the other patterns, the rod carrying the arm and body of the instrument must necessarily be made to slide with great stiffness in the box that rests on the centres.

The frontispiece to this volume represents a *first-class* binocular microscope, with the most recent improvements. The stage has a complete rotating motion, concentric with the axis of the microscope, worked by rack and pinion, as well as motions in two directions, at right angles to each other. The underneath part of the stage is shown, fitted with the iris diaphragm.

ACCESSORIES OF THE MICROSCOPE.

Numerous accessory instruments, besides those already described, are used with the microscope for variety and perfection of illumination for the exhibition of particular phenomena, and for holding objects of various kinds.

Direct Illumination of Transparent Objects.—To view transparent objects by means of light directly transmitted through them, either the concave mirror attached to the microscope is made use of, or the light is reflected from a mirror, or a rectangular prism, through an achromatic condenser. In either case the light must be brought to a focus at the place of the object, in order that the rays may continue their course as though diverging from points of the object. Now, with object-pieces of four-tenths of an inch focal length, and with lower power, the light is generally in excess; and since this cannot be remedied by altering the position of the mirror or condenser without impairing the distinctness of the image, recourse must be had to the diaphragm. The degree of light best suited for an object depends upon the degree of its transparency; and, consequently, it is generally desirable to try several of the holes in the diaphragm, some of which have stops in the middle shutting off the central rays of the pencils, and examine the effect produced in each case. To facilitate the operation, each hole, when concentric with the body of the instrument, is stopped by a weak spring. A great improvement, however, consists in continuously examining the object while the amount of light thrown on it is gradually varied. This is effected by a diaphragm with shutters, which by a screw action can be more or less closed; and the best form of such a diaphragm is *Brown's Iris Diaphragm*, in which the number of shutters is numerous, and the circular form of the aperture is consequently preserved during their expansion or contraction.

Oblique Illumination of Transparent Objects.—In many cases the structure of transparent objects is but imperfectly shown by rays thrown directly through them, and it becomes important to view them by oblique rays, traversing them either from all sides, or from one side only, and then passing through the object-piece; or it may be desirable to illuminate them so obliquely, that none of the illuminating rays shall pass through the objective, but that the image shall be entirely formed by rays irregularly reflected from the object, as though it were opaque, in which case it appears as an illuminated object on a dark ground: this is called dark-field illumination. To give oblique illumination on one side, the mirror attached to the microscope below the stage is mounted on a double arm, and the stage is made very thin. Another mode is by a prism, which reflects and condenses the light at the same time.

The Spot Lens.—To give oblique illumination on all sides, a plano-convex lens of great convexity is made use of, the central portion of which is covered by a circle of black varnish, so that only the peripheral rays pass through, and fall obliquely on the object; or a condenser of considerable aperture is employed, in which the central rays are stopped out in a similar manner.

Wenham's Parabolic Reflector.—The most perfect method of giving to the illuminating rays an obliquity so great that none of them shall pass directly on to the object-piece, is the intervention of the parabolic reflector.

Fig. 1 represents this instrument, the paraboloid being set in a tube, A, with a milled edge, B, for adjusting it, to focus on the object the light reflected from it. This form of setting is adapted for applying it to Beck's Popular Microscope: the form of setting must, however, be varied to suit the appliances of the stage to which it is to be attached.

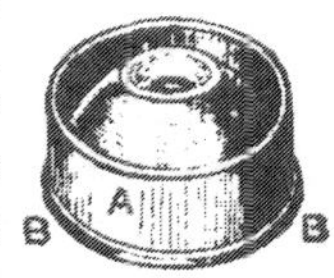

Fig. 1.

Fig. 2 represents a section of the paraboloid in its full size. C is the base, and A and B are the parabolic sides of the section,

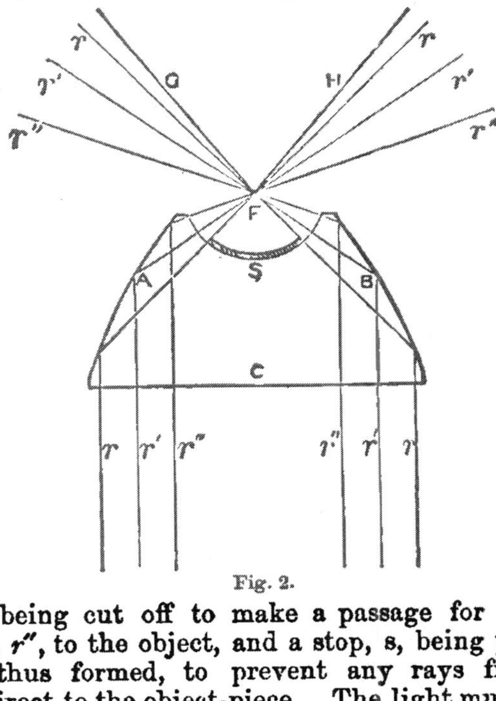

Fig. 2.

the apex being cut off to make a passage for the rays of light r, r', r'', to the object, and a stop, S, being placed in the aperture thus formed, to prevent any rays from passing through direct to the object-piece. The light must be thrown upon the parabolic surface A and B in rays parallel to the axis of the parabola, and will thus be reflected to the focus F of the parabola, which must be adjusted upon the object. The object-piece lying entirely within the angular space G F H, which admits no rays direct from the reflector, can, therefore, receive only the rays irregularly reflected from the object.

As the rays of light are to be thrown upon the reflector in parallel pencils, a plane mirror or prism must be used to reflect the daylight upon it; but, if the source of light be a lamp sending forth divergent pencils, these must be converted into parallel rays, either by the interposition of a condenser between the plane mirror and the parabolic reflector, or by direct reflection from a concave mirror. In either case the parallelism of the rays may be tested by placing a card or piece of white paper over the base c of the paraboloid: this should be equally illuminated over its whole surface.

Oblique Illumination of Opaque Objects.—The most perfect illuminator for the oblique illumination of opaque objects on one side only, with object-pieces of low power, is *Beck's Parabolic Illuminator.* It is fitted on to the object-piece by a ring clip, and reflects the light, which is thrown upon it

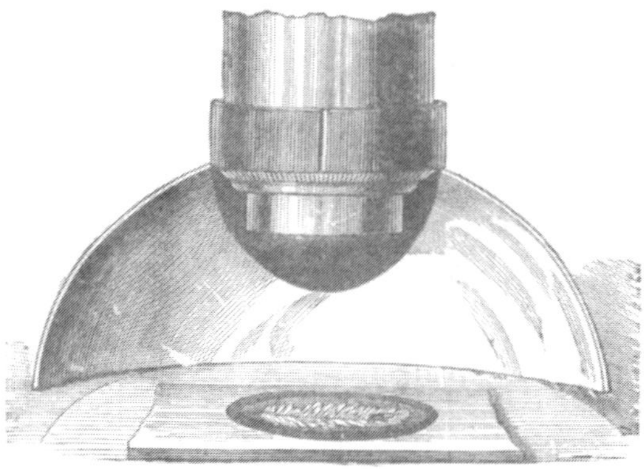

horizontally, down upon the object placed upon the stage. The ring clip is of course adapted to fit the object-pieces with which it is intended to be used; but Mr. Crouch has constructed a holder for this illuminator, by means of which it is attached to the body of the microscope, and can be used with any object-piece whatever.

Oblique Illumination of Opaque Objects on all Sides is afforded by the Lieberkuhn, already described; and by throwing the light from the mirror upon the Lieberkuhn in an oblique direction, oblique illumination of the object on one side only may also be obtained with object-pieces of high powers; or this result may be arrived at by covering one half of the Lieberkuhn with a non-reflecting substance—such as black paper. The disadvantages of the Lieberkuhn are that in order for its focus to agree with the distance from the object-piece to which the object must be brought, to be in the focus of the microscope, every different object-piece must have its own Lieberkuhn; and that the size of the objects which can be submitted to inspection with it, is limited by the necessity of allowing sufficient space beyond them for the passage of the light from the reflector to the Lieberkuhn.

Beck's Patent Illuminator.—The direct illumination of opaque objects under high powers, is effected by Beck's Illuminator, in which the light is thrown down upon the object through

Fig. 1.

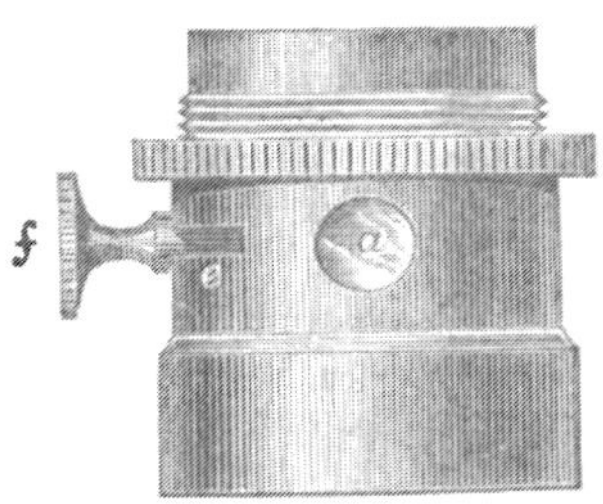

Fig. 2.

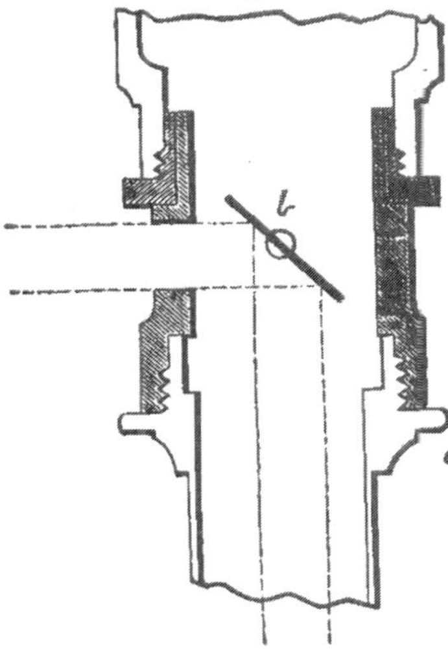

Fig. 3.

the object-piece itself. This object was first accomplished by a small metallic reflector, introduced into the body above the object-piece; this, of course, intercepted the central portion of the light returned from the object. Powell and Ireland substituted for the metallic reflector a plate of polished glass, which allowed the principal portion of the central rays from the object-piece to pass through it, and Mr. R. Beck further improved the instrument by using for the reflector a disc of very thin glass, which reflects more light than a polished surface.

Fig. 1 represents the instrument of its full size; Fig. 2 gives a sectional view of it, screwed into its position on the bottom, *c*, of the body of the microscope, and having the object-piece, *d*, screwed into itself; and Fig. 3 represents the disc, *b*, of thin glass, which forms the reflector. This reflector is introduced into the tube which adapts it to the microscope, through the slot *e* (Fig. 1), until it comes opposite the aperture, *a*, through which the light is admitted; when it is stopped by a shoulder on the stem to which it is attached; the milled head *f*, then serves to set it at any desired inclination. It is manifest that, by slightly varying the inclination, the rays reflected from the disc *b*, can be made to pass either through the centre of the object-piece, or through a portion of it more or less on one side; and thus the illumination of the object can be rendered either direct, or more or less one-sided and oblique.

Sorby's Spectroscope Eye-piece.—The increasing use of spectrum analysis, as a means of investigating the ultimate constitution of bodies, has led to the invention of this instrument for the comparison of the spectra formed by different substances. In the place of the ordinary diaphragm or stop of the Huyghenian eye-piece, is inserted a stop with a narrow slit in it, and in front of the eye-glass is placed a tube containing five prisms, two of flint and three of crown glass, in passing through which the light, admitted through the narrow slit, is refracted into a spectrum, and thus the spectra, formed by transmitting light through different substances placed on the stage of the microscope, can be observed; or the light may pass through the microscope direct from the mirror, and be refracted into the ordinary solar spectrum. To compare the spectrum thus formed, with a second spectrum, formed by light from another source, or transmitted through a different substance, a right-angled prism is placed so as to cover half the slit, and the light is reflected by a mirror through a tube, containing the substance whose spectrum is to be examined, and into the prism, by which it is reflected upwards, so that a portion of it passes through the half slit over the prism side by side with a similar portion of the light from the object-piece through the other half of the slit, and thus two spectra are formed side by side, whose slightest differences can be most accurately distinguished.

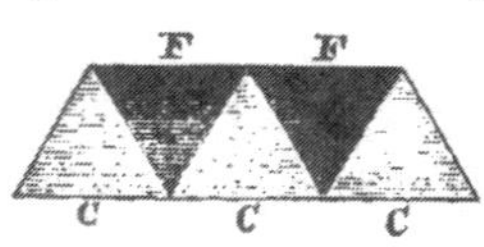

Polarizing Apparatus.—Most beautiful and brilliant results are obtained by viewing many microscopic objects with the assistance of polarised light; some objects which, when illuminated by ordinary light are scarcely visible, appear in the most varied and vivid colours under the influence of polarised light; and much knowledge is obtained by this means, of the structure of some substances, which could be arrived at in no other way. The peculiar property called polarisation may be imparted to light by various peculiar modes, both of reflection, and refraction; but the most simple apparatus to apply to the microscope for this purpose, is the following:—A prism, called a Nicol's Prism, of double refracting crystal, arranged so as to transmit only one portion of each refracted ray, forms the polariser A, and is fitted so that it can be turned round underneath the stage: and another prism, denominated the analyser, exhibits the peculiar effects of the polarised light after transmission through the substance under examination by it. This prism, mounted in a short tube, *e*, screws into the place of the back stop of the object-piece *f*, and the object-piece together with the prism, is attached to the nose-piece of the body of the microscope by the adapter H, which has a revolving fillet, *k*; or the short tube *e*, with the prism enclosed, may be screwed into an outer tube *c*, and then, having the cap *d*, screwed in over it, forms the piece B, which both slides and revolves in the place of the cap of the eye-piece. When the analysing prism is thus placed over the eye-piece, the field of view is contracted; and, when it is placed immediately above the object-piece, the definition is somewhat impaired: its position must, therefore, be regulated by the consideration of which of these defects is of the most importance in the examination of the particular object under view. G is a plate of selenite for the exhibition of colours arranged in a variety of beautiful forms.

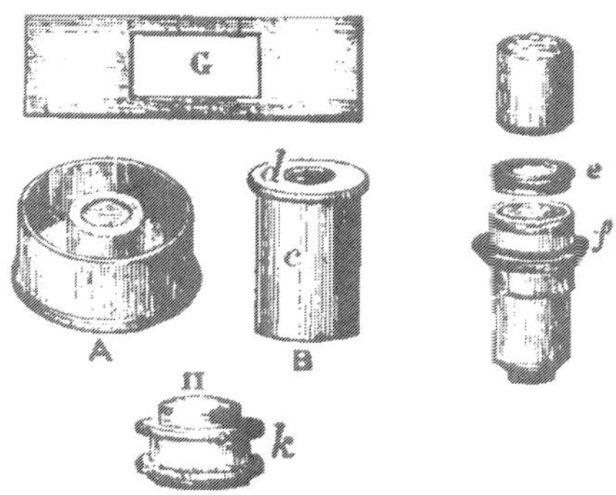

The Camera Lucida.—One of the most beautiful and useful applications of the camera lucida is its adaptation to the compound microscope, by means of which minute details and lineaments of objects are depicted with a precision and fidelity only surpassed by photography.

The microscope being placed in a horizontal position, the camera is fixed upon the eye-piece in place of its ordinary cap. When Wollaston's camera (p. 15) is made use of, the observer, looking upon it, sees the object by reflection, and the paper, placed on the table beneath it, directly, and the distance being so arranged that the object appears clearly delineated on the paper, an accurate drawing of it can be traced.

Since, in estimating the magnifying power of microscopes, it is usual to take ten inches as the distance of most distinct vision, it is advantageous to have the paper at this distance from the eye, and if the paper and pencil point are not clearly seen in coincidence with the image of the object, when the paper is thus placed, a lens is brought round under the prism, to produce the desired effect.

When Amici's camera (p. 18) is used the operator, looking in a horizontal direction, views the object directly, and the paper and pencil by reflection.

There is a knack in the use of either form of the instrument only to be acquired by practice, and the microscopist must not be discouraged by the failure of his first attempts, but persevere, until he can avail himself thoroughly of the great advantages to be derived from the use of this instrument.

The Stage Micrometer.—The stage micrometer is a glass slider having upon it, in the usual place for the object, lines ruled at the distances of 100ths and 1000ths of an inch. Being placed upon the stage, and having its image drawn on a sheet of paper at a distance of ten inches from the eye, the magnifying power of the microscope is determined by comparing the sketch thus made with an ordinary scale ruled into inches and tenths of an inch. Thus, if three divisions, at 1000th of an inch apart, occupy, when marked on the paper, a distance of $\frac{6}{10}$ of an inch, the magnifying power would be 200. Again, the micrometer being replaced on the stage by any object, and the image of this object being thrown by the camera lucida on the same paper on which the enlarged view of the micrometer has been delineated, the actual size of the object will be determined.

The Eye-piece Micrometer.—The most accurate means of measuring the dimension of an object is afforded by the eye-piece micrometer. Any errors in the accuracy of the divisions of the stage micrometer are multiplied by the whole power of the microscope; but in the eye-piece micrometer they are multiplied by the magnifying power of the eye-glass only.

The eye-piece micrometer was at first placed by Mr. Andrew Ross in the focus of the positive eye-piece; but, more recently, it has been adapted to the Huyghenian eye-piece. In its most perfect form, denominated the cobweb micrometer, it consists of two parallel threads passing across the diaphragm of the eye-piece, one of which is fixed, and the other is moved across the field by turning a screw with a circular head. The distance between the threads of the screw is 100th of an inch, and the rim of the circular head of the screw is divided into 100 equal parts, and, when turned, passes by a fixed index, so that a motion of the interval between two adjacent divisions, past this index, corresponds to a variation in the distance between the parallel cobweb threads of the 10,000th of an inch. A portion of the aperture of the diaphragm is cut off by a comb at right angles to the thread, the distances between the teeth of the comb being the same as those between the threads of the screw. To measure an object by this micrometer, one of its boundaries must be brought into contact with the fixed thread, and then, by turning the circular head, the other thread is to be moved until it comes into contact with the opposite boundary; the number of teeth of the comb that have been passed by will then show the complete number of turns of the screw, and the number on the circular head opposite the index will show the number of 100ths of a turn, that have been required to pass the thread across the whole breadth of the object. The value to be attached to the indications of this instrument must be determined by means of the stage micrometer; thus, if three complete turns and five 100ths of a turn be required to move the thread over the magnified image of the space between two adjacent lines on the stage micrometer, ruled at 1000th of an inch apart, a complete turn of the screw will indicate an actual dimension in the object of $\frac{1}{3050}$ of an inch, and a motion of one of the divisions on the circular head past the index, will denote a distance of $\frac{1}{305000}$ of an inch.

This form of micrometer is necessarily expensive, and a much cheaper form is Jackson's eye-piece micrometer, which consists of a thin plate of glass, ruled with lines like the stage micrometer, but set in a frame, so that when placed in the eye-piece in

the focus of the eye-glass, it can be moved across the field of view by turning a screw with a milled head, against which it is held by a spring. Every tenth line is ruled longer than the others, and one of these being brought into contact with one boundary of the object, the number of divisions between this and the opposite boundary gives the breadth of the object. When the breadth of the object does not occupy an exact number of divisions, a mode of estimating, with a great degree of accuracy, the parts of a division, beyond an exact number, contained in the breadth, is afforded by a line ruled across the divisions, and thus forming a diagonal scale (vol. i. p. 24).

The actual value of the divisions must be determined, as in the cobweb micrometer, by means of the stage micrometer; and in both cases it is advisable to eliminate the errors of the latter by taking the mean of several measurements on different parts of it.

Stage Forceps.—The stage forceps serve to hold objects, and

to turn them about in various directions. They are closed by a spring, and opened by pressing against a pin. They are attached to a short arm, which fits on a pin at some part of the stage, or object-holder; and by pushing the stem *b*, through the ring *a*, at the end of the arm, and turning it round in the ring by the milled head at the end of the stem, while the entire instrument is revolved upon the pin on which it fits, the object can be brought into any desired position whatever.

A very convenient form of mounting for a condensing lens, to illustrate opaque objects, is also represented in this figure.

The Live Box.—In the live box, B, the object is confined between the glass, *a*, of the lower part, and that of the cap, C: the distance between them can be varied by sliding the latter more or less on. As the thin glass is only dropped into a slight recess in the top of the cap, and held there by the heads of two screws, it can easily be taken out for wiping, or replaced by another when broken.

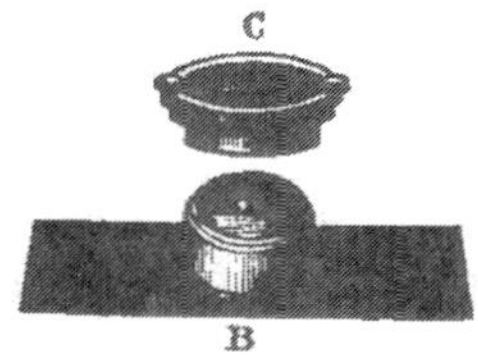

The Glass Trough.—This trough, A, is used for exhibiting objects in water, and must be adjusted with its thin plate of glass, *b*, in front. The object may be confined by placing a piece of glass, *e*, diagonally in the trough, its lower edge being kept in its place by a strip, *d*, at the bottom; then, if the object introduced be heavier than water, it will sink until stopped by the sloping plate. To regulate more particularly the position of the object, a very slight spring, *f*, may be applied behind the diagonal plate, and an ivory wedge, *g*, introduced in front of it, and pushed down to any desired extent.

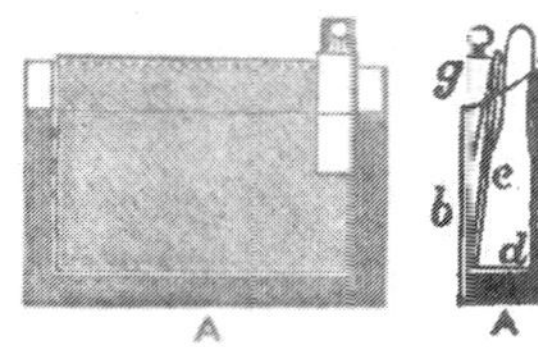

USE AND TESTING OF THE MICROSCOPE.

The Light.—The best position for the light is on the left of the observer, leaving his right-hand free to attend to the adjustments of the instrument without any danger of intercepting the illumination.

If daylight be made use of, that proceeding from a window whose aspect is away from the position of the sun should be selected; so that reflected, and not direct, solar rays should be made use of. The best light of all is that reflected in this manner from a white cloud. For artificial light, that from a wax or Palmer's candle may be used, if protected by a glass chimney to keep it steady; or a good oil or paraffin lamp, or an Argand gas-burner. If any of these artificial sources of illumination be used, they should be at a distance of between ten and twelve inches from the microscope, and means should be provided for raising and lowering the light, and for shading it from the eye.

Care of the Eyes.—In using a monocular instrument, the unemployed eye should be kept open, and, though at first some confusion is caused by the two distinct views thus presented to the two eyes, the power is soon acquired of regarding mentally only the view of the microscopic image. The constant use of the same eye is also injurious, and the young microscopist should therefore accustom himself to use either eye with equal facility, and employ them alternately. No expedients, however, with the monocular instrument can enable it to be used with the same amount of comfort and freedom from fatigue as the binocular, which is now, therefore, almost universally employed for general purposes; and in particular cases, when the whole of the light is required to be concentrated on a single image, the binocular, as now constructed, is converted into a monocular instrument by the simple withdrawal of the binocular prism or prisms.

Opaque Objects Uncovered.—In viewing opaque objects illuminated by light thrown upon them from one side, whether by a condenser or reflector, the light necessarily falls upon them very obliquely, and, consequently, if they were covered with glass, a great part of the light would be reflected from the surface of the glass. Such objects, therefore, should always be uncovered, when presented for examination.

The terms lower powers and higher powers have been on several occasions employed by us in speaking of object-pieces; and we shall need them again in discussing the properties of a good microscope, and the modes of testing them. It will, therefore, be advisable to state that by lower powers we mean object-pieces having focal lengths not less than half an inch; by medium powers, objectives with focal lengths less than half an inch, and exceeding one-sixth of an inch; and by higher powers, objectives having focal lengths of one-sixth of an inch and less.

Quality of Object-pieces.—There are four attributes to be considered in judging of the character of an object-piece:—(1.) The Defining Power; (2.) The Penetrating Power; (3.) The Resolving Power; (4.) Flatness of Field.

1. *Defining Power.*—This consists in showing well marked the principal features of the object, and especially its boundaries. This power depends upon the completeness of the corrections, and is an essential quality of every good objective. It is somewhat impaired by extending the aperture. It is to be tested by using a deep eye-piece, and still further by pulling out the draw-tube.

2. *Penetrating Power.*—This consists in the power of seeing clearly not merely the points of the object which are exactly in focus, but also points somewhat in advance of the focal

distance, or retired beyond it, and is essential to the examination of any object having inequalities of surface, or markings that penetrate to different depths. The penetrating power is inversely proportionate to the angle of aperture, and, consequently, in those object-pieces that are designed for a view of an object in its entirety, the angular aperture must not be too far increased. The disadvantage arising from excessive aperture, and the consequent want of penetration, is more especially felt when using the stereoscopic binocular instrument. Not only is the distinct view limited to a small depth, but the exaggeration of the stereoscopic effects (p. 33) by a large aperture must also be considered an imperfection.

3. *Resolving Power.*—This consists in the power of separating, and distinguishing, the finest and most delicate details of an object. It is only to be increased by increasing the angle of aperture, and is, consequently, inconsistent with great penetrating power, which is inversely proportionate to the aperture. The large apertures conferring this power should, therefore, only be bestowed on the higher powers, for which the task of resolving with the highest degree of perfection, may be reserved, while to the lower powers is allotted more especially that of correct definition and penetration.

4. *Flatness of Field.*—This quality, again, is not so important with the higher powers, intended to view the details of an object in succession, as with the lower powers, used for forming a correct view of its entire form and structure. It practically depends to a considerable degree on the extent of the field of view; and, consequently, the comparison of this quality in different object-pieces must be made with the same eye-piece.

For testing the completeness of the corrections, and the perfection of the centerings, a minute globule of mercury, called an artificial star, is made use of. Minute globules of mercury are spread on a black cloth, and the light from a window, reflected from one of them, is brought into focus in the centre of the field of view. The image thus formed should appear as a very small disc, free from colour, with the exception of a very pale green tint. Then, varying the position of the eye-piece, by drawing out and pushing in the draw-tube, the object should present the same appearances, at the same distances within and beyond the correct focus, and the disc in either case should enlarge concentrically.

A good test for defining power is afforded by the pollen grains of the hollyhock, viewed as an opaque object, and showing minute spines. A deep eye-piece should be used, which will exaggerate want of definition.

For resolving power, one of the best test-objects is afforded by the scales of the *Podura plumbea*, or skiptail, a small wingless insect, the size of a flea, found in damp cellars. These scales should appear covered with a great number of delicate marks, like notes of admiration.

For *testing flatness of field*, good objects are sections of wood, viewed with eye-pieces to give a field of from nine to twelve inches.

Care of the Instrument.—In using the higher powers, in which the object-piece is very close to the stage, great care should be taken not to bring it in contact with the object; and, when either the object, or the object-piece is changed, the body should be moved away from over the stage, in those instruments admitting of this arrangement, or should be raised well above the stage by means of the rack and pinion for quick adjustment.

The glasses of the eye-piece may be unscrewed and wiped, when necessary, with a piece of soft chamois leather, from which the dust has been well beaten out, and the outside surfaces of the objective may be also wiped in the same manner; but the objectives of English manufacture, when requiring to be unscrewed, must be sent to the manufacturer, who alone will be able to reset them without impairing their perfection. When only a little dust is on any of the glasses, it is best to remove it by a smart puff of the breath, and then to evaporate the moisture, deposited on the surface, by moving it briskly, backwards and forwards, through the atmosphere, as wiping, even with the softest material, will in time impair the high polish to which the surfaces are brought.

CHAPTER VI.

TELESCOPES.

REFRACTING TELESCOPES.

The Refracting Telescope consists of a convex object-glass, which forms an image of a distant object, and an eye-piece of one or more lenses, which performs the office of a microscope for viewing this image. The most simple form of the telescope is that called the *astronomical telescope*, and consists of

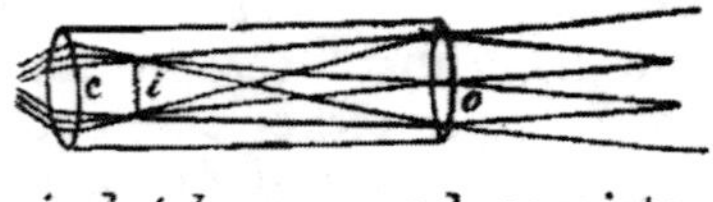

two convex lenses, the object-glass *o*, of as great focal length, and, consequently, low magnifying power, as the size of the telescope will permit, and the eye-glass *e*, of small focal length and, consequently, high magnifying power. When arranged for distinct vision of a distant object, the distance between the two lenses is equal to the sum of their focal lengths: an inverted image *i*, of the object is consequently formed in the common focus of the two lenses, and the pencils proceeding from the image consist, after refraction at the eye-glass, of parallel rays, which are the most favourable for distinct vision.

The magnifying power of this instrument is represented by the ratio of the focal length of the object-glass to that of the eye-glass, and may therefore be increased either by increasing the focal length of the object-glass, or by diminishing that of the eye-glass. The latter means, however, cannot be resorted to without increasing both the chromatic dispersion, and the spherical aberration. Hence, before the means were discovered of forming achromatic and aplanatic object-glasses, the only unobjectionable way of increasing the power of the telescope was by increasing the focal length of the object-glass, and astronomers used to attach the object-glass to the end of a long pole. This contrivance was called an aërial telescope. Huyghens used one of 123 feet in length, and Cassini one of 150 feet.

That the field of view should be as bright as possible, the image of the object-glass, formed by the eye-glass at the place of the eye, should not be larger than the pupil of the eye; and the brightness will then vary directly as the square of the diameter of the object-glass, and inversely as the square of the magnifying power. The brightness is also diminished by passing through the refracting media; and hence it is always an object to employ as few lenses as possible, consistently with the attainment of the other requisites of a good telescope.

Refracting telescopes for astronomical purposes are now constructed with achromatic object-glasses, and either Huyghenian or Ramsden's eye-pieces; these eye-pieces, however, when applied to telescopes, are called *celestial* eye-pieces.

The Huyghenian eye-piece is to be preferred, when we are only seeking to obtain the best defined and most distinct view of an object, and is the best eye-piece for all reflecting telescopes; but when it is necessary to place cross-wires or spider-lines at the place of the image in the field of view, for the purpose of accurately measuring the position of an object, at the time of observation, or to apply an apparatus, called a micrometer, for measuring the dimensions of an image, Ramsden's positive eye-piece is employed. With either of these eye-pieces the object appears inverted, which is no inconvenience when this object is one of the heavenly bodies; it is for this reason they are called celestial eye-pieces.

For the convenient observation of stars near the zenith, a plane reflector or prism is placed in the eye-piece, by which the directions of the pencils are turned, so that the axis of the eye-lens is at right angles to the axis of the instrument. Such an eye-piece is called a *diagonal eye-piece.*

When terrestrial objects are to be viewed, it is generally necessary that they should appear erect, for which purpose the inverted image formed by the object-glass must be again inverted by the eye-piece. The *terrestrial* or *erect* eye-piece, used for this purpose, is, in fact, a compound microscope, consisting of an object-lens, *l l*, a diaphragm, *d d*, an amplifying lens, *m m*, a field-lens, *f f*, and eye-lens, *e e*, the two latter forming either a negative or positive eye-piece, having the usual diaphragm at the place of the formation of the erect image. The pencils of rays from the inverted image of a

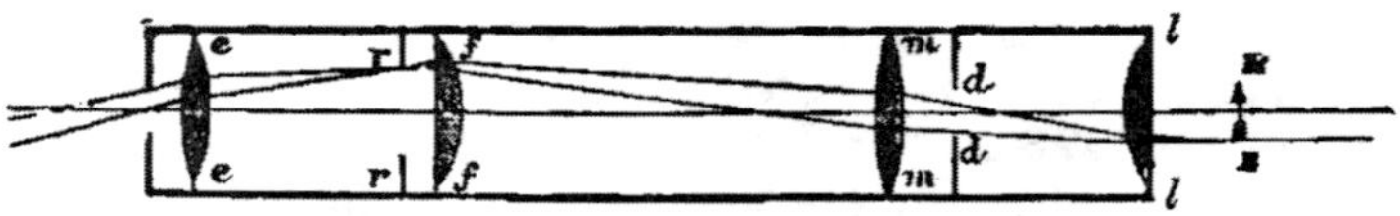

distant object formed by the object-glass of the telescope at E F, pass first through the object-lens, *l l*, of the eye-piece, crossing the axis of the instrument near the diaphragm *d d*, and thence passing through the amplifying-lens *m m*, form an image at *r r*, a second time reversed, and therefore erect, either before or after passing the field-lens *f f*, as the eye-piece is positive or negative; and this erect image is viewed, of magnified dimensions, through the eye-lens *e e*. An inspection of the figure in which the passage of a pencil from one extremity of the inverted image E F, to its exit from the eye-lens, is indicated, will show how the diaphragms, *d d*, and *r r*, cut off the extreme rays, which would not be sufficiently freed from spherical and chromatic aberration.

In consequence of the loss of light consequent upon this construction, portable telescopes with celestial eye-pieces were formerly much used by navigators for descrying objects at night, and these telescopes were, consequently, called night-glasses. They are now almost superseded by the great improvements which have been made in the binocular Galilean telescope, or double opera-glass.

By substituting for the convex eye-lens of the astronomical telescope a concave eye-lens of the same focal length, a simple telescope is formed with only two lenses, which shows objects erect. This is called the *Galilean telescope*, and is the construction used for opera glasses. When arranged for distinct vision of a distant object, the object-glass and eye-lens are separated by a distance equal to the difference of their focal lengths. The pencils of light proceeding from the object, after refraction at the object-glass *o*, tend to form an image of the object in the common focus of the two lenses; but, being intercepted by the concave eye-lens *e*, their rays are rendered parallel, and, consequently, adapted to produce distinct vision to an eye placed behind this lens.

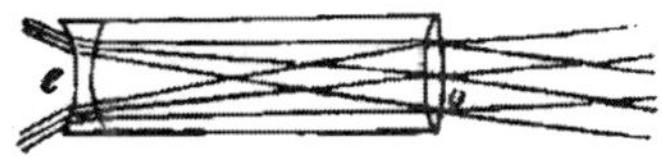

The magnifying power, as in the astronomical telescope, is represented by the ratio of the focal length of the object-glass to that of the eye-lens.

Great attention has been paid to this class of telescope during the last twenty or thirty years. They are made of the binocular form; that is, two telescopes combined together, one for each eye, the focus being adjusted by a screw and rack, which moves the eye-glasses of each telescope simultaneously. The object-glasses and eye-glasses are both compound lenses, forming achromatic combinations; and when made of the best quality, and larger in size than the ladies' opera glass, they are denominated race-glasses, or field-glasses, or rifle-glasses, according to the fancy of the manufacturer.

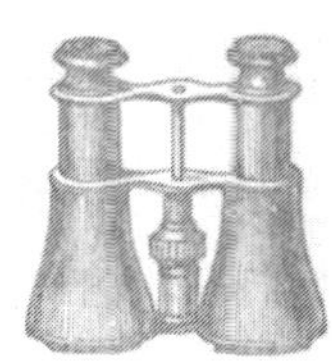

Messrs. Parkes and Son, of Birmingham, have two very fine glasses of this kind. One of them is fitted with twelve glasses, that is, each of the eye-glasses and object-glasses is composed of three pieces, forming a very perfect achromatic combination; and from this circumstance, and the beautiful quality of the crystal from which they are formed, I have seen nothing to surpass them in brightness, clearness, and distinctness of definition. The other is a Patent Tourist's Binocular, with three different magnifying powers; in fact, there are three eye-glasses in each telescope, which may be used successively by turning a milled head. The first adjustment contains a pair of tinted crystal lenses, and gives a power of $4\frac{1}{2}$ diameters; the second gives a power of about

6½ diameters; and the third gives a power of about 9½ diameters, which is the highest practicable power attainable. The object-glasses are triple, and 2⅛ inches in diameter; and the eye is protected from the irritation and fatigue produced by heat, by a pair of accurately-ground pebbles fitted in front of the eye-glasses. This is altogether a very superior instrument, and in dull or misty weather is as powerful a telescope as can be used with advantage for the observation of terrestrial objects.

In clear bright weather telescopes may be used having magnifying powers of from 10 to 40 diameters, the highest powers being only attainable when a large object-glass is used; but it must be borne in mind that whatever magnifies terrestrial objects, will also magnify the mist or other imperfections of the terrestrial atmosphere, through which they are seen. On the other hand, the Binocular Galilean Telescope enables a person to use both eyes in viewing objects, and, by transmitting to them a much greater amount of light from these objects, than they would otherwise receive, renders the objects brighter, clearer, and more distinct. The light from the objects is in fact caught upon the comparatively large object-glasses of the binocular, and transmitted to the eye, instead of the much smaller amount of light which would otherwise be caught by the pupil of the eye itself. The Binocular Galilean, as has been already stated above, is found to make the best night-glass, for viewing terrestrial objects by moonlight or in the twilight, from the very large amount of light that they transmit to the eyes.

For viewing terrestrial objects in bright weather, telescopes with erect eye-pieces are most advantageously used with powers varying generally from 15 to 25 diameters. Higher powers than this are not advantageous, unless the object-glasses be of extra size, or the weather unusually clear and bright. Telescopes are, however, constructed with eye-pieces of variable powers, so that the highest can be used upon objects that are brightly illuminated, and the lower powers in dull weather or for objects in shadow. Such eye-pieces are called *pancratic*.

It has already been stated that the erect eye-piece is in fact a compound microscope, consisting of an object-piece formed with two lenses and a diaphragm, and an eye-piece formed also of two lenses, and also generally containing a diaphragm. Now the object-piece and eye-piece are usually fitted into the opposite ends of a single tube, which slides in the tube carrying the object-glass of the telescope, and is pushed in or out to adjust the focus for distinct vision. In this form the magnifying power of the telescope is absolutely fixed; but if the eye-piece be fixed in one draw-tube, and the object-piece in another, so that the distance between them can be varied, while the focus can be re-adjusted by moving both together nearer to or further from the image formed by the object-glass, we have the *pancratic* eye-piece, which is analogous to the microscope with movable eye-piece or object-piece, in which the power can be considerably varied. The advantage of this arrangement is that a glass can be constructed, whose lowest powers shall be good for all ordinary occasions, while on favourable occasions a power may be used, which could only otherwise be obtained by telescopes with larger object-glasses, and of a more expensive character.

Many improvements have been made in the construction of terrestrial eye-pieces, the points aimed at being increase of extent, and luminosity, of the field, and more perfect achromatism. Several glasses have been forwarded to us, which for their size and price have attained these requisites to a remarkable degree, and our attention has been particularly called to a new achromatic eye-piece,* which, with an increase of magnifying power, gives also, what is unusual with such increase, a *more luminous and extended* field, combined with a remarkably clear and sharp definition. We have tested it, by trying it, and a good eye-piece of the ordinary construction, with the same telescope-body and object-glass.

REFLECTING TELESCOPES.

Since the discovery of the methods of forming achromatic and aplanatic object-glasses, the magnitude and available magnifying powers of refracting telescopes are theoretically unlimited; but the difficulty of procuring flint glass of even texture and free from flaws, in pieces of any considerable magnitude, has hitherto practically placed a limit upon the magnitude and available power of refracting telescopes. By the substitution, however, of reflectors, which are always achromatic, for the object-glasses, telescopes of colossal magnitude have been most successfully constructed. Of reflecting telescopes there are four kinds—the Newtonian, the Gregorian, the Cassegrainian, and the Herschelian.

* Forwarded by Messrs. Parkes.

The Newtonian Telescope consists of a concave object-speculum, *s*, a reflecting prism, or plane reflector, *m*, making

an angle of 45° with the axis of the telescope, placed between the object-speculum and its focus, and an eye-piece. The pencils of light proceeding from a distant object tend to form an image after reflection at the object-speculum, but are bent by the plane reflector, or prism, so that the image is formed at *i*, on the axis of the eye-piece, and in the focus of the eye-lens.

Nasmyth's Telescope is a combination of the Newtonian telescope and of the Cassegrainian telescope described below. The rays reflected from the concave object-speculum are received either upon a small speculum or prism placed in the axis of the tube, between the object-speculum and its focus. By this they are turned at right angles, and form an image in a tube inserted in one of the trunnions upon which the instrument turns. Into this tube also the eye-piece is inserted. The advantage derived from this arrangement is, that while the great tube is moved round in a complete circle, if necessary, the tube, in the trunnion, containing the eye-piece, remains fixed; and the observer can survey the whole meridian, or other circle in the heavens, without changing his position.

The Gregorian Telescope consists of a concave object-speculum, *s*, a small concave-speculum, *r*, whose focal length is short compared with that of the object-speculum, and an eye-

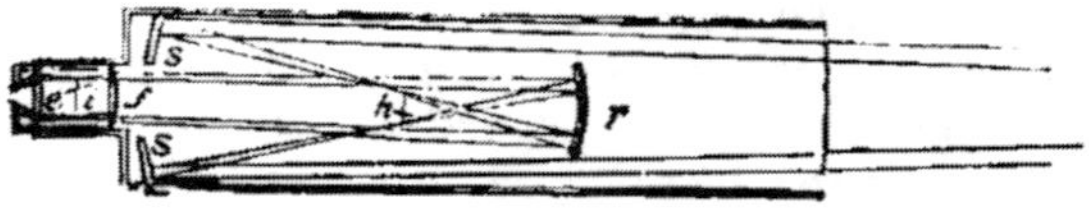

piece. The small speculum is placed so that its focus is near the focus of the object-speculum, but a little further from this speculum. The pencils of light proceeding from a distant object, after reflection at the object-speculum, form an inverted image, *h*, of the object at the focus of this speculum, and after reflection again at the small speculum form a second image, *i*, inverted with respect to the former, and, consequently, erect with respect to the object.

This telescope has, for terrestrial purposes, the advantage over the Newtonian telescope of showing objects erect, but yields to it both in the brightness and perfection of the image, because the second mirror increases the spherical aberration produced by the first, and it is extremely difficult to give the mirrors the proper curvature to remedy this evil.

The Cassegrainian Telescope consists of two specula and an eye-piece, like the Gregorian; but the second speculum is convex instead of concave, and is placed between the object-speculum and its principal focus, at a distance from this focus somewhat less than its own focal length. The pencils of light proceeding from a distant object, after reflection at the object-speculum, tend to form an inverted image of the object, but are intercepted, before doing so, by the convex speculum, and made to form the image, still inverted, in the focus of the eye-lens. Objects, therefore, are still inverted; but the spherical aberration of the convex speculum being opposite to that of the concave object-speculum, the whole spherical aberration is diminished. This telescope is also shorter than the Gregorian. It is, however, inferior to the Newtonian telescope for celestial observations, and not well adapted for terrestrial purposes on account of the inversion of the object.

When light is reflected at a mirror or speculum, there will always be a waste and dispersion; and in consequence of the two reflections, and also of the light intercepted by the plane mirror, or second speculum, the loss of light in all the reflecting telescopes hitherto described is considerable. Sir W Herschel, by a very simple contrivance, obtained what is called

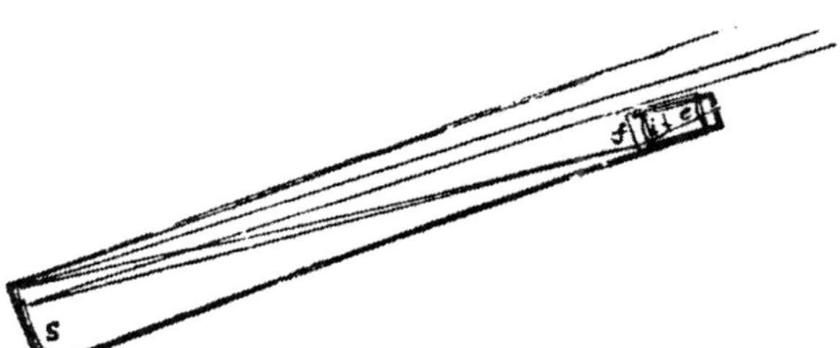

the front view; but this construction is only applicable to instruments of very large dimensions. In the Herschelian telescope the axis of the object-speculum *s* is slightly inclined to the axis of the tube, and the image *i*, being thus thrown to one side of the tube, is there viewed by the eye-piece.

ADJUSTMENT AND TESTING OF TELESCOPES.

Pouillet's Method of Determining the Magnifying Power of a Telescope.—It has been stated above that the magnifying powers of astronomical and Galilean telescopes may be deduced from the focal lengths of the lenses of which they are composed; but the results thus obtained are only approximate, and in fact the focal lengths of the lenses themselves are not always accurately known. Direct, accurate, and simple means of determining the magnifying power are therefore desirable. The following is the method adopted by Pouillet. At fifty or sixty yards distance a staff is set up, marked with white and black divisions, to which the telescope is directed: the bent tube T, containing two plane mirrors *m* and *m'*, is fitted on to the eye-piece of the telescope, and held there steadily by the screws *s*, *s*; the mirror *m* is pierced with a small hole about the twelfth of an inch in diameter;

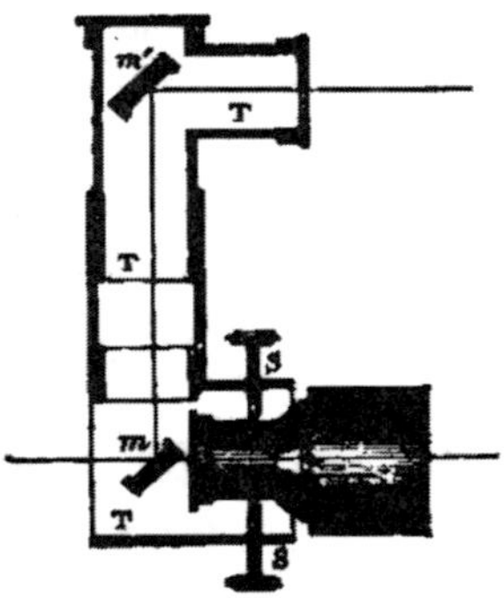

looking through this hole, the magnified image of the staff is seen in the telescope, while the staff itself is also seen by rays proceeding from it to the mirror *m'*, reflected thence to the mirror *m*, and thence to the eye. These two images are easily made to coincide with one another, and the number of the natural divisions overlying one of the magnified divisions, expresses the magnifying power of the telescope: thus, if fifty-three of the natural divisions overlie three of the magnified divisions, the number overlying one magnified division is seventeen and two-thirds, and this is an accurate measurement of the magnifying power of the telescope.

We shall now proceed to explain the best methods of adjusting and testing telescopes, as given by Pearson in his valuable work on Practical Astronomy.

Methods of Adjusting and Testing Refracting Telescopes.—Let us suppose that we have a refracting telescope of $3\frac{1}{2}$ feet focal length, and $3\frac{1}{2}$ inches aperture. Then, *to test the object-glass*, lay the tube of the telescope in a horizontal position upon some fixed support about the height of the eye, and place a printed card vertically, but for a celestial eye-piece in an inverted position, against some wall or pillar at thirty or forty yards' distance, so as to be exposed to a clear sky; then, when the telescope is directed to this object, and adjusted by the sliding tube for distinct vision, the letters on the card should appear clearly and sharply defined, without any colouration or mistiness; and, if very small points appear well defined, the object-glass may be deemed a pretty good one for viewing terrestrial objects. If the glass be intended for astronomical observations, fix at the same distance a black board, or one-half of a sheet of black paper, and a circular disc of white paper, about a quarter of an inch or less in diameter, upon the centre of the black ground; then having directed the telescope to this object, and adjusted for distinct vision, mark with a black-lead pencil the sliding eye-tube, at the end of the main tube, so that this position can always be known; and if this sliding-tube be gradually drawn out, or pushed in, while the eye beholds the disc, it will gradually enlarge and lose its colour, till its edges cease to be well defined. Now, if the enlarged misty circle is observed to be concentric with the disc itself, the object-glass is properly centered, as it has reference to the tube; but, if the misty circle goes to one side of the disc, the cell of the object-glass is not at right angles to the tube, and must have its screws removed, and its holes elongated by a rat-tailed file, small enough to enter the holes. When this has been done, replace the cell, and examine the disc a second time, and a slight stroke on the edge of the cell by a wooden mallet will show, by the alteration made in the position of the misty portion of the disc, how the adjustment is to be effected, which is known to be right when a motion in the sliding tube will make the disc enlarge in a circle concentric with the disc itself. When, then, the disc will enlarge so as to make a ring of diluted white light round its circumference, as the sliding tube holding the eye-piece is pushed in, or drawn out, the cell may be finally fixed by the screws passing through its elongated holes. When the object-glass is thus adjusted, we can proceed to ascertain whether the curves of the respective lenses composing the object-glass are well formed and suitable for each other. If a small motion of the sliding tube of about one-tenth of an inch from the point of distinct vision, in a 3½-feet telescope, will dilute the light of the disc and render the appearance confused, the figure of the object-glass is good; particularly if the same effect will take place at equal distances from the point of good vision, when the tube is alternately drawn out and pushed in. Such an object-glass is said to be aplanatic. A telescope that will admit of much motion in the sliding tube without affecting sensibly the distinctness of vision will not define an object well at any point of adjustment, and must be considered as having an imperfect object-glass in which the spherical aberration is not duly corrected. The achromatism of the object-glass is to be judged of by the absence of colouration round the enlarged disc. When an object-glass is free from imperfection both in respect of its aplanatism and achromatism, it may be considered a good glass for all terrestrial purposes.

How far an object-glass is good for astronomical observations can only be determined by actual observation of a heavenly body. When a good telescope is directed to the Moon, or to Jupiter, the achromatism may be judged of by alternately pushing in and drawing out the eye-piece from the place of distinct vision; in the former case a ring of purple will be formed round the edge; and, in the latter, a ring of light green, which is the central colour of the prismatic spectrum; for these appearances show that the extreme colours, red and violet, are corrected. Again, if one part of a lens employed have a different refractive power from another part of it, that is, if the glass, particularly flint glass, be more dense in one part than another, a star of the first or even of the second magnitude will point out the natural defect by the exhibition of an irradiation, or what opticians call a wing at one side, which no perfection of figure or adjustment will banish; and, the greater the aperture, the more liable is the evil to happen.

Another method of determining both the figure and quality of the object-glass is by first covering its centre by a circular piece of paper, as much as one-half of its diameter, and adjusting it for distinct vision of a given object, which may be the disc above mentioned, when the central rays are intercepted, and then trying if the focal length remains unaltered, when the paper is taken away, and an aperture of the same size applied, so that the extreme rays may in their turn be cut off. If the vision remains equally distinct in both cases, without any new adjustment for focal distance, the figure is good, and the spherical aberration cured; and it may be seen, by viewing a star of the first magnitude successively in both cases, whether the irradiation is produced more by the extreme or by the central parts of the glass; or, in case one half of the glass be faulty and the other good, a semicircular aperture, by being turned gradually round in trial, will detect what semicircle contains the defective portion of the glass; and, if such portion should be covered, the only inconvenience that would ensue would be the loss of so much light as is thus excluded.

The smaller a large star appears in any telescope, the better is the figure of the object-glass; but, if the image of the star be free from wings, the size of its disc is not an objection in practical observations, as it may be bisected by the small line by which the measure is to be taken. When, however, an object-glass produces radiations in a large star, it is unfit for the nicer purposes of astronomy. In testing a telescope, if a glass globe be placed at 40 yards' distance when the sun is shining, the speck of light reflected from this globe forms a good substitute for a large star, as an object to be viewed.

Whenever an object-glass is under examination, it will be proper to have the object examined by it in the centre of the field of view; and, when an object-glass is tested for astronomical purposes by the methods described above, it is necessary to employ a good negative eye-piece, which generally gives a better field of view than the positive.

If any fringes of red or yellow are observed on the edges of a white disc placed on a black ground, when the telescope is adjusted for distinct vision, and the disc carried too near the edges of the field, this species of colouration indicates that the *eye-piece* is not sufficiently free from spherical aberrations; and, if the curves of the lenses are suitable for each other, the cure is effected by an alteration in the distance between them, which must be finally adjusted by trial with a good object-glass.

Methods of Adjusting and Testing Reflecting Telescopes.— To adjust the specula of a Cassegrainian or Gregorian instrument, procure a Ramsden's eye-piece, which will render an object visible in the compound focus of the two lenses of which it is composed; then hold this eye-piece in front of the Huyghenian eye-piece of the telescope, and, by varying the distance, find the position in which the image of the large speculum is seen, well defined through both eye-pieces, and, if the image of the small speculum is seen precisely on the centre of the large one, the metals may be considered as rightly placed; but, if not, the proper screws must be used in succession, till the required position is determined. When the face of the large metal stands at right angles to the length of the tube, the adjustment may generally be finished without disturbing it; and, when the bed that receives it has once been properly finished, it will be advisable not to alter it, unless some accident should render such alteration indispensable.

To try whether the figures of the metals are adapted for each other: let the instrument be directed to some luminous point, as a white disc on a black ground, or, what is better, to a star: then having adjusted for distinct vision, firstly observe if the disc or star is well defined, and free from irradiations; secondly, carrying the small speculum short distances beyond, and short of, the place for distinct vision, examine if the disc or star enlarges alike in similar changes of position: if the result be satisfactory, the metals may be considered as well placed, and well adapted for each other.

To try whether the large speculum partake of the parabolic form, let the aperture be partially covered, first at the central part, and then round the circumference, by tin, pasteboard, or stiff paper; and if on trial the same adjustment for distinct vision be good in both these cases, and also when the speculum is all exposed, the figure may be considered good. If these effects be not produced, the instrument will be incompetent to perform several of the nicer observations in astronomy. When a mistiness appears in the field, it is a proof that the aberrations are not corrected, and that the figure of at least one of the specula is not perfect.

If a telescope is not good with its full aperture, its effect may be greatly improved, by putting a cover on the mouth, with a circular aperture, of about one-half the diameter that the tube has, in such a way that the diminished aperture may fall entirely at one side of the opening of the tube.

MOUNTING TELESCOPES.

For general purposes telescopes of moderate power are used by holding them in the hand, and they may be carried about in leather cases, fitted with straps for slinging them from the shoulders. Telescopes of moderate power are very extensively used at sea, and it is evident that from the rolling and pitching on the waves no stand could be of any use on board ship.

On land, a tripod stand forms a very simple mounting, and for telescopes of considerable power, stands with mechanical appliances for directing the telescope, may be employed. The accompanying illustration represents such a form of mounting. The handle H, is connected by a universal joint *f*, with an endless screw *e*, which acts upon a rack, attached to the pillar supporting the telescope, and turns the whole instrument round; it is raised or lowered by turning the milled heads s, s, which move in and out, one tube sliding within another. When the field of view of the telescope is small, in consequence of the high power of the telescope, a small telescope F, of lower power, is attached to it, the axes of the two telescopes being parallel. This small telescope is called the

finder, and when an object is brought into the centre of its field of view, it will also be seen in the field of view of the larger telescope.

The methods of mounting telescopes for particular purposes belong to the consideration of surveying and astronomical instruments, and will be described in Parts V. and VI., in the third volume.

CHAPTER VII.

INSTRUMENTS FOR EXHIBITING MAGNIFIED PICTURES OF OBJECTS TO AN ASSEMBLAGE OF SPECTATORS.

IN the microscope and telescope (Chapters V. and VI.) the images of objects formed by a convex lens or combination of lenses, have been confined within a tube, and viewed by an eye-piece; but by the omission of the eye-piece, and the employment of a screen to receive the image, a picture is formed which can be viewed at once by a whole assembly of spectators.

THE MAGIC LANTERN.

The magic lantern is an instrument for exhibiting on a screen an enlarged view of pictures painted on glass with transparent colours.

It has already been explained (p. 22), that when an object is placed at a distance from a lens greater than its focal length, a real image of this object is formed on the other side of the lens. When the object is at a distance from the lens equal to twice its focal length, the magnitude of the image is equal to that of the object; and as the object is moved nearer and nearer towards the principal focus, the magnitude of the image is increased more and more, to an unlimited extent. The extent to which the magnifying power may be made available is, however, limited by the amount of light which can be thrown on the object, and by the indistinctness arising from the spherical and chromatic aberration, which increases with the magnification of the image.

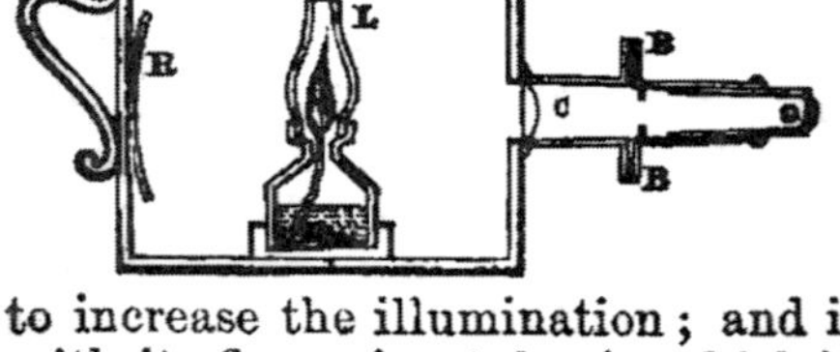

In its simplest and cheapest form the magic lantern consists of a dark lantern, with a bent chimney, to carry off the smoke and heated air with no emission of light.

Within the lantern a lamp L, with an Argand burner, is placed, to give a powerful light; behind the lamp is a reflector, R, to increase the illumination; and in front of it, and on a level with its flame, is a tube, in which is fixed a large plano-convex lens, C, to condense the light upon the picture, which slides through a groove, B B, formed upon the tube. The tube is continued beyond the groove carrying the picture, and in it is fitted a second tube, sliding in and out like the draw-tube in the body of a microscope, or the eye-tube in a telescope. In this tube is set another plano-convex lens, O, to receive the light radiated from the picture painted on the slide, and transmit it, to form an enlarged picture on a screen at an appropriate distance from the lantern. The position of the lens O, is so adjusted that the slide is just beyond its principal focus B, when the tube that carries the lens is pushed in as far as it will go. In this position the picture on the screen will be enlarged to its greatest extent, and its imperfections and want of brightness will also be the greatest; but by pulling out the draw-tube, and at the same time advancing the lantern nearer to the screen, the magnification will be diminished, while the brightness and distinctness will be increased; and after a few trials the operator will soon find the position that best suits his purpose.

The body of the lantern should be large, that it may not become inconveniently heated; the best oil should be burned; and the glass chimney of the lamp should be as tall as the lantern will permit.

The larger the picture painted on the slide, the less the magnification that will be required, to produce a picture of a given magnitude on the screen, and, consequently, the brighter and more perfect this picture will be.

There are two modes of exhibiting the pictures on the screen, and the construction of the screen must be suited to the mode intended to be adopted. One mode produces effects the most brilliant and distinct, the other the most illusory and amusing.

In the former, the lantern is placed in front of the screen, along with the spectators; and the picture is seen by the light projected from the lantern, and irregularly reflected from the screen.

The object in this case is to *avoid transparency* in the screen, as all light passing through the screen would be a loss to the picture, transmitted to the eye by reflection. The best screen in this case is made of strong white paper, pasted on canvas, and fixed on a stretching frame.

Phantasmagoria.—In the other mode of using it, the instrument is placed on the opposite side of the screen, to that occupied by the spectators. The screen in this case must be semi-transparent, and may be made of white muslin, or fine calico, stretched upon a frame, its transparency being increased by wetting it well with water.

The object of this arrangement is to produce an optical illusion, termed *phantasmagoria*, or the meeting of phantoms. The exhibitor and his apparatus being concealed from the spectators, the lantern is first placed close to the screen with the draw-tube drawn out as far as it will go, and a very small picture is thus formed on the screen. The exhibitor then moves the lantern slowly back from the screen, pushing in the draw-tube at the same time, so as to keep the focus of the projected rays upon the screen; the picture in consequence grows larger and larger; and the figures represented upon it, consequently, appear to approach the spectators, and to be actually coming into the midst of them. At length the exhibitor keeps the large and apparently near figures stationary for a brief period, and then reverses his former movements, until, the lantern being again close to the screen, the figure is apparently lost in the distance. At this moment he changes the slide without attracting the notice of the spectators, and proceeds to the production of a fresh illusion in the same manner.

Dissolving Views.—The concealment of the mode of operation is always an element of the interest produced by exhibitions of this kind, and another mode of disguising the change of slides, and, at the same time, of introducing highly entertaining effects, is that adopted in the production of dissolving views. For this purpose two similar lanterns are placed side by side, so as to throw pictures of exactly the same magnitude upon the same part of the same screen; to the front of the stand, upon which the lantern is placed, a box is attached, through which passes a bar that carries two shutters; on this bar a rack is formed, which is acted on by a pinion turned by a winch; and the shutters are so arranged that, when by one of them the nozzle of one of the lanterns is completely closed, the light from the other is entirely unobstructed, and *vice versâ.* A slide containing a picture being introduced into each of the lanterns, when the shutters are in this position, it is manifest that only one of the pictures will be exhibited on the screen. The winch being then turned slowly, the nozzle of the lantern containing this picture is gradually closed, while the nozzle of the other lantern is as gradually opened; the picture first exhibited grows fainter and fainter, and changes into the picture projected from the second lantern, which grows stronger and stronger until it is shown with the fullest brilliancy, and the picture first exhibited is entirely shut out. The slide containing this picture is then changed for another,

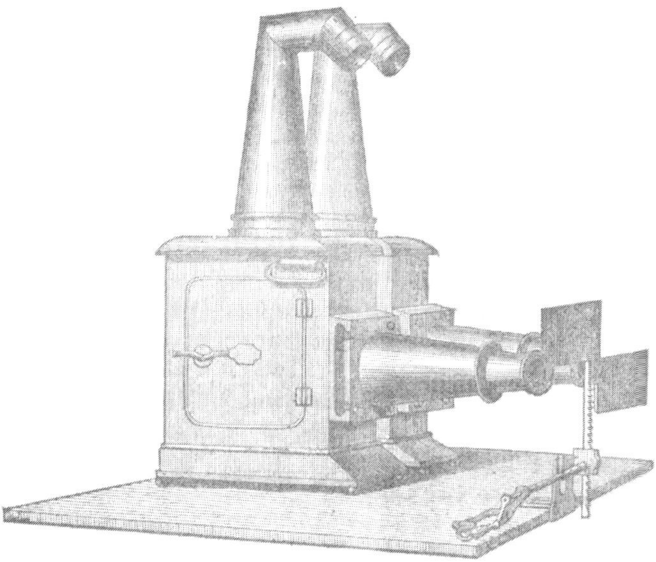

which is in like manner made to supersede that last exhibited; and in this manner a series of pictures may be made to glide from one into the other, to any extent that may be desirable.

A very pleasing effect is produced by exhibiting in this manner a summer scene changing into the same scene in its winter dress, an interior by day merging into the same brilliantly illuminated at night, &c.

The Oxy-hydrogen Light.—When it is desired to exhibit very large pictures before a public audience, the increased magnification requires a good achromatic combination of lenses, and a more intense illumination, than can be yielded by an ordinary lamp. The most brilliant artificial light that could be produced, before the discovery of the electric light, was the oxy-hydrogen, or Drummond light; and this was the light used for the illumination of dissolving views on a large scale at various exhibitions in London and elsewhere. This light is given off from a ball or cylinder of lime, heated to a white heat by the combustion of hydrogen and oxygen gases, supplied from two pipes, in the proportions in which they combine, to form water.

The Oxy-calcium Light.—The apparatus for the production of the gases, and for their application to the lantern, is expensive, and from the explosive nature of the hydrogen, its manipulation is attended with considerable danger in inexperienced hands. The oxy-calcium light is produced by forcing a jet of oxygen through the flame of a spirit-lamp, and placing a ball of lime so that the deflected flame may impinge upon it. This is a much more economical light than the oxy-hydrogen; its use is entirely unattended with danger; and the brilliancy of the light is constant and but little inferior to that of the oxy-hydrogen.

The Electric Light.—The most brilliant of all artificial lights, produced by the passage of a voltaic current from one to another of two charcoal points, has been applied to the magic lantern by M. Dubosc of Paris: and by throwing it into two lanterns, by means of two reflectors, it has been adapted to the exhibition of dissolving views.

THE SOLAR MICROSCOPE.

The solar microscope, by which effects are produced that may be reckoned among the most interesting and instructive of optical phenomena, consists of two distinct parts—a reflector and system of lenses, for strongly illuminating the object to be represented; and a combination of lenses, in which the chromatic and spherical aberrations are nicely counterbalanced, and a flat field produced, for the purpose of projecting a greatly magnified picture of the object upon a screen. It may, in short, be described as a magic lantern with the sun for an illuminator, and a compound object-piece for the more perfect projection of the image.

The Illuminating Apparatus.—A is a square copper plate, to one side of which the tube T, is attached, and on the other side is the mirror M; this plate screws on to a square board, not shown in the figure, which has a hole in the centre corresponding with the opening through the plate A, into the tube T. The mirror can be set at any inclination by turning the milled head *i*, and can be moved round into any desired position, by turning the milled head *r*, attached to a pinion that works upon a circular rack, formed on the plate A. The light reflected from the mirror M, is directed upon a large condensing lens, set in the larger end of the tube T, so as to pass through this lens in the direction of its axis: it is thus converted into a convergent pencil, proceeding along the tube T, into the smaller tube *t*, at the extremity of which it traverses a small focussing lens, and is thrown with a greater degree of convergence upon the object, placed upon a slider, which

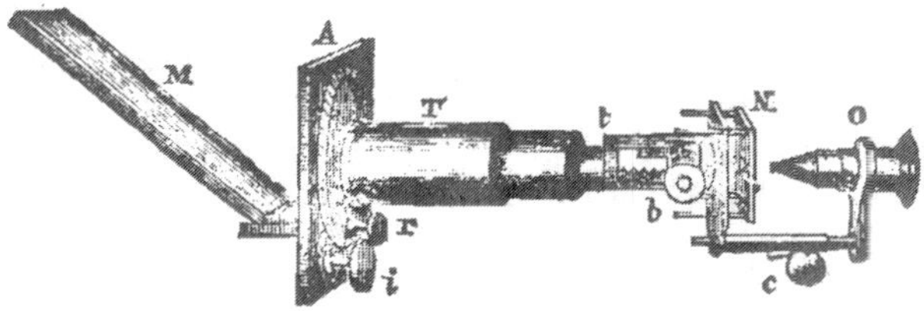

passes between two plates, N, held together by spiral springs. The tube *t* slides within the tube T, and affords the means for a rough adjustment to produce the desired degree of condensation on the object, and the milled head *b* serves to set the object with greater accuracy, so that the whole of it may be evenly and strongly illuminated.

In condensing the rays, proceeding from the sun, directly upon the object, not only is the light condensed, but the heat also; and, consequently, many objects would soon be seriously injured or destroyed. This inconvenience is to be obviated by the interposition of a medium, which is pervious to the sun's light, but impervious to its heat. Water possesses the desired property to a considerable degree, and is rendered almost completely impervious to heat by dissolving in it as much alum as it is capable of holding in solution. A cell, terminated by two parallel plates of glass about an inch asunder, and filled with a saturated solution of alum, is placed between the condensing apparatus and the slides; by this means the whole of the heat is kept back, while the light intercepted is altogether inconsiderable.

Amplifying Apparatus.—The objects for a magic lantern can be painted of a considerable size, so that the magnification required to exhibit them, is very much less than that necessary for minute microscopic objects; and again, the extreme accuracy of delineation required in the latter case, can be dispensed with in the former. A comparatively inexpensive apparatus will, therefore, suffice even for the best description of magic lantern; but for the microscope a combination of lenses, producing a flat field freed from chromatic and spherical aberrations, is indispensable.

This combination of lenses, or *compound objective*, is mounted in a brass ring *o*, so that its optical axis shall coincide with that of the illuminating apparatus; and, by turning the milled head *c*, it can be moved to or from the object, so as to adjust the focus to the distance of the screen, upon which the image is to be projected.

With the microscope the plan of showing the image through a transparent screen is very objectionable. The screen, therefore, should be quite opaque. Stout white paper pasted on canvas, or a smooth surface of plaster of Paris, will answer the purpose well.

The greater the distance at which the screen is placed, the greater the magnification of the object. It is not desirable, however, to push the amplifying power of the instrument too far; in that case the illumination becomes inconveniently faint, and the smallest amounts of uncorrected aberration perceptibly impair the result.

Method of Adjusting and Using the Instrument.—In a room of sufficient magnitude, a window must be selected, whose aspect will allow the sun's rays to be reflected in the direction of the axis of the tube. This window must be furnished with a shutter having an opening in it, as large as that in the plate and board to which the microscope is attached, and the board must be screwed to this shutter, so that the openings shall correspond, and admit a pencil of light to the large condensing lens, of a diameter fully equal to that of the lens itself. The mirror will thus be outside the shutter, while the microscope projects horizontally into the room; and in front of it the screen must now be placed in a vertical plane at right angles to the axis of the instrument. A distance of ten or twelve feet from the nozzle of the microscope will generally be found sufficient for the production of a clear, and sufficiently enlarged, image upon this screen. The room must next be rendered as dark as possible by closing the shutters, stopping up all interstices, and drawing the curtains, if the windows be provided with them.

These preliminaries having been arranged, remove the plates N, and vary the position and inclination of the mirror M, by turning the milled heads *i* and *r*, until a uniformly illuminated disc appears on the screen; then, by sliding the tube *t* in the tube T, a position will be found, in which this disc will be the most clear and distinct. Now, replace the plates N, and insert between them an object; adjust for focus by turning the milled head *c*, and for illumination by the milled head *b*, and a distinct image of the object upon a large scale will be seen upon the screen.

If the exhibition be continued for any considerable time, the position of the mirror must be occasionally altered, to suit the changed position of the sun; this is to be effected, as at first, by turning the milled heads *i* and *r*.

THE OXY-HYDROGEN AND PHOTO-ELECTRIC MICROSCOPES.

The chief inconvenience of the solar microscope is its dependence upon sunshine for illumination, so that it can only be used on fine sunshiny days, and then only during a limited number of hours, in which the sun is in a favourable position.

The invention of the *oxy-hydrogen light* supplied a means of illumination sufficiently intense for the microscope, and thus rendered it independent of the sun. The manner in which this light is produced has already been explained (p. 104); and the accompanying figure shows the manner in which the oxygen and hydrogen gases are supplied from gas bags. The microscope itself, which is here seen attached to the lantern, consists of condensing lenses, for concentrating the illumination upon the object, and a compound objective, for projecting its image on the screen, in all essential particulars identical with the same parts of the solar microscope, already described.

A still more perfect instrument has been produced by the substitution of the *electric*, for the *oxy-hydrogen* light. Great skill and ingenuity have been employed in order to counterbalance the gradual wasting of the charcoal points, and keep them by a self-regulating apparatus always at the same distance from one another, so as to ensure continuity and uniformity of illumination. These contrivances, however, belong to the domain of electricity and magnetism, and are foreign to the subject of this work: the optical instrument, the microscope, remains still identical with the solar microscope, the source only of the illumination being altèred, and not the manner of its condensation, or the objective for the projection of the image.

CHAPTER VIII.

THE CAMERA OBSCURA.

COPYING OF PLANS BY PHOTOGRAPHY.

THE *Camera Obscura* bears the same relation to the *magic lantern* and its microscopic variations, that the *telescope* bears to the *compound refracting microscope:* pencils of light, from objects more or less distant, are brought to a focus by an objective of large diameter; and the picture thus formed, instead of being viewed by an eye-piece, as in the telescope, is projected on a screen, as in the magic lantern, and solar and cognate microscopes already described.

If the distance between the objective and the screen of the camera be regulated to the focus, and the objective be directed to some object, the image of the object will be seen clearly defined on the screen in a reversed position. The image may be reinverted by the introduction of a prism, and extraneous light must be shut out from the screen, so that it receive only that proceeding from the object. If the camera be set up in the neighbourhood of a well-frequented thoroughfare, we have then an agreeable succession of distinct and vividly coloured pictures, differing from finely executed paintings only by exhibiting the actual motions of the objects viewed, men walking, horses trotting, soldiers marching, banners streaming, and foliage shaking in the breeze.

The camera obscura was always an instrument of great utility in the arts of design, since by it the process of drawing was reduced to that of mere tracing; but by the discovery of the photographic art, its use has been extended to a most unparalleled extent.*

The form of camera used for photography is most simple; but great care and expense is lavished upon the production of lenses, to give the most accurate results. The lens or lenses are set in a tube at the end of a rectangular box, in which another rectangular box slides, to produce a rough adjustment for focussing. A finer adjustment can then be attained by a slow motion, a screw and rack acting on the tube

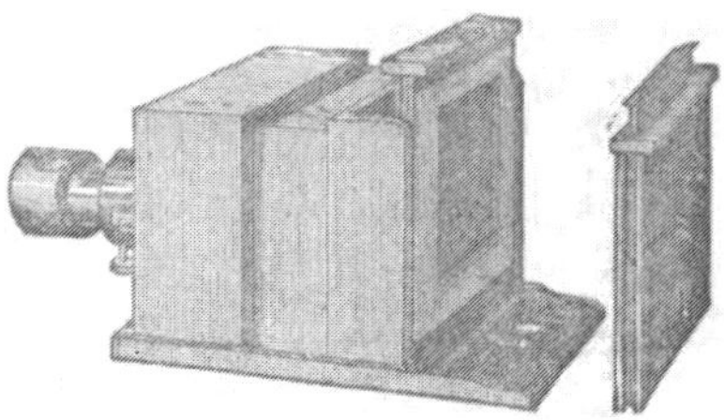

containing the lenses. The picture is first received on a plate of ground glass, inserted in grooves at the end of the rectangular sliding-box, and, after adjusting the focus, until the picture thus formed is satisfactory, the ground glass is removed, and replaced by the surface prepared to receive the photographic impression.†

The wonderful accuracy of photography in copying every minute detail, without the slightest strain on the hand or eye of the operator, gives it an immense advantage over the pantagraph or eidograph, and has led to its adoption by Sir H. James for the reduction of the maps and plans of the Ordnance Survey.

The only purpose for which the use of the pantagraph has been retained is for the reduction of plans to so small a scale, that it becomes necessary, firstly, at the sacrifice of truth,

* See also "A Popular Treatise on Photography, the Stereoscope, and Photographic Optics, from the French of D. von Monckoven." By W. H. Thornthwaite, price 1*s.* 6*d.* WEALE SERIES.

† For description of improved cameras and lenses see "A Treatise on Optics, or Light and Sight theoretically and practically treated." By E. Nugent, C.E. Second Edition, price 5*s.* Strahan and Co.

to enlarge such details as, were they drawn to scale, would be so minute as to be unintelligible, or fail to strike the eye; and secondly, to omit some details in order to avoid confusion.

We proceed to give, from a document compiled for the information of the officers entrusted with the superintendence of the Survey of India, an account of the methods employed at the Ordnance Survey Office, Southampton, for the reduction of plans by photography

APPARATUS USED FOR REDUCTION OF PLANS.

The reduction of plans from larger to smaller scales, is regulated by the law that the scale of the image formed by the lens is to the scale of the plan, as the distance of the image from the lens is to the distance of the plan from the

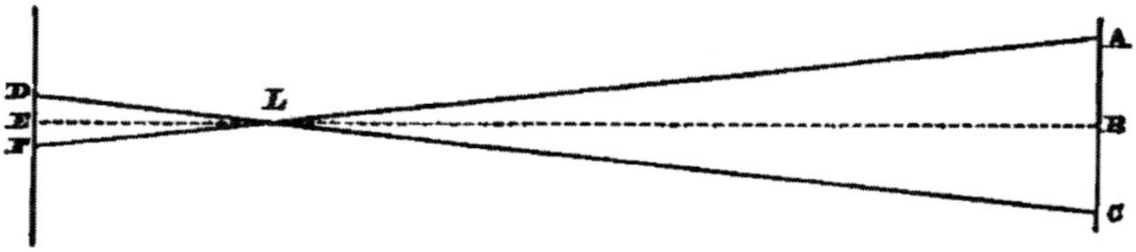

lens: that is to say, L being the lens, A C the plan, and D F the image; D F is to A C as E L to L B. It would be only necessary, therefore, when reducing plans from one scale to another, to place the lens in such a position that the distances E L and L B should be to one another as the scales.

Various circumstances modify this law as applied to the lenses actually used. Those which the photographer has principally to deal with, when working with the best achromatic lenses, are spherical aberration and curvature of the image. When reducing plans, these influencing causes are of particular importance. The usual remedy is to place a stop 4⅞ inches in front of the lens, and ¼ inch in diameter, which has the effect of diminishing the curvature to the extent of rendering it inappreciable.

The curvature increasing very rapidly with the obliquity of the pencils, there is a limit to the size of the plan which can be reduced from one scale to another without sensible distortion; and there is also a limit to the size of the image which can be taken free from sensible distortion. The largest plans reduced measure 38·016 inches by 25·344 inches, and no negative is taken of a greater size than 9 inches by 6 inches. When the reduction is not so great in degree as to allow of the whole plan coming within these limits, it is subdivided by pencil lines into rectangles, such that, when reduced, they shall occupy the area of 9 inches by 6 inches on the negative, and each rectangle is photographed separately, and the prints mounted together afterwards.

In practice the operation resolves itself into adjusting the camera so that the image, as seen on the ground glass, shall occupy the area which the plate to be photographed should, when reduced to the smaller scale. In order that this may be done, a rectangle is ruled on the ground glass of the camera, representing the area the plans should occupy when reduced to the smaller scale, and the plan and camera are so adjusted that the margin lines of the image of the plan lie exactly on the lines ruled on the ground glass.

As each plan successively photographed cannot well be fixed on the board in exactly the same place, and as the reductions are required to be made in different proportions, it is necessary to provide mechanical contrivances, to enable the operator to adjust each plan to be photographed both rapidly and accurately.

The following are the arrangements, which are found to answer remarkably well:—

The camera is made of mahogany, strongly bound with brass, and is provided with an endless screw to adjust the focal distance. The construction is shown in the Figs. 1, 2, and 3.

The lens is an achromatic meniscus, or concavo-convex, 3½ inches in diameter, 25 inches principal focal length, with a rack and pinion for adjusting the focus.

The stand on which the camera is supported is of simple construction. It is necessary that it should possess the means of giving to the camera a motion through small arcs of altitude and azimuth, and a sliding rectilineal motion in a direction at right angles to the optical axis of the lens. The construction is such as to admit of these movements. (See Figs. 1, 2, and 3.)

The top of the stand consists of two separate portions; the lower, a strong board working on a transverse axis, which rests on the frame of the stand, and is clamped at any required altitude by four screws, biting four hinged plates, which connect the top of the stand with the frame. The amount of motion is regulated by an elevating screw.

The upper board on which the camera is placed is connected

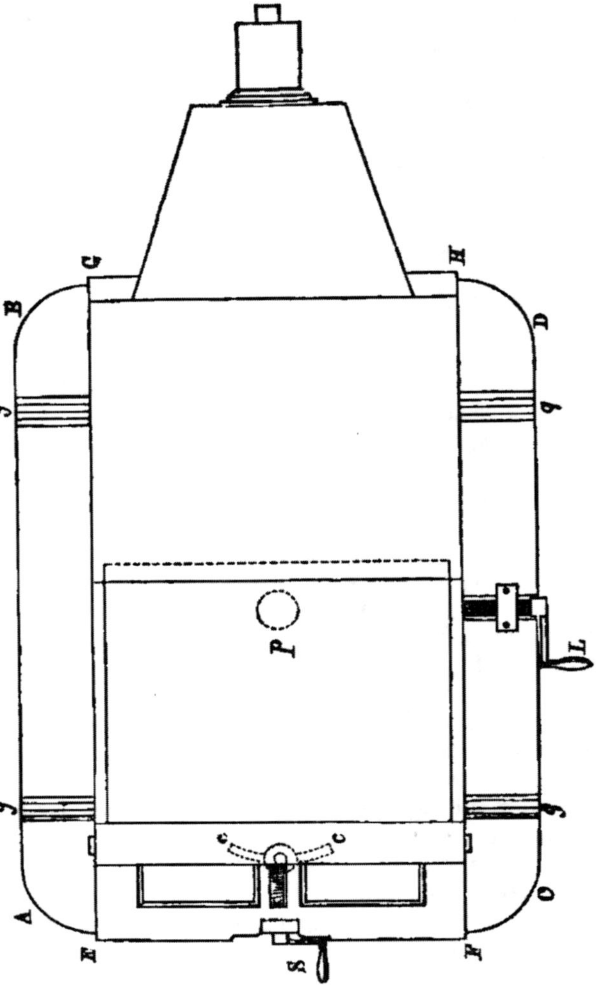

Fig. 1.—Plan of camera and stand. (Scale, 1 inch to a foot.)

with the lower one by a central pin, which passes through the top, and works in a socket in the lower board. (See Fig. 2.)

Through the latter an arc of a circle is cut, and a screw

passing upwards through the arc, and fastened to the lower surface of the upper board, gives the means of clamping the top board in any portion of the azimuthal arc it describes on

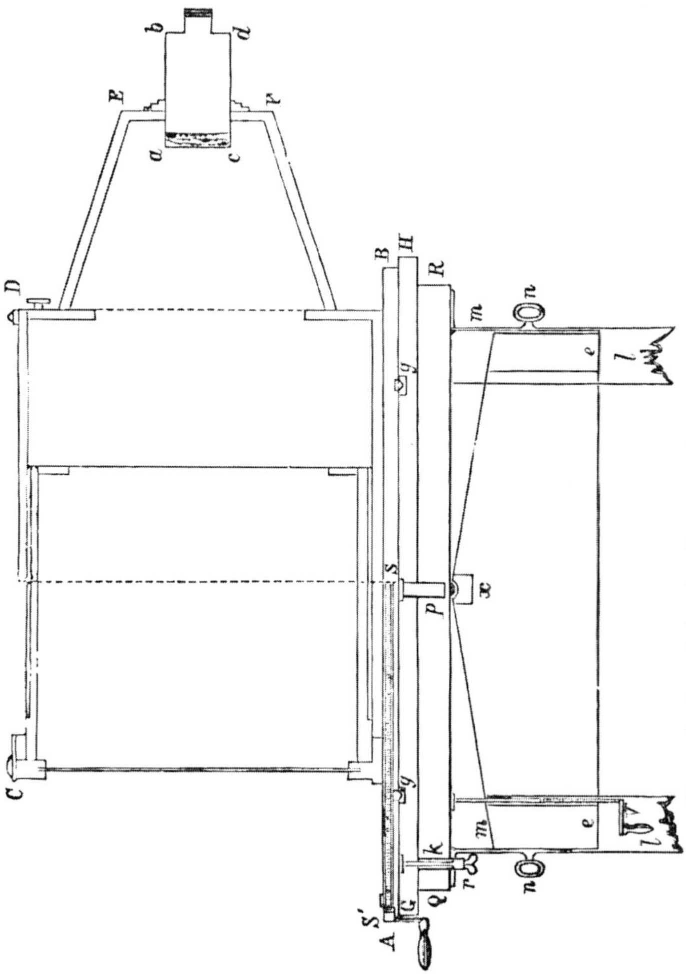

Fig. 2.—Section and elevation of camera and stand. (Scale, 1 inch to a foot.)

the centre pin, its motion being of course limited by the length of the arc cut in the lower board.

V-shaped grooved plates of brass are let transversely into the top surface of the stand, and to the bottom of the camera are screwed similarly shaped brass runners; so that when the runners are fitted into the grooves, the camera lies symmetrically on the stand, and can be made to slide with an easy motion in a transverse direction, the motion being given by a

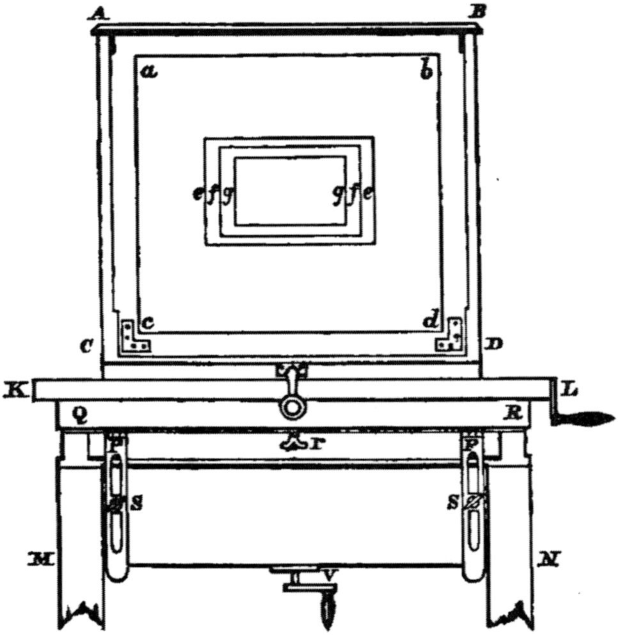

Fig. 3.—Elevation of camera and portion of stand. (Scale, 1 inch to a foot.)

male screw likewise let into the surface of the stand, and working in a small female-screw socket attached to the bottom of the camera.

Each of the legs of the stand has a groove at the bottom, to fit the convex surface of the rails on which it slides.

EXPLANATION OF THE PRECEDING FIGURES

Fig. 1.

A B C D Top of camera stand.
E F G H Camera.
g g g g Brass V grooves.
p Centre pin, on which top of stand turns.
c c Circular groove through lower portion of stand.
L Screw for giving lateral motion to camera.

Fig. 2.

A B C D E F The camera.
G H Q R Top of stand.

G H Upper board.
a b c d Lens and mounting.
g g V shaped runners, working in V grooves.
k Circular arc cut through lower board.
l l Portion of legs of stand.
m m Hinged plates.
n n Screws for clamping plates.
p Centre pin, on which upper board works.
r Screw passing through circular arc, clamping upper to lower board.
s s Endless screw for focussing.
v Elevating screw for giving top of stand motion on axis *x*.
x Axis of top of stand.

Fig. 3.

A B C D Movable slide.
a b c d Ground glass plate.
e f g Rectangles ruled on glass, and to be made to coincide with margin line of plan according to nature of reduction.
K L M N Stand for camera.
P P Plates to allow motion in altitude.
S S Clamping screws for do.
Q R Lower board.
r Screw passing through circular arc, clamping upper to lower board.
V Screw for giving top of stand motion in altitude on axis.

The stand on which the plan-board is fixed is shown in Figs 4, 5, 6, 7, and 8.

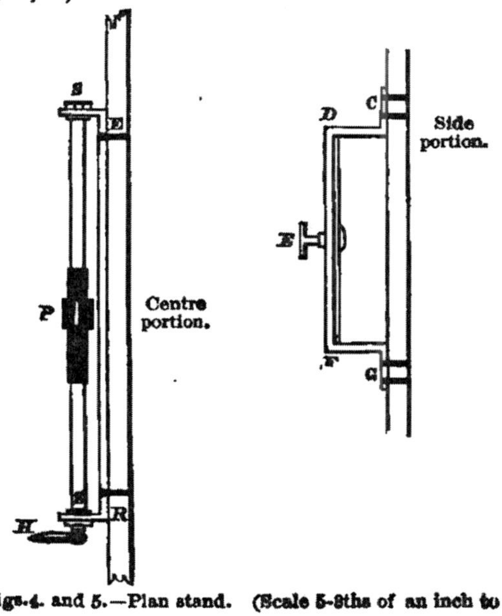

Figs. 4. and 5.—Plan stand. (Scale 5-8ths of an inch to a foot.)

It consists of a centre and two side pieces. The centre

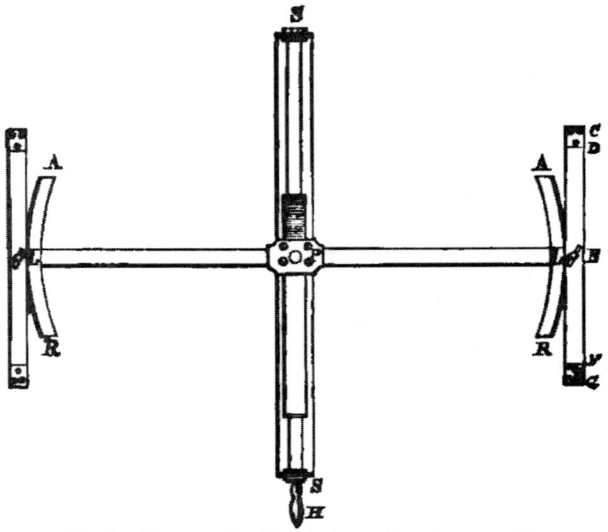

Fig. 6.—Plan-stand. (Front view. Board removed.)

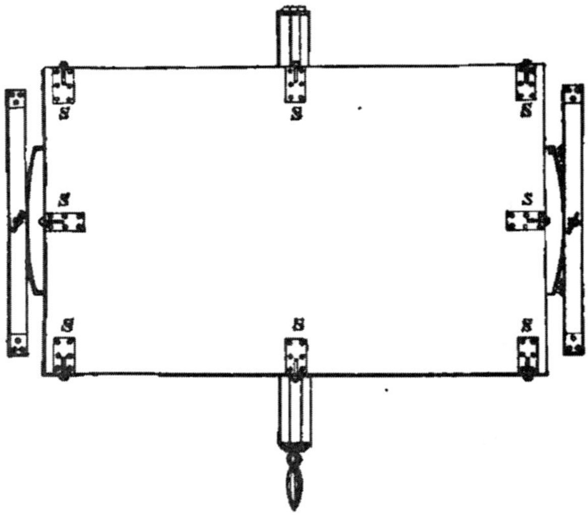

Fig. 7.—Plan-stand and board. (Front view.)

portion, Fig. 4, is composed of a frame, E R, screwed to the partition with projecting sockets, in which the screw S S, works, turned by the handle H. A centre pin, P, is attached

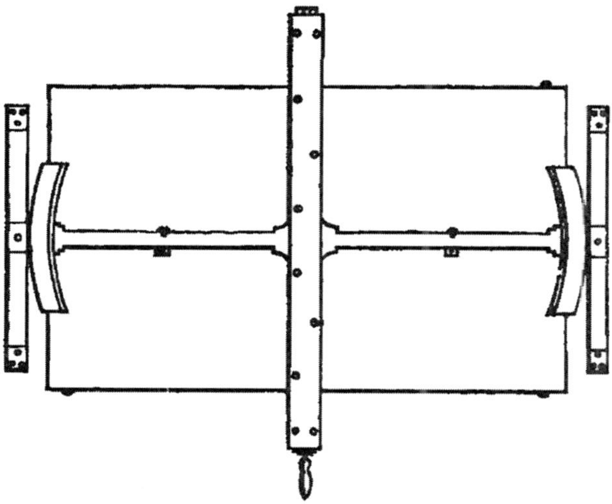

Fig. 8.—Plan-stand and board. (Back view. Partition removed. Scale, ⅛ of an inch to a foot.)

to the screw S S, by a female-screw socket. A transverse bar, L L, Fig. 6, is connected with the screw by this pin, on which it revolves; and at the extremity of the bar are two arcs of circles, thinned at the outer edge.

Each of the side portions consists of a bent iron frame, C D F G, Fig. 5, screwed to the partition at C and G. The portion D F of the frame is double; the interior thin piece is pinned to the exterior only at the ends, and can, by reason of its flexibility, be separated at the centre by relaxing the screw E.

The thinned portion of the circular arcs A B, Fig. 6, being, when the stand is set up, between the exterior and interior piece of the side-frames, the arm L L can be clamped in any part of the arc it describes on the pin P, within a certain limit, by tightening the screw E.

Figs. 7 and 8 show the front and the back of the stand, with the plan-board attached.

The board, Fig. 7, is provided with spring catches, S, by which the plans are attached to the board.

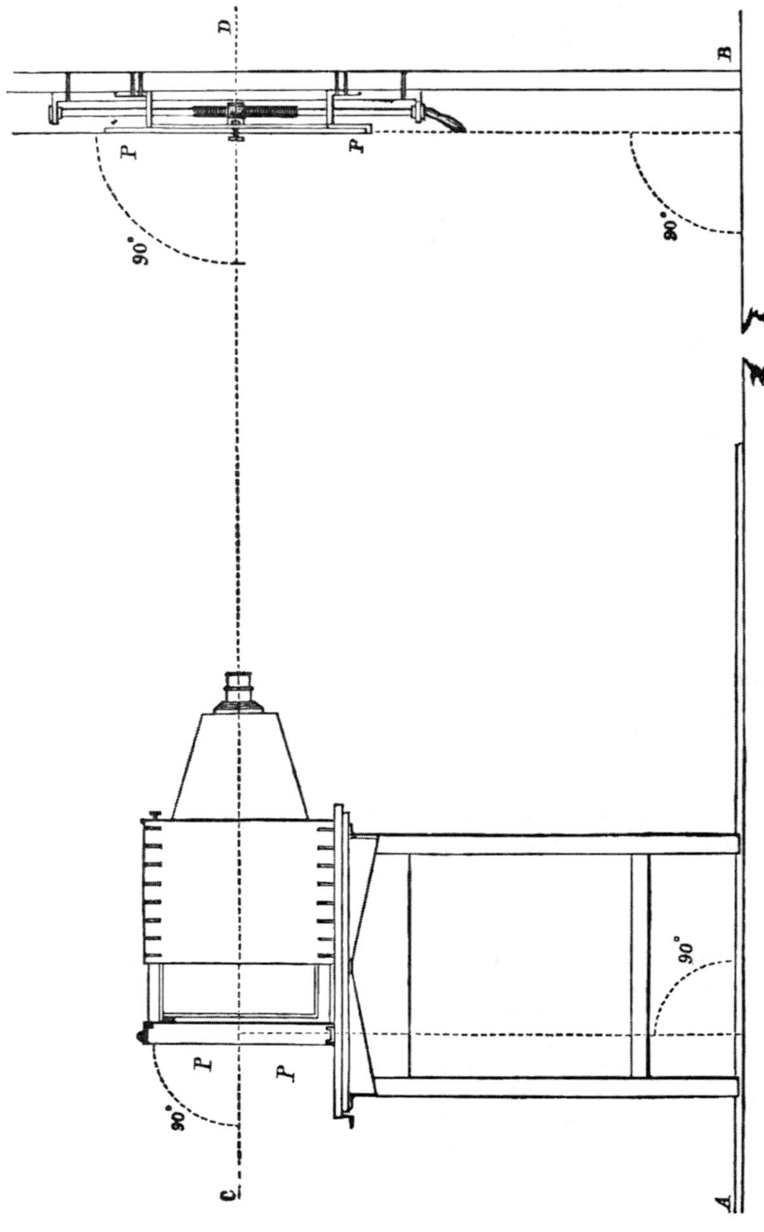

Fig. 9. Sectional Elevation on A B.

Fig. 10.—Plan, showing geometrical position of camera and plan for reduction.

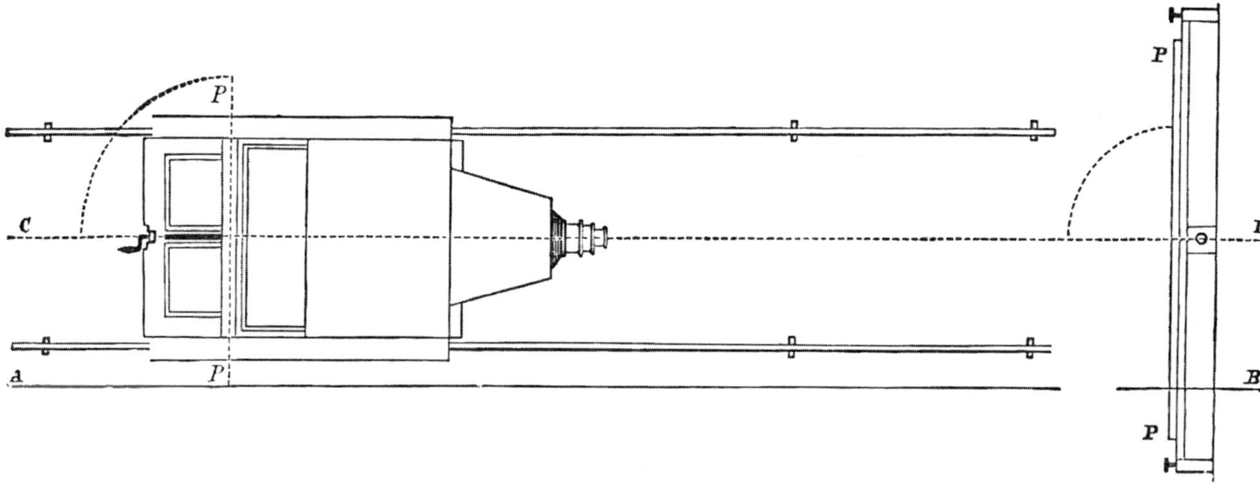

EXPLANATION OF FIGS 9 AND 10.

C D Line in plane of adjustment, passing through centre pin of Plan Stand, centre of sensitive plate in camera, and coinciding with optical axis of lens.

P P Position of plan on board.

v v Position of sensitive plate in camera.

The theoretical conditions that must be satisfied to reduce correctly to a given scale are, that the centre of the sensitive plate and the centre of the plan be in the optical axis of the lens, that the surface of the sensitive plate should be parallel to the surface of the plan, and in the plane of the image, and that the marginal lines of the image of the plan must coincide with the rectangle ruled on the ground glass, representing the rectangle the plan should occupy when reduced to the given scale.

Figs. 9 and 10 give a sectional elevation and general plan, showing the arrangements made to enable these conditions to be satisfied.

The necessary adjustments are made partly with the camera and its stand, and partly with the stand on which the plan is fixed.

The plan to be reduced is fastened to the board by the catches, and the arm of the stand is turned till it is as nearly horizontal as can be judged.

The photographer, looking on the ground-glass plate, draws the camera-stand backward or forward, till the image of the plan appears nearly to fit the rectangle ruled on the glass, at the same time bringing the glass into the plane of the image as it advances to or recedes from the lens.

The centre of the image of the plan (previously marked on the plan) is now made to coincide with the centre of the rectangle ruled on the glass, by moving the plan in a vertical direction by means of the screw of the plan-stand, and by moving the camera in a horizontal direction on the runners.

The parallelism of the surface of the plan and the ground glass will be indicated by the symmetrical position of the margin lines of the image and of the sides of the rectangle on the glass; or, in other words, when the centres coincide, if the surfaces are parallel, the margin of the image will either coincide with the margin of the rectangle on the ground glass, or it will be equally distant from it on every side, and conversely if the surfaces are *not* parallel.

The rails having been placed carefully at right angles to the surface of the plan and levelled, and the plan-stand being as nearly as possible vertical, the degree of inclination of the surfaces to each other will be very slight, and such as can easily be adjusted by the motions provided for the camera-stand.

A little consideration will enable the operator to judge of

and apply the movement required to produce parallelism, from the position already referred to, of the margin of the plan image relative to the margin of the rectangle on the glass. It has been already stated that a want of parallelism will be indicated by the margins not being at equal distances on all sides; and it follows from this that the mark representing the centre of the plan seen in the image will not be in the centre of the rectangular image. From the position of the margin with reference to the centre can be deduced the relative position of the surfaces of the plan and ground glass.

Fig. 11, is supposed to be either a plan or section of the two surfaces: G R being the surface of the ground glass;

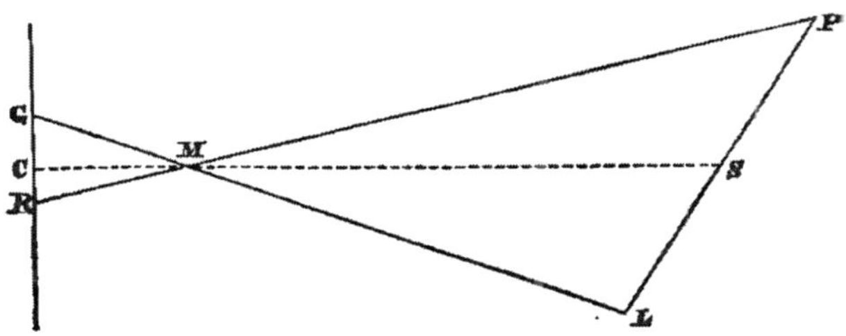

Fig. 11.

M the lens, the optical axis of which (C M) is at right angles to the surface, G R, and P L, the plan; S the centre of the plan, and C the centre of the rectangle, ruled on the ground glass; G R will be the image of the plan thrown on the glass, G and R being in the centre of opposite sides of the image.

Now it is evident, on inspection of the diagram, that when R P is greater than G L, G C is greater than C R; that is, the edge of the plan at P being farther, than the edge at L, from the lens M, the image of P at R, is nearer the centre of the ground glass, than the image of L at G.

By reasoning in this way, it will be at once suggested to the operator what movement in altitude or azimuth is to be given to the camera, to produce parallelism.

This movement will throw the centres of the plan-image and ground glass off each other; they are made again to coincide as before, and thus, by repeated adjustment, coincidence of the centres and parallelism of the surfaces are ensured.

It then only remains to produce coincidence of the sides of the rectangles by the rack-and-pinion motion of the lens and the screw for focusing the camera.

The adjustment is now complete, and it is only necessary to remove the focusing-glass, and to slip the sensitive plate into the same place, to receive a correct reduced image.

The time taken in this adjustment is about four minutes in the first instance, and one minute afterwards for each change of plan, provided the ratio of the reduction is the same.

THE PHOTOGRAPHIC PROCESS USED FOR THE PRODUCTION OF REDUCED COPIES.

The process used for the reduction of plans is that known by the name of the "wet collodion" process.

A general description of the different operations is all that is considered necessary in these memoranda. Detailed and minute directions in manipulating, as well as a complete synopsis of the chemistry of the subject, are given in Mr. Hardwick's "Manual of Photography."*

The operations for the production of the "negative" are—

1. The cleaning of the glass plate.
2. Coating the plate with collodion.
3. Immersion of plate in nitrate-of-silver bath, or sensitising plate.
4. Exposure of sensitive plate in camera.
5. Development of image on plate.
6. Fixing negative plate.
7. Intensifying negative, including immersion in solution of bichloride of mercury and washing with hydrosulphate of ammonia.
8. Varnishing negative.

1. *Cleaning the Glass Plates.*—The cleaning of the glasses used for photography is a point of considerable importance, and of more difficulty than would at first sight appear.

The delicacy of the chemical changes which unite in producing the negative is so great, that a very slight impurity in

* Also in Mr. Thornthwaite's adaptation of Professor Von Monckhoven's "Traité Populaire de Photographie sur Collodion," WEALE SERIES, No. 79.

the surface of the glass has an effect in controlling or influencing them; and there is a special difficulty in the present case from the effect produced on the glass by the bichloride of mercury used in process No. 7. This effect appears to be caused by the affinity of the mercury for the lead used in the manufacture of the glass. The consequence is a combination of the two, and the mercury becomes, as it were, a part of the glass. Unfortunately, it only remains combined with the lead till it is brought into contact with iodide of silver in the sensitive film; it then appears to decompose that salt, combining with the iodine, and rendering the film insensitive at those points where it existed previous to the immersion of the plate in the silver bath. The effect produced is that of a number of transparent spots on the negative.

In order, therefore, to render glass which has been once used again fit for photographic purposes, it is boiled in water for three or four hours, when the greater portion of the mercury passes off with the steam; it is then immersed in a dilute solution of iodine, in order that the iodine may form a more stable compound with it. The glass plate is then rubbed with Tripoli powder and ammonia, and placed in a holder (to avoid contact with the hand), dried, and polished with chamois leather. It is then ready for use; but, after all, when a little additional expense is no objection, it is better to use glass plates not more than twice, if they go through the intensifying process.

2. *Coating the Plate with Collodion.*—The plate, held in a pneumatic holder, receives a small quantity of collodion, such that a slight inclination of the plate may cause it to run over the whole surface, it is then drained off, and the evaporation of the ether leaves a thin film on the glass.

A full description of the mode of preparing collodion is given in Mr. Hardwick's "Manual of Photography," already referred to.*

3. *Immersion of the Plate in the Nitrate-of-silver Bath.*—The formulæ of the various solutions used, including that of the nitrate bath, are given page 130. The characteristics of a good negative are—

1st. Clearness of the detail of the plan photographed.

2nd. Intensity of the ground of the negative.

The condition of the bath which is calculated to produce

* Also in Mr. Thornthwaite's adaptation, &c., p. 20.

the best negatives may be defined to be such that the slightest decrease in quantity of the acid present will produce fogging in the detail of the negative.

The plate, when coated with the iodized collodion, is immersed in the bath for about five minutes, removed, drained, and placed in the ordinary dark slide.

4. *Exposure of the Plate.*—The time of exposure varies continually. The principal cause of the variation is the quantity, intensity, and actinic force of the light entering the lens; and the range is from eight minutes to twenty-five seconds.

When the means employed to estimate the proper exposure are the eye and a watch, the variations are so rapid, and the plate so sensitive to them, that on some days forty per cent. of the negatives may either be under or over exposed. These failures have been almost entirely obviated by the use of sensitive paper as a photometer. The following is the *modus operandi*:—

The paper is prepared as for "printing;" a box about 9 inches long, 2½ in. broad, and ⅛ in. deep, is divided by sixty interior partitions placed transversely across. Each of the portions thus made is closed by a sliding lid. This box is then hung up by the side of the plan-board, and in each compartment of the box is placed a slip of sensitive paper, the lids being then closed. At the instant the negative is exposed by the removal of the stop of the lens, one of the lids of the box is removed so as to uncover one of the slips of paper. When the paper acquires that shade, which we know from previous experience indicates the time at which the plate has been sufficiently exposed, the cap is placed on the lens, and the lid is closed on the box containing the sensitive paper.

If the negative appears to have been under-exposed (and the mistake may occur with the first negative taken on any day), the stop is not placed on the lens the next time, till the paper has received a slightly darker shade. While the next exposure is going on, the slip of paper first coloured is uncovered from time to time, in order that the eye may compare the tint with the new one now in use. If the negative first taken is found to have been rather over-exposed, the exposure is stopped before the second slip of paper has attained so dark a tint; and, finally, if the development shows that the negative received the proper amount of exposure, the exposure of each negative subsequently taken is stopped when the paper has attained that shade.

Clearness of detail in the negative is the first essential, and the time of exposure must be regulated so as most perfectly to produce this result.

5. *Development of the Negative.*—When the plate has been exposed, it is carried to the dark room, and a portion of the developing solution poured on it sufficient to cover it. The fluid is kept in motion over the surface of the plate; and if there is any appearance of undue exposure, a small quantity of nitrate of silver is added.

The amount of development is, like the exposure, regulated by the essential condition of clearness of detail, and is carried to the utmost extent compatible with that. In cold weather it is advantageous to warm the developing solution.

6. *Fixing the Negative.*—The negative, when sufficiently developed, is washed with water, and fixed with cyanide of potassium.

If the previous operations have been well performed, the process of fixing will dissolve from the detail every portion of the iodide of silver, and leave the lines perfectly clear. After fixing, the negative is washed with water.

The following general remarks apply to the above processes:—

It is found that when the plan is perfectly clean (as in the case of a new copper-plate impression), and the bath in good order, a negative can be produced (by development with pyrogallic acid alone), in good light, so dense in the ground, and clear in the detail, that the resulting print shall be only a slight shade darker in the ground than the paper of the impression, while the detail shall be jet black. If there is a failure in any of these conditions, viz., in the cleanliness of the plan, in the light, or in the bath, there is a perceptible difference in the prints.

7. *Intensifying the Negatives.*—The plans reduced are, for the most part, manuscripts, and pass through many hands. They are all considerably soiled in consequence, and thus one of the conditions for producing perfect impressions is wanting. It is found necessary, therefore, to intensify the negatives of MS. plans after they have been fixed. For this purpose, the negative is immersed in a saturated solution of bichloride of mercury (corrosive sublimate) till it is thoroughly whitened; it is then removed, and washed with water. A dilute solution

of hydrosulphate of ammonia, ten of water to one, is poured over the plate, so as to flow over the whole surface.

The ground of the negative is in this way converted into a sulphuret of silver and mercury, and the film is of a brown-yellow colour, and in great measure impervious to chemical rays.

It is impossible to produce good prints unless the negatives of the MS. plans are treated in this way.

8. *Varnishing the Negative.*—When dry, the negative is varnished with a solution of gum mastic in benzine. This varnish is only one-third the cost of the best (amber in chloroform) kind, but answers very well when negatives are not to be kept for any length of time. For work required to be preserved, the latter kind is used.

The operations for the production of the print, or positive, are—

1. Salting and albumenizing the paper,
2. Making the paper sensitive,
3. Printing,
4. Fixing and toning,
5. Washing the prints.

1. *Salting and albumenizing.*—The paper is floated on the solution for one minute and a half, then removed and hung up to dry.

2. *Making the Paper sensitive.*—The paper is floated on a solution of nitrate of silver in water for five minutes and then removed and dried.

This process is carried on in a dark room.

3. *Printing.*—The negative is placed in a printing-frame, in contact with a piece of sensitive paper, and exposed to the light. The exposure is carried on until the detail is sufficiently clearly seen, allowing for the reducing power of the toning bath.

If the negative is a good one, the print will be of an intense chocolate-colour, whilst the ground of the print will be hardly tinged; and after toning, the ground will be white, and the print a jet black.

4. *Toning and fixing the Prints.*—The fixing and toning are carried on simultaneously, in a bath of hyposulphate of soda with chloride of gold.

The detail, as before observed, should in the toning bath become a jet black; and when this stage has been attained, the print should be removed. Experience is the best guide in performing these operations: fixed rules can hardly be laid down for them; for if the printing is carried on for a considerable time, to bring out details which, having been in an unfavourable colour on the plan, have not photographed clearly, it will be at the expense of the general whiteness of the ground of the print. If, on the other hand, the printing is stopped early, in order to preserve the ground of the negative, it is probable that some of the details will not be sufficiently printed; and the same remarks apply to the toning.

Judgment must be exercised so as to carry each process to that extent from which the best general result will be secured.

5. *Washing the Prints.*—It is very important that the prints should be washed so thoroughly as to leave in the paper no trace of the hyposulphite of soda. A very minute portion of this salt is sufficient to cause the print to fade in a short time.

The prints are placed in a trough of water, through which a stream is constantly running from two pipes (see Figs, 13 and 14.) The mouth of one of the pipes is at the bottom

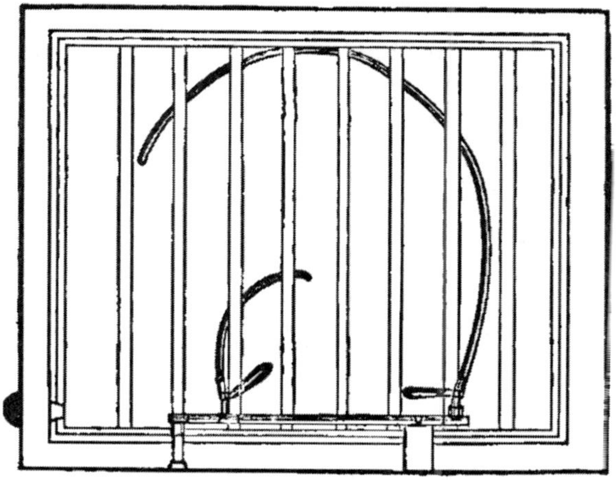

Fig. 13.—Plan.

of the trough, and that of the other halfway between the bottom and the top. Both pipes are curved, so that a rotatory motion is given to the whole body of water. A waste pipe at the top, and another at the bottom, carry off the surplus water as fast as it flows in. The prints are left in the water-trough from six to eight hours.

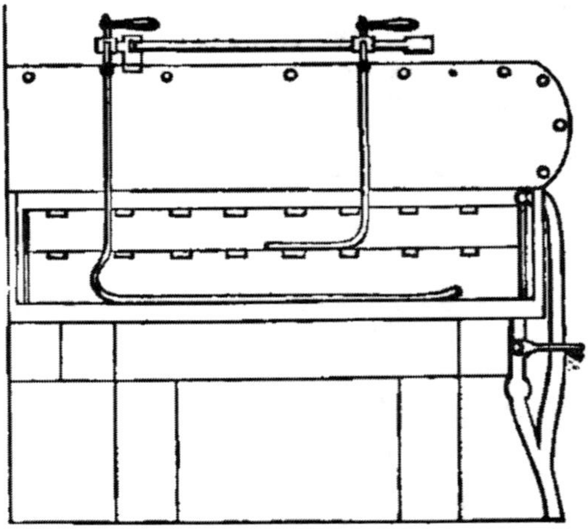

Fig. 14.—Section.

In order that the prints may be dispersed throughout the body of water in the trough, instead of floating at the top, the trough is divided horizontally into three compartments, by movable wooden trays; as soon as a certain number of prints have been placed in the trough, one of the trays is placed on top of them, and the whole are forced down to the lowest compartment in the trough. The next supply is treated in the same manner with the second tray, and the remainder are allowed to float on the top. The trays are made of battens nailed to a wooden frame at intervals of an inch apart, and the frame is loaded with lead.

FORMULÆ OF THE SOLUTIONS USED IN PHOTOGRAPHY.

Nitrate Baths for Collodion Negatives.

Nitrate of silver	1 ounce
Distilled water	2 ounces, dissolve
Iodide of potassium	3 grains
Distilled water	1 drachm, dissolve.

Mix these two solutions, the precipitate (iodide of silver) thus formed is by shaking entirely dissolved.

Now add fourteen ounces of distilled water, when the excess of iodide of silver is again thrown down, but in such a finely divided state as to render the saturation of the bath with iodide of silver perfect.

After half an hour, add alcohol 2 drachms, sulphuric ether 1 drachm, and then filter.

Developing Solution for Negatives.

Pyrogallic acid	20 grains
Glacial acetic acid	1 ounce
Alcohol	6 drachms
Distilled water	20 ounces.

Dissolve the pyrogallic acid in water, and then add the acetic acid and alcohol.

Fixing Solution.

Cyanide of potassium	15 grains
Distilled water	1 ounce, dissolve.

Solution for intensifying Negatives

Bichloride of mercury	30 grains
Distilled water	1 ounce (or saturated solution)
Hydrosulphuret of ammonia	1 ,,
Water	5 ounces.

Immerse the negative in the solution of bichloride of mercury until it has become well whitened; wash with water, and then pour quickly over the plate the dilute hydrosulphuret of ammonia, and again well wash with water.

Albumen and Salting Solution of Paper.

White of six eggs	
Distilled water	6 ounces
Chloride of ammonium	3 drachms.

When the ingredients are mixed, take a bundle of quills or a fork, and beat the whole into a perfect froth. As the froth forms, it is to be skimmed off and placed in a flat dish to subside. When the froth partially subsides, transfer it to a tall and narrow jar, and allow it to stand for several hours, that the membranous shreds may settle to the bottom; then pour off the upper clean portion, which is fit for use.

Nitrate Bath for Paper.

Nitrate of silver	1 ounce
Distilled water	7 ounces.

Fixing and Toning Bath.

Chloride of gold	8 grains
Nitrate of silver	16 „
Hyposulphite of soda	4 ounces
Water	8 „

Dissolve the hyposulphite of soda in four ounces of water, the chloride of gold in three ounces, and the nitrate of silver in the remaining ounce; then pour the diluted chloride by degrees into the hyposulphite, stirring with a glass rod, and afterwards the nitrate of silver in the same way.

CHAPTER IX.

INSTRUMENTS FOR MEASURING THE ANGLES OF CRYSTALS.

THE GONIOMETER.

This instrument is used for measuring the angles of crystals. The following description of the construction and method of using this instrument is extracted from the able article on Crystallography, in the "Encyclopædia Metropolitana:"—

"The instruments used for measuring the angles at which the planes of crystal incline to each other are called goniometers.

"The mutual inclination of any two planes, as of *a* and *b*,

Fig. 1, is indicated by the angle formed by two lines, *e d*, *e f*, drawn upon them from any point, *e*, on the edge at which they meet, and perpendicular to that edge.

"Now it is known that if two right lines, as *g f*, *d h*, Fig. 2, cross each other at any point, *e*, the opposite angles *d e f*, *g e h*, are equal. If, therefore, the lines, *g f*, *d h*, are supposed to be very thin and narrow plates, and to be attached together

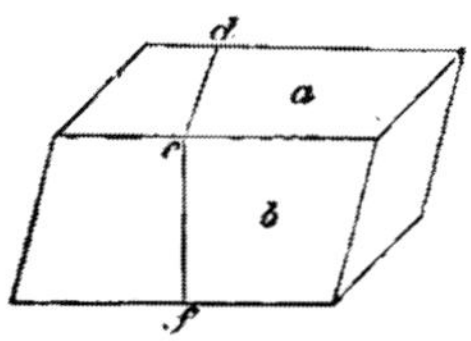

Fig. 1.

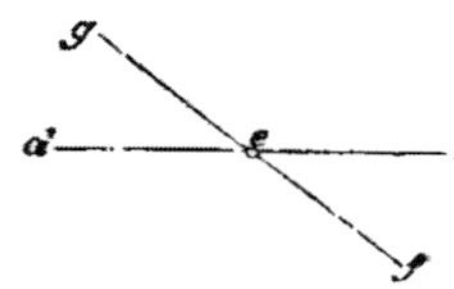

Fig. 2.

by a pin at *e*, serving as an axis to permit the point, *f*, to be brought nearer either to *d* or to *h*, and that the edges, *e d*, *e f*, of those plates are applied to the planes of the crystal, Fig. 1, so as to rest upon the lines, *e d*, *e f*, it is obvious that the angle, *g e h*, of the movable plates would be exactly equal to the angle, *d e f*, of the crystal.

"The common goniometer is a small instrument for measuring this angle, *g e h*, of the movable plates. It consists of a semicircle, Fig. 3, divided into 360 equal parts, or half degrees, and a pair of movable arms, *d h*, *g f*, Fig. 4, the semi-

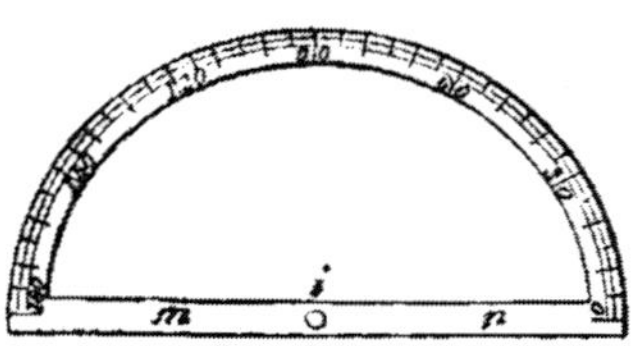

Fig. 3.

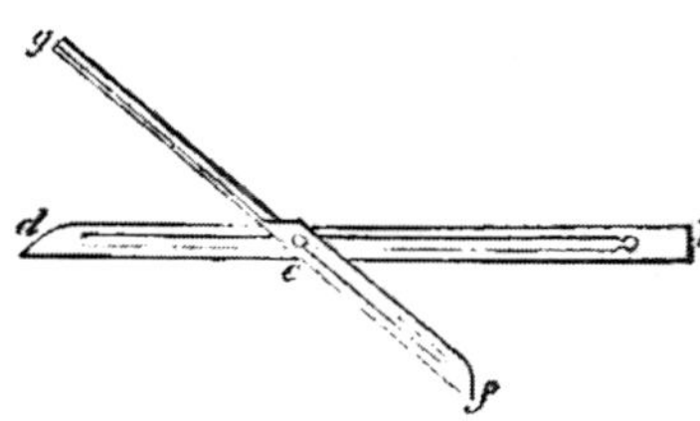

Fig. 4.

circle having a pin at *i*, which fits into a hole in the movable arms at *e*.

'The method of using this instrument is to apply the edges, *d e*, *e f*, of the movable arms to the two adjacent planes of any crystal, so that they shall actually touch or rest upon those planes in directions perpendicular to their edge. The arm, *d h*, is then to be laid on the plate, *m n*, of the semicircle, Fig. 3, the hole at *e* being suffered to drop on the pin at *i*, and the edge nearest to *h* of the arm will then indicate on the semicircle, as in Fig. 5, the number of degrees which the measured angle contains.

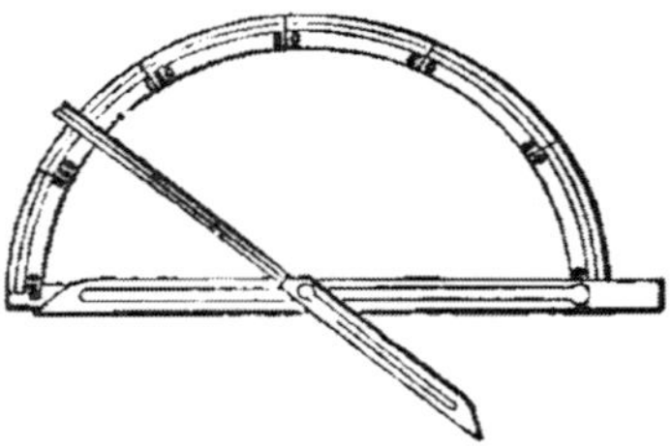

Fig. 5.

"When this instrument is applied to the planes of a crystal, the points, *d* and *f*, Fig. 4, should be previously brought sufficiently near together for the edges, *d e*, *e f*, to form a more acute angle than that about to be measured. The edges being then gently pressed upon the crystal, the points, *d* and *f*, will be gradually separated, until the edges coincide so accurately with the planes that no light can be perceived between them.

"The common goniometer is, however, incapable of affording very precise results, owing to the occasional imperfection of the planes of crystals, their frequent minuteness, and the difficulty of applying the instrument with the requisite degree of precision.

"The more perfect instrument, and one of the highest value to crystallography, is the reflecting goniometer, invented by Dr. Wollaston, which will give the inclination of planes whose area is less than $\frac{1}{100000}$ of an inch, to less than a minute of a degree.

"This instrument has been less resorted to than might, from its importance to the science, have been expected, owing, perhaps, to an opinion of its use being attended with some difficulty. But the observance of simple rules will render its application easy.

"The principle of the instrument may be thus explained:—

"Let *a b*, Fig. 6, represent a crystal, of which one plane only is visible in the figure, attached to a circle, graduated on its edge, and movable on its axis at *o*; and let *a* and *b* mark the position of the two planes whose mutual inclination is required.

"And let the lines, *o e*, *o g*, represent imaginary lines, resting on those planes in direction perpendicular to their common edge, and the dots at *i* and *h* some permament marks in a line with the centre, *o*.

"Let the circle be in such a position that the line, *o e*, would pass through the dot at *h*, if extended in that direction, as in Fig. 6.

"If the circle now be turned round with its attached

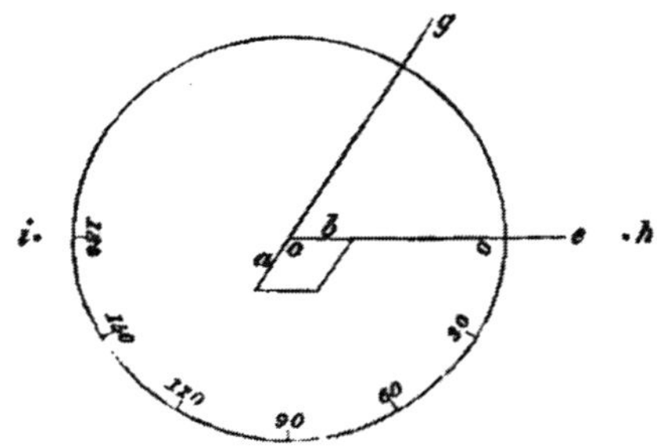

Fig. 6.

crystal, as in Fig. 7, until the imaginary line, *o g*, is brought into the position of the line, *o e*, in Fig. 6, the number 120 will stand opposite the dot at *i*. This is the number of degrees at which the planes *a* and *b* incline to each other. For if the line *o g* be extended in the direction *o i*, as in Fig. 7, it is obvious that the lines, *o e*, *o i*, which are perpendicular to the common edge of the planes, *a* and *b*, would intercept exactly 120° of the circle.

"Hence an instrument constructed upon the principle of these diagrams is capable of giving with accuracy the mutual inclination of any two planes which reflect objects with

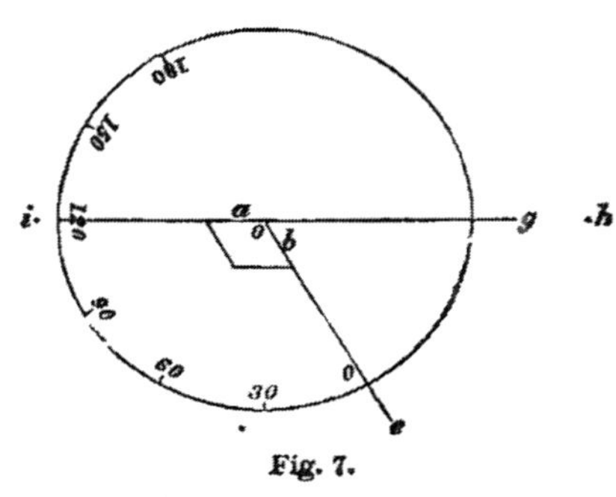

Fig. 7.

sufficient distinctness, if the means can be found for placing them successively in the relative positions shown in the two preceding figures.

"This purpose is effected by causing an object, as the line at *m*, Fig. 8, to be reflected successively from the two planes, *a* and *b*, at the same angle. It is well known that the images of objects are reflected from bright planes at the same angle

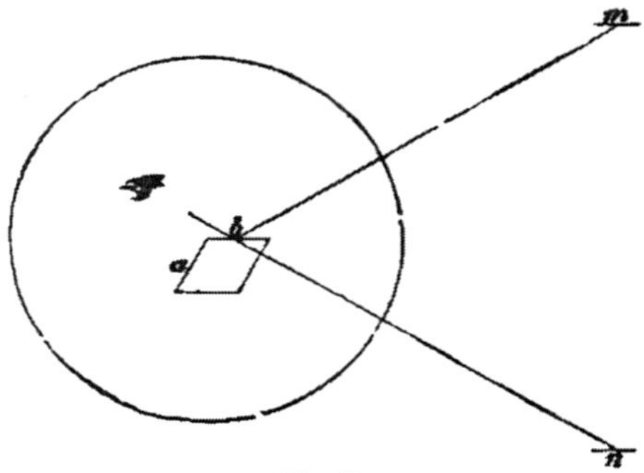

Fig. 8.

as that at which their rays fall on those planes; and that when the image of an object reflected from a horizontal plane is observed, it appears so much below the reflecting surface as the object itself is above.

"If, therefore, the planes *a* and *b*, Fig. 8, are successively brought into such positions as will cause the reflection of the line at *m*, from each plane, to appear to coincide with another line at *n*, both planes will be successively placed in the relative positions of the corresponding planes in Figs. 6 and 7.

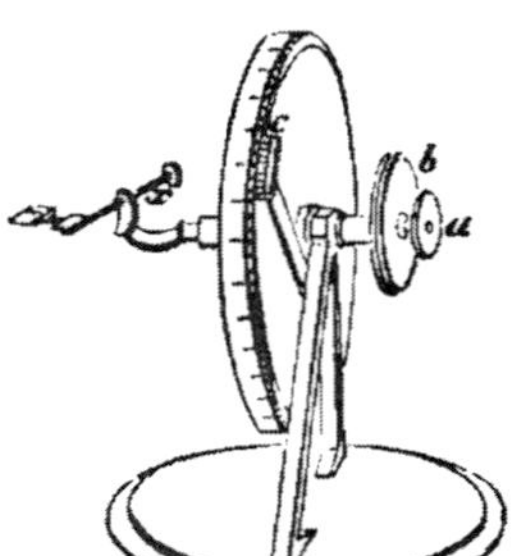

Fig. 9.

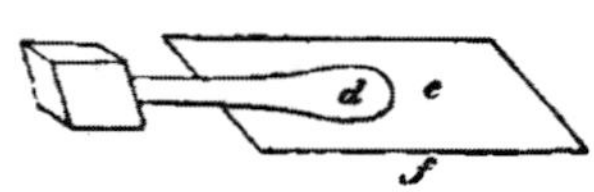

Fig. 10.

To bring the planes of any crystal successively into these relative positions, the following directions will be found useful.

"The instrument, as shown in the sketch, Fig. 9, should be first placed on a pyramidal stand and the stand on a small steady table, about 6 or 10 or 12 feet from a flat window. The graduated circular plate should stand perpendicularly from the window, the pin *x* being horizontal, not in the direction of the axis, as it is usually figured, but with the slit end nearest to the eye.

"Place the crystal which is to be measured on the table, resting on one of the two planes whose inclination is required, and with the edge at which those planes meet nearest and parallel to the window.

"Attach a portion of wax, about the size of *d*, to one side of a small brass plate, *e*, Fig. 10; lay the plate on the table with the edge, *f*, parallel to the window, the side to which the wax is attached being uppermost, and press the end of the wax against the crystal until it adheres; then lift the plate with its attached crystal, and place it in the slit of the pin, *x*, with that side uppermost which rested on the table.

"Bring the eye now so near the crystal, as, without perceiving the crystal itself, to permit the images of objects reflected from its planes to be distinctly observed, and raise or lower that end of the pin, *x*, which has the small circular plate on it, until one of the horizontal upper bars of the window is seen reflected from the upper or first plane of the crystal, corresponding with the plane *a*, Fig. 6, and until the image of the bar appears to touch some line below the window, as the edge of the skirting-board where it joins the floor.

"Turn the pin, *x*, on its own axis also, if necessary, until the reflected image of the bar of the window coincides accurately with the observed line below the window.

"Turn now the small circular handle, *a*, on its axis, until the same bar of the window appears reflected from the second plane of the crystal corresponding with plane *b*, Figs. 6 and 7, and until it appears to touch the line below; and having, in adjusting the *first* plane, turned the pin, *x*, *on its axis*, to bring the reflected image of the bar of the window to coincide with the line below, *now move the lower end of the pin laterally*, either towards or from the instrument, in order to make the image of the same bar, reflected from the second plane, coincide with the same line below.

"Having ascertained by repeatedly looking at, and adjusting both planes, that the image of the horizontal bar, reflected successively from each plane, coincides with the observed lower line, the crystal may be considered ready for measurement.

"Let the 180° on the graduated circle be now brought opposite the 0 of the vernier at *c*, by turning the handle *b*; and while the circle is retained accurately in this position, bring the reflected image of the bar from the *first* plane, to coincide with the line below, by turning the *small* circular handle *a*. Now turn the graduated circle, by means of the handle *b*, until the image of the bar, reflected from the *second* plane, is also observed to coincide with the same line below. In this state of the instrument the vernier at *c* will indicate the degrees and minutes at which the two planes are inclined to each other.

"The accuracy of the measurements taken with this instrument will depend upon the precision with which the image of the bar, reflected successively from both planes, is made to appear to coincide with the same line below; and also upon the 0, or the 180°, on the graduated circle, being made to stand precisely even with the lower line of the vernier, when the first plane of the crystal is adjusted for measurement. A wire being placed horizontally between two upper bars of the window, and a black line of the same thickness being drawn parallel to it below the window, will contribute to the exactness of the measurement by being used instead of the bar of the window and any other line.

"Persons beginning to use this instrument are recommended to apply it first to the measurement of fragments at least as large as that represented in Fig. 10, and of some substance whose planes are bright. Crystals of carbonate of lime will supply good fragments for this purpose, if they are merely broken by a slight blow of a small hammer.

"For accurate measurement, however, the fragments ought not, when the planes are bright, to exceed the size of that shown in Fig. 9, and they ought to be so placed on the instrument, that a line passing through its axis should also pass through the centre of the small minute fragment which is to be measured. This position on the instrument ought also to be attended to when the fragments of crystal are large. In which case the common edge of the two planes, whose inclination is required, should be brought very nearly to coincide with the axis of the goniometer; and it is frequently useful to blacken the whole of the planes to be measured, except a narrow stripe on each close to the edge over which the measurement is to be taken."

Wollaston's goniometer has been somewhat modified and

improved by the French optician, Charles, whose construction we proceed to describe.

This instrument consists of a graduated brass circle, *a*, supported by, and turning on, an axial pillar at right angles to its plane: this pillar rises from a tripod, standing on three screws, which serve to set the circle horizontal. An index-bar turns on a pivot at the centre of the circle, and over the pivot is a capstan-head, on which is fixed with *soft* wax the crystal whose angles are to be measured. The extremity of the index-bar carries a vernier, and has a clamp and tangent screw for setting it accurately.

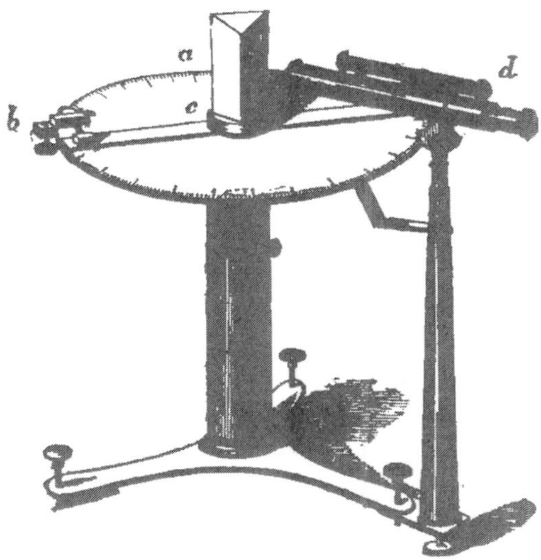

From one of the corners of the tripod rises a tube supporting a telescope, *d*, that has a motion in a horizontal, and also one in a vertical, plane.

The circle having first been adjusted for horizontality, it only remains to place the edge of the crystal, that forms the vertex of the dihedral* angle to be measured, vertical, and consequently parallel to the axis of rotation of the index-bar, without being too far from the centre of motion, the amount of eccentricity permissible depending upon the extent of the faces of the angle to be measured, as it is only necessary that

* *Dihedral angle*, angle between two plane surfaces.

the fixed axis of the telescope should always be directed to some point within the boundaries of these faces.

Adjustments.—At the focus of the telescope are placed several parallel and vertical cobweb threads, with means for adjusting their exact verticality: the telescope being directed to a distant vertical line, such as a lightning conductor, or the corner of a building, one of the micrometric threads must be brought into coincidence with this line: the telescope is then to be turned round, and directed towards the crystal, and the image of the same line is to be observed, reflected first from one of the faces of the dihedral angle, and then from the other: the inclination of the image can be varied by moving the crystal on the soft wax; and, when, after some trials, the reflected image, passed in succession across

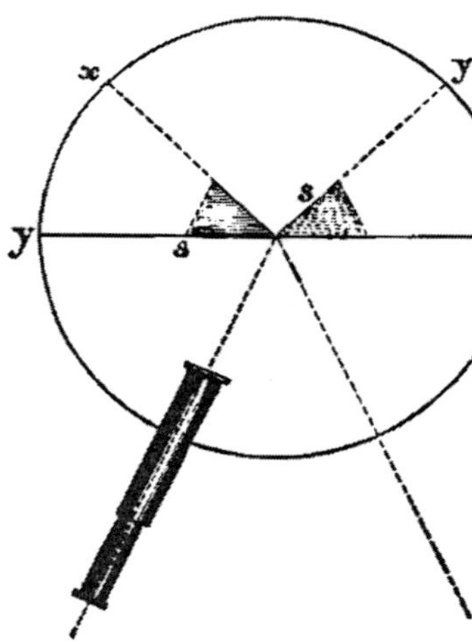

the field of view by turning the index-bar, fall exactly on the micrometric lines, the two faces of the dihedral angle will be both vertical, and so also will be its edge.

When this adjustment is perfected, the index of the vernier is to be brought to the zero division of the circle *a*, and the circle and index-bar are to be turned round together, until the image of the line reflected from one of the faces of the angle, falls exactly on the central micrometric thread; the circle is then to be fixed by the clamping screw, and the index-bar only to be turned, until the image reflected from the other face falls on the same thread; and the angle through which it has been necessary to move the index-bar is *the supplement* of the angle required. If, for instance, the first reflection has proceeded from the face $s\ y$, the second reflection from the face $s\ x$ will fall on the central micrometric thread, when this face has come into the position $s\ x'$, and consequently when it has described the angle $x\ s\ x'$, the supplement of the required angle $y\ s\ x$.

INDEX TO VOL. II.

Aberration, spherical, 20; chromatic, 21.
Accessories of microscope, 73.
Amici's camera lucida, 18.
Angle subtended by two objects measured, 5.
Angular aperture, 60.
Apparent images formed by plane reflectors, 2.
Astronomical telescope, 86.
Axis of pencil of light defined, 1.

Beck's improved single dissecting microscope, 43; binocular prisms, 45; popular microscope, 66; parabolic illuminator, 76; patent illuminator, 77.
Binocular body, Nachet's, 56; Wenham's, 57; Nachet's improved, 58.
Binocular parallax, 32.
Binocular prisms, Beck's, 45; Nachet's, 47.
Binocular telescope, 89.
Box, live, 83.
Brewster's lenticular stereoscope, 35.
Brown's Iris diaphragm, 74.

Camera lucida, Wollaston's, 15; Amici's, 18; applied to the microscope, 79.
Camera obscura, 108; photographic, 109.
Care of the eyes, 83.
Cassegrainian telescope, 93.
Changing object or object-piece, precautions in, 86.
Charles's goniometer, 137.
Chevalier's universal microscope, 69.
Chromatic dispersion, 21.
Coddington lens, 39.
Compound microscope, 47; magnifying apparatus of, 48; magnification produced by, and dimensions of, 49; modern improvements in, 50; illumination for, 59, 73, 76; stands for, 65; accessories of, 73.
Condenser, mirror and, 61; on stand, 75.
Crystal, Parkes and Son's patent, 25.

Darwin's dissecting single microscope, 41.
Defining power of microscope, 84.
Deviation and decomposition of light by prism, 12.
Diagonal eye-piece, 88.
Diaphragm, 61; Brown's Iris, 74.
Dimensions of compound microscope, 49.
Dissolving views, 102.
Distinct vision, conditions of, 20, 47.
Double vision, phenomena of, 33.
Doublets, 40.
Draw tube, 55.

Erector, 55.
Eye-pieces, 52; Huyghenian, 53; Ramsden's, 53; Kellner's, 54; Sorby's spectroscope, 78; micrometer, 80; diagonal, 88; terrestrial, 88; pancratic, 90.

Focal length of lens, 21.
Forceps, stage, 82.

Galilean telescope, 89.
Glass trough, 83.
Goniometer, 130; Wollaston's, 132; Charles's, 137.
Gregorian telescope, 92.

Herschelian telescope, 93.
Huyghenian eye-piece, 52.

Illumination of translucent objects, 60, 73, 74; of opaque objects, 61, 76.
Illuminator, Beck's parabolic, 76; Beck's patent, 77.
Images, apparent or virtual, formed by plane reflectors, 2; real and imaginary, formed by lenses, 22; formed on the retina, 24.
Inclined surfaces, reflection from, 4.
Iris diaphragm, 74.

Jackson's eye-piece micrometer, 81.

Kaleidoscope, 6.
Kellner's eye-piece, 54.

Lantern, magic, 100.
Lenses, defined, 18; different forms of, 19; focal lengths of, 21; real and imaginary images formed by, 22; magnifying powers of, 27; Coddington, 39; spot, 74.
Lieberkuhn, 62.
Light for microscope, best position and source of, 83; oxy-hydrogen, 103; oxy-calcium, 104; electric, 104.
Live box, 83.
Looking-glass, test of a good, 4.
Lower and higher powers, the terms defined, 84.

Magic lantern, 100.
Magnifying powers of lenses, 27; of compound microscope, 49; of telescopes, 87, 89, 90. 94.

Medium, defined, 1; refracting power of, determined, 14.
Micrometer, stage, 80; eye-piece, 80; Jackson's, 81.
Microscopes, simple, 38; Darwin's dissecting, 41; Quekett's dissecting, 43; Beck's improved dissecting, 43; compound, 47; magnification produced by, 49; modern improvements in, 50; stands for, 65; Beck's popular, 66; Chevalier's universal, 69; first class, the Ross, Powell and Ireland's, and R. and J. Beck's, 72; accessories of, 73; use and testing of, 83; care of, 86.
Mirror and condenser, 61.
Mirrors, inclined, repeated reflections from, 5.
Mounting of prisms, 15; spectacles, 26; microscopes, 41, 65; telescopes, 98.

Nachet's binocular prisms, 47; binocular body, 56; improved binocular body, 58.
Nasmyth's telescope, 92.
Newtonian telescope, 92.

Object-pieces or objectives, 51; precaution in changing, 86.
Opaque objects, illumination of, 61, 76; management of, 84.
Oxy-calcium light, 104.
Oxy-hydrogen light, 103; microscope, 107.

Parabolic illuminator, 76; reflectors, 9; reflector, Wenham's, 75.
Parallax, binocular, 32.
Parallel plane surfaces, reflections from, 2.
Pencil of light, defined, 1; eccentric, 20.
Penetrating power of microscopes, 85.
Periscopic spectacles, 27.
Phantasmagoria, 101.
Photo-electric microscope, 107.
Photographic, pictures for stereoscope, 38; camera, 109; process for the production of reduced copies, 122.
Plane surfaces, reflections from, 2.
Polarizing apparatus, 78.
Pouillet's method of determining the magnifying power of a telescope, 94.
Prism, edge, face, and refracting angle of, 12; deviation and decomposition of light by, 12; total reflection from, 14; mounting of, 15; binocular, 45, 47; Powell and Ireland's, 59.

Quekett's dissecting microscope, 43.

Ramsden's eye-piece, 53.
Reduction of plans, apparatus for, 110.
Reflecting stereoscope, 34; telescope, 91.
Reflections from parallel plane surfaces, 2; from inclined plane surfaces, 4; repeated, from inclined mirrors, 5; total, from prism, 14.
Reflectors, plane, apparent or virtual images formed by, 2; thickness of silvered glass determined, 3; spherical, parabolic, and elliptical, 9.
Refracting, power of medium determined, 14; telescope, 86.
Resolving power of microscope, 85.
Retina, images formed on, 24.

Short sight, cause of, 24; corrected by concave lenses, 25.
Sight, imperfections of, 24; corrected by lenses, 25.
Simple microscope, 38.
Sliders, 60.
Solar microscope, illuminating apparatus of, 104; amplifying apparatus of, 105; adjustment and use of, 106.
Sorby's spectroscope eye-piece, 78.
Spectacles, requisites of good, 25; Parkes and Son's patent crystal for, 25; mounting of, 26; periscopic, 27.
Spherical reflectors, 9; aberration, 20; confusion, 20.
Spot lens, 74.
Stage of microscope, 61; with delicate movements, 62; White's lever, 63; micrometer, 80; forceps, 82.
Stands for microscopes, 65.
Stereoscope, 31; Wheatstone's reflecting, 34; Brewster's lenticular, 35; photographic pictures for, 38.
Surfaces, plane, reflections from, 2; inclined reflections from, 4.
Symmetry, production of, in reflections from plane surfaces, 6.

Telescopes, astronomical, 86; aërial, 87; Galilean and binocular, 89; reflecting, 91; adjustment and testing of refracting, 94; adjustment and testing of reflecting, 97; mounting of, 98.
Tests, for a good looking-glass, 4; for microscopes, 85; for telescopes, 84, 87.
Thickness of silvered glass reflectors determined, 3.
Translucent objects, illumination of, 60, 73.
Trough, glass, 83.

Varying powers, use of, 54.
Vision, distinct conditions of, 20; double, phenomena of, 33.
Visometer, 27.

Weak, or long sight, cause of, 24; corrected by convex lenses, 25.
Wenham's binocular body, 57; parabolic reflector, 75.
Wheatstone's reflecting stereoscope, 34.
White's lever stage, 63.
Wollaston's camera lucida, 15; goniometer, 132.

PRINTED BY J. S. VIRTUE AND CO., LIMITED, CITY ROAD, LONDON.

Frontispiece.

TELESCOPE, MOUNTED EQUATORIALLY. (*See p.* 154.)

MATHEMATICAL INSTRUMENTS

VOL. III.

SURVEYING AND ASTRONOMICAL INSTRUMENTS

INCLUDING

I.—INSTRUMENTS USED FOR DETERMINING THE GEOMETRICAL FEATURES OF A PORTION OF GROUND

II.—INSTRUMENTS EMPLOYED IN ASTRONOMICAL OBSERVATIONS

PREFACE TO VOL. III.

The original one-volume edition of the Treatise on Mathematical Instruments having now been expanded into three volumes, this third volume is devoted entirely to the consideration of Surveying Instruments and of Astronomical Instruments, forming Parts IV. and V. of the work.

The additional space thus obtained has been employed by the Author, in the first place, in describing more fully the several varieties of the instruments treated of, and in giving more detailed and comprehensive illustrations of their application; and, in the second place, in introducing entirely new chapters upon instruments for the determination of distances by observation, and upon the measurement of altitudes by the barometer and thermometer. A chapter has also been added to Part V. upon the construction and use of the Equatorial, undoubtedly the form of instrument best adapted for the examination of those phenomena of the solar orb, which now demand so much attention from the scientific world.

The Author is again bound to mention his obligations to Messrs. Elliott, who have kindly supplied several of the woodcuts that illustrate the volume, and to Messrs. Parkes and Son of Birmingham, who sent him a large number of instruments for inspection, among which were the Drawing and Mining Levels, described at pp. 37, 38.

To these gentlemen, and to all from whom he has derived assistance in the preparation of this and the preceding volumes, the Author begs leave again to return his most sincere thanks, and ventures to repeat the hope already expressed in the preface to the first volume, "that the skill of the workman may not appear to have been altogether inadequate to the manipulation of the materials supplied him."

CONTENTS OF VOL. III.

PART IV.

SURVEYING INSTRUMENTS.

CHAPTER I.

INSTRUMENTS FOR MEASURING HORIZONTAL DISTANCES.

	PAGE
The Land Chain	2
Clinometers	7
The Offsetting Staff	8
Measuring Tape	9
Station Poles and Ranging Rods	9

CHAPTER II.

INSTRUMENTS FOR MEASURING VERTICAL DISTANCES.

The Spirit Level	10
The Y Level	12
Gravatt's Dumpy Level	20
Elliott's Improved Dumpy Level	22
The Improved Dumpy Y Level	24
Levelling Staves	24
Levelling, Remarks on	26
Water Level	33
Reflecting Level	34
Improved Reflecting Level	35
Draining and Mining Levels	36

CHAPTER III.

INSTRUMENTS FOR MEASURING ANGLES.

PAGE

The Prismatic Compass 39
The Box Sextant 47
Artificial Horizon 52
The Optical Square 53
The Cross Staff 53
The Line Ranger 54
The Theodolite 55
The Transit Theodolite 60
Everest's Theodolite 64
Lister's Patent Theodolite 66
The Theodolite of M. d'Abbadie 66
Measuring Angles with the Theodolite 67
Ranging Lines with the Theodolite 70
Traversing with the Theodolite 71
Plotting the Survey 74
Protracting by Chords 75
Northings, Southings, Eastings, and Westings 75

CHAPTER IV.

INSTRUMENTS FOR DETERMINING DISTANCES OF OBSERVED OBJECTS.

The Stadium 79
Edgeworth's Stadiometer 84
The Omnimeter 88

CHAPTER V.

INSTRUMENTS FOR THE DETERMINATION OF CONSIDERABLE HEIGHTS.

Preliminary Remarks on Hypsometrical Instruments . . 92
The Mercurial Thermometer 96
The Standard Barometer 98
The Syphon Barometer 99
The Portable or Mountain Barometer 99
The Wheel Barometer 103
The Aneroid Barometer 104
Application of the Barometer to the determination of differences of Altitude 108
Meteorological Phenomena denoted by the Barometer . . 118
Thermometric Hypsometry 124

PART V.

ASTRONOMICAL INSTRUMENTS.

CHAPTER I.

HADLEY SEXTANT.

PAGE
Description of the Sextant 127
Adjustment of the Sextant 130

CHAPTER II.

THE TRANSIT INSTRUMENT.

Description of the Transit Instrument 132
Adjustments of the Transit Instrument 135
Method of observing with the Transit 139
Use of the Portable Transit 141

CHAPTER III.

THE ALTITUDE AND AZIMUTH INSTRUMENT.

Description of the Altitude and Azimuth Instrument . . 142
Adjustments of this Instrument 144
Method of observing with this Instrument 146
Use of the Instrument 147

CHAPTER IV.

SUBSIDIARY INSTRUMENTS.

The Reading Microscope 149
The Collimator 151

CHAPTER V.

THE EQUATORIAL.

Description of the Equatorial 153
Adjustments of the Equatorial 154
Uses of the Equatorial 158
Mode of observing with the Equatorial 158

MATHEMATICAL INSTRUMENTS.

PART IV.

SURVEYING INSTRUMENTS.

Surveying consists in the measurement of such distances and angles, as shall most correctly, and conveniently, define the boundaries of a plot of ground, as well as the positions of the various objects that diversify its surface, whether natural, as hills, rivers, forests, &c., or artificial, as roads, railways, canals, &c.

From the measurement thus made, these boundaries and objects can be laid down on paper, so as to preserve their *relative* magnitudes and positions, the observed measurements being all reduced in the same proportion. The features of the ground thus laid down, form what is called a map or plan, according as the plot of ground represented is of greater or smaller extent; and the process of laying it down is called *plotting the survey*.

The instruments used in plotting, and also in the copying, reduction, and measurement, of maps and plans, are fully described in the volumes on Mathematical Drawing, Measuring, and Optical Instruments. It remains, then, to describe:—(1) The instruments used for measuring horizontal distances; (2) the instruments used for measuring vertical distances; and (3) the instruments used for measuring angles.

CHAPTER I.

INSTRUMENTS FOR MEASURING HORIZONTAL DISTANCES.

THE LAND CHAIN.

GUNTER's chain is the instrument used almost universally for measuring the distances required in a survey. For extensive and important surveys, however, such as those carried on under the Board of Ordnance, a base of about 5 or 6 miles in length is first measured by some more accurate instrument, and all the principal lines, and the distances of the extreme points, are calculated from triangles connecting them with this base. One of the instruments which has been used for this purpose was a steel chain 100 feet long, constructed by Ramsden, jointed like a watch chain. This chain was always stretched to the same tension, supported on troughs laid horizontally, and allowances were made for changes in its length made by temperature, at the rate of ·0075 of an inch for each degree of heat from 62° of Fahrenheit. More recently, compound rods have been used, so constructed that, by the unequal expansion of two metals, two points on cross-pieces, connecting the ends of the two bars composed of those metals, always remain at the same distance from each other.

To return, however, to Gunter's chain:—it is 66 feet, or four poles in length, and is divided into 100 links, which are joined together by rings. The length of each link, together with the rings connecting it with the next, is, consequently, $\frac{66 \times 12}{100}$ inches or 7·92 inches. To every tenth link are attached pieces of brass of different shapes for more readily counting the links in distances less than a chain. They are something like the fingers of the hand, one finger being placed at the tenth link, two fingers at the twentieth link, three at the thirtieth, and four at the fortieth; and at the fiftieth link, or centre of the chain, is placed a round mark. The same marks, in the reverse order, are placed beyond the centre; so that one finger may denote either ten links, or ninety links; two fingers, either twenty, or eighty links, &c.,

accordingly as they lie before, or beyond, the centre. At each end of the chain is a brass handle, by which to hold and pull it straight when in use.

The following tables exhibit the number of chains and links in the different units of lineal measure, and the number of square chains and links in the different units of square measure, made use of in this country :—

A TABLE OF LINEAR MEASURES

Links.	Feet.	Yards.	Poles.	Chains.	Furlongs.	Mile.
25	16½	5½	1			
100	66	22	4	1		
1,000	660	220	40	10	1	
8,000	5,280	1,760	320	80	8	1

A TABLE OF SQUARE MEASURES.

Sq. Links.	Sq. Feet.	Sq. Yards.	Sq. Pole, or Perch.	Sq. Ch.	Roods.	Acres.	Sq. Mile.
625	272¼	30¼	1				
10,000	4,356	484	16	1			
25,000	10,890	1,210	40	2½	1		
100,000	43,560	4,840	160	10	4	1	
64,000,000	27,878,400	3,097,600	102,400	6,400	2,560	640	1

As, then, an acre contains 100,000 square links, if the content of a survey, cast up in square links, be divided by 100,000, the quotient gives at once the content in acres, and decimals of an acre. But the division by 100,000 is performed by merely pointing off the five last figures towards the right hand for the decimals of an acre, and the remaining figures towards the left hand are the acres in the content required.

The decimals thus pointed off being then multiplied by 4, and the five last figures pointed off as before, the remaining figures are the roods; and the five decimals cut off from this product, multiplied by 40, give the poles or perches, and decimals of a pole, the same number, 5, of digits being again pointed off, including the zero which arises

```
     1175
     1175
    -----
     5875
    8225
  12925
 --------
 13·80625
        4
 --------
  3·22500
       40
 --------
  9·00000
```

from the multiplication by 40. Thus, if the side of a square field measured 11 chains, 75 links, or 1175 links, the area of the field would contain 1175 × 1175, or 1,380,625 square links, which is equivalent to 13·80625 acres. Then ·80625 acres is equivalent to ·80625 × 4, or 3·22500 roods; and, again, ·22500 roods is equivalent to ·22500 × 40, or 9·00000 poles. The field consequently would measure 13 acres, 3 roods, 9 poles.

Ten arrows must be provided with the chain, about 12 inches long, pointed at one end, so as to be easily pressed into the ground, and turned at the other end, so as to form a ring, to serve for a handle by which they may be hung on the finger.

In using the chain, marks are first to be set up at the extremities of the line to be measured. Two persons are then required to perform the measurement. The chain leader starts with the ten arrows in his left hand, and one end of the chain in his right, while the follower remains at the starting-point, and, looking at the mark or staff at the other extremity of the line to be measured, directs the leader to extend the chain in the direction of this mark. The leader then puts down one of his arrows, and proceeds a second chain's length towards the end of the line, while the follower comes up to the arrow first put down. A second arrow being now put down by the leader, the first is taken up by the follower, and the same operation is repeated till the leader has expended all his arrows. Ten chains, or 1000 links, have now been measured, and this measurement having been noted in the field book, the follower returns the ten arrows to the leader, and the same operations are repeated. When the leader arrives at the end of the line, the number of arrows in the follower's hand shows the number of chains measured since the last exchange of arrows noted in the field book, and the number of links extending from the last arrow to the mark, or staff, at the end of the line being also added, gives the entire measurement of the line. Thus, if the arrows have been exchanged 9 times, and if the follower have 4 arrows, and from the arrow last laid down to the end of the line measure 63 links, the whole measurement will be 9463 links.

$$\begin{array}{r} 9000 \\ 400 \\ 63 \\ \hline 9463 \end{array}$$

To assist, in preserving the line straight as well as to

serve for a check upon the number of chains measured, it is a good method to set up a staff at each ten chains, when the arrows are exchanged.

In using the chain care must be taken to stretch it always with the same tension. As it will give when much used, especially when new, it should constantly be examined, and shortened if necessary. For this purpose a chain's length should be accurately set out on a level piece of ground, or along the coping of a wall,* as a standard, and before taking the chains out for use, any links that may have got bent should first be straightened, and the chain then compared with the standard. The excess of length, if any, above the standard, should then be corrected by shortening equally two of the links, the first link, for instance, and the fifty-first, or first beyond the centre to begin with; on the next occasion, the twenty-fifth and the seventy-fifth; on the third, the forty-ninth and the hundredth; on the fourth, the eleventh and eighty-ninth; and on the fifth, the thirty-ninth and sixty-first. After this, the shortening, when necessary, may be proceeded with by taking in like order the links adjoining these. By this means the shortening is equally distributed throughout the chain; so that not only the entire length, but the lengths from any one point to the other, are kept as nearly as possible correct. The shortening should be made by the removal of a ring at the end of the link, when the elongation is sufficient to require it.

When the ground over which the measurement is taken rises or falls, or both alternately, the horizontal distances are what we require for plotting the survey, and not the actual distances measured along the line of the ground.

For many ordinary purposes the horizontal measurement may be obtained by holding one end of the chain up, so as to keep it, as nearly as can be judged, horizontal, the arrow being placed vertically under the end so held up; and, to sure this, both leader and follower should be provided with a line and weight at the end, to be used as a plummet. When, however, the measured line is of considerable length, and

* When making a standard chain cable for the Government base line at the Cape of Good Hope, Messrs. Elliott Brothers laid down very correctly, on the parapet of Somerset House, a distance of 100 feet. There are brasses inserted into the granite at every 10 feet, on which are cut the divisions. This information may be very useful to surveyors, as there is no other standard length laid down in London.

forms one of the principal lines of an extensive survey, the distances must be measured along the line of ground, and, the angles of elevation and depression of the several inclined parts of the line being afterwards taken with the theodolite (p. 55), or the vertical risings and fallings being taken by the process of levelling with the spirit level (p. 12) and staves (p. 25), the correct horizontal distances must thence be computed.

The following table shows the number of links to be subtracted from every chain, or 100 links, for the angles there set down, being in fact the versed sines of those angles to a radius of 100. The correction for each 100 links, for any angle whatever, may at once be taken from a table of natural versed sines, by considering the first two figures as integers. The correction may also be taken from a table of natural cosines, by subtracting each of the first four figures from 9, and reckoning the first two figures as integers, and the last two as decimals: thus, to find the correction for an inclination of 8° 19′, take the first four figures of the cosine of 8° 19′, which will be 9894, and subtracting each of these four figures from 9, we obtain 0105: then, considering the first two figures of this result as integers, and the last two as decimals, we have 1·05 for the correction, due to the inclination 8° 19′, for every 100 links. If the last figure in the correction thus found be increased by 1, whenever the fifth figure of the cosine is less than 5, the result will be more accurate.

TABLE showing the Reduction, in Links and Decimals of a Link, upon 100 Links, for every Half Degree of Inclination from 3° to 20° 30′.

Angle.	Reduction.	Angle.	Reduction.	Angle.	Reduction.
3° 0′	0·14	9° 0′	1·23	15° 0′	3·41
30	0·19	30	1·37	30	3·64
4 0	0·24	10 0	1·52	16 0	3·87
30	0·31	30	1·67	30	4·12
5 0	0·38	11 0	1·84	17 0	4·37
30	0·46	30	2·01	30	4·63
6 0	0·55	12 0	2·19	18 0	4·89
30	0·64	30	2·37	30	5·17
7 0	0·75	13 0	2·56	19 0	5·45
30	0·86	30	2·76	30	5·74
8 0	0·97	14 0	2·97	20 0	6·03
30	1·10	30	3·19	30	6·33

Recurring to measurements for short distances, if the inclination be considerable, although the method first mentioned may be still applied, an entire chain's length cannot be measured at once; but the length of the incline must be taken in several short steps denoted in the annexed figure; whence the method is called stepping. It is seen here that the required horizontal distance, A B, is equal to the sum of the distances *a c*, *d e*, and *f g*.

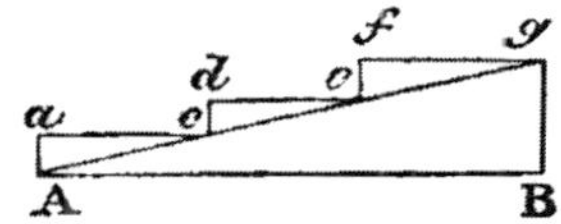

If, however, the length of the line should exceed three chains, with considerable inclination, the measurements should be taken along the ground, and the true *horizontal* distance deduced from the *measured* distance, either by some instrument graduated to measure, roughly, either the angle of inclination, or, what would be much better, the correction to be deducted from the measured distance, or by some simple levelling instrument, such as the reflecting level described hereafter (p. 34). To the latter method we should give the preference; but instruments of various forms, termed *clinometers*, for measuring angles of inclination with sufficient accuracy for the purpose now under consideration, are constructed by the instrument makers, being also found useful for mining and draining.

A simple form of clinometer suitable for the purpose may be constructed as follows: Describe on a card, A, a semicircle about four inches diameter, and divide it into degrees; a card already divided for this purpose may be procured from the instrument maker: we have one procured from Messrs. Elliott, who also supply a similarly divided semicircle, mounted in wood. Procure a box, B B, about half an inch deep, with a glass face to it; fix the semicircle against the back of the box; and attach to the centre of the semicircle a fine cord, c c, with a small bullet, *b*, at the end, to act as a pendulum.

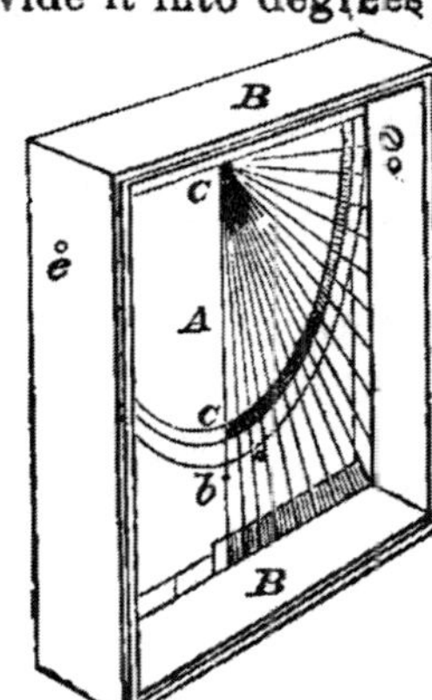

Make two circular apertures in the rim of the box, having their centres exactly in line with the diameter of the semicircle. The opening at *o*, to form the object-end of the instrument, should be about a quarter of an inch in diameter, and should have a fine wire stretched across it horizontally, exactly in line with the diameter of the semicircle; and the aperture at *e*, to form the eye-end of the instrument, should be about the twentieth of an inch in diameter. In using it the height of the observer's eye from the ground should be conspicuously marked on an offsetting staff, described below, and a man being sent forward the required distance, to plant the staff on the ground and hold it in a vertical position, the sight is to be directed to the mark on the staff, and the pendulum, as soon as it is at rest, points to the degree of inclination, and to the correction to be made, if this also be marked on the card, as it is advisable that it should be.

The advantage of Gunter's chain consists in its adaptation to the superficial measure of land in acres, &c.; but when a survey is to be made for the purpose of linear measurements only, or when it may be more convenient to compute the area in square feet, a chain 100 feet long, divided into links of a foot long, is to be preferred. Such a chain is the best for military surveying.

Offsets, perpendicular to the main line, to hedges and remarkable objects on either side of it, are measured from the chain as it lies stretched upon the ground, by means of an offsetting staff. This staff should be 10 links in length, and divided into links. With Gunter's chain the staff, then, will be 6·6 feet, or 6 feet 7·2 inches long, while with the 100 feet chain it will be 10 feet in length.

The offsetting staff may be made of hickory, ash, or red pine. It should be about an inch and a quarter in diameter in the centre, tapering until it is about three-quarters of an inch diameter at the ends. One end should have a ferrule with steel point for sticking in the ground; and at the other end should be another ferrule with a strong hook fastened to the side, for the purpose of taking hold of one of the brass handles of the chain, when it is necessary to drag it through a hedge. The links should be marked with rings drawn round the rod, and cut in, and the centre should have some special mark, as a peculiar coloured ring, or a V, to denote five links. A surveyor might also keep by him one particular offsetting staff, marked conspicuously at a length denoting the height of his eye. Offsets are measured by laying the staff flat on the ground, pointing to the spot to be measured to, and turning it over and forward, always pointing to the same spot.

Offsets may also be measured with a measuring tape very conveniently and accurately, and it is particularly convenient for the measurement of buildings, in surveying through a town. It requires, however, two persons to attend to it, and is liable to fray out and get dirty. Messrs. Parkes have taken out a patent for a tape, which they call the inflexible tape. It is woven from a strong fibre; and being saturated with a waterproof coating, is impervious to the effects of wet.

Surveys of small extent can be made with the chain only, and the mode of proceeding consists in chaining lines to form triangles, either included within, or inclosing, the area to be surveyed. The triangle, in fact, is the only figure the angular points of which can be determined by the lengths of the sides alone. The best form for the triangles is the equilateral, and they should, therefore, be made to approach this form as nearly as may be consistent with other important considerations. These considerations are that the chain lines should be as few as possible, and the offsets short. Lines across the triangles should be measured, thus forming new triangles, and checking the position of the angular points, common to them with the original triangles. Such lines are called *tie lines.*

It is best to have the principal triangles as large as possible; and, as it is of the greatest importance that the long lines necessary for this purpose should be made accurately straight, they must be marked out by station poles and ranging rods.

The station poles to be fixed at the ends of the principal lines should be fifteen or twenty feet long; they should be clean, straight, natural spars, and should be carefully fixed in a vertical position, by trying them with a line and plummet on every side.

The ranging rods should be about eight or nine feet long, and perfectly straight, shod with iron, and tipped with steel. They should be painted white; and both they and the station poles should have small flags fastened to their tops.

CHAPTER II.

INSTRUMENTS FOR THE MEASUREMENT OF VERTICAL DISTANCES.

THE SPIRIT LEVEL.

Certain parts of the principal instruments used in surveying, and in astronomical observations, require to be adjusted in truly horizontal positions; and, to arrive at this adjustment, one or more subsidiary instruments, called *spirit levels*, are attached to such principal instruments. The spirit level attached to a good telescope, furnished with a compass, and such means of correct adjustment as we shall presently describe, becomes also itself an important instrument, being used in the department of surveying termed levelling, which consists in measuring the vertical distances between various stations.

The spirit level consists of a glass tube, differing from the cylindrical form by having its diameter largest in the middle, and decreasing slightly and with great regularity from the middle to the ends. The tube is nearly, but not quite, filled with spirits of wine, thus leaving in it a bubble of air, *b b*, which rises to the highest part of the tube, so as to have its two ends equally distant from the middle when the instrument is in adjustment, as represented in the annexed figure.

The tube is generally fitted into another tube of metal, and attached to a frame terminating in angular bearings, by which the level can either be suspended from, or else be supported on, cylindrical pivots. When, however, the level forms a permanent part of any instrument, the manner of attaching it is modified to suit the particular form of the instrument to which it is attached. A small and accurately-divided scale is attached to the best instruments, or otherwise a scale is scratched upon the glass tube itself, as represented in the figure given above.

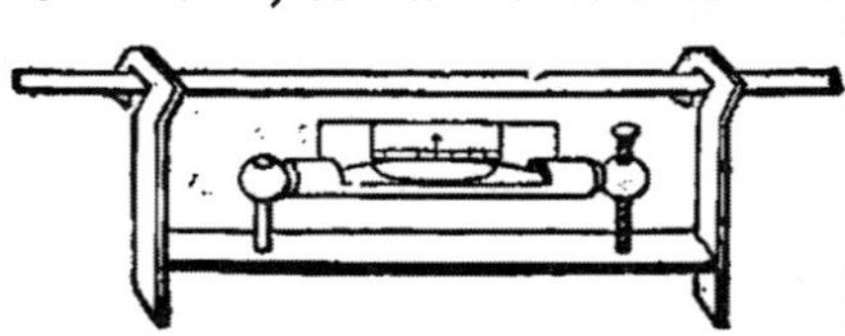

The annexed figure is a representation of such a level as is used for levelling the axis of the best astronomical instruments. It is provided with a fixed scale, seen in the figure, and is suspended by means of accurately constructed angular bearings.

The following criteria of a good level are extracted from Dr. Pearson's valuable work on Practical Astronomy, before referred to.

"Firstly, the bubble must be long enough, compared with the whole tube, to admit of quick displacement, and yet not too long to admit of its proper elongation by low temperature.

"Secondly, the curve must be such, that the sensibility and uniform run of the bubble will indicate quantities sufficiently minute, while those quantities correspond exactly to the changes of inclination, as read on the graduated limb of the instrument of which it forms a part.

"Thirdly, the bubble must keep its station when the angles are moved a little round the pivots of suspension.

"Fourthly, the opposite ends of the bubble must vary alike in all changes of temperature, or, in other words, the ends of the bubble must elongate or contract alike in opposite directions, so that the middle point may always be stationary.

"Fifthly, the angles of the metallic end-pieces must be so nicely adjusted that reversion on horizontal pivots that are equal will not alter the place of the bubble.

"Sixthly, the distance between the two zeros of a fixed scale, when such a graduated scale is used, should be equal to the length of the bubble at the temperature of 60° of Fahrenheit's scale, and should be marked at equal distances from the visible ends of the glass tube. Then, as the bubble lengthens by cold or shortens by heat, its extreme ends may always be referred to these fixed marks, 0 0, on the scale, and will fall either within, upon, or beyond them, according to the existing temperature. The number of subdivisions of the scale that each end of the bubble is standing at, counted from the fixed zero marks, at the instant of finishing an observation, must always be noted, that an allowance may be made for the value of the deviation in seconds, or as the case may require.

"Seventhly, when the two ends of the bubble are not alike affected by a change of temperature, the scale should be a detached one, and adjustable to the new zero points, by an inversion of the level.

"Eighthly, when the scale has only one zero at its centre,

which is a mode of dividing the least liable to misapprehension, the positions must be reversed at each observation, and both ends of the bubble read in each position; for in this case, if any change has taken place in the true position of this zero, the resulting error will merge in the reduction of the observation. This mode of graduating is generally practised on the Continent."

We proceed now to the description of the most accurate instruments for measuring the differences of level, or vertical distances, between different stations.

Since the introduction of railways, the astonishing rate at which they have progressed, and the great amount of levelling work involved in their construction, have necessarily directed the attention, both of surveyors and instrument makers, to the production of an instrument which shall combine the advantages of expedition in its use, accuracy in the results obtained, and durability under the effect of constant wear and tear.

The best form is most undoubtedly the Y level; and the best construction of Y level has resulted from the adaptation of the Dumpy level to this form. We shall, therefore, proceed first to the complete description of the Y level with its adjustments, premising that, in other forms of level, not possessing the same facility for correcting the adjustment at any time, the adjustment must be made by the maker, in the manner here described, before finally fixing the parts of the instrument; and dependence must then be placed upon its solidity, and the accurate and close construction of all the joints, for the permanency of correct adjustment.

THE Y LEVEL.

The following figure represents this instrument. A is an achromatic telescope, resting upon two supporters, which in shape resemble the letter Y, and are consequently called the Y's. The lower ends of these supporters are let perpendicularly into a strong bar, which carries a compass box C. This compass box is convenient for taking bearings, and has a contrivance for throwing the needle off its centre, when not in use. One of the Y supporters is fitted into a socket, and can be raised or lowered by the screw B.

Beneath the compass box, which is generally in one piece with the bar, is a conical axis passing through the upper of two parallel plates, and terminating in a ball supported in a socket. Immediately above this upper parallel plate is a

collar, which can be made to embrace the conical axis tightly by turning the clamping screw E, and a slow horizontal motion may then be given to the instrument by means of the tangent screw D. The two parallel plates are connected together by the ball and socket already mentioned, and are set firm by four mill-headed screws, which turn in sockets fixed to the lower plate, while their heads press against the under side of the upper plate, and thus serve the purpose of setting the instrument up truly level.

Beneath the lower parallel plate is a female screw, adapted to the staff-head, which is connected by brass joints with three mahogany legs, so constructed as, when shut together, to form one round staff, a very convenient form for portability, and, when opened out, to make a very firm stand, be the ground ever so uneven.

The spirit level *l l* is fixed to the telescope by a joint at one end, and a capstan-headed screw at the other, to raise or depress it for adjustment.

In looking through a telescope a considerable field of view is embraced; but the measurements, indicated by any instrument, of which the telescope may form a part, will only have reference to one particular point, which particular point is considered as the centre of this field of view. We must therefore place some fixed point in the field of view, at the focus of the eye-piece, and the point to which the measurement will have reference will be that point of the object viewed, which appears to be coincident with this fixed point, or which, as the technical phrase is, is bisected by the fixed point.

The intersection of two fixed lines will furnish us with such a fixed point, and, consequently, two lines of spider's thread are fixed at right angles to each other in the focus of the eye-piece. They are attached by a little gum to a brass ring of smaller dimensions than the tube of the telescope, and which is fixed to the tube by four small screws, *a*, *b*, *c*, *d*. If the screw *d* be eased, while at the same time *c* is tightened, the ring will be moved to the right; but if *c* be eased, and *d* tightened, the ring will be moved to the left; and, in a like manner, it may be moved up or down by means of the screws *a* and *b*.

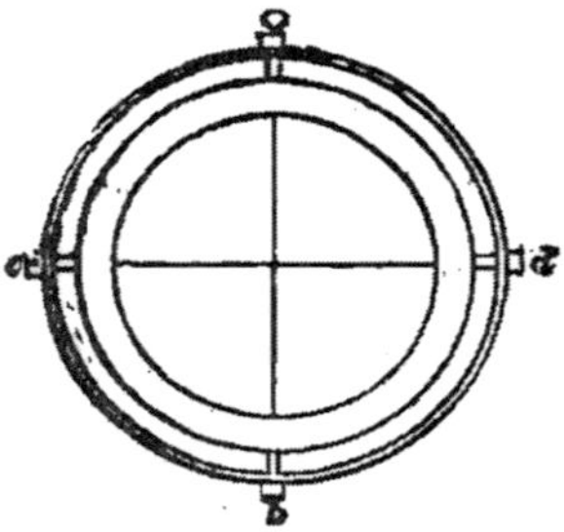

When the instrument is in adjustment, the axis of the tube of the telescope is set truly horizontal by means of the level beneath it, and the line of observation ought consequently to be parallel to this axis. Let A represent the proper position of the intersection of the cross wires, and *o* A the direction of the axis of a pencil of light passing through the object-glass and coming to its focus at A. Then, the axis of the tube of the telescope being set truly horizontal, the line A *o* is also truly horizontal, and every point bisected by the intersection of the cross wires will be situated on the prolongation of the horizontal line A *o*.

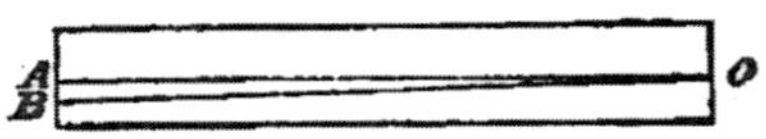

Suppose now the position of the diaphragm carrying the cross wires to have become deranged, so that the point of intersection is moved to B, then every point bisected by the intersection of the cross wires will be on the prolongation of the line B O, and will, consequently, be below the true level point on the line A O.

Let now the telescope be turned half round in the Y's, and

let the annexed figure represent it in its new position; then, in this new position of the telescope, the prolongation of the line B *o* will rise above the prolongation of the level line

A *o*, and, at the same distance from the telescope, the point now bisected by the intersection of the cross wires will be as much above the true level point on the line A *o*, as the point before bisected by them was below it. The true level point is, therefore, midway between the two points observed in the two positions of the telescope, and the diaphragm carrying the cross wires is to be moved by means of the screws, *a*, *b*, *c*, *d*, till their point of intersection coincides with that true level point. The telescope is then to be again turned round upon the Y's, and if the same point be still bisected by the intersection of the cross wires, they are in their proper position; but if not, the same method of adjustment must be repeated, till the same point is bisected by the intersection of the cross wires in every position of the telescope.

The error from misplacement of the spider-lines has a technical denomination. The line *o* A, or *o* B, from *o* to the point of intersection of the cross wires, is called the *line of collimation*, and the error arising from their derangement, which we have shown the method of detecting, and correcting, is called the *error of collimation.*

When the image of the object viewed, formed by the object-glass, falls either short of, or beyond, the place of the cross wires, the error arising from this cause is called *parallax.* The existence of parallax is determined by moving the eye about when looking through the telescope, and observing whether the cross wires change their position, and are flittering and undefined.

To correct this error, first adjust the eye-piece by means of the movable eye-piece tube, till you can perceive the cross wire clearly defined and sharply marked against any white object. Then by moving the milled-headed screw A, at the side of the telescope, the internal tube, *a a*, is to be thrust outwards or drawn inwards, until you obtain the proper focus, according to the distance of the object, and you are enabled to see, at the same time, the object clearly, and the intersection of the wires, clearly and sharply defined, before it. The existence of parallax is very inconvenient, and, where disregarded, has frequently been productive of serious error. It will not always be found sufficient to set the eye-piece first, and the object-glass afterwards. The setting of the object-glass, by introducing more distant rays of light, will affect the focus of the eye-piece, and produce parallax, or indistinctness of the wires, when there was none before: the eye-piece must, in this case, be adjusted again.

Generally, when once set for the day, there is no occasion for altering the *eye-piece*, but the *object-glass* will of course have to be altered at every change of distance of the object.

In adjusting the instrument, the parallax should be first corrected, and then the error of collimation. The line of collimation being thus brought to coincide with the axis of the tube of the telescope, two further adjustments are necessary: the first to adjust the bubble-tube, so that it may truly indicate when the axis of the telescope is horizontal; and the second to set the axis of the telescope perpendicular to the vertical axis round which the instrument turns.

To adjust the Bubble-Tube.—Move the telescope till it lies in the direction of two of the parallel plate screws, and by giving motion to these screws bring the air-bubble to the centre of its run. Now reverse the telescope carefully in the Y's, that is, turn it end for end; and should the bubble not settle at the same point of the tube as before, it shows that the bubble-tube is out of adjustment, and requires correcting. The end to which the bubble retires must then be noticed, and the bubble made to return one-half the distance by turning the parallel plate screws, and the other half by turning the capstan-headed screw at the end of the bubble-tube. The telescope must now again be reversed, and the operation be repeated, until the bubble settles at the same point of the tube, in the centre of its run, in both positions of the instrument. The adjustment is then perfect, and the clips which serve to confine the telescope in the Y's should be made fast.

The bubble-tube has sometimes three vertical screws at one end instead of a single screw to adjust it. By this means the tube can be adjusted to horizontal as well as vertical parallelism with the axis of the telescope. This adjustment is tested by gently rocking the telescope round in the Y's, through a small angle on either side of the position it occupies. When the level-tube is properly placed, the correction, if required, is made with the three vertical screws, by pushing or pulling the level-tube from the end the bubble travels towards, when brought round a small portion of a turn; a very slight touch of the screws will suffice.

Lastly, to set the Axis of the Telescope perpendicular to the Vertical Axis round which the Instrument turns.—Place the telescope over two of the parallel plate screws, and move them, unscrewing one while screwing up the other, until the bubble of the level settles in the centre of its run; then turn the instrument half round upon the vertical axis, so that the contrary ends of the telescope may be over the same two screws, and, if the bubble does not again settle at the same point as before, half the error must be corrected by turning the screw B, and the other half by turning the two parallel plate screws over which the telescope is placed. Next turn the telescope a quarter round, that it may lie over the other two screws, and repeat the process to bring these two screws also into adjustment; and when, after a few trials, the bubble maintains exactly the same position in the centre of its run, while the telescope is turned all round upon the axis, this axis will be truly vertical, and the axis of the telescope being horizontal by reason of the previous adjustment of the bubble-tube, will be perpendicular to that vertical axis, and remain truly horizontal while the telescope is turned completely round upon the stand. The adjustment is therefore perfect.

The object of the above adjustments is to make the line of collimation move round in a horizontal plane, when the instrument is turned round its vertical axis, and the methods above explained suppose that the telescope itself is constructed with the utmost perfection, so that the axis of the tube carrying the object-glass is always in the same straight line with the axis of the tube which carries the diaphragm with the cross wires. If this perfection in the construction of the instrument does not exist, the line of collimation will vary as the tube carrying the object-glass is thrust out and drawn in, to adjust the focus for objects of different distances.

In the best constructed instruments, however, of whatever pattern, the collimation, having been adjusted by the maker by the methods detailed above, is scarcely liable to derangement afterwards. We do not, therefore, think much advantage would be derived from any plan of collimating a badly-constructed instrument. If the axis of the tube which carries the diaphragm moves accurately in a straight line, although this line should not be the same with the axis of the tube carrying the object-glass, the points viewed, when the focus is correctly adjusted for each observation, will lie in a straight line, which can be made horizontal, and when from a dropping of the tube as it is drawn out, or a curvilinear motion produced either by inferior workmanship originally, or from having received some subsequent injury, we know of no means of collimation by which such a defect can be remedied, and it can only be put right by the hands of the instrument maker.

Mr. Gravatt published a method of collimating with the view of removing the error arising from imperfection in the slides of the telescope, and at the same time eliminating the small errors arising from the curvature of the earth and the horizontal refraction. The error, however, arising from the imperfection of the slide cannot be removed, unless the maker can improve the motion of the tubes; and the most that Gravatt's mode of collimating can do is so to divide the error arising from the readings at different distances as to diminish its importance. With respect to the errors arising from the curvature of the earth and from refraction, as they are almost, if not entirely, eliminated by the method of taking the levels from positions nearly equi-distant from the station observed, as usually practised, the advantage of considering them in any mode of collimating is extremely problematical.

We subjoin, however, Mr. Gravatt's mode of collimating, which may be used to test the motion of the diaphragm, if we have reason to fear that it does not move in a straight line; but for levels in which, as in the Dumpy, the telescope cannot be reversed on its bearings, a simpler mode of testing, and, if necessary, of adjusting the collimation is to be preferred.

To examine and correct the Collimation by Mr. Gravatt's Method.—"On a tolerably level piece of ground, drive in three stakes, at intervals of about four or five chains, calling the first stake *a*, the second *b*, and the third *c*.

"Place the instrument half way between the stakes *a* and *b*, and read the staff A, placed on the stake *a*, and also the staff B, placed on the stake *b*; call the two readings A and B;

then, although the instrument be out of adjustment,* yet the points read off will be equi-distant from the earth's centre, and consequently level.

"Now remove the instrument to a point half way between *b* and *c*. Again read off the staff B, and read also a staff placed on the stake *c*, which call staff C (the one before called A being removed into that situation). Now, by adding the difference of the readings on B (with its proper sign) to the reading on C, we get three points, say A′, B′, and C′, equi-distant from the earth's centre, or in the same true level.

"Place the instrument at any short distance, say half a chain beyond it, and, using the bubble merely to see that you do not disturb the instrument, read all three staffs, or, to speak more correctly, get a reading from each of the stakes, *a*, *b*, *c*; call these three readings A″, B″, C″. Now, if the stake *b* be half way between *a* and *c*,† then ought C″ − C′ − (A″ − A) to be equal to 2 [B″ − B′ − (A″ − A′)]; but if not, alter the screws which adjust the diaphragm, and consequently the horizontal spider line, or wire, until such be the case; and then the instrument will be adjusted for collimation.

"To adjust the spirit-bubble without removing the instrument, read the staff A, say it reads A‴, then adding (A‴ − A′) with its proper sign to B′ we get a value, say B‴.

"Adjust the instrument by means of the parallel plate screws,‡ to read B‴ on the staff B.

"Now, by the screws attached to the bubble-tube, bring the bubble into the centre of its run.

"The instrument will now be in complete practical adjustment for level, curvature, and horizontal refraction, for any distance not exceeding ten chains, the maximum error being only $\frac{1}{1000}$th of a foot."

Gravatt's method of collimating is objectionable from its

* The axis of the instrument is to be set vertical by means of the parallel plate screws, by placing it over each pair alternately, and moving them until the air-bubble remains in the same position, while the instrument is turned half round upon its axis.

† Whatever be the distances between the stakes *a*, *b*, *c*, the following proportions ought to hold, viz.:—

The distance from *a* : *b* : the distance *a* to *c* : : B″ — B′ — (A″ — A′) : C″ — C′ — (A″ — A′) : if not the diaphragm does not move in a straight line.

‡ If this adjustment be made by the screw B, instead of the parallel plate screws, the line of collimation will be brought into its proper position with respect to the vertical axis.

needless complexity in the case of a good instrument, and from its failing to correct a bad one, since the curvilinear motion of the diaphragm, if any, which it indicated, can only be set right by the instrument maker.

To test and adjust, if necessary, the Collimation of a Level, the following simple method is the best:—If a pool of water large enough be at hand, drive in two stakes, three or four chains apart, exactly level with the surface of the water. If no such pool of water be convenient, select a piece of ground as nearly level as possible, and drive in two stakes, three or four chains apart, until the readings obtained from a point midway between them be exactly the same for each: then, if the instrument have been set up with its axis truly vertical, the tops of the two stakes will be exactly on the same level, as is the case with the stakes in the pool of water.

Set up the instrument in the line of the stakes, about half a chain beyond one of them; then if the readings of the staff, placed alternately on each of the stakes, be the same, the line of collimation is correct; but if not, alter the position of the diaphragm, until exactly the same reading is obtained from both stakes, and the collimation will be adjusted.

The instrument having been carefully examined and rectified, in using the instrument the screw B should never be touched; but at each station the parallel plate screws alone should be used for setting the axis round which the instrument turns truly vertical, when, in consequence of the adjustments previously made, the line of collimation will be truly level. For this purpose the telescope must be placed over each pair of the parallel plate screws alternately, and they must be moved till the air-bubble settles in the middle of the level, and the operation being repeated till the telescope can be turned quite round upon the stand, without any change taking place in the position of the bubble, the instrument will be ready to read off the graduations upon the levelling staves, which we proceed to describe.

GRAVATT'S DUMPY LEVEL.

To Mr. Gravatt is undoubtedly due the credit of having, in the construction of the Dumpy level, led the way to all the improvements which have recently been made in this instrument, terminating in the improved Y level, than which we can imagine nothing better adapted to satisfy all the requisites of a perfect instrument.

We shall here merely mention the characteristic points of Gravatt's level; and its construction, and mode of adjustment by the maker, as well as the method of using it, will be fully understood from the detailed description of the Y level (p. 12).

This instrument is furnished with an object-glass of large aperture and short focal length; and, sufficient light being thus obtained to admit of a high magnifying power in the eye-piece, the advantages of a much larger instrument are

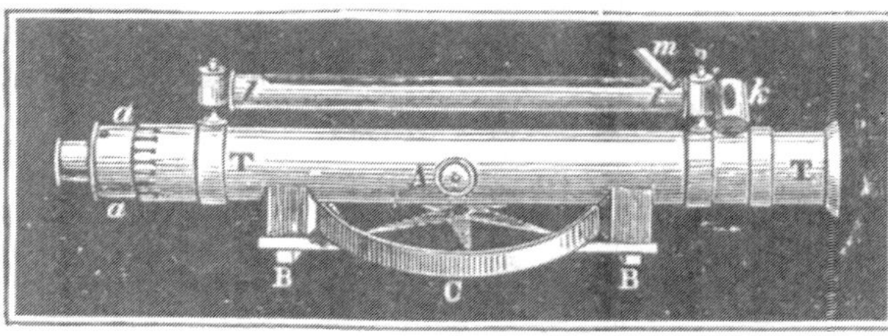

secured, without the inconvenience of its length: hence has arisen the name of Dumpy, given to the instrument. The diaphragm is carried by the internal tube, *a a*, which is nearly equal in length to the external tube. The external tube, T T, is sprung at its aperture, and gives a steady and even motion to the internal tube, *a*, *a*, which is thrust out, and drawn in, to adjust the focus for objects at different distances, by means of the milled-headed screw A. The spirit level is placed above the telescope, and attached to it by capstan-headed screws, one at either end, by means of which the bubble can be brought to the centre of its run.

The telescope is attached to a horizontal bar, but room is just left between the telescope and the bar for the compass-box. By this means the instrument is rendered very firm and compact, in comparison with the old instrument, in which the level was placed below the telescope.

A cross level, *k*, is placed upon the telescope at right angles to the principal level, *l*, *l*, by which we are enabled to set the instrument up at once with the axis nearly vertical. This is intended to give the advantages of expedition in setting the instrument, and saving in the wear and tear of the parallel plate screws. From its extreme shortness, however, it does not perform its duty well, and to give it greater length, to a sufficient extent to be really useful, would be very inconvenient, and it would be then very liable to disarrangement.

A mirror, *m*, mounted upon a hinge-joint, is placed at the end of the level *l*, *l*, so that the observer, while reading the staff, can at the same time see that the instrument retains its proper position—a precaution by no means unnecessary in windy weather, or on bad springy ground.

The telescope is attached to the horizontal bar by capstan-headed screws, B, B, by which the line of collimation is set perpendicular to the vertical axis; and the instrument is set up upon parallel plates, as will be understood from the description of the Y level.

ELLIOTT'S IMPROVED DUMPY LEVEL.

In Gravatt's, as in levels previously constructed, the level, telescope, and horizontal bar were made in one piece, and the parallel plate, with conical axis, formed a head-piece to which the former was screwed when in use. Now a tighter or slacker connection between the telescope and the parallel plate portion of the instrument would produce a disparagement, which would manifest itself in the impossibility of making the bubble maintain the centre of its run, whilst the instrument was turned completely round on its axis. For this reason it is imperative always, in levels so fixed to the parallel plates, to turn them constantly to the right, and never to the left, in adjusting them and pointing them for use. If they were often turned to the left, not only would their correctness be certainly injured, but they might even get screwed off, and fall.

In the level now under consideration, the horizontal bar and conical axis are in one piece, and cannot be disunited, so that the level, telescope, and parallel plates are not adapted for separation, or liable to the disarrangement which accompanies it. Secondly, the horizontal bar is made so as to have its principal thickness and strength in the vertical direction, and not in the horizontal, where there is neither any strain nor any material injury that could be produced by a strain. Thirdly, the cross-level *k* is removed, and a circular level is placed below the telescope instead of the compass box.

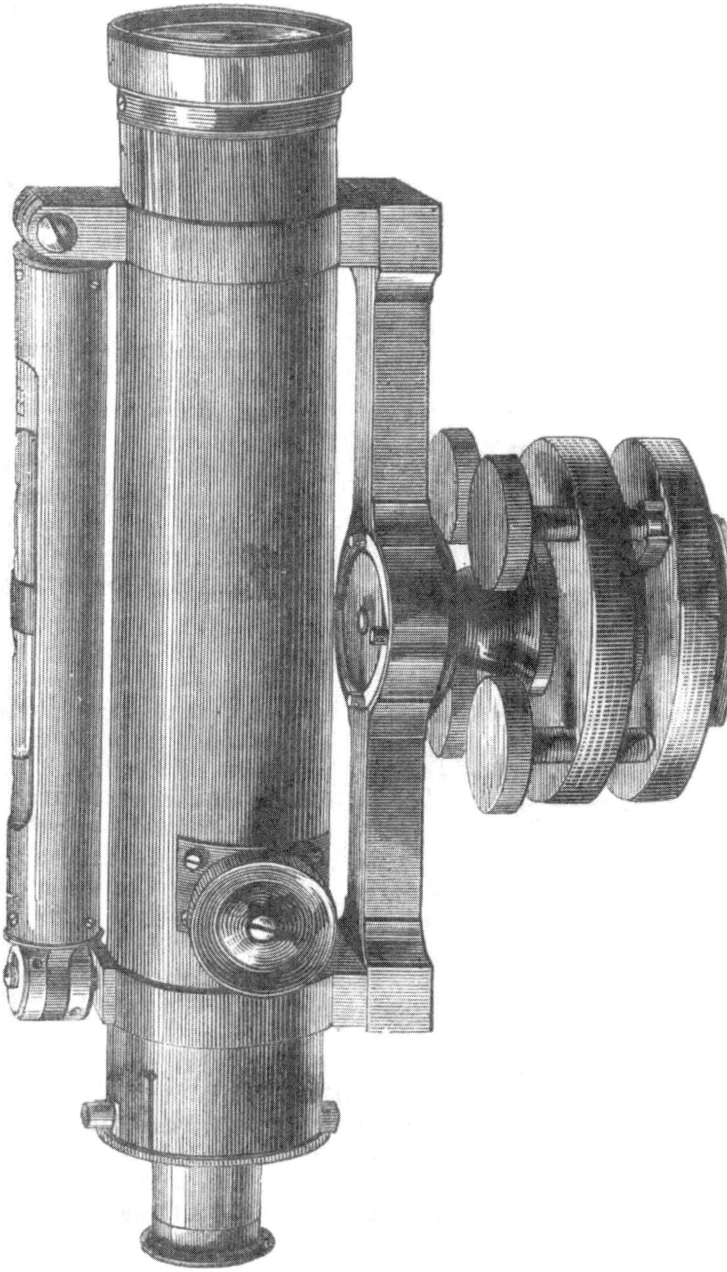

Elliott's Dumpy Level.

This adds considerably to the simplicity and compactness of the instrument; and this form of level, being equally sensitive in every direction, is admirably adapted for the first rough adjustment of the instrument by moving the legs of the tripod stand.

In levelling, the ground passed over has always been previously surveyed, and, besides, the bearing, if required, can be taken with a pocket compass of the ordinary construction, or with a prismatic compass, better than with one attached to the level.

THE IMPROVED DUMPY Y LEVEL.

This instrument differs from the improved Dumpy level last described only in resting the telescope upon Y-shaped bearings attached to the horizontal bar, instead of being permanently fixed to this bar; and when in adjustment, the telescope is held in position by clips attached to the Y's. One of the Y supporters may be fitted in a socket, as in the ordinary Y level, already described, and may be raised or lowered by a capstan-headed screw; or the required effect may be produced by carefully rubbing off slight portions of the surface of one of the Y's.

LEVELLING STAVES.

The best constructed levelling staff consists of three parts, which pack together by sliding, in a neat manner, and, when drawn out for use, make a staff from 14 to 17 feet long; the bottom joint is about four inches wide; and the whole length is divided into hundredths of a foot, coloured black and white alternately. The feet and tenths are numbered, the figures for the tenths being always painted black, while those for the feet are distinguished by their larger size, and often also by being painted red. In the annexed figures, Fig. 1 shows the whole of one foot, reduced to *one-fourth* of the actual size, and Fig. 2 shows a length of sixteen-hundredths of a foot, *full* size. Every alternate tenth is numbered with a figure, the length of which exactly occupies a tenth: the top of the figure 1, for instance, being exactly in line with the top of the black division, which measures one-tenth of a foot from the base of the staff, and the bottom of the figure 3 being in line with the top of the black division, which measures two-tenths of a foot from the base. It will also be seen that all the horizontal lines in the figures are also in line with some particular hundredths on the scale: the bottom of the top stroke of the 3 with twenty-eight hundredths, the top of central portion with twenty-six hundredths, the bottom of this portion with twenty-five, and the top of the bottom portion with twenty-two hundredths. This arrangement is an assistance in reading the division off, and, when long sights are required, every assistance of this kind is valuable. All the even hundredths reach to the top of a black stroke, and all the odd hundredths, to the

Fig. 1.

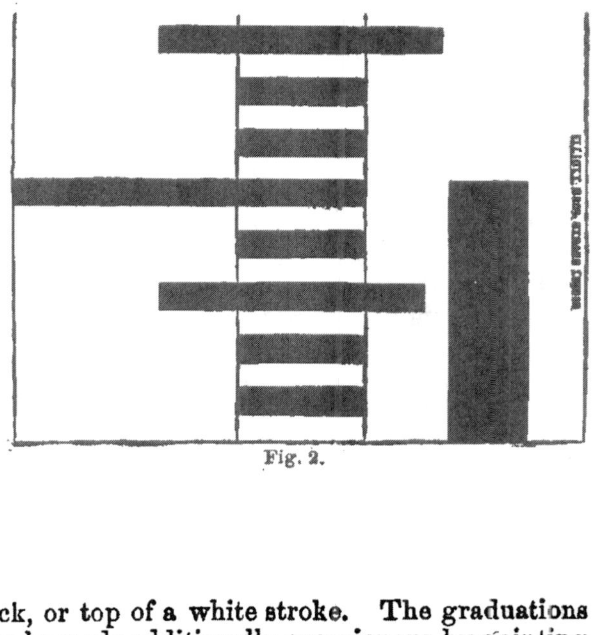

Fig. 2.

bottom of a black, or top of a white stroke. The graduations for feet may also be made additionally conspicuous by painting a border on the left hand side of the staff, alternately black and white, the black and white divisions each occupying the length of a foot, and a vertical white band, about three-eighths of an inch wide, being left between the border and the column of graduations.

It is to be observed that, when viewed through the telescope, the staff appears inverted, as well as the figure upon it, the telescope having an astronomical eye-piece, to avoid the loss of light that would be incurred by the use of a terrestrial, or erecting, eye-piece. This produces no inconvenience, the eye soon becoming familiar with their appearance in this form, and readily reading the figures downwards, as in fact an upward measurement from the ground.

Levelling.—The operation of determining the difference of level between two stations, by observations made at a single station, is called *simple levelling*, and is performed as follows:—

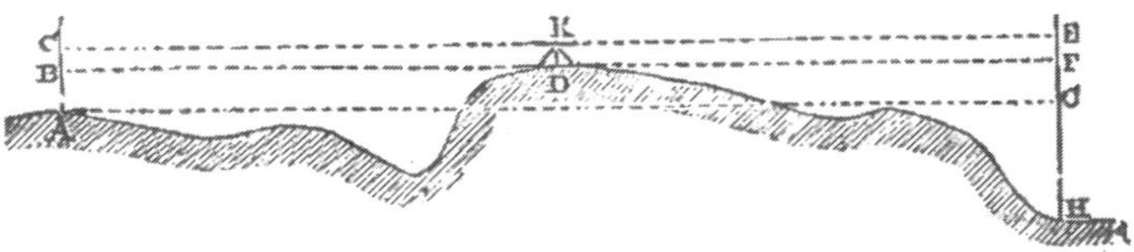

Let A and H be two points whose difference of level is required. Plant the instrument at D, and adjust it to a horizontal position. Read off A C, the height of a staff held at A, then turn the telescope round and read off E H, the height of a similar staff, or of the same staff held at H. Then A C is the height of K, the axis of the telescope above the point A, and E H is the height of K above another point H; and it is clear that E H − A C = G H, the difference of level between A and H.

In this operation the station A, at which the staff is first read off, is called the back station, and the station H is called the fore station; and, if the reading of the staff at the back station be greater than that at the fore station, the difference of level is called a rise; but if the reading at the back station be less than that at the fore, as in the example just given, the difference of level is called a fall.

When from the nature of the ground, or the great distance between the two points, they cannot both be observed from a single spot, a series of simple levels must be taken, the fore station at each operation being made the back station at the next operation; and from the combination of all the results thus obtained the required difference of level is found. In these operations, care must be taken, in going over soft ground, lest the staff at the fore station, when turned round to be read as the staff at the back station in the next operation, should sink farther into the ground; and, to prevent this, the foot of the staff must be placed upon a flat, hard substance as a piece of slate or tile. There is a simple instrument called a tripod, sold for this purpose by the instrument makers, being simply a plate of iron with a small rounded projection in the centre, two small spikes at the side to fix it in its place, and a short chain to lift it by, when the staff-holder wishes to remove from his place.

In determining by this method the difference of level between two distant points, it is immaterial by what route we proceed from one to another, so that such spots may be selected for the intermediate stations as are most convenient for the purpose. The bearings of the stations from the instrument are also matter of indifference; but the more nearly the instrument is equidistant from the two stations observed at each operation, the more correct will be the result obtained, the errors in the back readings compensating, for the most part, the errors in the fore readings, whether the errors arise from refraction* and curvature,† or from the imperfect adjustment of the instrument. The sum of the distances from the instrument to the fore stations should also be nearly equal to the sum of those to the back stations.

If, then, the object be only to obtain the difference of level of two points, we have only to record in two separate columns the readings of the staff at the back stations and fore stations respectively, and the difference of the sums of these readings will be the difference of level required. Thus, if the difference of level between two points A and B be required, and if the readings at A and B, and three intermediate stations ⊙ 1, ⊙ 2, ⊙ 3, be recorded as follows, viz.:—

* The error of refraction is that arising from the bending of the rays of light during their passage through the atmosphere, and makes all objects appear higher than they really are.

† The object of levelling is to determine points upon a spherical surface or equally distant from the earth's centre, or to determine the differences of the distances of a series of points from the earth's centre. The line of sight, or prolongation of the line of collimation, however, is a tangent to the spherical surface, and therefore the points observed upon this line are really above the level of the point of observation. The correction for curvature is therefore additive, while that for refraction is subtractive.

		Back ⊙	Fore ⊙		
		Feet.	Feet.		
Reading of staff at	A .	3·65	5·80	Reading of staff at	⊙ 1
„	⊙ 1 .	2·05	8·50	„	⊙ 2
„	⊙ 2 .	3·89	8·40	„	⊙ 3
„	⊙ 3 .	5·28	14·35	„	B
		14·87	37·05		
			14·87		
			22·18		

Then 22·18 feet is the fall from A to B, or A is 22·18 feet above B.

When, however, it is required not only to find the difference of level between two distant points, but to make such observations as shall enable us to draw a section, exhibiting the undulations of the ground along some specified route from the one point to the other, then the stations must be so chosen, that one of them shall be at the commencement of each change in the inclination of the ground; the distances between the stations must also be carefully measured; and it is further advisable to note the distances and bearings of the stations from the instrument, which it will be more convenient now to place on a point in the specified route between the stations. The bearings may be best taken with a prismatic compass, described hereafter (p. 40); but the compass attached to the instrument will do, if there be one, or any pocket compass may be used.

In drawing the section, it is the horizontal distances between the several stations that must be laid down. For short distances, or over very irregular ground, such horizontal measurements may be obtained by bidding an assistant hold one end of a measuring tape close to the ground at the highest end of the distance, and holding the other end above the ground, stretching the tape in a horizontal line, a stone let fall from this end then marking upon the ground the point to which the measurement reaches. But, when the ground rises and falls in long regular slopes, the measurements should be taken along the slopes, and then be reduced to horizontal distances by calculation. If the rise or fall is but slight, this reduction may be altogether disregarded, the difference between the horizontal and hypothenusal measurements not exceeding the limits of error in the measurements themselves

Care should be taken to record all the observations in a clear and intelligible form, and for this purpose a field book may be prepared of the accompanying form.

	Distance to stations.			Staff readings.					
	From starting point.	From instruments.	Bearings.	Back.	Fore.	Heights above datum.	Reduction.	Reduced horizontal distances.	REMARKS.
	feet.	feet.	°	100·00		feet.			
Back ⊙		210	30·000	3·65		100·00			Back ⊙ 300 feet from hedge, windmill bearing 125° from instrument, church spire bearing 223·5°.
Fore ⊙	460	250	120·10		5·80	3·65 103·65 5·80 97·85			
Back ⊙	460	180	300·00	2·05		97·85			Road to lime kilns.
Fore ⊙	320 780	140	120·00		8·50	2·05 99 90 8·50 91·40			
Back ⊙	780	180	299·75	3·89		91·40			
Fore ⊙	380 1160	200	119·40		8·40	3·89 95·29 8·40 86·89			
Back ⊙	1160	180	300·25	5·28		86·89			
Fore ⊙	360 1520	180	120·00		14·35	5·28 92·17 14·35 77·82			
Back ⊙	1520	300	300·00	12·25		77·82			
Fore ⊙					15·78	12·25 90·07 15·78 74·29			Bottom of canal distant 150 feet.
Back ⊙				15·78		74·29			
Fore ⊙	580 2100	280	120·00		9·21	15·78 90·07 9·21 80·86			
Back ⊙	2100	205	300·15	11·05		80·86			
Fore ⊙	400 2500	195	120·00		1·12	11·05 91·91 1·12 90·79			
		2500		153·95 63·16 90·79	63·16				

In the first column are entered the distances between the several stations, which, being successively added to the preceding total, give the total distances of each station from the starting-point; in the next column are entered the distances of the stations from the instrument; and in the third, are entered the bearings of the stations from the instrument. In the fourth and fifth columns are entered the readings of the staves; and in the sixth column the heights above datum of the several stations are computed, by adding the back reading to the height last found, and subtracting the fore reading from the sum. The seventh and eighth columns are added for performing the reduction of the measured distances to horizontal distances, when the slope is sufficient to render this reduction necessary. In carrying forward the distances to the next page of the book, the total reduced horizontal distance should be carried to the top of the first and second columns, instead of the total measured distance along the slope; but such substitutions should not be made at any other part of the page, as it would interfere with the proof of the distances by adding up the second column, which ought to produce the last distance entered in the first. The levels are proved by subtracting the sum of the numbers in the fifth column from the sum of the numbers in the fourth, when the remainder should be the height above datum of the last station, recorded at the bottom of the sixth column.

To facilitate the reduction of the measured distances to the corresponding horizontal distances, the following table, showing the reduction, upon each 100 feet, for each foot difference of level, should be inserted in the field book:—

Difference of Level in 100 feet distance.	Reduction upon 100 feet of distance.	Difference of Level in 100 feet distance.	Reduction upon 100 feet of distance.	Difference of Level in 100 feet distance.	Reduction upon 100 feet of distance.
4	0·08	13	0·85	22	2·45
5	0·13	14	0·98	23	2·68
6	0·18	15	1·13	24	2·92
7	0·25	16	1·29	25	3·18
8	0·32	17	1·46	26	3·44
9	0·41	18	1·63	27	3·71
10	0·50	19	1·82	28	4·00
11	0·61	20	2·02	29	4·30
12	0·72	21	2·23	30	4·61

When it is required to plot the section on a large scale,

and to show every undulation of the surface, it is not necessary to remove and re-set the instrument, to obtain the height above datum of every point necessary to be known for this purpose; but, besides reading the staff at the back and fore stations, it may be read off, from the same place of the instrument, at as many intermediate points as may be deemed desirable; and these readings, being entered both as back and fore readings, will produce the same effect, as back and fore readings of the same points, obtained in different positions of the instrument. The distances of these points from the instrument should be omitted from the second column; but, the distances between them being entered successively in the first column, their respective distances from the instrument may at any time be determined, if required. The height of the instrument itself may be entered in this way as an intermediate sight; and as the same height that is added as a back reading, is subtracted again as a fore reading, any error in this reading will not at all affect the levels afterwards taken, and, provided it be not greater than the limit within which distances can be laid down and estimated upon the plot, is of no moment. Now, in taking the section of a line of any considerable extent, the scale is seldom sufficiently large to admit of less than six inches being laid down, or estimated, upon the plot, and, consequently, an error of two or three inches in the intermediate sights would be immaterial. When observations are made out of the line to be levelled, in order, for instance, to obtain the height of this line above neighbouring rivers, canals, roads, &c., the readings are to be entered in the same manner as for other intermediate sights; and, the columns of bearing and distance being left blank, no mistake can be made in drawing the section. The bearing and distance of such points, if desirable to be noted, must be entered in the space left for remarks.

For the purpose of reference on any future occasion, in order either to check the accuracy of the levels already obtained, or for the convenience of commencing a new series in some other direction, marks should be left upon some convenient fixed points, upon which the staff has been held, and the reading noted with the greatest possible care. These bench marks, as they are called, should ordinarily be left at about every half-mile of distance, and may be either on or off the line. In the latter case the readings are to be recorded in the manner already explained for points out of the line. The hooks and tops of gates, copings, sills, or steps of doors, &c., are commonly used for bench marks, and the mark must be made exactly on the point upon which the staff has been held. A stout stake may be driven into the ground for a bench mark, and is by many persons preferred to any other.

When a section of considerable length is to be plotted, the horizontal distances cannot be laid down on as large a scale as is necessary for the vertical heights above datum, in order that the section may be of any practical use, without making the plot of most unwieldy dimensions. It is therefore usual

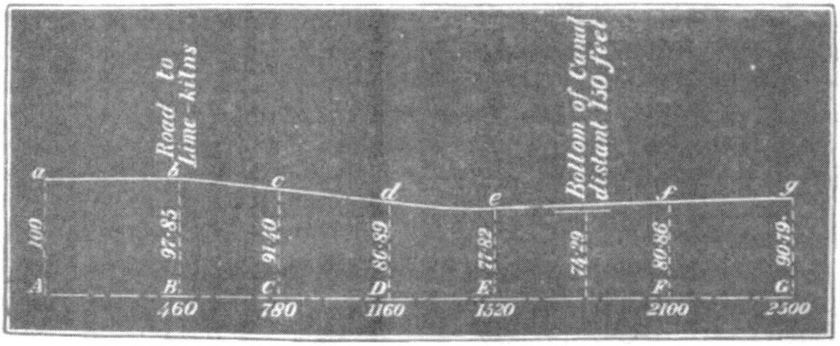

to make the vertical scale much larger than the horizontal one: thus 4 inches to a mile for the horizontal distances, with one inch to 100 feet for the vertical distances, is a usual combination. In the accompanying figure we have drawn the portion of a section, from the portion of the field book at page 29, making use of a scale of 1 inch to 800 feet for the horizontal distances, and of a scale of 1 inch to 200 feet for the vertical distances.

A G is ruled for the datum line; on it are set off from A the horizontal distances at the points B, C, D, E, F, G, according to the horizontal scale of 1 inch for each 800 feet, and through the points A, B, C, D, E, F, and G, are drawn lines A*a*, B*b*, &c., perpendicular to A G; on these lines are set off the vertical distances to the points *a*, *b*, *c*, &c.; according to the vertical scale of 1 inch for each 200 feet; and the line *a g*, passing through all the points *a*, *b*, *c*, &c., will represent the required section. A line is drawn between the stations E and F at the proper distance from the datum line, to represent the level of the canal; and proceeding in this manner, and making any remarks that may seem desirable, opposite the corresponding points of the section, the work will be completed.

LEVELS FOR CONTOURING, MINING, DRAINING, ETC.

Having now explained the construction and use of the most perfect instrument for tracing the level of any portion of country, we proceed to notice instruments adapted for use in cases where rapidity of execution is of greater importance than minute accuracy.

THE WATER LEVEL.

A very simple instrument of this kind is the *water level.* It can be made anywhere, and by any workman, costs but a few shillings, and requires no adjustment when using it.

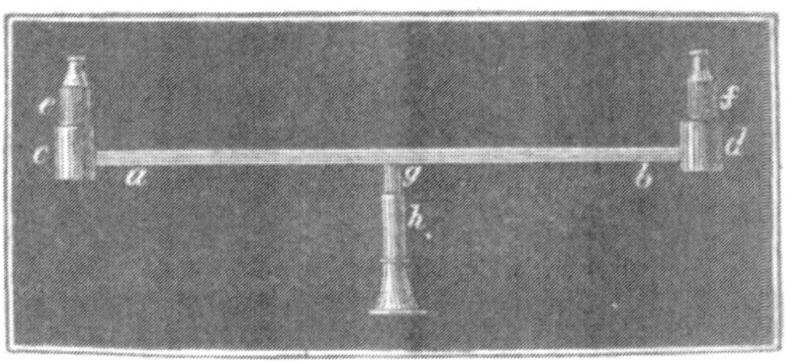

a b is a hollow tube of brass, about half an inch in diameter, and about 3 feet long; *c* and *d* are short pieces of brass tube, of larger diameter, into which the long tube is soldered, and are for the purpose of receiving the two small bottles, *e* and *f*, the ends of which, after the bottoms have been cut off, by tying a piece of string round them when heated, are fixed in their positions by putty or white lead; the projecting short axis, *g*, works, in the instrument from which the sketch was taken, in a hollow brass cylinder, *h*, which forms the top of a stand, used for observing with a repeating circle; but it may be made in a variety of ways, so as to revolve on any light portable stand. The tube, when required for use, is filled with water, coloured with lake or indigo, till it nearly reaches to the necks of the bottles, which are then corked for the convenience of carriage. On setting the stand tolerably level by the eye, these corks are both withdrawn, which must be done carefully, and when the tube is nearly level, or the water will be ejected with violence; and the surface of the water in the bottles, being necessarily on the same level, gives a horizontal line in whatever direction the tube is turned.

THE REFLECTING LEVEL.

An instrument, however, with which observations upon the level of a country may be more expeditiously made, and generally with greater correctness, than with the water level, is the reflecting level. This instrument consists merely of a piece of common looking-glass, *l l*, 1 inch square, set in a frame fixed against a plate of metal weighing about a pound, and suspended from a ring, *r*, by a twisted wire, *w*, so that it may swing freely, but not turn round on its axis of suspension. A fine silk thread, *t t*, is stretched across the centre of the mirror, and a small opening, *o*, is at one side of the mirror.

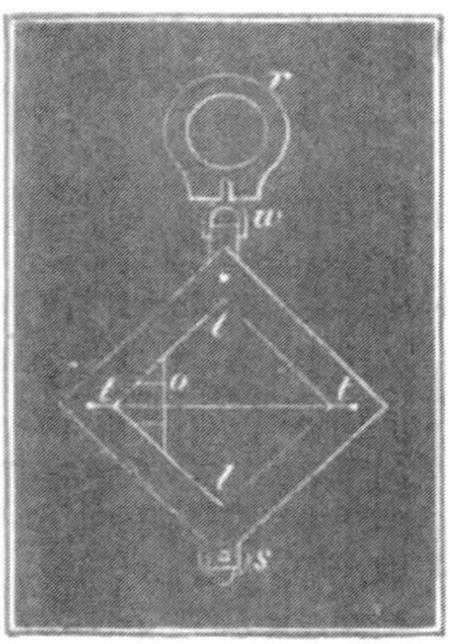

The instrument is adjusted as follows. It is suspended at about 50 yards in front of a wall. The observer looks into the mirror, and brings his eye into such a position that its image is bisected by the silk thread *t t*; and the point upon the wall, seen through the opening, *o*, which coincides with the silk thread, is marked upon the wall. The mirror is then turned round, and when the silk thread again bisects the image of the observer's eye, the point, the reflection of which in the mirror now coincides with this thread, is also marked upon the wall. Lastly, the middle point, between the two thus found, is marked upon the wall; and, by turning a screw, *s*, the centre of gravity of the instrument is altered, till the mirror hangs so that the reflection of the last mark comes upon the thread, when the observer's eye is also bisected by it. The instrument will now be in perfect adjustment, and, when the image of the eye is brought upon the thread, all points bisected by the thread, whether seen by reflection, or directly through the opening *o*, will be on the same level with the eye of the observer.

The observations may be made either by holding the

instrument at arm's length, or by suspending it from the branch of a tree, or from any post or rail of a convenient height. Greater accuracy is obtained by suspending it by means of a frame fitting on a three-legged stand, such as already described, as used for supporting the more accurate instruments; but it must not be forgotten that this instrument is not to be at all compared with them for minute accuracy; but that its advantages are the great rapidity with which it can be used, whether in a very confined space, or in an open country.

IMPROVED REFLECTING LEVEL.

All the advantages of the instrument just described are possessed in a higher degree by the improved reflecting level, which resembles it in nothing but the name, and the fact that

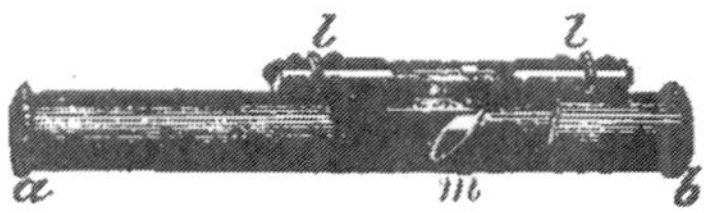

a reflector is used in its construction. It is a small portable instrument, which requires no adjustment, can be taken from the pocket at any time, and used immediately.

It consists of a tube, *a b*, about six inches long, and three-quarters of an inch in diameter; one end, *a*, forming the eye-end, being closed by a disc, pierced with a sight-hole of about the twentieth of an inch in diameter. A spirit-level, *l l*, is placed over a slit made in the tube, so that the bubble can be reflected from a mirror placed within the tube. This mirror, *m*, fills up half of the tube, and being inclined to the axis of the tube, at an angle of 45 degrees, reflects along the axis, a ray falling upon it at right angles to this axis. The level, then, being so placed that a line drawn from the centre of its run to the edge of the mirror is at right angles to the axis of the tube, the bubble, being at the centre of its run when the axis of the tube is horizontal, is then seen in the direction of the axis by an eye applied to the eye-end of the instrument; and consequently, upon looking through the tube, that point of any object, which is in coincidence with the image of the bubble, must be on the same level as the eye of the observer.

We have received from Messrs. Elliott an instrument, in

which the mirror is made to fill half the tube on one side, so that its edge appears as a vertical diameter, instead of being horizontal. A fine wire is then placed across the tube at right angles to the edge of the mirror; and, when the

image of the bubble is bisected by this wire, any point, seen through the tube in coincidence with the wire, is on the same level as the eye of the observer. This mode of construction is a very great improvement.

The reflecting level is a very useful instrument for contouring, that is, for determining lines upon the ground, which would form the boundaries of horizontal sections of its surface. For this purpose the observer has the height of his eye marked upon a staff, and an assistant, carrying this staff, plants it upon the ground at short intervals, and moving to the right or left, according to signals from the observer, until the mark is seen on a level with the observer's eye, a series of points on the desired contour line are determined. A second mark being made upon the staff at a known distance below the first, a point differing in level from the contour already determined by this distance, or by any multiple of it, may be determined, and, the observer proceeding to this point, another contour is determined passing through it; and so a series of contours is ultimately found at equal differences of level, which, being laid down on a plan previously constructed, delineates all the undulations of the plot represented by that plan. The distances to the points at which the observations are made, from points previously determined, should be taken with a measuring tape, and the points can thence be easily laid down on the paper.

Draining and Mining Levels.—For the purposes of mining and draining, instruments adapted to measure the rise or fall for each 100 feet of distance are found convenient. The illustrations annexed, taken from instruments forwarded to us by Messrs. Parkes, give the forms of two such levels.

In Fig. 1, the telescope A B is attached at the object-end to the bar C D by the hinge-joint H, and this bar is attached to a second bar F, by the spring *k*, and pivot *p*. The bar F screws on to an axis G, fitted to the head of a tripod stand, and the bar C D, being approximately levelled by moving the legs of the stand, is brought truly horizontal by turning the screw S until the level comes to the centre of its run. The eye-end, A, of the telescope is attached to the bar C D by the arc R R, and is moved along this arc by means of a rack and pinion, moved by the screw T. The arc R R is graduated, to show the rise and fall in every 100 feet of distance, and is read by means of an index seen through the small circular opening *e*. The bubble is adjusted by the capstan-headed screw X.

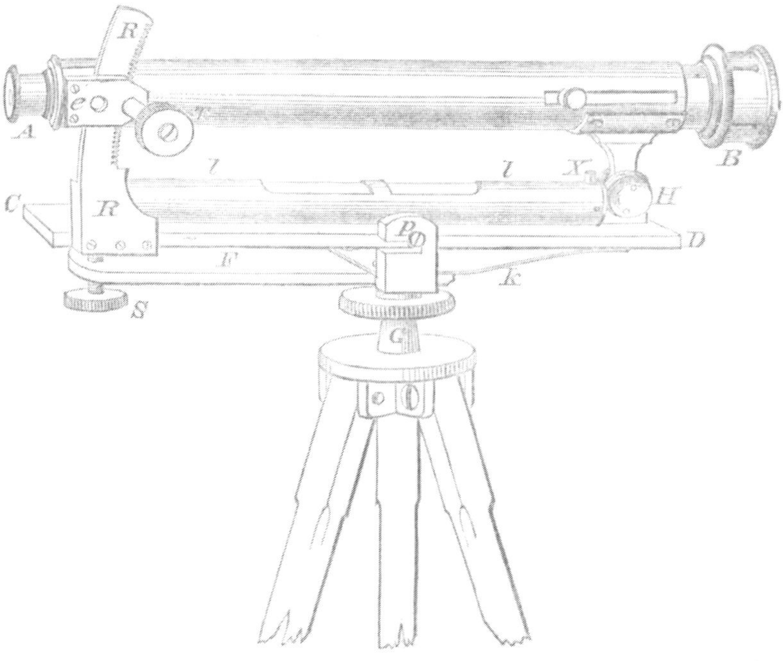

Fig. 1.

To adjust the Bubble—Set the telescope with the index at 0; then, planting the instrument about half a chain beyond one of two stakes, driven into the ground until their tops are on the same level, turn the screw S, until the same reading is obtained from a levelling staff, placed alternately on each of the stakes; if then the bubble be in the centre of its run, the adjustment is perfect; but if not, make it come to the centre of its run by turning the screw x.

This is an inexpensive instrument, and is used with great facility. For instance, to construct a drain with a given fall per 100 feet, set the index to the figure indicating the given fall, and then the differences of the readings on a staff, placed on different stations along the route of the drain, will indicate the

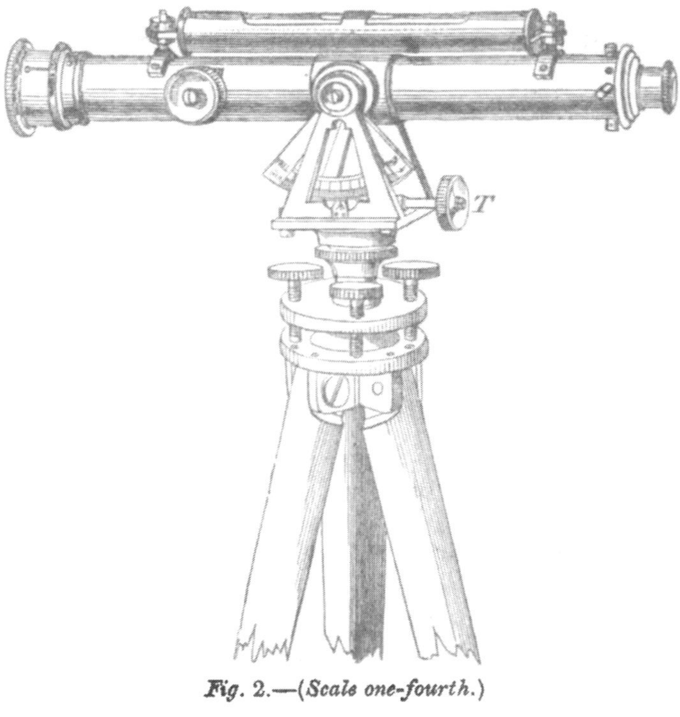

Fig. 2.—(Scale one-fourth.)

differences of the depths to which to go at these points, to come to the position for the bottom of the drain.

Fig. 2 represents a more expensive instrument. It can be used to take both back sights and fore sights, or sights in any direction, since it is mounted on parallel plates, and can be used in the same manner as a Dumpy, or other surveying level: in fact, the index being first set to zero on the arc beneath the telescope, the adjustments are then the same for this instrument as for a Dumpy level. To determine a line of rise or fall, the graduated arc must be moved by means of the tangent screw T, until the index points to the rise or fall desired. This instrument can then be used in the same manner as the one last described. In that instrument, however, there is no means of setting the axis vertical, and, consequently, that instrument must be adjusted, by bringing the bubble to the centre of its run, at each observation, though made from the same station.

CHAPTER III.

INSTRUMENTS FOR MEASURING ANGLES.

In every map and plan the distances and angles laid down are not the actual distances and angles between the points, the relative positions of which are intended to be represented; but they are the distances and angles between the projections* of those points upon the same horizontal plane, and are called the horizontal angles and distances between the points. Now, if our surveying instruments were constructed to measure the actual angles subtended by different objects, the process of calculating all the horizontal angles from these observed angles would be very laborious; but, by having such instruments as will at once determine by observation the horizontal angles, we are saved a vast amount of labour, and also from errors which might creep into the calculations.

THE PRISMATIC COMPASS.

With this instrument horizontal angles can be observed with great rapidity, and, when used with a tripod stand, with a considerable degree of accuracy. It is, consequently, a very valuable instrument to the military surveyor, who can make

* The projection of a point upon a horizontal plane is the point in which a vertical line through that point meets the horizontal plane.

his observations with it, while holding it in his hand, with all the accuracy necessary for a military sketch. It is also a useful instrument for filling in the detail of an extensive survey,* after the principal points have been laid down by means of observations made with the theodolite, hereafter to be described, and for any purpose, in short, in which the portability of the instrument and rapidity of execution are of more importance than extreme accuracy.

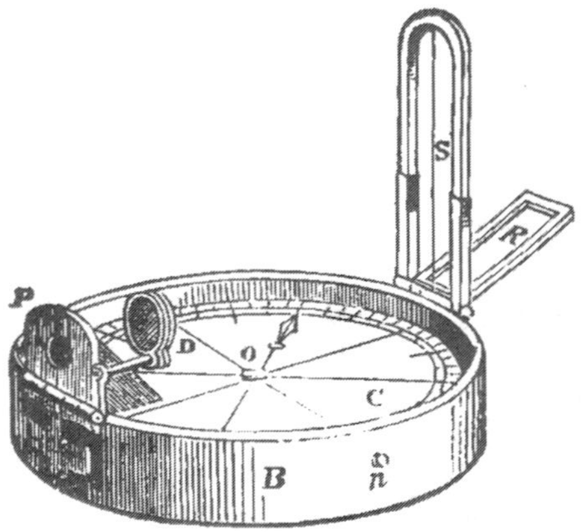

C is a compass card divided usually to every 20′, or third part of a degree, and having attached to its under side a magnetic needle, which turns upon an agate centre, O, fixed in the box B ; *n* is a spring, which, being touched by the finger, acts upon the card, and checks its vibrations, so as to bring it sooner to rest, when making an observation. S is the sight-vane, having a fine thread stretched along its opening, by which the point to be observed with the instrument is to be bisected. The sight-vane is mounted upon a hinge-joint, so that it can be turned down flat into the box when not in use. P is the prism, attached to a plate sliding in a socket, and thus admitting of being raised or lowered at pleasure, and also supplied with a hinge-joint, so that it can be turned down into the box when not in use. In the plate to which the prism is attached, and which projects beyond the prism, is a narrow slit, forming the sight through which the vision is directed when making an observation. On looking through this slit, and raising or lowering the prism in its socket, distinct vision of the divisions on the compass card immediately under the sight-vane is soon obtained, and these divisions, seen through the prism,

* The prismatic compass was used for this purpose by the gentlemen engaged in making the Ordnance surveys.

all appear, as each is successively brought into coincidence with the thread of the sight-vane by turning the instrument round, as continuations of the thread, which is seen directly through the part of the slit that projects beyond the prism.

The method of using the instrument is as follows:—The sight-vane S, and the prism P, being turned up upon their hinge-joints as represented in our figure, hold the instrument as nearly in a horizontal position as you can judge, or, if it be used with a tripod stand, set it, as nearly as you can, in a horizontal position by moving the legs of the stand, so that the card may play freely. Raise the prism in its socket, till the divisions upon the card are seen distinctly through the prism, and, turning the instrument round, until the object to be observed is seen, through the portion of the slit projecting beyond the prism, in exact coincidence with the thread of the sight-vane, bring the card to rest by touching the spring *n*; and then the reading at the division upon the card, which appears in coincidence with the prolongation of the thread, gives the magnetic azimuth of the object observed, or the angle which a straight line, drawn from the eye to the object, makes with the magnetic meridian.* By repeating each observation two or three times, and taking the mean of the readings, greater accuracy in the measurement of the bearings will be obtained; and, to ensure the most correct results, the observer, after observing an object two or three times, should proceed to the place of that object, and again observe an object at his first station the same number of times, taking the readings from the north in the first case, and from the south in the second, or *vice versâ*: the mean of all the results in these to be taken as the correct reading. This method will also detect any local attraction

* The magnetic meridian now makes an angle of about 24° with the true meridian, at London, the north point of the compass being 24° west of the true north point. This angle is called the variation of the compass, and is different at different places, and also at the same place at different times. Since this variation will affect equally, or nearly so, all azimuths observed within a limited extent, and during a limited time, the angles subtended by any two of the objects observed, being the difference of their azimuths, will not be affected by the variation, and hence the map, or plan, may be constructed with all the objects in their proper relative positions; but the true meridian must be laid down, if required, by observations made for the purpose.

on the needle, that may be acting at the one station, and not at the other. The magnetic azimuth of a second object being obtained in the same manner, the difference between these two azimuths is the angle subtended by the objects at the place of the eye, and, which is an important point, is independent of any error in the azimuths, arising from the slit in the prism not being diametrically opposite to the thread of the sight-vane.

For the purpose of taking the bearings of objects much above or below the level of the observer, a mirror, R, is supplied with the instrument, which slides on and off the sight-vane, S, with sufficient friction to remain at any part of the vane that may be desired. It can be put on with its face either upwards or downwards, so as to reflect the images of objects, either considerably above or below the horizontal plane, to the eye of the observer; and, if the instrument be used for obtaining the magnetic azimuth of the sun, it must be supplied with dark glasses, D, to be interposed between the sun's image and the eye.

There is a stop in the side of the box, not shown in our figure, by touching which a little lever is raised, and the card thrown off its centre, as it always should be when not in use, or the constant playing of the needle would wear the fine agate point upon which it is balanced, and the sensibility of the instrument would be thereby impaired. The sight-vane and prism being turned down, a cover fits on to the box, which is about three inches in diameter, and one inch deep; and the whole, being packed in a leather case, may be carried in the pocket without inconvenience.*

The compass card being divided into 360°, the degrees are numbered, in some instruments, from the north point round by the east, south, and west, to the north again, from 0° up to 360°, consecutively, the 0° not being figured, since it coincides with the 360°; there is, consequently, no need to mention more than the angular reading, to describe the exact bearing, which is all easterly. Thus, an object bearing exactly south-east, would be fully described by the angular

* For much valuable information respecting the use of the prismatic compass, especially in military surveying and sketching, we can refer our readers to a "Treatise on Military Surveying," &c., by Lieutenant-Colonel Basil Jackson, in which the subject is handled with great ability.

reading 135°, that is, 135° from the north round by the east, and an object bearing south-west, would be described by the angular reading 225°, that is, 225° from the north round by the east. In other instruments, the degrees are figured from the north, round by the east, to the south, up to 180°, and then, commencing again at 0°, from the south, round by the west, up to 180° again at the north. The bearing of an object may then be described by giving the angular reading, with the direction east or west, since all bearings reading east, go progressively from north to south, and all bearings reading west, go progressively from south to north.

The following illustrations of the use of the prismatic compass are taken from Haskoll's "Practice of Engineering Field Works."

"Let A B and C D be portions of two main chain lines of a survey; it is required to fill in the fencing along *a b c d e f*. From the point *a*, take the bearings of *a* B and of *a f*, *f* being a station on the line C D, previously fixed on for this purpose. Fix on some object in the fence near some point, for instance, *d*, in line with *a f*. Take also the bearing of *a b*, and chain to *b*, offsetting as you go along, and at *b* take the bearing point of *b a*; and if *a b* bear 136° East, then *b a* should bear 136° West. If there is a difference of a few minutes between the readings, take the mean for the correct reading. At *b* and at *c* repeat the same kind of work. On reaching *d*, find this station on the line *a f*; take the bearings of *d a* and of *d f*; and whatever number of degrees one is West from South, the other will be East from North. In the four-sided figure *a b c d*, the four angles will be equal to four right angles, or 360°; and as, having the bearings, we can obtain the angles from them, we can ascertain how nearly our bearings make these angles equal to four right angles. If, as is almost sure to be the case with this instrument, we get some few minutes, either over or under the four right angles, divide this difference among the bearings. It is also possible, that from *d* we may be able to get a bearing of some known point on the line A B, by which, when we come to plot, we may check the position of *d*; for if *d g* reads 130° West, then, the plotting from *g* on the plan of a line bearing 130° East, will run through the station *d*, when this has also been plotted. For the continuation of the work from *d* to *e* and *f*, we have a triangle *d e f*, the three internal angles of which should be equal to two right angles, by which we may check the angles at *d*, *e*, and *f*. It cannot fail to be observed that all the bearings may, and should, be plotted from the station *a*, without once moving the protractor.

"As we consider it very desirable that the student should make himself thoroughly master of the prismatic compass, before he engages with the principal surveying instrument, the theodolite, we will go a little further into this matter, premising, however, that it will be more as an introduction,

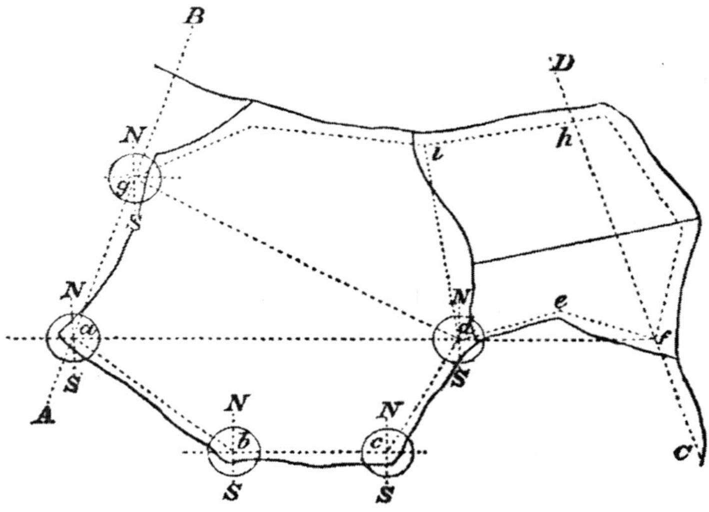

than anything else, to what is termed surveying by traverse, to which we shall come at a future page. Probably, also, it will make the subject more familiar to the reader, when he comes to it.

"At *a*, let the bearing of *a d* be 91° 5′ East, and the bearing of *a b*, 136° East; note this, at the commencement of the line *a b*, before you commence chaining. On reaching the point *b*, take the bearing of *b a*, and let this be 136° 5′: note this at the end of the line, and take 136° 2½′ for the mean, or corrected bearing, east and west, of the line, *a b*. In the same manner, let the mean bearing of *b c* equal 89°, and that of *c d*, equal 30°. At *d*, take the bearing of *d a*, and let this be 91° 10′; then the mean or corrected bearing of *a d*, or *d f*, will be equal to 91° 7½′. The angle *d a b* is equal to 136° 2½′, minus 1° 7½′, or to 44° 55′. The angle *a b c* is made up of the angles *a b* N, and N *b c*; *a b* N is equal to 180° minus 136*b* 2½′, or to 43° 57½′; and, N *b c* being 89°, the angle *a* ₒ *c* is equal to 43° 57½′, plus 89, or to 132° 57½′. The angle *b c d* is made up of the angles *b c* N, and N *c d*; *b c* N is equal to 180° minus 89°,* or to 91°; and, N *c d* being 30°, *b c d* is equal to 91°, plus 30°, or 121°. Lastly, *c d a* is equal to S *d a* minus S *d c*; S *d a* is equal to 91° 7½′; and, S *d c*, being 30°, *c d a* is equal to 91° 7½′ minus 30°, or to 61° 7½′."

Now 44° 55′ + 132° 57½′ + 121° + 61° 7½′ = 360°.

In the same manner we may check the triangle *d e f*, the three angles of a triangle being equal to two right angles.

Observe that three more lines for filling in have been run, from *g*, on A B, to *h*, on C D; and that a tie line has been chained from *d* to *i*, forming a further check upon the work; for, if in plotting, *d* or *i* lean one way or the other out of position, then the line *d i* will not plot truly.

For the operation of filling in, the large 5-inch prismatic compass with silver ring is very useful, and, when once the surveyor has become familiar and handy with the use of the instrument, it will give very close approximations.

To plot the bearings, place the protractor at *a*, so that A B shall have its due bearing with regard to the magnetic meridian N S, the north and south being each represented on the protractor by 180°; prick off, from *a*, the bearings of *a f*, *a b*, *b c*, *c d*, *d e*, *e f*; from *a* draw *a f* and *a b*, making the latter the length of the measured chain line; and from *b*, *c*, *d*, and *e*, draw *b c*, *c d*, *d e*, and *e f*, parallel to their bearings pricked off from *a*, and make them of lengths corresponding to the measured chain distances.

With due attention this instrument may also be used for a road survey. The bearings along the bends of the road are taken at the same time that the lengths are chained, care being

* A straight line falling on two parallel straight lines makes the two internal angles on the same side of it together equal to two right angles; it also makes the alternate angles equal to one another —Euc. Bk. I. Prop. 29.

taken, also, to get bearings from several stations to one or two objects, likely to be visible from different parts of the survey. The work is plotted as before, by laying off the bearings all from one point. The figure illustrates a similar method of getting a plan of a stream; but here a visual line. A B C, has been set off, passing through two prominent objects, selected for the purpose. As the bearings of the chain lines along the banks of the stream are measured, a few bearings

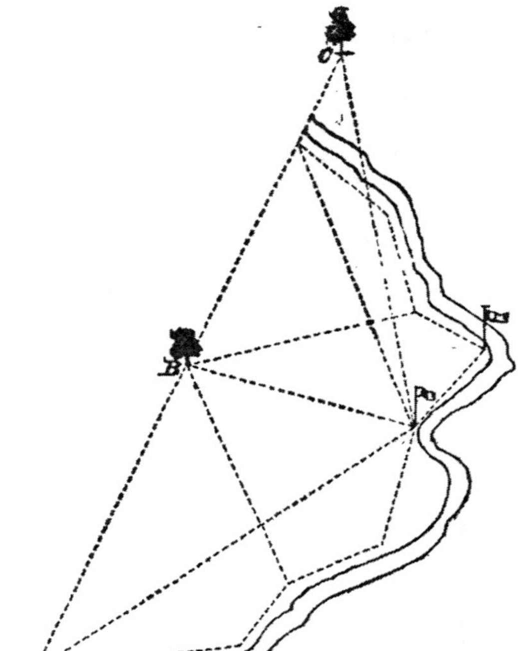

are taken here and there, at different stations, upon the prominent objects, and if convenient, one or two back sights upon the starting-point. If the chain lines have been correctly measured, and the bearings correctly taken, these bearings upon the prominent objects will all intersect very near the objects, and prove the degree of accuracy of the work; and though, it must be remembered, we cannot expect the same amount of accuracy as with a theodolite, yet, if the prismatic compass be applied in the manner we have been pointing out, errors will be inappreciable. The measurement of the bases A B and B C will completely check the work.

There is yet another service to which the prismatic compass is well adapted, which is, from their bearings, to ascertain the angles that fences make with chain lines, more particularly those straight fences which do not require offsetting. If, in the course of filling in, such a fence be intersected by two chain lines, the intersections, of course, fix the direction of the fence upon the plan; and a check upon the chaining is afforded by the use of the prismatic compass in the manner here mentioned. It may also be rendered very useful in testing the accuracy of a plan, the correctness of which we have any reason to doubt. Draw two or three lines across the plan, and marking on it the angles they make with each other; then in the field set out the corresponding lines with ranging rods, take their bearings, and see if the angles between them agree with those measured upon the plan. This may also be done without rods, by drawing any line on the plan through two or three junctions of fences, setting off the corresponding line on the ground, taking the angles that adjacent fences make with this line, and then comparing these angles with those similarly made on the paper.

THE BOX SEXTANT.

This instrument, which is equally portable with the prismatic compass, forming, when shut up, a box of about three inches in diameter, and an inch and a half deep, will measure the actual angle between any two objects to a single minute. It requires no support but the hand, is easily adjusted, and, when once adjusted, but seldom requires re-adjusting.

When the sextant is to be used, the lid, E, of the box is taken off, and screwed on to the bottom, where it makes a convenient handle for holding the instrument. The telescope, T, being then drawn out, the instrument appears as represented in our figure. A, is an index arm, having at its extremity a vernier, of which 30 divisions coincide with 29 of the divisions upon the graduated limb *l l*; and the divided spaces upon the limb denoting, each, 30 minutes, or half a degree, the angles observed are read off, by means of the vernier, to a single minute. The index is moved by turning the milled head, B, which acts upon a rack and pinion within the box. To the index arm is attached a mirror, called the index glass, which moves with the index arm, and is firmly fixed upon it by the maker, so as to have its plane accurately perpendicular to the plane in which the motion of the index arm takes place, and which is called the plane of the instrument. This plane is evidently the same as the plane of the face of the instrument, or of the graduated limb *l l*. In the line of sight of the telescope is placed a second glass, called the horizon glass, having only half its surface silvered, and which must be so adjusted that its plane may be perpendicular to the plane of the instrument, and parallel to the plane of the index glass when the index is at zero. The instrument is provided with two dark glasses, which can be raised or lowered by means of little levers seen at *d*, so as to be interposed, when necessary, between the mirrors, and any object too bright to be otherwise conveniently observed, as the sun. The eye-end of the telescope is also furnished with a dark glass, to be used when necessary.

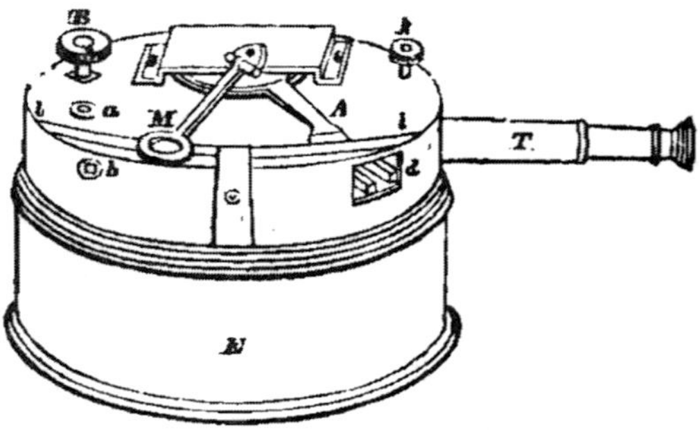

The principle upon which the sextant is constructed has been proved (vol. ii. p. 4), viz.: that the total deviation of a ray of light, after reflection successively at the index glass and horizon glass, is double the inclination of the two glasses. Now, the limb *l l*, being divided into spaces, each of 15′ extent, and these spaces being figured at 30′ each, the reading of the limb gives double the angle moved over by the index arm from the position in which the reading is zero, or double the angle of inclination of the two mirrors, I and H, if these mirrors be parallel when the reading is zero. If, then, the instrument be in perfect adjustment, and any object be viewed by it after reflection at both the mirrors, the reading of the instrument gives the total deviation of the rays of light, by which the vision is produced, or the angle S P H between the bearing of the object, S, from the centre of the index mirror, I, and the bearing of the reflected image, S′, from E, the place of the eye, that is, between lines, S I and S′ E, drawn, respectively, from the object to the centre of the index glass, and from the reflected image in the horizon glass to the eye. This angle is very nearly equal to the angle S E H, subtended by the object and its image at the place of the eye, differing from it only by the small angle I S E, subtended at the object by the place of the eye, and the centre of the index glass. This small angle is called the parallax of the instrument, and is scarcely perceptible at the distance of a quarter of a mile, while for distances greater than that it is so small that it may be considered to vanish. It also varies with the amount of deviation, and vanishes altogether whenever the centre of the index glass is in a direct line between the object and the eye.*

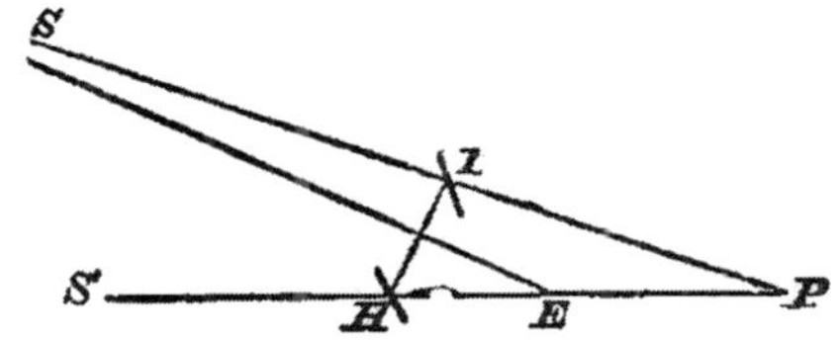

To see if the instrument be in perfect adjustment, place the dark glass before the eye-end of the telescope, and looking at the sun, and moving the index backwards and forwards a little distance on either side of zero, the sun's reflected image will be seen to pass over its disc, as seen directly through the horizon glass, and if, in its passage, the reflected image completely covers the direct image, so that but one perfect orb is seen, the horizon glass is perpendicular to the plane of the instrument; but if not, the screw at *a* must be

* We have seen a method given for what is called correcting the parallax when an observation is made at a short distance, by finding the deviation at this distance, when the angle between the object and its image is equal to zero; this deviation being given by the reading of the instrument, when the reflected image of the object observed exactly coincides with the object itself, seen through the unsilvered part of the horizon glass. This deviation, however, is not the parallax, even for a small angle between the object and its image, and, if the angle be not very small, the error introduced by the method will be greater than the parallax itself.

turned by the key *k* till such is the case. The key *k* fits the square heads of both the screws seen at *a* and *b*, and also fits into a spare part of the face of the instrument, so as to be at hand when wanted. This adjustment being perfected, bring the reflected image of the sun's upper limb in exact contact with the direct image of his lower limb, and note the reading of the vernier; then move the index back beyond the zero division of the limb, till the reflected image of the sun's lower limb is in exact contact with the direct image of his upper limb, and, if the zero of the vernier be now exactly as far behind the zero of the limb, as it was, at the former reading, in front of it, so that the reading now on the part of the limb called the arc of excess, behind its zero division,* is the same as the former reading, the instrument is in perfect adjustment; but, if not, half the difference of the two readings is the amount of the error, and is called the index error, being a constant error, for all angles observed by the instrument, of excess, if the first reading be the greatest, and of defect, if the second reading on the arc of excess be the greatest.

In the former case, then, the true angle will be found by subtracting the index error from, and in the latter, by adding it to, the reading of the instrument at every observation.

This method of correcting for the index error is to be used with the larger instruments, hereafter to be described under the head of Astronomical Instruments; but in the box sextant, this error should be removed by applying the key *k* to the screw at *b*, and turning it gently, till both readings are alike, each being made equal to half the sum of the two readings first obtained. When this adjustment is perfected, if the zeros of the vernier and limb are made exactly to coincide, the reflected and direct image of the sun will exactly coincide, so as to form but one perfect orb, and the reflected

* In reading an angle upon the arc of excess, the division to read on the limb is that next in front of the zero of the vernier, or between the zero of the vernier and the zero of the limb, and the divisions of the vernier itself are to be read from the end division marked 30, and not, as usually, from the zero division: thus, if the zero division of the vernier were a little farther from the zero division of the limb, than the first division on the arc of excess; and if the twenty-seventh division on the vernier, or the third from the end division, marked 30, coincided with a division upon the limb, then the reading would be 33.

and direct image of any line, sufficiently distant not to be sensibly affected by parallax, as the distant horizon, or the top or end of a wall more than half a mile off, will coincide so as to form one unbroken line.

To obtain the angle subtended by two objects, situated nearly or quite in the same vertical plane, hold the instrument in the right hand, and bring down the reflected image of the upper object, by turning the milled head B till it exactly coincides with the direct image of the lower object, and the readings of the instrument will give the angle between the two objects.

To obtain the angle subtended by two objects nearly in the same horizontal plane, hold the sextant in the left hand, and bring the reflected image of the right-hand object into coincidence with the direct image of the left-hand object.

It will be seldom that the surveyor need pay any attention to the small error arising from parallax; but, should great accuracy be desirable, and one of the objects be distant, while the other is near, the parallax will be eliminated by observing the distant object by reflection, and the near one by direct vision, holding the instrument for this purpose with its face downwards, if the distant object be on the left hand. If both objects be near, the reflected image of a distant object, in a direct line with one of the objects, must be brought into coincidence with the direct image of the other object, and the parallax will thus be eliminated.

For the purposes of surveying, the horizontal angles, between different objects, are required, and the reduction of these angles from the actual oblique angles, subtended by the objects, would be a troublesome and laborious process. If the angle subtended by two objects be large, and one be not much higher than the other, the actual angle observed will be, however, a sufficient approximation to the horizontal angle required; and, if the angle between the two objects be small, the horizontal angle will be obtained, with sufficient accuracy, by taking the difference of the angles observed between each of the objects, and a third object, at a considerable angular distance from them. With a little practice the eye will be able to select an object in the same direction as one of the objects, and nearly on a level with the other object, and the angle between this object, and the object selected, will be the horizontal angle required.

At sea the altitude of an object may be determined by observing the angle subtended by it and the verge of the horizon; but upon land a contrivance, called *an artificial horizon*, becomes necessary, for correctly determining altitudes. The best kind of artificial horizon consists of an oblong trough, *t t*, filled with mercury, and protected from the wind by a roof, *r r*, having in either slope a plate of glass with its two surfaces ground into perfectly parallel planes. The angle, S E S′, between the object, and its reflected image seen in the mercury, is double the angle of elevation S E H; and, the angle S E S′ being observed, its half will, consequently, be the angle of elevation required. If the angle of elevation be greater than 60°, the angle S E S′ will be greater than 120°, and cannot be observed with the sextant we have been describing.

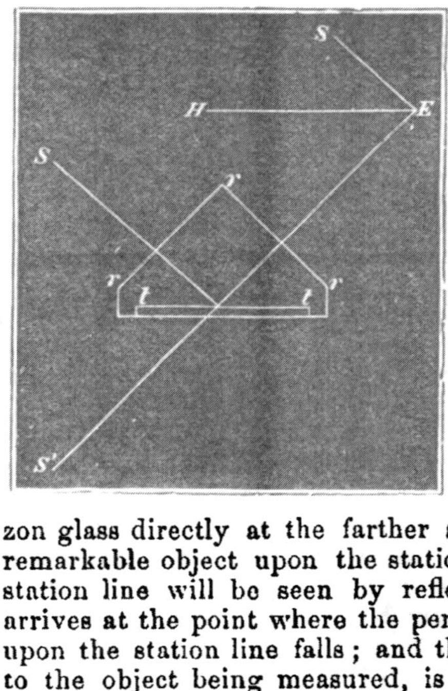

The box sextant is a most convenient instrument for laying off offsets, or perpendicular distances from a station line; for, by setting the index at 90°, and walking along the station line, looking through the horizon glass directly at the farther station staff, or any other remarkable object upon the station line, any object off the station line will be seen by reflection, when the observer arrives at the point where the perpendicular from this object upon the station line falls; and the distance from this point to the object being measured, is its perpendicular distance from the station line.

For the mere purpose of measuring offsets an instrument called an *optical square* is now very generally employed, which consists of the two glasses of the sextant fixed permanently at an angle of 45°, so that any two objects seen in it, the one by direct vision, and the other by reflection, subtend at the place of the observer an angle of 90°.

THE OPTICAL SQUARE.

The mirrors A and B are fixed in a shallow circular box, so that the vertex E of the isosceles right-angled triangle A B E comes to the circumference of the box, where the aperture for the eye is placed. By this means parallax is eliminated, and a ray of light from the object F, proceeding in the direction F E, and falling upon the mirror A, is reflected in the direction A B, meets the mirror B, is again reflected in the direction B E, and seen by the eye at E, in the direction E D, at right angles to E F. The mirror B is unsilvered in the lower part, so that the eye, looking along the line E D through the aperture in the rim of the box, sees an object D in this line, by direct vision, in coincidence with the reflected image of the object F, on the line E F, at right angles to E D.

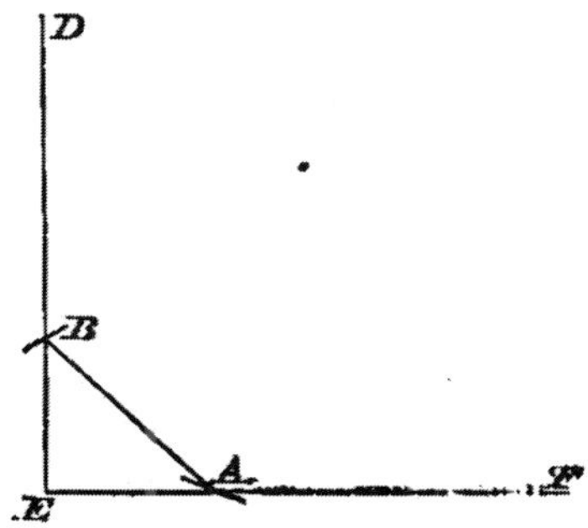

This instrument cannot be excelled for facility and rapidity of use. To set off a perpendicular from a given point on a line, the observer, standing at this point, and holding the square in his hand, directs the motions of an assistant, who walks out from the line with a rod, so as to keep the reflection of the rod in the mirror B, in coincidence with an object, on the line, seen through the unsilvered portion of the same mirror: the line from the observer's eye to the rod, is, then, the perpendicular required.

Again, if a perpendicular is to be made to fall from a given point upon a given line, the observer walks along the line, until he comes to a point from which he sees an object on the line, coinciding with the image, in the mirror B, of an object at the given point: the observer is then standing at the spot, on which the perpendicular will fall.

THE CROSS STAFF.

Another instrument for setting out lines at right angles to other lines, is the cross staff, which, though old fashioned and cumbrous, is still very commonly used for the purpose.

There are two forms of cross staff. One form consists of four sights fixed at right angles upon a brass cross, which, when in use, is fixed on the top of a staff. The second form consists of a hollow brass cylinder, about three or four inches in diameter, and as many in depth, through which are pierced sights at right angles to each other: this also, when in use, is affixed to a staff.

To set out with the cross staff a perpendicular to a line, from a given point on it, thrust the staff into the ground at this point, so that through one pair of sights you can see both ends of your main line, by looking first through one of these sights, and then moving round and looking through the opposite one. Having thus ensured a point exactly on the line, and the proper position of the cross, now move round, and, looking across the instrument, through the other pair of sights, have a pole set up in a line with them; and the perpendicular will be set off.

THE LINE RANGER.

We may here notice another application of the law of

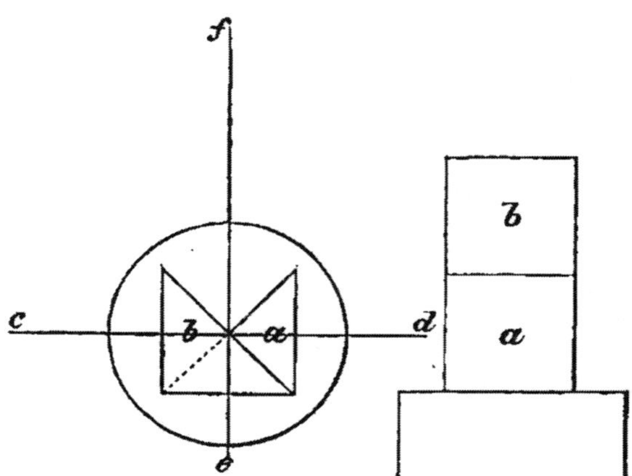

Fig. 1. *Fig.* 2.

reflected light, to the determination of the relative angular position of objects, in Adie's line ranger, a very simple pocket instrument, for setting out straight lines. It is shown in plan in Fig. 1, and in elevation in Fig. 2. *a* and *b* are two totally reflecting glass prisms, placed one over the other, on a cylindrical base, which serves as a handle. One of the faces of the right angle in the upper prism, is placed flush with one of the faces of the right angle in the lower prism, the other faces containing the right angles being parallel. The observer, holding the instrument in his hand, and looking into the prisms in the direction *e f*, sees directly before him, in the prism *a*, the reflected image of a pole at *d*, on his right hand, and in the prism *b*, that of another at *c*, on his left hand; and, when these images are in the same straight line, the instrument is also exactly in the same straight line with the objects *c* and *d*.

THE THEODOLITE.

The theodolite is the most important instrument used by surveyors, and measures, at the same time, both the horizontal angles subtended by each two of the points observed with it, and the angles of elevation of these points from the point of observation.

This instrument may be considered as consisting of three parts: the parallel plates with adjusting screws fitting on to the staff head, of exactly the same construction as already described for supporting the Y and other levels; the horizontal limb, for measuring the horizontal angles; and the vertical limb, for measuring the vertical angles, or angles of elevation.

The horizontal limb is composed of two circular plates, L and V, which fit accurately one upon the other. The lower plate projects beyond the other, and its projecting edge is sloped off, or chamfered, as it is called, and graduated at every half degree. The upper plate is called the vernier plate, and has portions of its edge chamfered off, so as to form with the chamfered edge of the lower plate continued portions of the same conical surface.* These chamfered portions of the upper plate are graduated to form the verniers,

Fig. 1.

* In the ordinary mode of construction, represented in section in Fig. 1, the perfect coincidence of the junction of the upper and lower plates cannot be effected; but the edge of the vernier plate, slightly overlapping the lower, may produce a small error in the reading, from the

by which the limb is subdivided to single minutes. The six-inch theodolite represented in our figure has two such verniers, 180° apart. The lower plate of the horizontal limb is attached to a conical axis passing through the upper parallel

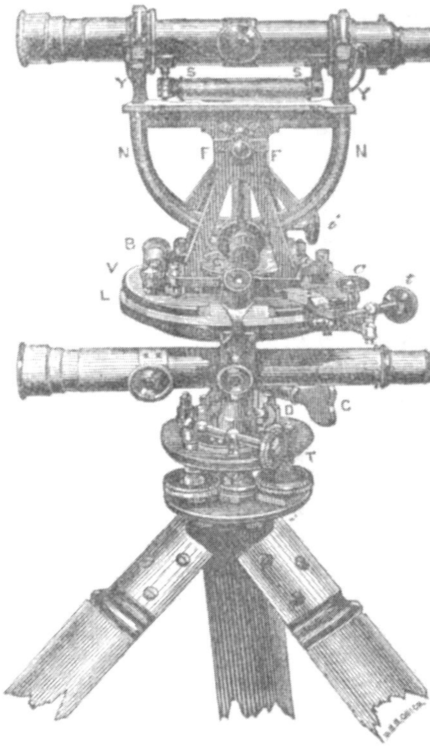

A six-inch Theodolite.

plate, and terminating in a ball fitting in a socket upon the lower parallel plate, exactly as the vertical axis of the Y level already described. This axis is, however, hollowed to receive

introduction of parallax. This defect has been remedied in some theodolites by squaring the edges, at the junction between the two plates, as represented in Fig. 2.

Fig. 2.

a similar conical axis ground accurately to fit it, so that the axes of the two cones may be exactly coincident or parallel.* To the internal axis the upper, or vernier, plate of the horizontal limb is attached, and thus, while the whole limb can be moved through any horizontal angle desired, the upper plate only can also be moved through any desired angle, when the lower plate is fixed by means of the clamping screw, C, which tightens the collar D. T is a slow-motion screw, which moves the whole limb through a small space, to adjust it more perfectly, after tightening the collar, D, by the clamping screw C. There is also a clamping screw, c, for fixing the upper, or vernier, plate to the lower plate, and a tangent screw, *t*, for giving the vernier plate a slow motion upon the lower plate, when so clamped. Two spirit levels, B, B, are placed upon the horizontal limb, at right angles to each other, and a compass, G, is also placed upon it in the centre, between the supports, F, F, for the vertical limb.

The vertical limb, N, N, is divided upon one side, at every 30 minutes, each way, from 0° to 90°, and subdivided by the vernier, which is fixed to the compass box, to single minutes. Upon the other side are marked the number of links to be deducted from each chain, for various angles of inclination, in order to reduce the distances, as measured along ground, rising or falling at these angles, to the corresponding horizontal distances. The axis of this limb must rest, in a position truly parallel to the horizontal limb, upon the supports F, F, so as to be horizontal, when the horizontal limb is set truly level; and the plane of the vertical limb should be accurately perpendicular to its axis. To the top of the limb N, N, is attached a bar, which carries two Y's for supporting a telescope, of the same construction as that before described for the Y spirit level; and underneath the telescope is a spirit level, S, S, attached to it at one end by a joint, and at the other end by a capstan-headed screw, as in the Y level. The horizontal axis can be fixed by a clamping screw, and the vertical limb can then be moved through a small space by a slow-motion screw *i*.

* Upon this depends, in a great measure, the perfection of the instrument, as far as the horizontal measurements are concerned; and, when we describe presently the adjustments of the instrument, we shall explain the method of detecting an inaccuracy in the grinding of the axes.

Before commencing observations with this instrument, the following adjustments must be attended to:—

1. Adjustments of the telescope, viz.: the adjustment for parallax; the adjustment for collimation.
2. Adjustment of the horizontal limb, viz.: to set the levels on the horizontal limb to indicate the verticality of the azimuthal axis.
3. Adjustment of the vertical limb, viz.: to set the level beneath the telescope to indicate the horizontality of the line of collimation.

1. *Parallax and Collimation.*—These adjustments have already been described (p. 18) under the head of the Y level.

2. *Adjustment of the Horizontal Limb.*—Set the instrument up as accurately as you can by the eye, by moving the legs of the stand. Tighten the collar, D, by the clamping screw C, and, unclamping the vernier plate, turn it round till the telescope is over two of the parallel plate screws. Bring the bubble of the level, S S, beneath the telescope to the centre of its run by turning the tangent screw *i*. Turn the vernier plate half round, bringing the telescope again over the same pair of the parallel plate screws; and, if the bubble of the level be not still in the centre of its run, bring it back to the centre, half way, by turning the parallel plate screws over which it is placed, and half way by turning the tangent screw *i*. Repeat this operation, till the bubble remains accurately in the centre of its run in both positions of the telescope; and then turning the vernier plate round till the telescope is over the other pair of parallel plate screws, bring the bubble again to the centre of its run by turning these screws. The bubble will now retain its position, while the vernier plate is turned completely round, showing that the internal azimuthal axis, about which it turns, is truly vertical. The bubbles of the levels on the vernier plate being now, therefore, brought to the centres of their tubes, will be adjusted, to show the verticality of the internal azimuthal axis. Now, having clamped the vernier plate, loosen the collar D, by turning back the screw C, and move the whole instrument slowly round upon the external azimuthal axis, and, if the bubble of the level S, S, beneath the telescope, maintains its position during a complete revolution, the external azimuthal axis is truly parallel with the internal, and both are vertical at the

same time; but, if the bubble does not maintain its position, it shows that the two parts of the axis have been inaccurately ground, and the fault can only be remedied by the instrument maker.

3. *Adjustment of the Vertical Limb.*—The bubble of the level, s, s, being in the centre of its run, reverse the telescope, end for end, in the Y's, and, if the bubble does not remain in the same position, correct for one half the error by the capstan-headed adjusting screw at one end of the level, and for the other half, by the vertical tangent screw *i*. Repeat the operation till the result is perfectly satisfactory. Next turn the telescope round a little, both to the right and to the left, and if the bubble does not still remain in the centre of its run, the level, s s, must be adjusted laterally by means of the screw at its other end. This adjustment will probably disturb the first, and the whole operation must then be carefully repeated. By means of a small screw, fastening the vernier of the vertical limb to the vernier plate over the compass box, the zero of this vernier may now be set to the zero of the limb, and the vertical limb will be in perfect adjustment.

The second telescope, placed beneath the horizontal limb, serves to detect any accidental derangement of the instrument during an observation, by noting whether it is directed to the same point of a distant object at the end of the observation to which it has been set at the commencement of the observation. This telescope is now very generally omitted, the advantage derived from it being considered to be counterbalanced by the additional length which must in consequence be given to the vertical axis, and which detracts from the compactness and firmness of the instrument.

In the larger instruments the vertical limb admits of an adjustment to make it move accurately in a vertical plane, when the horizontal limb has been first set in perfect adjustment. This adjustment is important, and should be examined with great care; and in the small theodolites, in which the vertical limb is permanently fixed to the horizontal limb, an instrument, which will not bear the test of the examination that we proceed to describe, must be condemned, till set in better adjustment by the maker. The azimuthal axis having been set truly vertical, direct the telescope to some well-defined angle of a building, and, making the intersection of the wires exactly coincide with this angle, near the ground, elevate the telescope by giving motion to the vertical limb, and, if the adjustment be perfect, the intersection of the cross wires will move accurately along the angle of the building, still continuing in coincidence with it. A still more perfect test will be, to make the intersection of the cross wires coincide with the reflected image of a star in an artificial horizon, and, elevating the telescope, if the adjustment be perfect, the direct image of the star itself will again be bisected by the cross wires.

THE TRANSIT THEODOLITE.

The transit theodolite is so called from the telescope being mounted in the same manner as in the transit instrument described hereafter (p. 132). In surveying, the difference of level being determined by the level in the manner already explained, the vertical limb, attached to the telescope of the theodolite, may for most purposes to which it is applied be dispensed with, which makes the instrument both simpler and lighter. When supplied with a vertical limb, that is a complete circle, it becomes at the same time a theodolite, an altitude and azimuth instrument, and a transit instrument, and may be used for astronomical, as well as for terrestrial purposes.

The tripod stand, the parallel plates, and the horizontal graduated limb and vernier plate, differ in no particular from those of the theodolite already described, and, like that, the vernier plate of the transit theodolite carries a compass in its centre between the supports of the telescope, and two spirit evels, *l*, *l*, to serve as a guide for setting the instrument level. As in the ordinary theodolite too, a small microscope for reading the verniers is adapted to slide round the horizontal limb.

The two frames, Y, Y, resting on the vernier plate, carry th bearings for the telescope with its spirit level, and the graduated circle M, called the *vertical* limb with its two verniers, V, and microscopes, for measuring vertical angles, called angles of elevation, or angular altitudes. Upon these bearings, which are usually of the ordinary Y pattern, called therefore the Y's, the horizontal axis of the telescope, with its appendages, is made to rest. The horizontal axis is formed of two cones, the larger ends of which are attached to the middle of the tube of the telescope, while the small ends are ground into two *perfectly equal* cylinders, called *pivots*; and it is these pivots that bear and turn upon the Y's. The common axis of the cylinders and cones must be accurately at right angles to the axis, or rather, to the line of collimation of the telescope, and the adjustment to produce

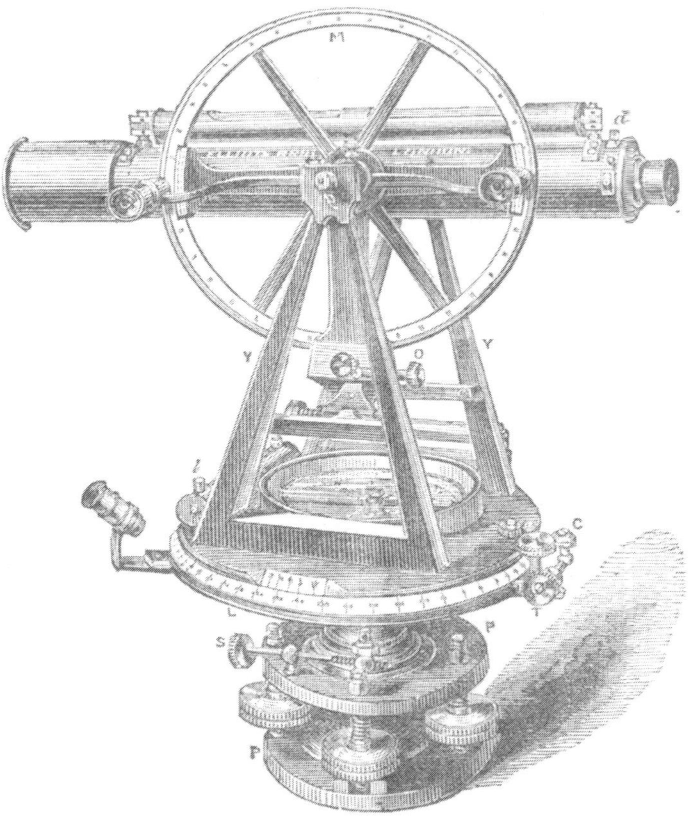

this is called the *adjustment for azimuth*. This mode of supporting the telescope and vertical limb gives to the transit theodolite its superiority over other modes of construction, as, whenever it is desirable to reverse the ends of the telescope, this is immediately accomplished by merely revolving it on its axis, instead of having recourse to the troublesome operation of opening clips, lifting the telescope out, and replacing it. It also serves to eliminate any error arising from imperfect adjustment, by reversing both the horizontal limb and the telescope, and taking the mean of the two readings thus obtained.

The vertical limb is graduated, from 0° to 90°, through one quadrant, and then again, in the same manner, throughout each of the other quadrants, going round from left to right.

At either end of the spirit level are capstan-headed screws, to adjust it parallel to the optical axis of the telescope, or line of collimation, by tightening the one and loosening the other. The vertical limb is also provided with a tangent screw, o, for slow motion.

The following are the adjustments:—

1. The adjustment for parallax.
2. Adjustment of the horizontal limb, and consequent verticality of the azimuthal axis.
3. Adjustment for horizontality of the axis on which the telescope turns.
4. Adjustment for collimation in azimuth.
5. Adjustment for collimation in altitude.

1. *Parallax.*—This adjustment, which has been already described (p. 19), must always be tested by the observer, and corrected, if necessary, as it will differ according as he is long or short sighted. The other adjustments are all made by the maker with great care, for the sake of his own reputation; and in well-constructed instruments derangements are but little likely to occur afterwards, if ordinary care be taken of them; but still they may accidentally occur, and it is desirable, therefore, to examine the instrument, both when new, and from time to time afterwards.

2. *Adjustment of the Horizontal Limb, l.*—This adjustment also has been explained (p. 58).

3. *Horizontality of the Axis on which the Telescope turns.*—This is to be tested by observing a long plumb line; first, making the intersection of the cross wires coincide with this line at the bottom, then, if this intersection moves accurately along the line, as the telescope is raised by turning it on its bearings, this adjustment is perfect; but, if not, the bearing, on the side opposite to that on which the intersection departs from the plumb line, is too high, and should be

lowered, by rubbing the face of the Y with slate pencil and water, or by some equally delicate process. Before proceeding, however, to do this, the next adjustment should be examined, and the test should then be verified, by observing if a star and its image, reflected in an artificial horizon, are both coincident with the intersection of the cross wires, when the telescope is turned upon its bearings.

4. *Adjustment for Collimation in Azimuth.*—Direct the telescope to some distant, small, and well defined, object, and, bringing its centre into exact coincidence with the intersection of the cross wires, clamp the horizontal limb. Lift the telescope off the Y's, and reverse it, so that the bubble will be downwards, and the ends of the horizontal axis, on opposite bearings to what they were before. Now examine if the centre of the object again comes into exact coincidence with the intersection of the cross wires, as the telescope is turned on its bearings; if so, the collimation in azimuth is perfect, and the line of collimation is accurately at right angles to the axis; but if not, correct one half the deviation by turning the two small side screws that adjust the diaphragm, and the other half, by moving the horizontal limb by means of the tangent screw, s. If the half distances have been correctly estimated, the centre of the object will again come into coincidence with the intersection of the cross wires, when the telescope is restored to its first position, and the adjustment will be accomplished; but if not, repeat the operation, until the centre of the object is coincident with the intersection of the cross wires in both positions of the instrument, and the adjustment will be perfected.

If a star and its reflected image be now observed, as directed for the preceding adjustment, and the test be satisfactorily satisfied, we may consider that both adjustments are perfected.

5. *Adjustment for Collimation in Altitude.*—Point the telescope to a very distant object, or a star, and, bringing the cross wires into coincidence with the object, read off the angle on the vertical circle, as indicated by the verniers. Turn the horizontal line through 180, and again direct the telescope to the same object, and read off the angle on the vertical limb. If the two readings be alike, the adjustment for collimation in altitude is perfect; but, if the readings be different, half their sum will give the true reading. The

verniers should then be adjusted to give the true reading, and, the vertical limb being brought down to zero, the bubble should be brought to the centre of its run: the adjustment will then be complete.

EVEREST'S THEODOLITE.

This instrument has distinctive features, as shown by the figure. The horizontal limb, L, consists of one plate only, on which the degrees are graduated; the verniers, V, are at

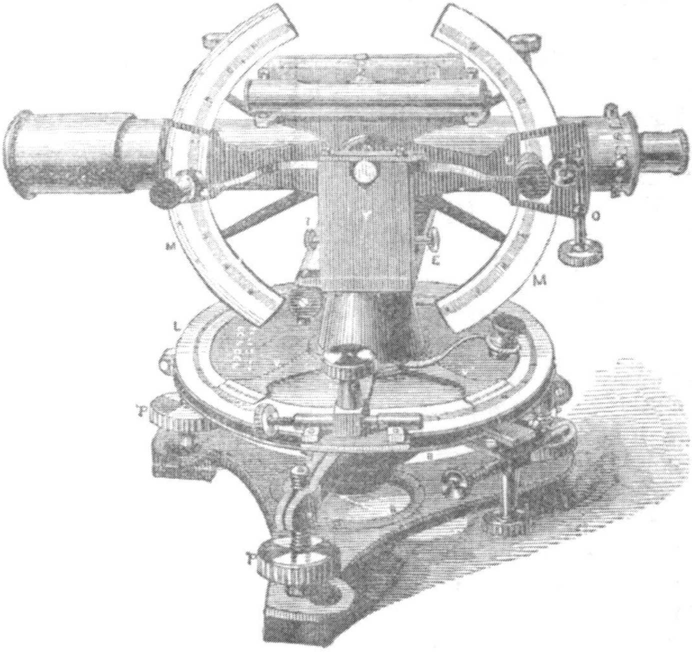

the end of bars radiating from the centre; and another bar carries the clamp, C, and the tangent screw for the verniers. At S are shown the clamp and slow-motion screw of the horizontal limb. The vernier bars are connected with the upper portion of the instrument carrying the telescope, vertical limb, M, &c., and, turning upon the same centre, they show the angle traversed by the telescope.

The tripod support, B, is provided with parallel plate screws, *p*, *p*, *p*. One advantage with this instrument is, that it may be used with a three-legged stand, the same as those already described, or be disengaged from the top of this staff, and placed upon the top of a wall, or in other situations where the ordinary tripod could not be used. It will be seen by the figure, that the telescope and vertical limb are supported in a manner very similar to the transit theodolite, the horizontal axis connected with the telescope resting on two supports, only one of which is seen in the drawing at Y. These are supported by a flat horizontal bar, E, to which is attached a spirit level, only one end of which is seen in the drawing. This level is for adjusting the axis horizontally; and, this being accomplished, the vertical arc, M M, attached to the telescope, moves with it in a vertical plane.

Everest's theodolite is much used in India, but has not met with a very favourable reception in this country. It labours under the disadvantage of not affording means of efficiently testing its adjustments; since the telescope can be turned neither round, nor end for end, as in the ordinary theodolite, nor over on opposite bearings, as in the transit theodolite; neither can it be reversed, like the telescope of the transit theodolite, by turning round on its bearings.

Many persons also dislike the mode of supporting the horizontal limb on three screws only, as these may be shaken by having no opposite bearing to steady them. With the parallel plates, and four screws, each opposite pair bear against each other, and no shake is possible. On the other hand, the advocates of a three-screw bearing object to the parallel plates, and four screws, that, by pressing the upper parallel plate unequally against the azimuthal axis, the centre of this plate, in which the axis turns, may become ovalled. The annexed figure represents a plan for preventing this ovalling, by forming the upper parallel plate of two concentric rings, A and B, joined together at one part only by the bridge C, placed midway between two of the screws S, S. The outer ring will thus admit of a slight bending, and the pressure of the inner ring against the axis being thereby relieved, scarcely any tendency to ovalling can occur.

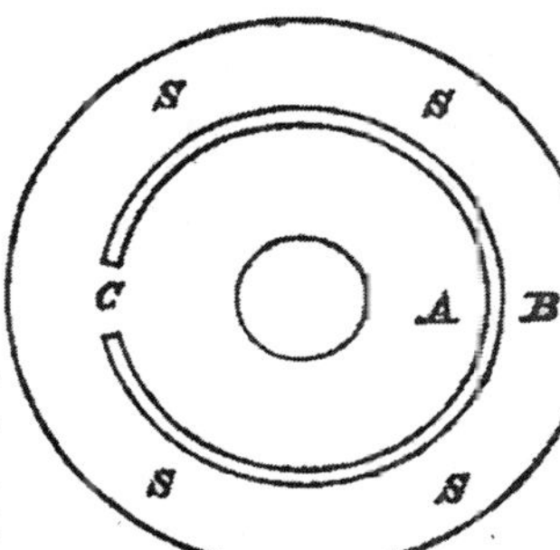

LISTER'S PATENT THEODOLITE.

The most important peculiarity of this theodolite is an adaptation for setting out mechanically, without recourse to measurement or calculation, the surface widths for batter pegs required in all railway earthworks, to indicate the tops of slopes of cuttings, and bottoms of slopes of embankments. The ordinary methods of performing this work involve much tedious calculation, and liability to error, particularly on rough and on sloping ground. By the use of Lister's theodolite this is entirely avoided, and a great saving of time effected, inasmuch as, when the instrument is once fixed in position, the widths may be put out on either side of it as far as the range of the telescope will admit, and almost as quickly as the pegs can be driven. The boundary fence, being parallel with the surface line of the slope, may be set out at the same time. A further advantage possessed by this instrument, consists in its superior adaptation to levelling by vertical angles, particularly where cross sections are required to be taken over very sidling ground. With the ordinary theodolite, the telescope must be reversed, and the vertical angle reset, to take the back sights, and each cross section must be taken separately; but this instrument, being so constructed that the telescope will revolve in any vertical plane, these defects are avoided, and a series of sections may be taken without moving the instrument.

Ordinary theodolites are readily altered to embrace these improvements.

An entirely new form of theodolite has been designed by M. d'Abbadie, and is made in Paris. The telescope is fixed with its line of collimation parallel to the plane of the horizontal circles, which are provided with crossed spirit levels, and divided in the usual way, for the reading of horizontal angles by the aid of a second telescope, fixed to the lower plate. Vertical angles are read off by turning the telescope upon its own line of collimation as an axis. The telescope is provided with a vertical circle, whose plane is fixed truly at right angles to its line of collimation. To measure the vertical angle subtended by two objects—one, suppose the lower one, is observed, as seen reflected in a totally reflecting prism in the telescope tube, whose plane of reflection is parallel to the line of collimation. The vertical circle being then brought to zero and clamped to the telescope, they are turned round together on the line of collimation, as an axis, until the higher of the two objects is seen reflected in the same prism, the precise position being fixed by cross wires in the field. The vertical angle between the two objects is now read off upon the vertical circle, whose centre is on the line of collimation. This instrument is very original in conception, and is said to work well, and with perfect accuracy. It presents the advantages of not being easily thrown out of adjustment by carriage or rough usage, is light, even when strongly made, and not as costly as the older forms.

MEASURING ANGLES WITH THE THEODOLITE..

It may be as well to observe to the beginner, that but little anxiety need be felt as to the adjustments of an instrument supplied by a good maker, although the student may make an examination of the value and condition of the instrument, as soon as he feels confident he has mastered the principles and details of its construction. What he has really to apprehend, is the damage the instrument is liable to, when brought into use—and this, not from his own hands, but from those of others, particularly *chainmen*, whose hands it is difficult to keep off the instrument, when his back is turned for a short time. Let him, therefore, be advised never to leave an instrument on the ground, without its being under the care of some person, on whom he can thoroughly depend; for the curiosity of an ignorant man, and his desire to handle that of which he has no conception, are often unrestrainable, where there is, as he thinks, no fear of detection, and have often led to serious consequences. When the day's work is over, restore the instrument to its case, and keep the key: the sense of security thus ensured, more than repays the few minutes of trouble incurred.

In measuring angles with the theodolite, supposing the instrument (p. 61) in adjustment, and secured on its three-legged support, ready for use, the first thing to be done is to plant the instrument somewhat about level, exactly over a station from which observations are to be made, giving the three legs a wide spread, and fixing them as firmly in the ground as circumstances will admit. The tripod is made of such a length, that, when a sufficient spread is given, the height of the eye-piece of the telescope will be most convenient to the observer. The exact position of the centre of the instrument over the centre of the station is ensured by the plummet and line, that hang from the centre of the instrument, being close over a peg in the ground, or dropping into the hole itself, from which a flag-pole has been removed, to make place for the theodolite. Now set it level, by turning the parallel plate screws, *p*, *p*, until the bubbles of the two spirit levels on the vernier plate keep the centre of their run, whilst the telescope is turned steadily quite round on its centre; and the instrument will be ready for measuring angles.

Let A be the station; and let it be required to measure the angle subtended by B and C. Clamp the zero of the vernier plate to zero, or 360°, on the limb, and make the coincidence perfect by means of the tangent screw T. Examine the zero of the other vernier, and see that it coincides with

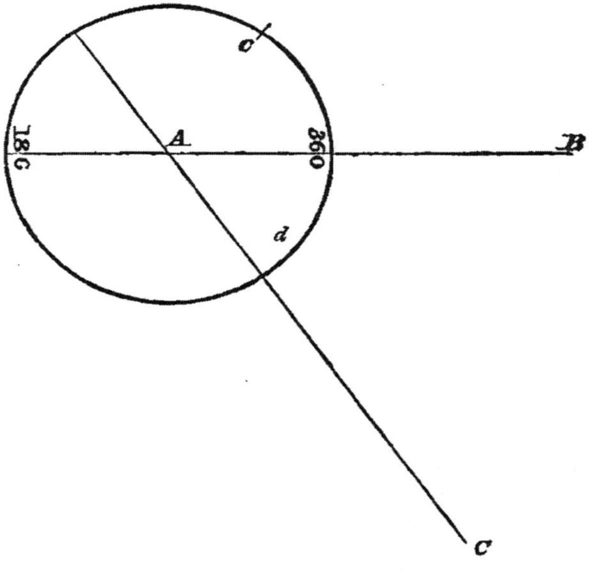

180°. Move the instrument bodily round, until the intersection of the wires in the telescope is in contact with the object at B. Clamp the instrument by means of the screw attached to the collar round the axis, and make the *bisection* of the object at B, by the wires, perfect by means of the slow-motion screw S. This exact bisection can only be secured, when clear and distinct vision has been obtained by means of the milled head in the side of the telescope, for adjusting the object-glass. The line of sight having thus been made to pass from A exactly through zero, or 360° on the limb, and the centre of the object at B, release the vernier plate, by loosening its clamp, and turn it round, until the intersection of the wires is again in contact with the object at C; then tighten the clamp of the vernier plate, and, by means of its tangent screw, T, make perfect the bisection, by the wires, of the object at C. If there is any considerable difference between the distances of B and C from A, the milled head of the telescope will have again to be turned, to ensure distinct vision. *The angle is now measured*, and the measurement is to be read off from *both* verniers, and the mean of the readings taken: thus let the fore vernier pointing to C read 62° 37′, and the back vernier show 36′ 30″ beyond a degree; then the mean of the minutes and seconds will be 36′ 45″, and the corrected measure of the angle will be 62° 36′ 45″.

In ordinary cases, with a 6-inch theodolite, the above measurement of an angle, read off with *both* verniers, will be found sufficiently accurate; but, where a considerable area is triangulated, or where distant observations are taken from a base line, as well as in traverses of several miles circuit, the angles should be *repeated*, and the mean of all the observations taken as the real measure.

The operation of repeating an angle is performed as follows:—Having taken the first measurement in the manner explained above, loosen the clamp of the lower plate, or horizontal limb, turn the theodolite bodily round, until the telescope is directed upon B, and, again clamping the instrument, perfect the bisection of the object by the cross wires, by means of the slow-motion screw S. The index of the vernier, together with the coincident division on the limb, will thus have been brought from *d* to the previous position of the zero, or division marked 360°, while this will have been carried through an equal arc to *c*. Now release the upper, or vernier plate; turn it until the telescope is directed upon C; clamp it again; and perfect the bisection by means of the tangent screw T. The means of the readings, indicated by the two verniers, will now denote the magnitude of an angle twice as large as B A C, and half the amount may generally be taken as a very accurate measurement of the angle. Should this measurement differ considerably, as half a minute or more, from that first obtained by single observation, the operation may again be repeated, and one-third of the amount, indicated by the readings of the verniers, be taken for the corrected magnitude of the angle.

RANGING LINES WITH THE THEODOLITE.

The theodolite is used, in engineering field-work, for ranging long lines, to form a base line for a system of triangulation, or for setting out a line of railway or canal. This is performed by means of a series of back and fore sights, taken at prominent stations along the line; and, as the back and fore sights, at each station, must be taken without the slightest variation in the position of the vernier plate upon the horizontal limb, this plate should be firmly clamped to the limb, and remain so throughout the whole operation, the indices of the verniers being fixed at zero, or 360°, and 180°, or any two equally marked divisions, and examined at each step, to see that the coincidence remains perfect. After having examined the adjustments, and clamped together the vernier plate and horizontal limb, as above directed, plant the instrument, by means of the plummet and line, exactly over a point selected for the first station, through which the line is to be ranged, and level it by means of the parallel-plate screws. The pole, set up at the place selected for the starting-point, or commencement of the base line, will here form the back object. Direct the telescope to a point on this pole as near the ground as possible; clamp the instrument, and perfect the bisection of the object by the cross wires, by means of the slow-motion screw. As many ranging rods as may be desirable may now be set out, in the prolongation of the line of collimation of the telescope, between the starting-point and the station. This being done, examine the verniers to see that they have not moved upon the limb, and, if the result be satisfactory, reverse the telescope, and again set out ranging rods as far as it commands the ground, in every case directing the intersection of the cross wires upon a point as low down upon the rod as possible. Now direct the telescope to the next commanding situation on the base, and set up a pole there, to mark the spot for the next station. Move on to this station, and proceed in the same way, and so from station to station, until the entire line is set out.

In the performance of this work the superiority of the transit theodolite is immediately perceived, the reversal of the telescope being accomplished by the simple operation of revolving it on its axis. With the ordinary theodolite, the reversal must be performed by opening the clips, lifting out the telescope, turning it round, and resetting it again. The Everest theodolite, in which the telescope can only be reversed by turning it round in azimuth through 180°, is still less adapted for this purpose.

TRAVERSING WITH THE THEODOLITE.

In the conduct of an extensive survey, the two principal desiderata are accuracy and dispatch, neither of which should be unduly sacrificed to the other. To obtain both these ends, the principal points of the survey should be determined by a system of triangles, proceeding from an accurately-measured base of considerable length. The angles of these triangles should be observed with a large and perfect theodolite, constructed for the purpose, or with an altitude and azimuth instrument; and numerous corrections should be applied, for the spherical form of the earth, the refraction of the atmosphere, the errors due to the imperfect graduation of the instrument, &c.

The boundaries of the entire country to be surveyed being thus determined with the greatest possible accuracy, and a series of stations laid down throughout, the spaces included between these stations may be subdivided into spaces of smaller extent, the boundaries of which may be surveyed with considerable dispatch by means of the chain, and a portable theodolite, such as we have been describing above; and lastly, the details of the country within these spaces may be sketched with still greater rapidity by means of the prismatic compass.

The boundaries of the spaces to be surveyed by the chain and small theodolite should not exceed three or four miles in extent, and the following is the manner of proceeding.

Let *a*, *b*, *c*, *d*, *e*, *f*, represent the boundary to be surveyed, and let A, B, and C, be three stations, that have been accurately laid down by the previous triangulation, of which both B and C can be seen from A, and A can be seen from C. First measure with the chain the lengths of the several lines *a b*, *b c*, *c d*, &c., taking offsets to all remarkable points on either side of these lines, in the usual manner, and driving pickets at *a*, *b*, *c*, &c. Measure also the distance from A to *a*, and from *d* to C. These measurements having been made, set up the theodolite at A, level it, and clamp the vernier plate to the lower plate of the horizontal limb, at zero, or so that the readings of the two verniers may be 360° and 180° respectively, this adjustment being perfected by the slow-motion screw T. Next move the whole instrument round upon the azimuthal axis, till the object B is accurately bisected by the cross wires, clamp it firmly in this position by the screw, tightening the collar, and enter in the field book the reading of the compass. Now release the vernier plate, and turning it round, bisect the object C, by the cross wires, and enter the readings of both verniers in the field book. Observing, in like manner, the bearings of any other remarkable objects, and, entering the readings in the same way, direct the telescope lastly to *a*, at which station an assistant must be placed, with a staff held upon the picket there driven into the ground, and, entering the reading of the vernier as before, clamp the vernier plate carefully, and remove the instrument to *a*. Level the instrument at *a*, unclamp the collar, and, turning round the whole instrument upon the azimuthal axis, direct the telescope to the last station A, tighten the collar again, and perfect the adjustment, if necessary, by the slow-motion screw S. Now release the vernier plate, and, bringing it back to zero, if the reading of the compass be the same* as the reading previously entered in the field book, we assume our work, as far as it has gone, to be correct; but, if not, we must go back to A, and go over the work again.† Next, with the lower plate still clamped, and the vernier plate released, enter the readings

* If the horizontal limb be figured throughout one semicircle up to 180°, and then, commencing again from 0°, be figured up to 180 at the original starting-point, the readings of the compass will be the same, as stated in the text; but if, as in the transit theodolite, the horizontal limb be figured all round, from 0 up to 360°, the readings would differ by 180°. To make the readings the same in this case, the observations at the second station should be made with the telescope reversed, by revolving it on its bearings.

† If the same result be again arrived at, we may presume that the compass is acted on by some local attraction, and proceed with the work; and the accuracy of this presumption will be further tested as we go on.

when the telescope is directed to the several remarkable points visible from *a*, and lastly direct the telescope to the next forward station *b*, as before, clamping the vernier plate at each observation, and perfecting the adjustment with the tangent screw. In the same manner, proceed from *b* to *c*, *c* to *d*, and *d* to C; and, having directed the telescope at C to the last back station *d*, and released the vernier plate, direct the telescope to A; and, if all the angles have been correctly measured up to this time, the readings of the verniers will now be the same as when the telescope was directed to C from the point A. If then we have not been able to make all the compass readings agree at the previous stations, after going twice over the work at such stations, we may now consider that our work was correct, and that the errors in the compass reading arose from some local attraction, or extraordinary variation of the needle. This verification of the work at C, is called *closing the work*.

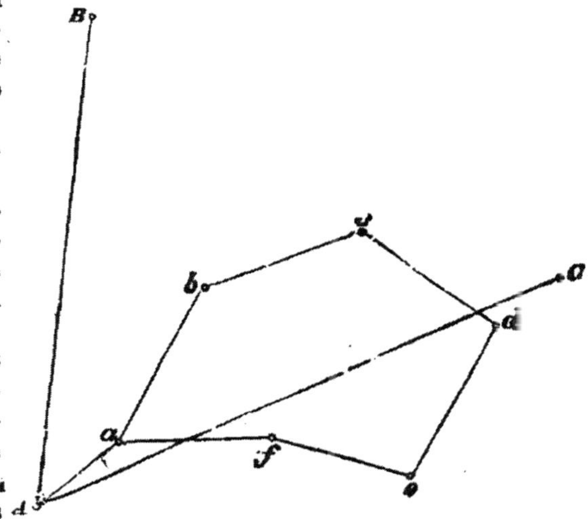

We now come back again to *d*, and proceed from *d* to *e*, and so on, as before, till we come to some other station, which has been observed, either from A or C, and which we again close upon, and at last arriving at *f*, if the readings agree within a minute or two with the readings for *f*, previously observed at *a*, the whole work may be considered to have been performed with a sufficient degree of accuracy; but, if the error amount to more than a minute or two, we must proceed back again from *f* to *e*, and so on till we find out the station at which the error has occurred.

If the ground along any of the lines *a b*, *b c*, &c., rise or fall, suppose for instance, along *b c*, then we must direct the telescope from *b*, so as to make the cross wires bisect, upon the staff, held upon the picket at *c*, a point at the same distance from the ground as the centre of the telescope, and then upon one side of the vertical limb is pointed out the number of links to be deducted for each chain, from the measured distance *b c*, to reduce it to the required horizontal distance This reduction is then to be entered in the field book.*

PLOTTING THE SURVEY.

Having now explained the instruments to be used for the measurement of both linear and angular distances, it remains to consider the means to be used for laying down on paper the results obtained from them. This is called *plotting the survey*. The instruments to be used in plotting are described fully in Vol. I. Part I., on Drawing Instruments.

If the survey has been made entirely with the chain, the chain lines will all be sides of triangles, which have been measured, and the survey is to be plotted by laying down their lengths with the ordinary or beam compasses, according as they are short or long; and the offsets from them may be pricked off by means of a plotting and offset scale.

Care must be taken that the sheet of paper is large enough to contain the work, and, also, to lay down the first line in such a manner that the whole of the survey shall come conveniently into the paper. For this purpose a plot of the outline, to a small scale, should first be made, or a rough sketch made on the ground will frequently suffice.

Having selected the position for the first line, and laid it down of its proper length, the lengths of the two other sides of the triangle, of which it forms the base, are to be taken with the compasses; arcs with these distances being struck from either end of the base, the intersection of these arcs will give the vertex of the first triangle to be laid down; and the sides must then be drawn from the vertex to either end of the base. Proceeding in a similar manner with the other measured lines, all the triangles will be filled in, and the detail may then be set off from the principal lines thus drawn.

When the theodolite has been used, and it is desired to protract the angles observed, a circular protractor of some

* The method of surveying with the chain and theodolite, explained above, is called surveying by traverse.

kind must be used. The different forms of this instrument are described in "Mathematical Instruments," Part I., and their respective advantages pointed out.

In plotting a traverse, unless great care be used, both as to the lengths of the lines, and their angular bearings, the work will not close accurately; and, before protractors were so perfectly constructed as at present, either the method of northings and southings, eastings and westings, hereafter described, had to be employed, or the angles had to be protracted by chords.

To *protract by chords*, find in a table of sines, the sine of half the angle to be protracted, and double it. The number thus found, multiplied into a convenient length, will give the chord of the angle to a radius of the assumed length, and an arc being described with this radius, the chord, laid in it, will subtend at the centre the required angle. Thus, to set off an angle of 34°. In the table of sines, the sine of $\frac{34}{2}$°, or 17°, is ·29237, and, multiplying this by 200, we have 58·5, for the chord

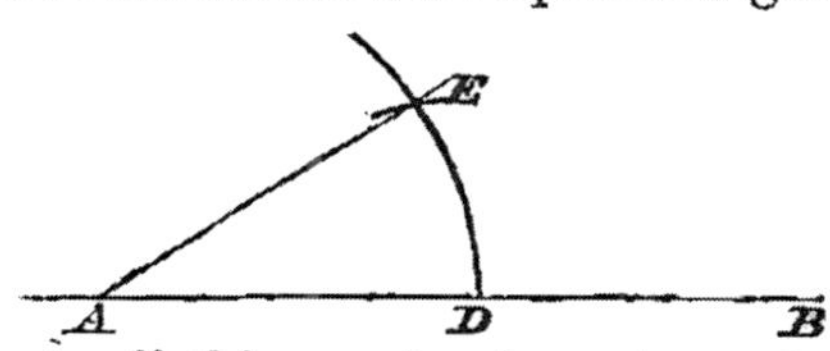

of 34° to radius 100. To set off this angle, then, from the point A of the line A B; with radius 100 describe the arc D E, intersecting A B in D, and with radius 58·5, from centre D cut the arc D E in E: then D E will be the chord, and D A E will be the required angle, of 34°. In practice, it is not necessary to draw a long arc D E, but only to mark the position of D, and describe two small arcs at E with radii A D, and D E, respectively.

The process of laying down angles by the system of chords, just explained, gives results far more accurate than can be obtained by any other means, since they can be protracted by the beam compass with radii as large as we please. This mode is necessarily used, therefore, for laying down the principal angles of a trigonometrical survey on a large scale. In fact, for all large triangles, protracting by chords is the only trustworthy method, and should always be resorted to when the sides of the triangles exceed two feet in length.

The method of northings, southings, eastings, and westings, is another most admirable system of plotting, especially useful when it is a principal point to obtain the area of an estate or parish, &c. The first step is to procure, or form, a table of northings, southings, eastings, and westings*, for all angles made with a meridian line, and for all distances from 1 to 100. These distances may be either links, feet, chains, or estimated in any denomination whatever, and the corresponding northings, southings, eastings, and westings, will be in the same denomination. This table is, in fact, nothing more than the products of the sines and cosines of the angles, made with the meridian line, multiplied by the several distances, and the following is the method of using it. Take out from this table the northings, southings, eastings, and westings, made on each of the lines of the survey, the line from which the angles have been measured being for this purpose assumed as the meridian, no matter in what direction it may lie, and place them in a table, which we may call a traverse table, in four separate columns, being the third, fourth, fifth, and sixth columns of the table†, headed N., S., E., and W. respectively. Add up these several columns, and, if the work is so far correct, the sum of the northings will equal the sum of the southings, and the sum of the eastings will equal the sum of the westings. Then, in two additional columns, enter the whole quantities of northing and easting, made at the termination of each of the several bounding lines of the survey. These quantities will be determined by putting zero for the greatest southing, and adding, or subtracting, the northing, or southing, made on each particular line, to, or from, the whole quantity of northing, made at the beginning of this line, or at the termination of the preceding line; and again, by putting zero for the greatest westing, and adding, or subtracting, the easting, or westing, made on each line, to, or from, the whole quantity of easting at the beginning of the line.

The preparatory table having been formed, the plot may be laid down with great accuracy; and, for an extensive traverse of a mile or two across, no method will give better results, although it is a little tedious. It is, however, applied with greater facility, if a rough sketch of the work be first plotted as a guide. This rough sketch will enable a point on

* This table is the same as the table given in books on navigation, and then called a table of latitude and departure.

† The first and second columns of the traverse table contain the courses and distances.

the paper to be selected for the starting-point, through which a line should be drawn to represent a meridian. The northings and southings may be marked off on this line, measuring each distance, taken from columns 3 and 4, from the point last found, and proceeding upwards for the northings, and downwards for the southings. The accuracy of the divisions may then be checked, by measuring their respective distances from the lowest division thus obtained, as these distances ought to agree with the total northings in column 7. Lines, perpendicular to the meridian, are then to be drawn through the divisions thus found, and on these perpendiculars are to be set off the eastings and westings. Mark them off, first, on one of the east and west lines just drawn, measuring each distance, taken from columns 5 and 6, from the point last found, to the right for eastings, and to the left for westings; then, opposite the division thus found, upon their proper perpendiculars to the meridian, will be found the correct position for plotting the points of the survey. Much labour may be saved by making the plot on paper previously ruled with two sets of equidistant parallel lines, at right angles to each other, the distances apart of the lines according with the scale to which the plot is to be made.

To compute the Area of the Plot.—Rule six additional columns. In the first of these, or ninth column of the traverse table, set the sums of the total northings, and in the tenth, the sums of the total eastings, at the beginning and end of each line in the survey; which sums will be found by adding together each pair of succeeding numbers in the two preceding columns. In the eleventh column set the products of the eastings, made on the respective lines of the survey, found in the fifth column, multiplied by the corresponding sums of the total northings, in the ninth column; and in the twelfth column set the products of the westings, found in the sixth column, and the corresponding sums of the total northings in the ninth column. Sum up the eleventh and twelfth columns, and the difference of the totals thus found will be twice the area of the plot. Again, in the thirteenth and fourteenth columns, set the products of the northings and southings in the third and fourth columns, multiplied by the corresponding sums of the total eastings in the tenth column, and the difference of the totals of the thirteenth and fourteenth columns will again be twice the area of the plot. If it agree nearly with the double area before found, it shows

1	2	3	4	5	6	7	8	9	10	11	12	13	14
Courses.	Distances.	Northings.	Southings.	Eastings.	Westings.	Total Northings.	Total Eastings.	Sums of Total Northings.	Sums of Total Eastings.	East Products.	West Products.	North Products.	South Products.
	ch. lin.	ch. lin.	ch. lin.	ch. lin.	ch. lin.	ch. lin.	ch. lin.	ch. lin.	ch. lin.	acres.	acres.	acres.	
	0 00	...	...	...	...	13 76	0 84						
N. 31° 20′ E.	10 16	8 68	...	5 29	...	22 44	6 13	36 20	6 97	19·14980	...	6·04996	
N. 87 15 E.	9 23	0 44	..	9 22	...	22 88	15 35	45 32	21 48	41·78504	...	·94512	
S. 14 00 E.	6 66	...	6 46	1 61	...	16 42	16 96	39 30	32 31	6·32730	...	...	20·87226
S. 12 25 W.	11 00	...	10 74	...	2 37	5 68	14 59	22 10	31 55	...	5·23770	...	33·88470
S. 64 10 W.	13 4	...	5 68	...	11 74	0 00	2 85	5 68	17 44	...	6·66832	...	9·90592
N. 44 17 W.	4 9	2 93	...	...	2 87	2 93	0 00	2 93	2 85	...	·84091	·83505	
N. 4 28 E.	10 86	10 83	...	0 84	...	13 76	0 84	16 69	0 84	1·40196	...	·90972	
		22 88	22 88	16 96	16 98					68·66410	12·74693	8·73985	64·66288
										12·74693			8·73985
										55·91717			55·92303
													55·91717
												4)	111·84020
												Acres	27·96005
													4
												Roods.........	3·84020
													40
												Perches	33·60800

the calculations to have been correctly performed. (We give an example in the preceding page.)

The near agreements of the sums of the third and fourth, and of the fifth and sixth columns, is a test of the accuracy of the survey; in columns 7 and 8, we have the distances to be set off by the plotting scale; in column 9, we have the multipliers, by which the east and west products, in columns 11 and 12, are found; and in column 10, we have the multipliers, for finding the north and south products, in columns 13 and 14. The difference of the sums of the eleventh and twelfth columns gives double the area; the difference of the sums of the thirteenth and fourteenth gives again double the area; and, taking the mean of these results, by adding them together, and dividing by 4, we obtain the area, most probably to within a quarter of a perch, since the two double areas differ by less than a perch.

In exploring a wild country, of which we have only a very imperfect map, as a guide, or perhaps none at all, the computations of the northings, southings, eastings, and westings, become indispensable, that the observer may determine his position, in the same way as the sailor, who makes use of the same distances under the names of latitude and departure.

CHAPTER IV.

INSTRUMENTS FOR DETERMINING AT ONCE THE DISTANCE OF AN OBSERVED OBJECT.

THE STADIUM FOR MEASURING DISTANCES IN RIFLE PRACTICE.

This instrument, used throughout the army for the rapid measurement of distances with sufficient accuracy for the purpose of rifle-practice, was originally designed by Major A. C. Cooke, R.E., and introduced in its present form by Major-General Hay, the Inspector-General of Musketry.

A B C (Fig. 1), represents the stadium.
A is a side sight.
B „ front ditto.
C „ back ditto.
H „ movable ditto.
D E is a tape or chain.
D H is the distance to be measured.

The principle of the stadium depends on the well-known

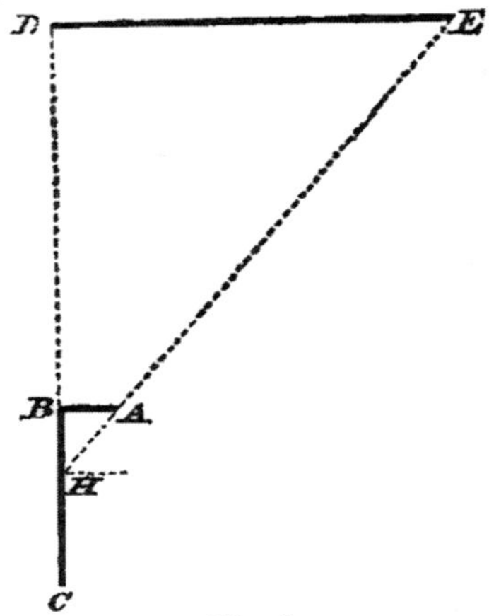

Fig. 1.

mathematical proposition, that in similar triangles the similar sides are proportional. In the two triangles E D H, A H B, the right angle E D H is equal to the right angle A B H, and the angle E H D is common to both triangles: the two triangles are, therefore, similar, and their sides proportional: H D has therefore the same proportion to H B, that E D has to A B. The length of D E is fixed, and that of B A being also fixed, their proportion is known: suppose it to be as 120 to 1, then D H will also be 120 times B H.

To construct the stadium, then, it is only necessary to fix on a suitable proportion between D E and B A, say 120 to 1, and then to mark the divisions on B C 120 times their real length; that is, if B H is really 12 inches, it is marked 120 feet, and indicates the distance from D to H. In fact, every inch on B H corresponds to a distance of 10 feet and is marked accordingly.

The absolute correctness of the result obtained depends upon the two triangles A B H, E D H being strictly similar, and this again depends upon the tape, D E, being held exactly at right angles to D H; for if it is not, the angle E D H will not be equal to the angle A B H. It is desirable, therefore, to ascertain what error in the determination of the distance may be expected from this cause.

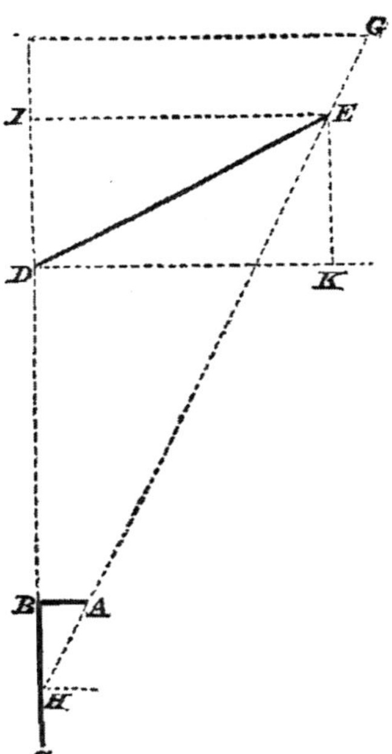

Fig. 2.

Suppose D E to be held as much out of the perpendicular as is shown in Fig. 2. The real distance D H remaining the same, the distance indicated, by B H, on the stadium, is evidently very different in Fig. 2, to what it is in Fig. 1. Let F G be an imaginary tape held at right angles to F H, the end of which, G, falls on the prolongation of H E; it is evident that the distance indicated on the stadium will be that of F H, and the amount of error, or the difference between the real and the indicated distance, will be F D.

Put F D $= x =$ error to be determined.

E K $= e =$ I D $=$ the number of yards that the man at the end, E, of the tape is out of the perpendicular.

H F $= d =$ apparent distance;

D E $= t =$ F G $=$ length of tape, in yards;

then $\text{IH} = \text{HF}\,\dfrac{\text{IE}}{\text{FG}} = d\,\dfrac{\sqrt{t^2-e^2}}{t}$;

$\text{HD} = \text{IH} - \text{ID} = d\,\dfrac{\sqrt{t^2-e^2}}{t} - e$; when the angle E D H is greater than a right angle;*

$= \text{IH} + \text{HD} = d\,\dfrac{\sqrt{t^2-e^2}}{t} + e$; when the angle E D H is less than a right angle.†

and $x = \text{HD} - \text{HF} = -\,d\left(1 - \dfrac{\sqrt{t^2-e^2}}{t}\right) - e$; when the angle E D H is greater than a right angle.*

$= -\,d\left(1 - \dfrac{\sqrt{t^2-e^2}}{t}\right) + e$; when the angle E D H is less than a right angle.†

By substituting values of t, e, and d, in these general equations, the amount of error in the distance indicated on the stadium can be calculated under all conditions.

* In this case the correction is always negative, or to be subtracted from the distance indicated.

† In this case e is sometimes greater than $d\left(\dfrac{\sqrt{t^2-e^2}}{t}\right)$, and sometimes less: in the former case the correction is positive, or to be added; in the latter negative, or to be subtracted.

The following tables have been calculated from these equations :—

TABLES *showing the error arising in the indications of the stadium when the tape or chain is not held at right angles to the line to be measured.*

t = 40 = length of tape, in yards.
e = number of yards that the end of the tape or chain is out of the perpendicular.
d = distance indicated by the stadium, in yards.
x = error in distance indicated: when the sign + is annexed, the value of x is to be added to the reading of the stadium, and when the sign — is annexed, it is to be subtracted.

	The Angle E D H less than a right angle.							The Angle E D H greater than a right angle.						
Value of d, in yards.	e=1	e=3	e=5	e=7	e=9	e=11	e=15	e=1	e=3	e=5	e=7	e=9	e=11	e=15
	Values of x, in yards.							Values of x, in yards.						
100	+1	+2¾	+4¼	+5½	+6½	+7¼	+7¾	—1	—3¼	—5¾	—8½	—11½	—14¾	—22¼
200	+⅞	+2½	+3½	+4	+3¾	+3¼	+½	—1⅛	—3½	—6½	—10	—14¼	—18¾	—29½
300	+⅞	+2¼	+2¾	+2½	+1¼	—½	—7	—1⅛	—3¾	—7¼	—11½	—16¾	—22½	—37
400	+⅞	+1¾	+1¾	+¾	—1¼	—4½	—14¼	—1⅛	—4¼	—8¼	—13¼	—19¼	—26½	—44¼
500	+⅞	+1½	+1	—¾	—3¾	—8¼	—21½	—1⅛	—4½	—9	—14¾	—21¾	—30¼	—51½
600	+¾	+1¼	+¼	—2¼	—6½	—12	—28¾	—1¼	—4¾	—9¾	—16¼	—24½	—34	—58¾
700	+¾	+1	—½	—3¾	—9	—16	—36	—1¼	—5	—10½	—17¾	—27	—38	—66
800	+¾	+¾	—1¼	—5¼	—11½	—19¾	—43½	—1¼	—5¼	—11¼	—19½	—29½	—41¾	—73½
900	+¾	+½	—2	—7	—14	—23½	—50¾	—1¼	—5½	—12	—21	—32	—45½	—80¾
1000	+⅝	+¼	—2¾	—8½	—16¾	—27½	—58	—1⅜	—5¾	—12¾	—22½	—34¾	—49½	—88

The indications of the stadium are also liable to error from a second cause, namely, the tape not being held horizontal. In this case the effect will be the same as if the tape were too short. Using the same notation as before, e representing the difference of level between the two ends of the tape, the equation, obtained from Fig. 3, will be

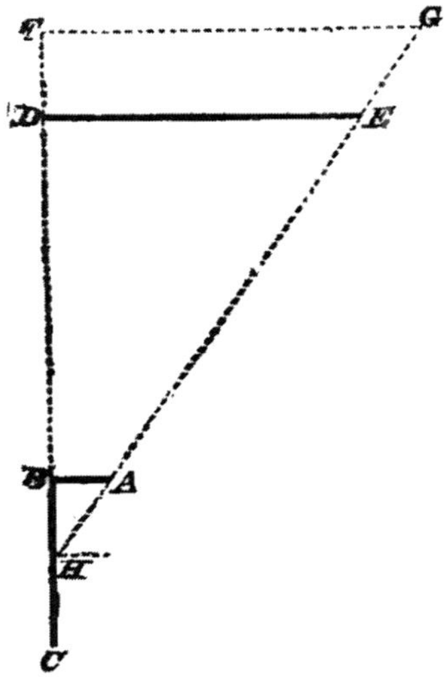

Fig. 3.

$$x = -d\left(1 - \frac{\sqrt{t^2 - e^2}}{t}\right)^*$$

The following table has been calculated from this equation :—

TABLE *showing the error in the indications of the stadium when the two ends of the tape or chain are not horizontal.*

t = 40 = length of tape, in yards.
e = difference of level between the two ends of the tape or chain.
d = distance indicated by the stadium, in yards.
x = error in distance indicated, to be subtracted from the reading of the stadium.

---

* This correction is always negative. If, however, the instrument instead of being used with B A horizontal, were held with B A parallel to the tape, no error could arise from want of horizontality in the tape.

Value of d, in yards.	$e = 1$	$e = 3$	$e = 5$	$e = 7$	$e = 9$	$e = 11$	$e = 15$
	Value of x, in yards.						
100	—	$\frac{1}{4}$	$\frac{3}{4}$	$1\frac{1}{2}$	$2\frac{1}{2}$	$3\frac{3}{4}$	$7\frac{1}{4}$
200	$\frac{1}{8}$	$\frac{1}{2}$	$1\frac{1}{2}$	3	$5\frac{1}{4}$	$7\frac{3}{4}$	$14\frac{1}{2}$
300	$\frac{1}{8}$	$\frac{3}{4}$	$2\frac{1}{4}$	$4\frac{1}{2}$	$7\frac{3}{4}$	$11\frac{1}{2}$	22
400	$\frac{1}{8}$	$1\frac{1}{4}$	$3\frac{1}{4}$	$6\frac{1}{4}$	$10\frac{1}{4}$	$15\frac{1}{2}$	$29\frac{1}{4}$
500	$\frac{1}{8}$	$1\frac{1}{2}$	4	$7\frac{3}{4}$	$12\frac{3}{4}$	$19\frac{1}{4}$	$36\frac{1}{2}$
600	$\frac{1}{4}$	$1\frac{3}{4}$	$4\frac{3}{4}$	$9\frac{1}{4}$	$15\frac{1}{2}$	23	$43\frac{3}{4}$
700	$\frac{1}{4}$	2	$5\frac{1}{2}$	$10\frac{3}{4}$	18	27	51
800	$\frac{1}{4}$	$2\frac{1}{4}$	$6\frac{1}{4}$	$12\frac{1}{4}$	$20\frac{1}{2}$	$30\frac{3}{4}$	$58\frac{1}{2}$
900	$\frac{1}{4}$	$2\frac{1}{2}$	7	14	23	$34\frac{1}{2}$	$65\frac{3}{4}$
1000	$\frac{3}{8}$	$2\frac{3}{4}$	$7\frac{3}{4}$	$15\frac{1}{2}$	$25\frac{3}{4}$	$38\frac{1}{2}$	73

EDGEWORTH'S STADIOMETER OR SURVEYING INSTRUMENT.

This instrument is mounted on an ordinary tripod stand, and consists of a set of parallel plates and screws, marked G′ (Fig. 1); a round disc, marked A, on which a sheet of paper can be fastened by means of the little clips C C′; a telescope M B, fitted on an arc O, which has a vertical motion in a frame P G; and this frame P G has a circular and independent motion round the disc or table A. There is a scale D D′, fastened in the frame P G, the centre of which corresponds with the centre of the instrument, and which is graduated to the scale to which the surveyor wishes to have his survey plotted. The telescope, M B, is fitted with a diaphragm, with two horizontal hairs, distant from one another a hundredth part of the focal length of the object-glass. From this proportion it follows that, when an ordinary levelling staff is held on any distant point, and the telescope brought to bear upon it, if the readings, in feet and decimals of a foot, of the intersections of the hairs on the staff be observed, their difference, multiplied by a hundred, gives the true distance, in feet, of the staff from the instrument. The hairs can be adjusted by means of the screws R and R′, from observations on a staff at a known distance.

We have thus a simple and speedy method of measuring the distances around the instrument.

Some correction, however, of this method must be made in obtaining distances in precipitous grounds: in such cases it is necessary that the staff should be held at right angles to the

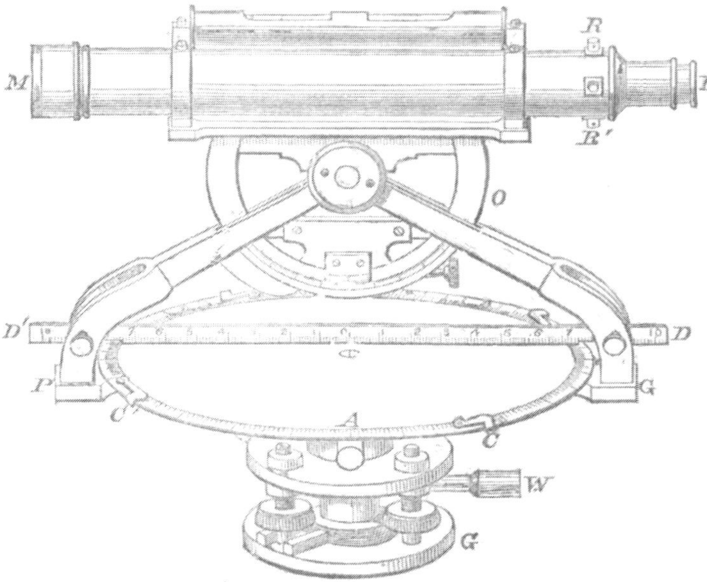

Fig. 1. (*About one-fourth full-size.*)

axis of the telescope observing it, and not upright, as the staff man would feel naturally disposed to hold it. Thus the operation of determining a distance in precipitous ground is as follows:—

The instrument is at the point A (Fig. 2), and the horizontal distance A F is required to be known. The man holding the staff sights the instrument through a hole, bored square through the staff at C, or other device, so that when he sights the instrument, the staff must necessarily be at right angles to the line of sight. He then signals to the observer, who takes the readings at E and D, the difference of which multiplied by 100, is the true distance from C to A. This, corrected for the hypothenusal excess, gives the horizontal distance A F. It is to be observed that these corrections are only necessary in precipitous places; and that, as a general rule, when the obliquity of the ground is less than 15°, there is no appreciable error introduced by taking the readings of the staff held in the ordinary way, and no necessity for any correction on account of the excess of hypothenusal over base length.

The correction for hypothenusal over base distances is marked on the limb O, which is moved by a thumb-screw, as in a theodolite.

The operation of surveying with this instrument is as follows:—The surveyor first sets out his base lines and lines of triangulation, which he can do with the telescope. The base line should, of course, be chained, if it is at all an extensive survey, and stakes driven at suitable and known points. The instrument is then by means of a

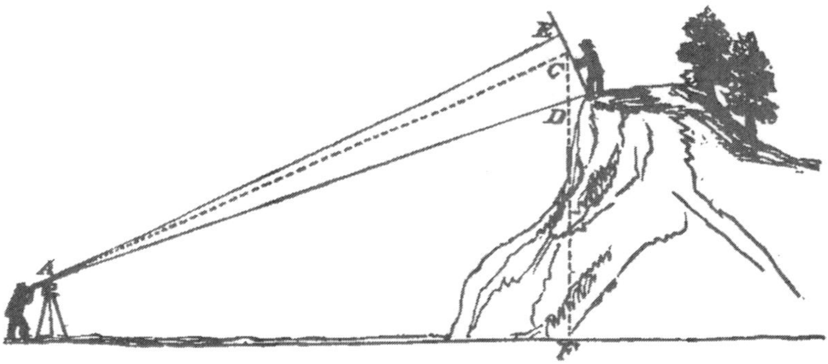

Fig. 2.

plumb bob, set vertically over any of these points. A sheet of paper is put on the table, and a line to correspond with the base line of the survey marked on the sheet. This line, by means of the telescope, is brought to coincide in direction with the base line, and the table is then clamped by the screw W. The staff men now take their staves to the various salient points in the surrounding ground, such as A, B, C, D. (Fig. 3.)

The distances of these points from the instrument are measured by means of the hairs in the eye-piece of the telescope; and their directions correspond with the directions of the telescope. When, then, the distances, corrected, if necessary, for hypothenusal excess, are taken along the scale D, D′, and marked on the paper, the points thus shown are in their true positions on the plan; and a line, drawn through these points with a pencil, completes the plotting of the rough sketch. When the surrounding fences, houses, roads, ponds, &c., have been observed, the instrument is moved on to the next station, and another paper put on, a note having been previously made on the old paper of what portion of the survey it referred to. The surveyor, in the evening, transfers all these detached portions of the survey to his office plan, by means of tracing, pricking through, or other draughtsman's device.

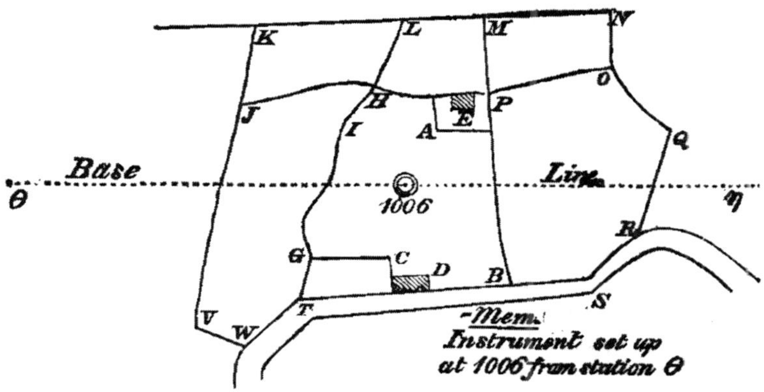

Fig. 3.

The advantages claimed by this system over the ordinary methods of chaining distances, and taking offsets, are these :—

1. A far increased rate of speed in the field.

2. The reduction to the minimum of office operations, which are at present so tedious, and occupy nearly as much time as the field work.

3. In places where waters, or inaccessible heights intervene, and where one cannot chain, but must triangulate, this method of obtaining distances, and their position, is obviously most advantageous.

4. The physical exertion of the surveyor is considerably reduced ; for, instead of following the chainman, he has only to stay by his instrument.

5. The chances of error are reduced, because an error in one observation is confined to that one operation, and not brought forward as in the ordinary method ; and the surveyor has an opportunity of comparing his plotted survey with the ground.

6. The levels of the ground, or contour lines, can be ascertained during the operations, and marked on the survey.

It is to be observed, that the scale D, D′ is capable of being changed for any scale that the surveyor may wish to use; and that the plate A comes off; which admits of the instrument being fitted into a very handy flat box. The plate A is eight inches in diameter, and would thus command, to a scale of 200 feet to the inch, 800 feet on both sides of the base line.

Moreover, the disc is divided into degrees, and may be used, as in a theolodite, for taking horizontal angles. A compass, also, can be attached to this instrument when necessary.

This instrument has been used with considerable success in several railway surveys, for which it is particularly suited.

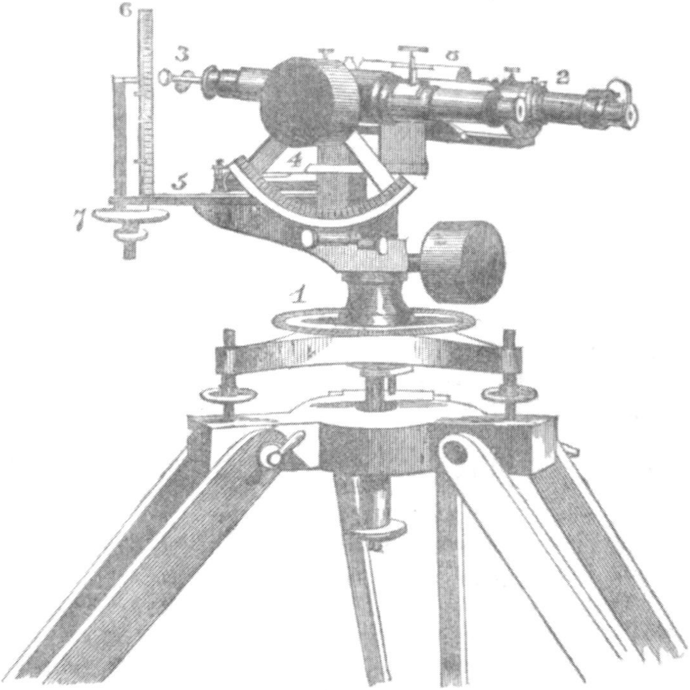

Fig. 1.

THE OMNIMETER.

An instrument called an omnimeter has recently been invented by M. Eckhold, a German engineer. It is intended, like the stadiometer just described, to measure base lines and distances, without chaining, and also differences of altitude and angles. It is, however, a far more perfect instrument than the stadiometer; and we are assured by Messrs. Elliott, that it gives wonderfully accurate results.

The instrument, represented in Fig. 1, on the opposite page, consists of the following parts:—

A graduated limb (1), reading by means of a vernier to ten seconds, for the measurement of horizontal angles;

A good telescope (), revolving in a plane perpendicular to the limb;

A powerful microscope (3), closely united to the telescope;

A highly sensitive level (4) lying upon a rule (5), which has a fixed length (of twenty centimètres, for example);

A scale (6), fixed vertically at the extremity of the said rule, in the plane of the optical axis of the microscope, and divided into half millimètres, the millimètres only being indicated by figures from 1 to 100;

A micrometrical screw (7), attached to the base of the scale, and giving the correct reading of the scale to the $\frac{2}{10000}$ of a millimètre, denoted on the graduated circle of the screw;

A second extremely sensitive level (8), capable of being applied to the telescope, and of determining, in case of need, its horizontality;

Further, of all the necessary screws, keys, and other matters, required to secure the efficient working of the instrument.

As the necessary complement of the omnimeter, there is a staff, Fig. 2, not divided, having an invariable length (as, for example, three mètres), defined on it by two white lines on a black ground, one at the upper, and the other at the lower extremity of the staff.

The instrument is used in the following manner:—

Supposing that the distance between any two points is required, the first thing to be done is to place the staff, which must be held in a truly vertical position, at one of the points; the omnimeter is set up at the other point, and levelled; the telescope is then directed to the upper white line, *m*, of the staff and clamped; and the inclination of the telescope is then read off on the scale by means of the microscope. On account of the figures being magnified by the microscope, only one number at a time, of the 100 on the scale, can be seen; suppose the reading to be 67 (millimètres), plus the, as yet unascertained, fraction comprised

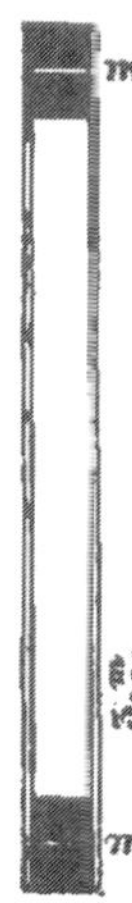

Fig. 2.

between this number, and the horizontal thread of the microscope. The value of this fraction is ascertained exactly by means of the micrometer; say that it reads 2,085 on the circle of the micrometer.

This is noted down in the field book to the right of the number 67, already obtained, thus . .	672,085
The operation is repeated in every particular for the lower line, m', of the staff, and we obtain, say	609,400
Difference	62,685

The operation with the instrument is now complete, as we possess the data required for calculating the distance. In fact, this is found by dividing the constant 6,000,000, by the difference 62,685; giving as a result 95·79 m. for the horizontal distance between the points in question.

For let us suppose the staff in position at $m\,m'$, Fig. 3, O to be the point round which the telescope revolves, and A B to represent the observed difference 62,685. From the point O

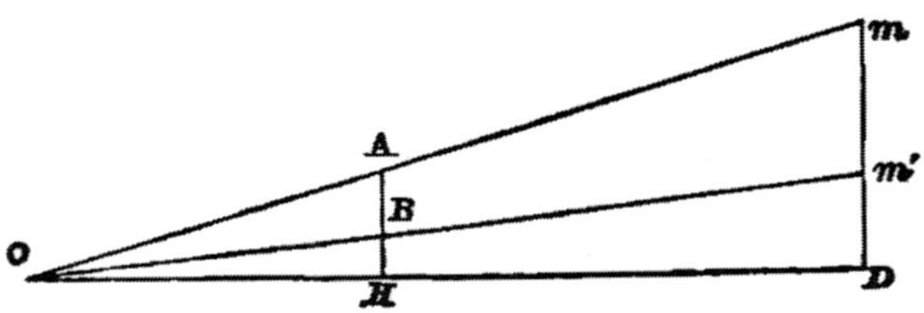

Fig. 3.

let a horizontal line O D be drawn, and let A B be produced to H, and $m\,m'$ to D. Then A B H and $m\,m'$ D are perpendicular to O D; and in the similar triangles O A H and O m D:—

$$\frac{\text{OH}}{\text{AB}} = \frac{\text{OD}}{mm'};\ \text{and, therefore OD} = \frac{\text{OH} \times mm'}{\text{AB}}$$

But O H = 0·20 m. by the construction of the instrument;

$m\,m'$ = 8·00 m., the invariable length of the staff;

A B = 62,685, the quantity obtained, in our example, by the scale.

The last quantity, estimated in mètres, is 0·0062685 m.; and if we substitute for the letters their respective values, we have

$$\text{OD} = \frac{0{\cdot}20 \times 3}{0{\cdot}0062685} = \frac{0{\cdot}6}{0{\cdot}0062685} = \frac{6{,}000{,}000}{62{,}685} = 95{\cdot}79 \text{ mètres.}$$

In the preceding example, the staff was supposed to be placed at one end of the line, and the omnimeter at the other. This mode of operating is not, however, the only one. There are two others; one, by placing the omnimeter in line with the two points, the other, by fixing it at some point out of this line. In either of these modes a staff must be set up at each of the points. In the first case, the total distance is obtained directly, by adding together the back and foresights. In the second case, the distance is obtained by taking the measure of the horizontal angle from the limb, and calculating the base of the triangle, trigonometrically, by means of the angle, thus obtained, and the two sides.

For the establishment of a base of triangulation, the latter method is that which is recommended for adoption; for if we proceed thus by small distances, from ten to twenty mètres on each side or thereabouts, a base line of 1000 mètres may be easily measured within an approximation of 0·004 m.; a result which is quite satisfactory. In fact, if for a distance of ten mètres, the scale gives us the difference 0·06 m., it will give for a distance of 9·99996 m. the difference 0·0600002 m., a quantity appreciable on the circle of the micrometer. The nature of this operation requiring extreme exactness, the staff should have both faces similar to that already described, should be provided with contrivances to ensure its verticality, and its base should be armed with a steel point, to rest on iron pickets driven into the ground.

For the measurement of the slope, or inclined distance, $\text{O}\,m'$, the operation requires further the reading off of the height BH, on the scale. We have then a new proportion:

$$\frac{\text{AB}}{\text{BH}} = \frac{m\,m'}{m'\,\text{D}}\text{; whence } m'\,\text{D} = \frac{m\,m' \times \text{BH}}{\text{AB}};$$

and, then,

$$\text{O}\,m' = \sqrt{m'.\,\text{D}^2 + \text{O}\,\text{D}^2}.$$

In using the omnimeter it has been found that, where extreme accuracy of distance is required, it is not prudent to make the points more than 200 mètres apart. At this distance an approximation within 0·4 m. is obtained. At 500 mètres, by making use of a staff four mètres high, the approximation will be within 0·17 m.; a sufficiently satisfactory result; as, by one fore and one back sight, we can determine a kilomètre within 0·34 m.

In levelling, the operation is identical with that just described for the measurement of inclined distances. The difference of altitude is given by the formula

$$m' \text{D} = \frac{m\, m' \times \text{B H}}{\text{A B}};$$

that is, three (mètres), multiplied by the reading at the lower sight divided by the difference of the reading at the two sights.

For distances not exceeding 100 mètres the differences of altitude are obtained within a fraction of a millimètre, and for greater distances, at which the instruments at present in use are inoperative, the measurement can be obtained within a millimètre. This degree of precision is attained because:

1. In placing the instrument it is not necessary to centralise the optical axis.

2. Because the operator has always to point his telescope on the same lines of the staff, which is generally of an invariable length, viz., three mètres.

3. There cannot be a doubt in the readings; as in those of the graduated levelling staves, for which recourse must be had to estimation.

In difficult and hilly ground, levelling with any of the ordinary levels becomes a long and expensive operation, because the sights are necessarily short, and the number of them multiplied, in proportion to the steepness of the incline. The omnimeter allows of levelling points placed at great distances apart, and which make considerable angles with the horizon.

CHAPTER V.

INSTRUMENTS FOR THE DETERMINATION OF CONSIDERABLE HEIGHTS.

In an earlier part of this book the instruments necessary to be used, to obtain an accurate knowledge of the undulations of any portion of ground, have been described, and the method of using them explained. The levels there described, however, were only capable of measuring differences of altitude of a few feet, from each station, or place of observation, and a continued series of stations had to be selected, and a very large number of observations to be taken, to determine the difference of altitude along a line of considerable length. If the object in view were only to obtain the difference of altitude of two points, where this difference was large, the process of levelling would be altogether inexpedient. The result may then be arrived at by taking angles of elevation of the object, of which the altitude is desired, from different points on a base line ; and if this operation be well conducted, it will give the most accurate results ; but this mode will occupy considerable time, and involve much labour in computation. For observing, therefore, the height of hills or mountains, barometers and thermometers are much used and, when adapted to such use, are called *hypsometrical instruments.*

The determination of altitudes by the barometer is deduced from the difference in the weights of the superincumbent columns of atmosphere, having the same sectional area, at the two stations at which the observations are made.

The barometer is, in fact, an instrument for measuring the weight of the atmosphere. It was invented in 1643, by Torricelli. Up to his time various phenomena, now known to be caused by the pressure produced by the atmosphere in consequence of its weight, were considered to be accounted for by the statement that Nature abhorred a vacuum. Torricelli, in investigating the phenomenon of water ascending in a pump, found that it could not be raised beyond a certain distance, and consequently, Nature's abhorrence of a vacuum had a limit. What, then, could be the cause of this limitation ? To determine this, he tried what would take place with a fluid much heavier than water,—mercury, in fact. He took a glass tube about 4 feet long, sealed at one end and open at the other, and, having filled it with mercury, closed the open end with his finger ; he then inverted the tube, and immersing the open end in a basin containing mercury, withdrew his finger, and raised the tube into a vertical position. He found the mercury descend in the tube, and flow out into the basin, until its surface stood 27½ inches above the surface of the mercury in the basin : it then remained stationary. Comparing the height of this column of mercury with the much greater height of water raised in the most perfect pump, he arrived at the conclusion that these heights were in the inverse ratio of their specific gravities, and, therefore, the difference in the altitudes of the columns depended upon the difference in their weights. He also observed that, in both cases, the phenomenon was produced by cutting off the communication between the atmosphere and the upper extremities of the columns of water, and mercury, while the surface of the fluid, into which their lower ends was immersed, was in direct communication with the atmosphere. He therefore considered that the weight of the atmosphere maintained the elevation of the columns in the tubes, and that the weight of these columns in feet measured the weight of the atmosphere on the same sectional area.

Pascal, a few years later, repeated and extended Torricelli's experiments, making use, not only of mercury, but also of a variety of other liquids, and found that, in all cases, the lighter the liquid, the higher it ascended in the tube. Thoroughly convinced himself of the correctness of Torricelli's theory, he considered what would be the most convincing evidence of its truth to others; and concluded that carrying the tube to a greater height, the column of atmosphere above it would be diminished, and consequently the column in the tube would descend; while, by carrying it to a greater depth the superincumbent column of atmosphere would be increased, and the column would rise in the tube. Experiments made by him upon the mountain Pay de Dôme, near Clermont, in Auvergne, completely proved the correctness of his views. Two tubes, filled with mercury, were carried to the foot of the mountain, their open ends being immersed in mercury, as in Torricelli's experiments; they both stood at 28 inches. One was left at the foot of the mountain, and the other carried up to its top; the mercury in this one kept sinking more and more, the higher it was taken, until at the top it had fallen to 24·7 inches; while upon being brought down again the phenomena were reversed, the mercury rising higher and higher, until, when at the bottom, it was again compared with the tube which had been left there, the heights of the two columns exactly coincided, as at first. Hence it appeared that, in ascending the mountain, the column of atmosphere above the surface of the basin being diminished by that portion of it between the levels of the top and bottom of the mountain, the pressure arising from the weight of this portion was removed from the surface of mercury in the basin, and the mercury descended in the tube, until its weight was just again counterpoised by the weight of the column of atmosphere above the level of the top of the mountain; and conversely in descending, the column of atmosphere above the basin being increased by a length equal to the height of the mountain, the pressure arising from the weight of this additional length of column, acting on the surface of the mercury in the basin, forced it to ascend the tube, until the weight of the mercury in the tube was again balanced by the weight of the atmospheric column.

Pascal conceived that the lower parts of the atmosphere would be more compressed than the upper, owing to the greater weight of the mass of air above them, and that, from this cause, the rise and fall of the mercury in the tube would not be equal, for equal differences of altitude. This view proved to be correct; the atmosphere, in fact, being compressible and elastic, and its density, at a constant temperature, being in exact proportion to the weight with which it is compressed. This relation between the pressure and density of the atmosphere, was proved by Mariott, and is therefore called Mariott's law. It follows, then, that the stratum of air nearest the earth's surface is the densest, from having to support the weight of all the air above it; and in ascending the strata becomes rarer, from being pressed by lesser quantities of air above them. In fact, for heights in arithmetical progression, the density of the atmosphere would decrease in geometrical progression, if the temperature at the different heights were all the same; but as this is not the case, the relation between the densities at different heights, is considerably modified by the differences of temperature at those heights. Air, under a constant pressure, expands about $\frac{1}{490}$ of its bulk at 32° Fahrenheit for every degree of temperature above that point. The warmer the air too, the greater the amount of aqueous vapour contained in it, and this vapour exerts an ever-varying influence on the mercurial column. The correction due to this cause, however, is very trifling. Sir George Shuckburg made numerous experiments upon the effects of temperature upon the weight of a given column of atmosphere. He found that when the mercurial column, measuring the total atmospheric pressure, was 30 inches, and the temperature 32° Fahrenheit, a rise of 86·85 feet produced a fall in the mercurial column of $\frac{1}{10}$th of an inch, and that for every additional degree of temperature, the column of atmosphere equivalent to $\frac{1}{10}$th of an inch of mercury, was increased by 0·211 of an inch. From these data the heights of mountains can be obtained, from observation of the barometer and thermometer, in the manner that will be pointed out, after describing the construction of these instruments.

THE MERCURIAL THERMOMETER.

The common mercurial thermometer consists of a glass tube of uniform bore, terminating in a hollow bulb. By holding the bulb over the flame of a spirit-lamp, a considerable portion of air is expelled from the bulb and tube, and the end of the tube, which is open, being immersed in a cup of mercury, as the air within the tube and bulb cools and condenses, the external atmospheric pressure drives a portion of mercury into the tube and bulb. A paper funnel filled with mercury is next to be tied round the open end of the tube, and, the mercury in the bulb being boiled over the spirit-lamp, the whole of the air remaining in the tube will soon be expelled, and its place supplied by mercurial vapour. The instrument being again allowed to cool, the mercurial vapour will be all condensed, and its place supplied by mercury, driven down the funnel, till both bulb and tube are completely filled with mercury. Lastly, when the mercury has cooled down nearly to the highest temperature intended to be measured by the instrument, the end of the tube, hitherto open, is to be perfectly sealed by means of the blowpipe, and as the mercury afterwards continues to cool, it will be considerably condensed, and, sinking down, will leave a vacuum in the upper part of the tube.

The thermometer has now to be graduated, and, for this purpose, it must first be immersed in melted snow, and, when the mercury has sunk as low as it will, a graduation must be marked opposite the extremity of the mercurial column for the freezing point. The thermometer must next be immersed in the vapour of water, boiling under a given atmospheric pressure, and, when the mercury is again stationary, another graduation must be marked, opposite the extremity of the mercurial column, for the boiling point. The distance between these two graduations is then to be divided into a determinate number of equal parts, and, continuing to mark off divisions of the same extent, in both directions, to the extremities of the tube, the instrument will be completly graduated.

In the centigrade thermometer, the freezing point is 0°, and the boiling point 100°; in Reaumur's, the freezing point is marked 0°, and the boiling point 80°; and in Fahrenheit's, which is the thermometer in common use in this country, the freezing point is marked 32°, and the boiling point 212°. Assuming, then, the boiling point to have been determined under the same pressure for all three, and consequently to indicate the same absolute temperature, we have—

$$\tfrac{1}{5}\,\mathrm{C}^\circ = \tfrac{1}{4}\,\mathrm{R}^\circ = \tfrac{1}{9}\,(\mathrm{F}^\circ - 32^\circ)\,;$$

if C,° R,° and F,° represent respectively the degrees of the centigrade thermometer of Reaumur's, and of Fahrenheit's, denoting the same absolute temperature.

The mercurial thermometer was constructed under the idea, that equal differences of temperature would cause the bulk of a body to vary by equal differences. Now, except in the case of gases, this assumption is not borne out by the facts of the case; but by a fortunate coincidence, the increasing rate of the expansion of the mercury is exactly compensated by the comparatively small expansion of the glass, so that the indications of the temperature given by the ordinary mercurial thermometer are strictly accurate.

The mere fact of speaking of the indications of a thermometer as accurate implies, that there is some mode of testing this accuracy, and, consequently, some idea of comparing temperatures independently of the observation of the variation in the bulk of a body, produced by a difference of temperature; and, in fact, the comparison of temperatures depends on the following considerations:—

1°. Two bodies are said to have the same temperature, if when placed in contact the temperature of either remains unaltered by the other.

2°. When bodies of different temperature are placed in contact, the temperature of the hotter body decreases, and that of the colder increases, till both become of the same temperature.

3°. If the bodies thus placed in contact be both of the same material, and be equal to each other in weight, the increase of temperature in one will be equal to its decrease in the other, and the product of the whole mass into the temperature will remain the same.

In order to test the accuracy of the common thermometer, Dr. Brooke Taylor heated two equal weights of water, till their temperatures, as indicated by the thermometer, were respectively 100° and 200°; and on mingling these bodies of water together, the thermometer indicated correctly 150°, as the temperature of the mixture.

THE BAROMETER.

There are various forms of the barometer, to suit the various circumstances under which it is to be used; but in all of them there is a straight tube, about 33 inches long, closed at one end, for holding the column of mercury, of which the altitude is to be observed. This is called the barometer tube, and, being filled with mercury, the open end of it is placed in communication either with an open cistern containing mercury, or with a bent tube containing mercury, which is exposed to the action of the atmosphere.

Fig. 1. represents a form of barometer, called the standard barometer, adapted for use in fixed observatories. The bore of the barometer tube, A C, should not be less than one-fourth of an inch diameter, and if it be one-third of an inch it is better, in order to avoid any considerable error from the capillary depression. The diameter of the open end, A B, of the barometer should be as large as it can conveniently be made, in order that the variation in the altitude of the mercury in A C should affect as little as possible the altitude of the surface A B. In order, however, to obtain the true altitude of the mercury within this barometer, the position of the surface A B must be taken into account, as well as that of the extremity of the mercurial column A C, the height of which above an assumed zero point is given by the scale attached to the barometer. This is done by making the scale movable, and terminating in an ivory point: when this point and its reflection appear to touch one another, the height indicated is correct.

Fig. 1.

Tubes of small diameters require correction for capillarity. The mercury is depressed in the glass tube from the repulsion between the two substances, and the following corrections must be added :—

Diameter of tube.	Correction.
inch.	
0·25	0·020
0·30	0·015
0·40	0·007
0·45	0·005
0·60	0·002

The syphon barometer, Fig. 2, gives the altitude of the mercurial column independently of errors from capillary depression; for, on account of the equal size of the bore throughout the tube, the extremities of the columns A and C are equally depressed.

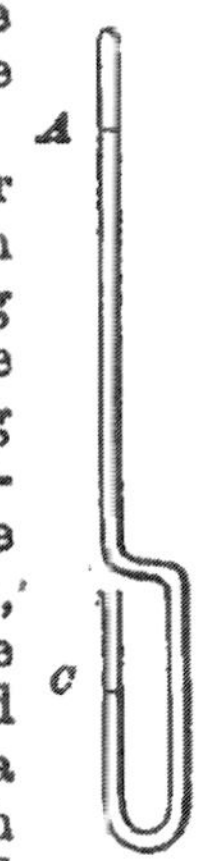

Fig. 2.

In the ordinary form of the portable barometer the tube is immersed in a reservoir with a leathern bottom, and furnished with a screw *f*, by turning which the mercury can be forced up, and made entirely to fill the tube, to prevent it from jerking about, and injuring the instrument during its carriage from place to place. This screw can be made to regulate the surface of the mercury in the cistern, and always bring it into line with the zero of the scale. For this purpose the instrument is furnished with a gauge or float *h*. The gauge consists of a small ivory cylinder, fixed to a float of cork, which rests on the top of the mercury. The upper end of this cylinder works in a groove, hollowed out of a second piece of ivory, which is fixed to the instrument, and has a fine line upon it denoting the zero of the scale: the floating cylinder has a corresponding fine line cut upon it; and, before observing the reading of the instrument, the two lines must be brought into exact coincidence by turning the screw *f*.

Let *a b*, Fig. 3, be the glass tube plunged into the

mercury in the cistern C, and D be the surface-line of the fluid in the cistern, made level with the commencement of the scale, in the construction of the instrument: then the extremity of the mercurial column in the tube at this time is called the *neutral point*. When the mercury rises in the tube, a portion equal to the rise leaves the cistern, and the surface-line falls towards the dotted line *e*; and, being lower than the surface from which its neutral point was measured, the indicated variation in the atmosphere is too little; but turn the screw *f* forwards, until the lines on the float *h* coincide, and the mercury then records the exact change; when a depression occurs, the mercury, sinking from the tube into the cistern, raises the surface-line towards *g*; in this case the screw *f* must be turned back, until the leather at the bottom of the cistern be sufficiently loosened to allow the mercury to assume its proper level at the surface D.

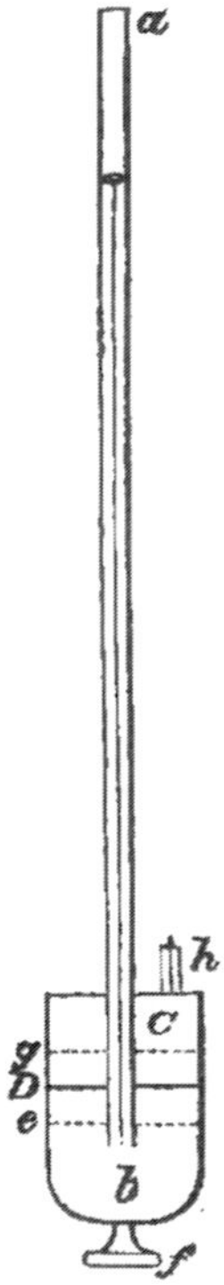

Fig. 3.

When there is not a gauge to the barometer, the relative capacities of the cistern and tube are ascertained by experiment, in the construction of the instrument, and marked thereon, as is also its neutral point.

The correction to be applied to the reading of the scale, to obtain the correct height, may be computed as follows:—

Let h be the neutral point altitude of the extremity of the mercurial column, when the zero point of the scale coincides with the surface D, and let h' be its apparent altitude in any other case; let also K represent the sectional area of D, and k that of the tube; so that $\frac{K}{k}$ is the capacity of the cistern relatively to the tube, which is marked on the instrument; then the distances, through which the surfaces in the tube, and cistern, will rise and fall respectively, will be manifestly inversely proportional to the areas of their surfaces; if, therefore, $\delta h'$ represent the required correction, viz., the distance through which the mercury falls in the cistern,

$$h' - h : \delta h' :: K : k.$$

$$\therefore \delta h' = \frac{k}{K}(h' - h),$$

and the true altitude of the barometer is given by the equation

$$h' + \delta h' = h' + \frac{k}{K}(h' - h).$$

When the indicated height, h', is above the neutral point h, $h'-h$ is positive, and the correction $\delta h'$ is to be added, to give the correct height; but when the mercury is below the neutral point, $h'-h$ is negative, and the correction $\delta h'$ is to be subtracted, to give the correct height. Hence the following rule:—

When the mercury in the tube is *above* the neutral point, the difference between it and the neutral point is to be divided by the capacity, and the quotient *added* to the observed height, to give the correct height; and, when the mercury is *below* the neutral point, the difference is to be divided as before, and the quotient *subtracted* from the observed height will give the correct height.

Ex.—Let $\frac{K}{k}$, or the capacity of the cistern relatively to the tube, be 40, and the neutral point altitude be 30 inches.

Ex. 1.—Observed height		30·480
Neutral point . .	30·000	
Height above neutral point	+·480	
Correction		+.012
Correct height		30·492
Ex. 2.—Observed height		28·720
Neutral point . .	30·000	
Height above neutral point	—1·280	
Correction		—·032
Correct height		28·688

The portable, or mountain barometer, has received great improvements. It is now produced in a form, in which the mercury is confined wholly by glass and iron, and can be made absolutely free from shake in the tube, so as to ensure safety in rough transit. The total height of the mercurial column is read directly against the light at both ends, and with no *structural* corrections for expansion of instrument, except as between brass and glass, the co-efficients of both being well-known.

The greatest range of the barometer in any one spot does not exceed 3½ inches, and consequently, in barometers intended for meteorological observations, it is unnecessary to graduate more than a few inches of the tube: the divisions, therefore, commence at 27 inches, and are continued to 31 inches. The graduations on the mountain barometer, for measuring great heights, commence at 15 inches, and are carried on to 33 inches.

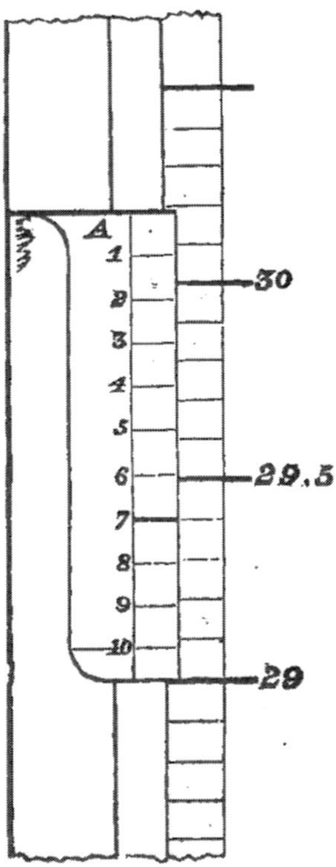

In the common, or meteorological barometer, each inch is divided into ten equal parts, and on the vernier, eleven tenths of an inch are divided into ten equal parts, so that each part is equal to eleven hundredths of an inch, or one tenth and one hundredth. By means of this vernier, then, the height of the column is indicated to the hundredth part of an inch.

The annexed figure shows the arrangement and mode of reading the barometer.

Here the pointer or index, A, of the vernier is above the division, indicating 30·1 inches on the barometer scale, and the division, marked 7 on the vernier, coincides with a division on the barometer scale: the reading is therefore 30·17.

In the mountain and standard barometers, 51 tenths of an inch on the vernier are divided into 50 equal parts, so that each part is equal to 51 five-hundredths of an inch, or one tenth, and one five-hundredth. By means of this vernier, then, the height of the column is read off to the five-hundredth part of an inch.

A thermometer is attached to the barometer, to indicate the temperature of the mercury in the cistern. The expansion of mercury is $\frac{1}{9990}$ of its bulk, for each degree of Fahrenheit between 32° and 212°; and, in making observations for the determination of altitudes, it is indispensable that the attached thermometer should be read off for the temperature of the mercury, and a detached thermometer for the determination of the temperature of the atmosphere, simultaneously with the observation of the height of mercury in the barometer.

Before taking an observation the instrument should be gently tapped, to prevent any adhesion of the mercury to the tube, and if the instrument be provided with a gauge, this should be adjusted, to bring the surface of the mercury in the cistern in a line with the zero of the scale; the vernier should then be brought level with the top of the mercury, and if light is admitted from behind, this arrangement can be perfected by making the lower part of the pointer tangential to the convex part of the mercury in the tube. In reading off, care must be taken to place the eye on a level with the top of the mercurial column, as by placing it above, the reading would be too low, and by placing it below, the reading would be too high, or, in other words, there would be an error from parallax.

THE WHEEL BAROMETER.

In the wheel barometer, a weight, *f*, floats on the surface of the mercury, in the open end of the bent tube A, and a string attached to the float *f* passes over a small wheel *w*, and is then attached to a second weight *p*. As the mercury rises and falls, the float *f* moves with it, and the string turns the wheel, the axis of which passes through a dial-plate and carries an index, like the hand of a clock, which points to the altitudes figured on the dial-plate. The wheel barometer is by its construction endowed with so many sources of error, that

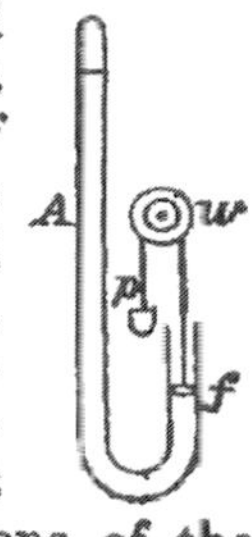

it cannot be depended upon for correct indications of the height of the barometric column; although it readily shows if the mercury be in a rising or falling state, as a small variation in the barometric column produces a considerable movement of the index. It may rather be considered as an ornamental piece of furniture, than as having the slightest pretensions to be called a scientific instrument.

THE ANEROID BAROMETER.

The action of the *aneroid* barometer, invented by M. Vidi, of Paris, for ascertaining the variations of the atmosphere, depends on the effect produced by the pressure of the atmosphere on a metallic box hermetically sealed, from which the air has been previously exhausted. It has already been explained, that the weight of the column of the mercurial barometer is counterpoised by the weight of the atmosphere, and that the variations in the weight of the atmosphere are shown by the variations in the *length* of this

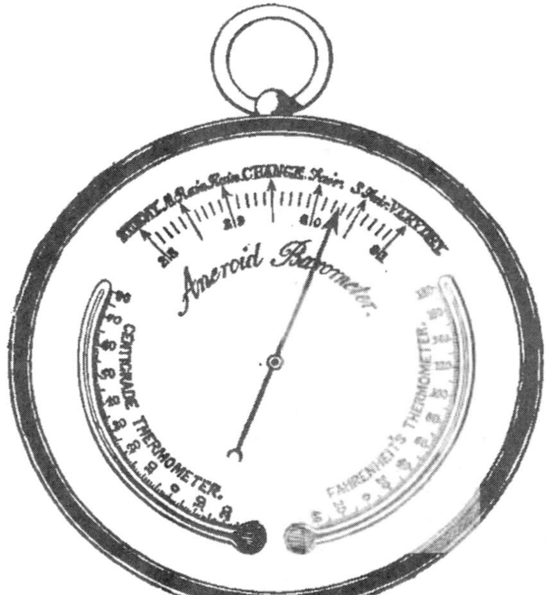

Fig. 1.

column, and measured in inches and tenths; but in the aneroid, an index traversing a dial records the changes in the weight or *pressure* of the atmosphere on a *given surface*, suppose a square inch; and it would therefore. have greatly facilitated the comprehension of the action of the instrument, had the dial been graduated to show the difference of the atmospheric pressure in absolute weight or pounds.

Simultaneous observations of the Aneroid and Mercurial Barometer for the month of March, 1848.

Date.	9 A.M.		Thermometer.	3 P.M.		Thermometer.
	Aneroid barometer.	Standard barometer.		Aneroid barometer.	Standard barometer.	
	in.	in.	degrees.	in.	in.	degrees.
1	28·66	28·67	50	28·80	28·80	50
2	29·15	29·15	50	29·29	29·29	50
3	29·88	29·90	48			
4	30·12	30·14	46	30·11	30·12	51
5	29·82	29·83	46	29·77	29·77	46
6	29·87	29·88	46	29·84	29·85	47
7	29·81	29·82	45			
8	30·28	30·29	44	30·22	30·25	46
9	29·98	29·99	49	29·89	29·90	52
10	29·44	29·45	51	29·41	29·42	51
11	28·91	28·93	50	28·84	28·85	50
12	28·69	28·70	48	28·79	28·80	48
14	29·76	29·78	47	29·85	29·88	49
15	29·76	29·78	46	29·64	29·65	49
16	29·49	29·50	48	29·49	29·49	49
17	29·34	29·35	49	29·34	29·34	46
18	29·44	29·45	46	29·37	29·37	52
19	29·18	29·20	48	29·12	29·12	51
20	28·98	28·99	48	28·97	28·98	49
21	28·80	28·81	49	29·13	29·13	49
22	29·60	29·60	47	29·67	29·68	51
23	29·67	29·70	54	29·80	29·80	54
24	30·02	30·02	55	30·10	30·10	55
25	30·16	30·16	52	30·11	30·11	54
26	29·89	29·90	53	29·80	29·80	54
27	29·70	29·70	53	29·70	29·70	56
29	29·91	29·91	54	29·91	29·90	56
30	29·81	29·80	55	29·81	29·80	58
31	29·98	29·98	58	30·00	30·00	65

Though for purely scientific purposes the aneroid is at present far removed from competition with the mercurial barometer, it nevertheless has some advantages in its extreme sensibility and its portability. Much has been urged against its variations from temperature; but in a range from 28° to 80°, these seldom exceed a tenth of an inch; and it must be borne in mind, that if the mercurial barometer be subjected to the same range, it will be equally affected; only in the latter case, the cause of the variation is satisfactorily established, and its exact amount for every degree of temperature accurately determined.

The observations recorded in the foregoing table show that, in the popular use of the aneroid, the same corrections for temperature may be taken as for the mercurial barometer.

Fig. 1 represents the external appearance of the instrument.

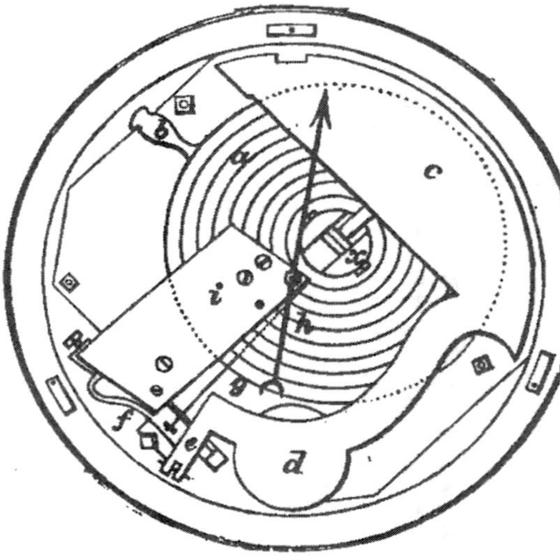

Fig. 2.

It is four inches and three-quarters in diameter across the face, and one inch and three-quarters in thickness. The pressure of the atmosphere is indicated by a hand pointing to a scale, which is graduated to correspond with the common barometer: thermometers are placed on the face, one of which is essential.

Fig 2 represents the internal construction, as seen when the face is removed, but with the hand still attached. *a* is a flat circular box, made of some white metal, exhausted of air through the short tube *b*, which is subsequently made air-tight by soldering: the upper and lower surfaces of the box are corrugated in concentric circles, which gives it greater elasticity; and the box is fixed to the bottom of a metallic case, which encloses the mechanism of the whole instrument. In the centre of the *upper* surface of the elastic box is a solid cylindrical socket *x*, about half an inch high, to the top of which the *principal lever, c d e*, is attached; and this lever, which brings the box into a state of tension by separating the surfaces, rests partly upon a spiral spring *d*, and partly on two fulcra having knife edges, with perfect freedom of motion. The end, *e*, of the large or principal lever is attached to a second lever *f*, from which a fine watch-chain *g* extends to *h*, where it works on a drum, attached to the arbour of the hand. A hair-spring at *h*, the attachments of which are made to the metallic plate *i*, regulates the motion of the hand.

As the weight of the atmosphere is increased or diminished, so is the surface of the corrugated elastic box depressed or elevated, as is also at the same time the spiral spring *d*, upon which the principal lever rests; and this motion is communicated through the levers to the arbour of the hand at *h*. The tension of the box in its construction is equal to 44 lbs. At the back of the aneroid is a screw to adjust the hand to the height of any standard mercurial barometer: for comparative observations the aneroid must be placed in the position for which the adjustment is made.

A perspective view of the interior of the Aneroid.

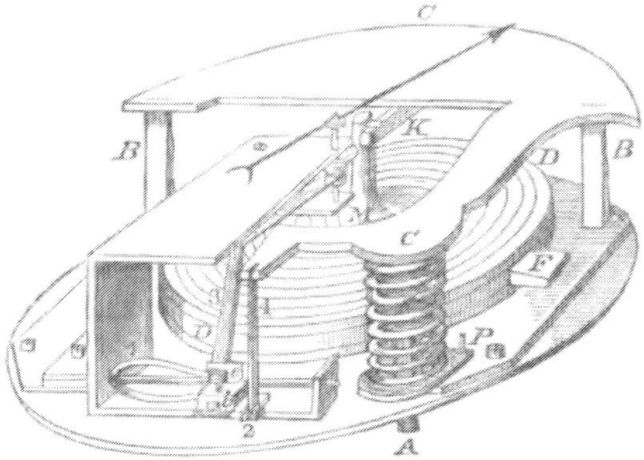

Fig. 3.

A. Screw adjusting the hand.
B B. Fulcrums.

C C. Principal lever.
D D. Vacuum vase.
1. Vertical rod connecting lever C C with levers 2 and 3.
a b. Adjusting screws for leverage.
S. Spiral spring.
M. Socket in vacuum vase.
K. Pin attached to socket.

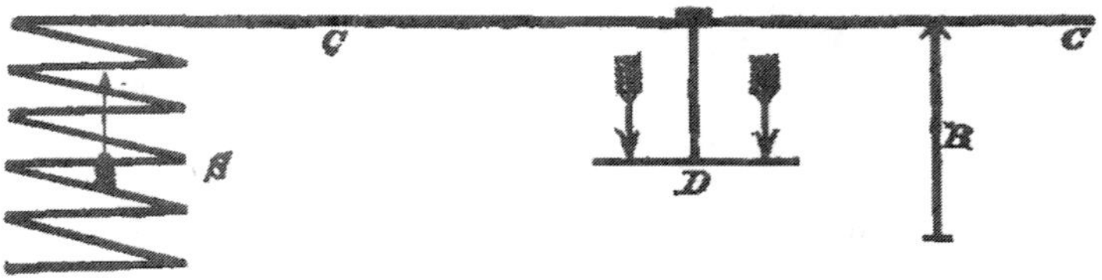

D. Vacuum vase (the arrows indicate the downward pressure of the atmosphere).
C. Principal lever.
B. Fulcrum.
S. Spring.

APPLICATION OF THE BAROMETER TO THE DETERMINATION OF DIFFERENCES OF ALTITUDE.

It has already been stated (p. 95) the density of the atmosphere would decrease in geometrical progression, for altitudes in arithmetical progression, and, since this density also varies directly as the pressure to which it is subjected, and which is measured by the height of the barometric column, it follows that, if at different altitudes these columns decrease in geometric progression, the altitudes will increase in arithmetical progression, and will, therefore, be proportional to the logarithms of the barometric columns. Hence, if the temperature remained constant, the difference of two altitudes would vary as the difference of the logs. of the barometric columns at those altitudes; so that, if h be taken to represent the height of a higher station above a lower one, and if B be the height of the barometer at the *lower* station, and b the height at the *higher* station, we should have

$$h = k \log. \frac{B}{b};$$

where k is a constant quantity, to be determined by experiment.

The following Table exhibits the results of Sir George

Shuckburgh's experimental determination of the number of feet in a column of the atmosphere, equivalent in weight to a like column of mercury, $\frac{1}{10}$ of an inch high for every five degrees of temperature, ranging from 32° to 80°, when the barometer stands at 30 inches.

Thermometer.	Feet.
32°	86·85
35	87·49
40	88·54
45	89·60
50	90·66
55	91·72
60	92·77
65	93·82
70	94·88
75	95·93
80	96·99

From an examination of this table, we see that the value of h varies with the temperature by $\frac{1}{440}$ of its value at 55°, for each degree of Fahrenheit above or below 55°: hence, if K be the value of k for temperature of 55°, for any other temperature, t, we shall have $k = K\left(1 + \frac{t - 55°}{440}\right)$

$$\text{and } h = K \log. \frac{B}{b}\left(1 + \frac{t - 55°}{440}\right)$$

Now the value of $\log \frac{B}{b}$ is given by the equation

$$\log. \frac{B}{b} = 2\left(\frac{B - b}{B + b}\right) + \left(\frac{B - b}{B + b}\right)^3 + \&c.$$

and, when $\frac{B - b}{B + b}$ is a small fraction, all the terms in this series, except the first, may be neglected as inconsiderable, and we have log. $\frac{B}{b} = 2\,\frac{B-b}{B+b}$. In this case, then,

$$h = 2\,K\,\frac{B-b}{B+b}.\left(1 + \frac{t-55^\circ}{440}\right)$$

t representing the temperature in degrees Fahrenheit.

Also, when $B - b = \frac{1}{10}$ of an inch,
$B = 30$ inches,
and $t = 55$,

the same table gives 91·72 feet for the value of h.

To determine the value of 2 K we have, therefore, the equation

$$91{\cdot}72 = 2\,K\;\frac{\frac{1}{10}}{59{\cdot}9},$$
$$= \frac{2\,K}{600}\text{ nearly;}$$
$$\therefore 2\,K = 600 \times 91{\cdot}72,$$
$$= 55032 \text{ nearly},$$

and, this value being in excess, we may take for the value of 2 K, the more simple number 55,000; we have then the value of h given by the equation,

$$h = 55{,}000\,\frac{B-b}{B+b}\left(1 + \frac{t^\circ - 55^\circ}{440}\right).$$

If the fall exceed 5 inches, $\frac{B-b}{B+b}$ should be replaced by log. $\frac{B}{b}$; but for any amount of fall less than this, the above formula will give results sufficiently accurate.

It has already been stated, that the barometer has a thermometer attached for the purpose of indicating the temperature of the instrument itself. In the determination of height when the difference of two stations at which observations are made is considerable, there is generally also a considerable difference in the temperatures at the two stations, and it is important to take account of the indication of the attached thermometer. The barometric heights should, therefore, be corrected for the variation of temperature, before applying the formula for the computation of the difference of altitude.

The following tables show the correction to be applied to

the reading of the barometer, for the temperatures indicated by the attached thermometer.

In applying the correction for the attached thermometer it is to be observed, that it is only the difference in the temperature at the two stations which will apparently affect the result, and that consequently, as long as the barometric heights are reduced to the *same* temperature, it matters not what that temperature may be.

The law of variation of the aneroid barometer for difference of temperature is not known, but from the closeness of its indications with those of the standard barometer, the correction may be assumed to be the same with that for the mercurial barometer. In many cases, too, the difference of temperature at the different stations is so small that the corrections for them may be neglected.

TABLE I.

Corrections for Temperature to be applied to Barometers mounted on Brass Scales.

Temp. of Fahr.	Inches. 24·5	Inches. 25·0	Inches. 25·5	Inches. 26·0	Inches. 26·5	Inches. 27·0	Inches. 27·5	Inches. 28·0	Inches. 28·5	Inches. 29·0	Inches. 29·5	Inches. 30·0	Inches. 30·5	Temp. of Fahr.
°														°
24	+ ·010	+ ·010	+ ·010	+ ·011	+ ·011	+ ·011	+ ·011	+ ·011	+ ·012	+ ·012	+ ·012	+ ·012	+ ·012	24
26	·006	·006	·006	·006	·006	·006	·006	·006	·006	·007	·007	·007	·007	26
28	+ ·001	+ ·001	+ ·001	+ ·001	+ ·001	+ ·001	+ ·001	+ ·001	+ ·001	+ ·001	+ ·001	+ ·001	+ ·001	28
30	− ·003	− ·003	− ·004	− ·004	− ·004	− ·004	− ·004	− ·004	− ·004	− ·004	− ·004	− ·004	− ·004	30
32	·008	·008	·008	·008	·008	·009	·009	·009	·009	·009	·009	·009	·010	32
34	·012	·012	·013	·013	·013	·013	·014	·014	·014	·014	·015	·015	·015	34
36	·017	·017	·017	·017	·018	·018	·019	·019	·019	·020	·020	·020	·020	36
38	·021	·021	·022	·022	·023	·023	·023	·024	·024	·025	·025	·026	·026	38
40	·025	·026	·026	·027	·027	·028	·028	·029	·029	·030	·030	·031	·031	40
42	·030	·030	·031	·031	.032	·033	·033	·034	·034	·035	·036	·036	·037	42
44	·034	·035	·035	·036	·037	·037	·038	·039	·040	·040	·041	·042	·042	44
46	·038	·039	·040	·041	·042	·042	·043	·044	·045	·045	·046	·047	·048	46
48	·043	·044	·045	·045	·046	·047	·048	·049	·050	·051	·052	·052	·053	48
50	·047	·048	·049	·050	·051	·052	·053	·054	·055	·056	·057	·058	·059	50
52	·052	·053	·054	·055	·056	·057	·058	·059	·060	·061	·062	·063	·064	52
54	·056	·057	·058	·059	·060	·062	·063	·064	·065	·066	·067	·068	·070	54
56	·060	·061	·063	·064	·065	·066	·068	·069	·070	·071	·073	·074	·075	56
58	·064	·066	·067	·068	·070	·071	·073	·074	·075	·077	·078	·079	·081	58
60	·069	·070	·072	·073	·075	·076	·077	·079	·080	·082	·083	·085	·086	60
62	·073	·075	·076	·078	·079	·081	·082	·084	·085	·087	·088	·090	·091	62
64	·078	·079	·081	·083	·084	·086	·087	·089	·090	·092	·094	·095	·097	64
66	·082	·084	·085	·087	·089	·090	·092	·094	·096	·097	·099	·101	·102	66
68	·086	·088	·090	·091	·094	·095	·097	·099	·101	·102	·104	·106	·108	68
70	·091	·093	·095	·096	·098	·100	·102	·104	·106	·108	·109	·111	·113	70
72	·095	·097	·099	·101	·103	·105	·107	·109	·111	·113	·115	·117	·119	72
74	·099	·102	·104	·106	·108	·110	·112	·114	·116	·118	·120	·122	·124	74
76	·103	·106	·108	·110	·112	·114	·117	·119	·121	·123	·125	·127	·129	76
78	·108	·111	·113	·115	·117	·119	·122	·124	·126	·128	·130	·133	·135	78
80	− ·114	− ·115	− ·117	− ·119	− ·122	− ·124	− ·126	− ·129	− ·131	− ·133	− ·136	− ·138	− ·140	80

Enter with approximate height of the barometer at the top of the table, and the degree of the thermometer on the side of the page; then take out the correction with its proper sign.

Let barometer **read** 29·000 Thermometer attached = 50°

Correction .... − ·056

Correct height for 32° 28·944

TABLE II.

*Corrections for Temperature to be applied to Barometers mounted in wood.**

Temperature.	Inches. 28·5	Inches. 29·0	Inches. 29·5	Inches. 30·0	Inches. 30·5	Temperature.	Inches. 28·5	Inches. 29·0	Inches. 29·5	Inches. 30·0	Inches. 30·5
°						°					
25	+ ·017	+ ·017	+ ·018	+ ·018	+ ·018	53	— ·053	— ·053	— ·054	— ·054	— ·055
26	·015	·015	·015	·015	·015	54	·055	·055	·056	·057	·058
27	·012	·012	·012	·012	·012	55	·056	·057	·058	·059	·060
28	·009	·010	·010	·010	·010	56	·059	·059	·060	·062	·063
29	·007	·007	·007	·007	·007	57	·062	·062	·062	·064	·065
30	·005	·005	·005	·005	·005	58	·064	·064	·065	·066	·067
31	+ ·002	+ ·003	+ ·003	+ ·003	+ ·003	59	·066	·067	·068	·069	·071
32	·000	·000	·000	·000	·000	60	·068	·069	·071	·072	·073
33	— ·002	— ·002	— ·002	— ·002	— ·003	61	·072	·072	·073	·074	·075
34	·005	·005	·005	·005	·005	62	·074	·074	·076	·077	·078
35	·007	·007	·008	·008	·008	63	·077	·077	·078	·080	·081
36	·010	·010	·011	·011	·011	64	·080	·080	·081	·083	·084
37	·013	·013	·013	·014	·014	65	·081	·082	·083	·085	·086
38	·015	·015	·015	·015	·015	66	·084	·085	·086	·087	·088
39	·018	·018	·018	·019	·019	67	·088	·088	·089	·090	·091
40	·020	·020	·020	·021	·021	68	·090	·090	·091	·093	·094
41	·023	·023	·023	·024	·024	69	·092	·092	·093	·095	·097
42	·025	·025	·025	·026	·026	70	·093	·094	·096	·098	·100
43	·028	·028	·028	·029	·030	71	·096	·097	·098	·100	·102
44	·030	·030	·030	·031	·032	72	·099	·099	·101	·103	·105
45	·032	·032	·033	·033	·034	73	·102	·102	·103	·106	·108
46	·035	·035	·036	·036	·036	74	·104	·105	·106	·108	·111
47	·037	·037	·038	·038	·038	75	·105	·107	·109	·112	·114
48	·039	·039	·040	·040	·040	76	·107	·109	·112	·115	·117
49	·041	·041	·043	·043	·043	77	·110	·112	·114	·117	·120
50	·044	·045	·046	·047	·047	78	·112	·114	·117	·120	·123
51	·047	·047	·049	·049	·050	79	·115	·117	·120	·123	·127
52	— ·050	— ·050	— ·051	— ·051	— ·052	80	— ·117	— ·119	— ·122	— ·126	— ·130

* The corrections in the above Table are due to the expansion of mercury only.

Enter with the approximate height of the barometer at the top of the table, and the degree of the thermometer on the left side of the page; then take out the correction with its proper sign.

Let barometer read...... 29·500 Thermometer attached — 68°
Correction by table...... ·091

Correct height for 32° 29·409

The following table gives the value of the factor K corrected for temperature:—

TABLE III.

Thermometer.	Factor.	Thermometer.	Factor.	Thermometer.	Factor.
degrees.		degrees.		degrees.	
30	25,928	47	27,006	64	28,083
31	25,992	48	27,069	65	28,146
32	26,055	49	27,132	66	28,210
33	26,118	50	27,196	67	28,273
34	26,182	51	27,259	68	28,336
35	26,245	52	27,322	69	28,400
36	26,308	53	27,386	70	28,463
37	26,372	54	27,449	71	28,527
38	26,435	55	27,513	72	28,590
39	26,499	56	27,576	73	28,653
40	26,562	57	27,639	74	28,717
41	26,625	58	27,703	75	28,780
42	26,689	59	27,766	76	28,843
43	26,752	60	27,829	77	28,907
44	26,815	61	27,893	78	28,970
45	26,879	62	27,956	79	29,034
46	26,942	63	28,020	80	29,097

APPLICATION OF THE PRECEDING TABLE.

Let m = the mean height of the two barometers, or $\frac{B + b}{2}$
d = the difference of the two heights,
f = the number corresponding to the mean of the readings of the detached thermometer,
x = difference of altitude at the two stations

then, $x = f\frac{d}{m}$.

EXAMPLE.

At the lower station—

Let the barometer stand at	30·08 inches,
attached thermometer	59°,
detached „	60°.

At the higher station—

Let the barometer stand at	26·40 inches,
attached thermometer	47°,
detached „	46°.

The corrections for the barometer from Table I. are ·08 and ·04, respectively. The corrected heights, therefore, are 30 inches, and 26·36 inches; their mean height is 28·18 inches, and their difference 3·64 inches. The mean of the readings of the detached thermometer is 53°, and opposite this number in Table III. is 27,386; therefore,

$$x = 27{,}386 \frac{3{\cdot}64}{28{\cdot}18} = 3{,}537 \text{ feet.}$$

The following table, made by computing the value of $\frac{m}{f}$, may be much more easily applied, to give results differing but slightly in their correctness :—

Mean of barometric readings.	Mean Temperature.					
	30°	40°	50°	60°	70°	80°
25	·00096	·00994	·00092	·00090	·00088	·00086
26	·00100	·00098	·00096	·00093	·00091	·00089
27	·00104	·00102	·00099	·00097	·00095	·00093
28	·00107	·00105	·00103	·00101	·00098	·00096
29	·00112	·00109	·00107	·00104	·00102	·00100
30	·00115	·00113	·00110	·00108	·00105	·00103

The difference of the barometric readings, divided by the appropriate number from this table, gives at once the difference of height sought.

EXAMPLE.

Barometer at lower station .	30·00	Mean temperature .	53°
" higher " .	26·36	" barometer . .	28·18
Difference .	3·64		

Corresponding tabular number, ·00103.

```
·00103)3·64(3534
       309
       ---
        550
        515
        ---
         350
         309
         ---
         410
```

The result 3534 differs by only 3 feet from that already obtained by the table of factors.

Perhaps, however, the simplest mode of all is to form a table of elevations corresponding to the heights of mercury in the barometer, assuming the zero point to correspond to 30 inches of the barometer, with the thermometer at 55°.

The altitudes corresponding to the readings of the barometer at any two stations being taken from the table, the difference between their altitudes gives the difference of altitude between the two stations, subject only to the correction for the temperature.

The following table is computed from the formula :—

$$h = 55{,}000\,\frac{30-b}{30+b}$$

the temperature being 55°.

Height of barometer. *b*.	Altitude of station. *h*.	Height of barometer. *b*.	Altitude of station. *h*.
in.	feet.	in.	feet.
31·0	—902	28·5	1411
30·9	813	28·4	1508
30·8	724	28·3	1605
30·7	634	28·2	1702
30·6	545	28·1	1799
30·5	455	28·0	1897
30·4	364	27·9	1996
30·3	274	27·8	2095
30·2	183	27·7	2194
30·1	91	27·6	2293
30·0	0	27·5	2392
29·9	+92	27·4	2491
29·8	184	27·3	2592
29·7	276	27·2	2692
29·6	368	27·1	2793
29·5	462	27·0	2895
29·4	556	26·9	2997
29·3	650	26·8	3099
29·2	744	26·7	3201
29·1	838	26·6	3304
29·0	932	26·5	3407
28·9	1028	26·4	3511
28·8	1123	26·3	3615
28·7	1219	26·2	3719
28·6	1315	26·1	3824
—	—	26·0	3929

EXAMPLES OF THE USE OF THE PRECEDING TABLE.

	in.		ft.
1. Barometer at higher station	29·5	Corresponding altitude in table . . .	462
″ lower ″	30·1	″ ″ . . .	−91
		Difference of altitudes	553
Assumed temperature . .	55°		
Mean temperature . . .	44°	Correction for temperature, $\frac{11}{440} \times 553 =$	14
	11		539

Hence the upper station is 539 feet above the lower.

Computed by the factor for 44° in Table III., at p. 114, the difference of the altitudes = 539·9.

	in.		ft.
2. Barometer at higher station	26·66	Corresponding altitude	3242
″ lower ″	29·89	″ ″	101
		Difference of altitudes	3141
Assumed temperature . . .	55°		
Mean temperature	50	Correction for temperature . $\frac{5}{440} \times 3141 =$	35
	5	Corrected difference of altitudes . . .	3106

Computed by the factor for 50° in Table III., at p. 114, the difference of altitudes is 3106·7.

The following modification of the above table gives the barometric readings for each 100 feet of altitude:—

Feet	in.	Feet	in.
−1000	31·11	1600	28·31
900	31·00	1700	28·20
800	30·88	1800	28·10
700	30·77	1900	28·00
600	30·66	2000	27·90
500	30·55	2100	27·79
400	30·44	2200	27·69
300	30·33	2300	27·59
200	30·22	2400	27·49
100	30·11	2500	27·39
0	30·00	2600	27·29
+ 100	29·89	2700	27·19
200	29·78	2800	27·09
300	29·67	2900	27·00
400	29·57	3000	26·90
500	29·46	3100	26·80
600	29·35	3200	26·70
700	29·25	3300	26·60
800	29·14	3400	26·51
900	29·03	3500	26·41
1000	28·93	3600	26·31
1100	28·82	3700	26·22
1200	28·72	3800	26·12
1300	28·62	3900	26·03
1400	28·51	4000	25·93
1500	28·41		

In the table thus modified, if the distance between the two readings of the barometer be taken off with compasses, or measured on a slip of paper, it will extend on the column of heights from zero to the difference of altitude, at the assumed temperature of 55°.

If the column of altitudes be arranged on a slide, adapted to a groove, in a rule having the column of barometric readings marked upon it, then by merely setting the zero of the slide to the barometric reading of any station at starting, the altitudes of all points passed over, at the assumed temperature of 55°, will at once be read on the slide, opposite the observed height of the barometer at those points. A slide-rule of this kind is manufactured by Mr. West, of Charing Cross, from the design of its inventor, Mr. Whitley, of Penarth, Truro.

Mr. Whitley makes a length of one inch on the slide correspond to 200 feet of altitude, and, dividing the inch into 20 equal parts, each division indicates 10 feet difference of altitude: the corresponding barometric heights on the fixed scale then differ, on the average, by about ·01′, so that the difference of altitude corresponding to each variation of $\frac{1}{100}$ of an inch of the barometer, is at once read on the slide.

A slide-rule thus constructed, 25 inches long, would extend from 31·11 inches of the barometer, corresponding to a depression of 1,000 feet, to 25·93 inches, corresponding to an altitude of 4,000.

Nothing can be more simple, than to set out with such a rule, and an aneroid barometer, and note the altitudes passed over, even while travelling on a railway.

METEOROLOGICAL PHENOMENA OF THE BAROMETER.

The following account of the meteorological phenomena indicated by the barometer, is extracted from a manual of the barometer by Mr. Belville, of the Royal Observatory, Greenwich, from which we have derived much assistance in this part of our work.

Strong winds in the winter from the west with a steady high pressure, invariably bring a high temperature and very little rain, winds from the east, a low temperature and sharp frosts.

If the mercury fall during a high wind from the south-west, south-south-west, or west-south-west, an increasing storm is probable; if the fall be rapid, the wind will be violent, but of short duration; if the fall be slow, the wind will be less violent, but of longer continuance; the disturbing cause is probably the same in each case, but its intensity unequal: nearly all our high winds from the south-west come with a falling barometer.

If the depression of the mercury be sudden and considerable with the wind due west, a violent storm may be expected from the north-west or north, during which the mercury will rise to its former height. If the mercury fall with the wind at north-west, or north, a great reduction of temperature will follow, in the winter severe frosts, in the summer cold rains.

A steady and considerable fall of the mercury during an east wind denotes that the wind will soon go round to the south, unless a heavy fall of snow or rain immediately follow; in this case the *upper* clouds usually come up from the south. The deep snow of the severe winter of 1814 was a notable instance.

The lowest depressions occur with the wind at south and south-east, when much rain falls, and frequently short and severe gales blow from these points. In the winter months, sudden depressions of the mercury with the wind in these quarters are attended with electrical phenomena.

A fall of the mercury with a south wind is invariably followed by rain in greater or less quantities.

A falling barometer with the wind at north brings the worst weather: in the summer, rain and storm follow; in the winter and spring, deep snows and severe frosts. This case is of rare occurrence.

A great depression of the mercury during a frosty period brings on a thaw: if the wind be south or south-east, the thaw will continue; if the wind be south-west, the frost will be likely to return with a rising barometer and northerly wind.

In the winter season, a rapid rise of the mercury immediately after a gale from the south-west with rain (the wind going round to north-west or north) is usually attended with clear sky and sharp white frosts.

Great depressions in the summer months are attended with storms of wind and rain with thunder and hail: cold, unseasonable weather generally succeeds the depressions.

During a period of broken cold weather in the winter months, with the wind at north or north-north-west, a sudden rise of the mercury denotes the approach of rain and a southerly wind.*

During a steady frost with the wind at north, north-east, or east, a continued slow rising of the mercury indicates snow and cloudy weather.

If the mercury rise with the wind at south-west, south, or even south-east, the temperature is generally high.

Observation does not show that *extremes of temperature* are contemporaneous with the greatest elevations and least depressions of the mercurial column.

Meteors are not prevalent during very low pressures : the *Aurora Borealis* has been noticed at all heights of the barometer. Small flashes of lightning are of frequent occurrence during stormy weather in the winter season when the mercury stands low.

Great elevations in the summer are generally attended with dry, warm weather.

Great depressions at all seasons are followed by change of wind, and by much rain.

A rising barometer with a southerly wind is usually followed by fine weather. In the summer it is dry and warm ; in the winter, dry with moderate frosts. This is of rare occurrence.

When the mercury is very unsteady during calm rainy weather, it denotes that the air is in an electrical state, and that thunder will follow.

In the summer months, if a depression of two or three tenths of the mercury occur in a hot period, it is attended with rain and thunder, and succeeded by a cool atmosphere. Sometimes heavy thunder-storms take place overhead without any fall of the mercury ; in this case a reduction of temperature does not usually follow.

Rain in some quantity may fall with a high pressure, provided the wind be in any of the northerly points ; and when much rain falls with a steady rising barometer and the mercury attains a great elevation, a long period of fine weather usually succeeds.

If after a storm of wind and rain, the mercury remains steady at the point to which it had fallen, serene weather may

* Thaws also commonly set in during the night.

follow without a change of wind; but on the rising of the mercury, rain and a change of wind may be expected.

During a series of stormy weather the mercury is in constant agitation, falling and rising twice or thrice in the space of twenty-four hours, the wind changing alternately from south to west, and backing again to the south: this alternation of winds continues until the mercury rises to a bold elevation, when it ceases, and the weather becomes settled.

Storms of wind, especially when accompanied with much rain, produce the greatest depressions of the mercury. No storm of wind on record has blown without some rain falling, although the time of its falling and its amount have been variable: sometimes the rain has increased with the increasing storm and sinking mercury; at other times the rain has fallen suddenly at the close of the storm, or at the time of the *minimum* pressure.

No great storm ever sets in with a steady rising barometer.

As far as regards the locality of Greenwich, the most violent gusts of wind come from due south, and those next in violence from due north; in both instances the mercury remains stationary at its *minimum* point during the greatest *horizontal pressure:* the winds from these quarters are of short duration, and limited in their extent. The ordinary south-west gales will blow unremittingly for twenty-four hours, and will sweep over the whole of the British Isles.

Note.—Although a rising mercury attends a northerly wind, great depressions occur previously to a great storm coming from that quarter.

In England, the winds which blow for the greatest number of days together without intermission are the west and west-south-west: they blow chiefly during the winter months, and are the principal cause of our mild winters.

The east and east-north-east are the winds the next most prevalent. The great antagonist winds, the north and south, are the origin of our most violent storms.

The westerly winds surge mostly by night, and their average force is twice that of the easterly winds.

The easterly winds are generally calm at night, but blow with some power during the day.

On an average, sunrise and sunset are the periods of the twenty-four hours in which there is the least wind. An hour or two after noon is the period when the wind is the highest.

As a general rule when the wind turns against the sun, or *retrogrades* from west to south, it is attended with a falling mercury; when it goes in the *same direction* as the sun, or turns direct from west to north, the mercury rises, and there is a probability of fine weather.

It never hails in calm weather. When hail falls, it is during sudden gusts of wind, and the mercury rises while the hail is actually falling.

If the weather during harvest-time has been generally fine, and a fall of the mercury with a shower occur,—if the wind turn a few points to the north and the barometer rises above 30 inches, the weather may be expected to be fair for some days.

The finest and most beneficial state of the atmosphere, more especially as regards the health of man, is with a uniform pressure at a mean height of the barometer varying from 29·80 to 30·00.

When there is only one current of air subsisting in the atmosphere, there is seldom much variation in the height of the mercurial column. It is when two or more *strata* of the air are in motion in different directions at the same time, that great fluctuations of the mercury occur.

In high pressures, the *upper* current usually sets from the northward; in low pressures it sets from the south and south-west.

The variations of the barometer are always greater in the winter than in the summer.

In accounting for the different currents of the atmosphere it must be remarked that the great heat of the torrid zone causes a constant ascent of air over it, which passes northward and southward; while an under current of cold air flows from the poles to supply its place; the diurnal rotation of the earth combined with these currents causes the trade-winds, whose direction is from east to west: these currents would from the same causes become in the north temperate zone north-east and south-west winds, and in the south temperate zone south-east and north-west winds; but the great irregularities of the temperature from the seasons, the large tracts of ocean, and the different geographical formations of the land, subject them to interruptions, and give to every country its prevailing winds, derived from local causes. In England, the south-south-west, south-west, and west-south-west winds set in towards the end of October, and blow with their greatest strength during November, December, and February, and are even powerful in June and July: the winds from the westerly quarters prevail in March, but they then veer more towards the north, whence they blow with great violence: in April, the east and north-east, and the west and north-west winds balance each other, and their comparative strength is nearly equal; in May, the east, north-east, and north-north-east winds preponderate; the latter blows the less frequently, but with the greatest violence; in this month the average of the winds from the westerly quarters ranges low; their average strength also decreases, with the exception of that from the west-south-west, which ranges higher than in April. In August the west and west-south-west winds prevail, but their power is moderate; the stormy winds of this month blow from the west-south-west and north-north-west. September is the calmest period of the year; in this month the north and south winds, and the east and west winds, balance each other; in January the east and west winds upon an average are nearly equal, both as regards the number of times they blow, and their average strength; the winds from the south-south-west, west-south-west, and the north-westerly quarters are more rare, but they blow with great violence. As the winds from these opposite quarters predominate, so is the character of our winters determined as to mildness or severity.

Sudden depressions of the barometer sometimes occur in weather apparently calm. It is almost an established fact that storms have a circular motion; and if, when an exhaustion or sudden diminution of the atmosphere takes place, the mercurial column happen to be in the partial vacuum or centre of motion, the air will be at rest; while the surrounding air at a greater distance from the centre will be violently agitated with a less fall of the barometer. This circular motion of the atmosphere is not confined to one spot where the storm may commence and expend its violence; but it has a progressive cycloidal movement onwards, changing constantly the situation of its centre of motion and, as it advances, enlarging its circumference, until, having traversed many hundred miles, it becomes exhausted as the air recovers its equilibrium. These great rarefactions of the atmosphere are probably the effects of electricity; they are common in their most terrific form in the Indian Ocean, on the western coast of Africa, and in the West Indies.

In our own climate the approach of thunder-clouds produces violent squalls of wind; and dense and highly electrified clouds will sometimes raise miniature whirlwinds as they pass overhead.

THERMOMETRIC HYPSOMETRY.

Water, and even ice, constantly give off steam or aqueous vapour at all temperatures, when exposed to the air; thus we know that if a glass of water be left in a room for some days, the whole of the water will gradually evaporate. This power of water to rise in vapour at all temperatures is called the *elastic force*, or *tension*, of aqueous vapour. It may be measured, when a small quantity of water is placed above the mercury in a barometer, by the depression which the tension of the vapour thus given off is capable of producing in the mercurial column. If we gradually heat the drops of water thus placed in the barometer, we shall notice that the column of mercury gradually sinks, and when the water is heated to the boiling point, the mercury in the barometer tube is found to stand at the same level as that in the trough—showing that *the elastic force of the vapour*, at that temperature, *is equal to the atmospheric pressure.*

Hence, *water boils when the tension of its vapour is equal to the pressure of the superincumbent column of atmosphere;* and, consequently, on the tops of mountains, where the atmospheric pressure is less than at the sea-level, water boils at a temperature below the 212° of common experience.

A simple experiment to illustrate this fact, consists in boiling water in a globular flask, into the neck of which a stop-cock is fitted. As soon as the air is expelled, the stop-cock is closed, and the flask removed from the source of heat. The boiling then ceases. But, on immersing the flask in *cold* water, the ebullition recommences briskly—owing to the *reduction of the pressure*, consequent upon the condensation of the steam—the tension of the vapour at the temperature of the water in the flask being greater than the diminished pressure.

Founded on this principle a thermometric instrument has been constructed for determining heights, by noticing the temperatures at which water boils.*

* All other liquids obey a similar law of ebullition; but as the tensions of their vapours are very different, their *boiling points* vary considerably.

Wollaston's Boiling-point Hypsometer.—This is simply a cylindrical copper vessel of small size, with or without a spirit lamp, closed at the top, in which there is a cork; and through the cork is passed a thermometer, the bulb of which descends to half an inch above a small quantity of water in the vessel. Of course there is a vent-hole for the steam at the side of the rim.

Mr. Casella has specially identified himself with thermometric hypsometry. His instrument is constructed generally on Dr. Wollaston's principle, but arranged and modified by Colonel Sykes, F.R.S., Regnault, and others, but with considerable simplification and improvement. The thermometer, strong and sensitive, with small bulb, is divided and figured on the stem, and sheltered from cold when in use by a double telescopic chamber, into which it is passed to any required depth, through a loose piece of india-rubber, which rests on the top; the chamber being filled with vapour from the boiling water beneath. The inner chamber being thus completely filled, the vapour descends into the outer chamber, and escapes by the outlet. By this means the mercury, both in the bulb and stem, is immersed in pure vapour, *whatever kind of water may be employed.* A portable leather case contains the whole, when packed for travelling.

The Alpine Hypsometer.—A still smaller apparatus on the same principle has been much used by the members of the Alpine Club, and other distinguished travellers.* It is fitted in a small tin case, in which it may be used in windy weather. Size 5½ inches long by 2⅓ diameter; weight, 13 ozs.; and it may be carried in the trowsers-pocket. It is used by many instead of a barometer, and by those a little experienced in its use it is often preferred, for its simplicity and certainty, to the mountain barometer.

"For the elevation of great mountain-masses," says Dr. I. H. Hooker, "and continuously elevated areas, I consider that hypsometrical results are as good as barometrical ones; for the general purposes of botanical geography, the *boiling-point thermometer* supersedes the barometer in point of practical utility, for under every advantage the transport of a glass tube full of mercury, nearly 3 feet long, and cased

* Among the rest by Baker, "Falls of the Nyanai."

in metal, is a great drawback to the unrestrained motion of the traveller."*

A convenient approximate rule has been given by Lefroy for ascertaining altitude by the boiling point of water, viz.:—

Allow for each degree below 212° Fahr. that water boils in a metal vessel, in mean state of barometer (*i.e.* a little above or below 30 inches in height)—

511 feet for first degree;
513 ,, for second ,,
515 ,, for third ,, &c.

But the errors of thermometric measurements of height are always large, except in the tropics.

* "Himalayan Journals," vol. ii.

PART V.

ASTRONOMICAL INSTRUMENTS.

PRELIMINARY.

In a rudimentary work of this kind, a description of the large fixed instruments used in observatories would hardly be in its proper place, as but few persons can become practically acquainted with their use. Our descriptions will therefore be confined to such instruments as are of a portable character, and may be set up and used by private individuals for the purposes of astronomical observation, or for the most important observations in geodetical surveys. Hadley's sextant is, in fact, an instrument indispensable to the navigator and to the marine surveyor, and to this our attention will be first directed, and we shall then proceed to describe the portable transit, and altitude and azimuth instruments, with the collimator, an instrument which may be used in their adjustment, and the equatorial.

CHAPTER I.

HADLEY'S SEXTANT.

This instrument differs from the pocket sextant, already described—in its appearance, from the absence of the box in which the pocket sextant is fixed—in its size varying usually from 4 inches to 6 inches radius—and in its requiring and admitting of more perfect and minute adjustment.

L is the limb of the instrument, graduated from 0° to 140°, at every 10′ or 20′, according to the size of the instrument, and subdivided by the vernier, V, to 10″ or 20″, thus enabling us to read off angles by estimation to 5″. The limb is also graduated through a small space, called the arc of excess, on the other side of the zero point. T is the tangent screw for giving a slow motion to the index bar after it has been clamped by the screw O. M is a microscope, attached to the index bar, movable by a screw, V, so as to command a view of the vernier throughout its entire length.

I is the index glass, or first reflector, attached to, and moving with, the index bar; and H is the horizon glass, having its lower half silvered, to form the second reflector, and its upper half transparent. Four dark glasses are placed at C, any one

or more of which can be turned down between the index glass and horizon glass, to moderate the intensity of the light from any very bright object, viewed by reflection; and at G are three dark glasses, any one or more of which can be turned up, to moderate the intensity of the light from any bright object, viewed directly through the transparent part of the horizon glass. D is a ring for carrying the telescope, attached to a stem S, called the up-and-down piece, which can be raised or lowered by means of a milled-headed screw. The use of this up-and-down piece is to raise or lower the telescope, till the objects seen directly and by reflection appear of the same brightness. K is the handle by which the instrument is held.

In selecting an instrument, care must be taken that all the joints of the frame are close, without the least opening or looseness, and that all the screws act well, and remain steady, while the instrument is shaken by being carried from place to place. All the divisions on the limb and vernier, when viewed through the microscope, must appear exceedingly fine and distinct, and the inlaid plates upon which the divisions are marked must be perfectly level with the surface of the instrument. The index, or zero, of the vernier should also be brought into exact coincidence successively with each division of the limb, till the last division upon the vernier reaches the last division upon the limb; and *if* the last division of the vernier do not in each case also exactly coincide with a division upon the limb, the instrument is badly graduated, and should be rejected. All the glass used in the instrument should be of the best quality, and the glasses of the reflectors should each have their faces ground and polished perfectly parallel to each other, to avoid refraction. Look, therefore, into each reflector, separately, in a very oblique direction, and observe the image of some distant object; and if the image appears clear and distinct in every part of the reflector, the glass is of good quality; but if the image appear notched, or drawn with small lines, the glass is veiny, and must be rejected. Again, if the image appears singly, and well defined about the edges, the two surfaces of the glass are truly parallel; but if the edge of the image appears misty, or separated like two images, the two surfaces are inclined to one another. The examination will be more perfect if the image be examined with a small telescope.

A plain tube and two telescopes, one showing objects inverted and the other erect, are usually supplied with the sextant. The manner of testing the telescopes has already been explained in the part of the work devoted to optical instruments. A dark glass is also supplied to fit on to the eye-end of the telescope, and a key for turning the adjusting screws.

To examine the Error arising from the Imperfection of the Dark Glasses.—Fit the dark glass to the eye-end of the telescope, and all the shades being removed, bring the reflected image of the sun into contact with its image seen directly through the unsilvered part of the horizon glass. Then remove the dark glass from the eye-end of the telescope, and setting up first each shade separately, and then their various combinations, if the two images do not in any case remain in contact, the angle through which the index must be moved to restore the contact is the error of the dark glass, or combination of dark glasses, used in the observation, and which error should be recorded for each glass and each combination of the glasses.

The adjustments of the instrument consist in setting the horizon glass perpendicular to the plane of the instrument, and in setting the line of collimation of the telescope parallel to the plane of the instrument.

To adjust the Horizon Glass.—While looking steadily at any convenient object, sweep the index slowly along the limb, and if the reflected image do not pass exactly over the direct image, but one projects laterally beyond the other, then the reflectors are not both perpendicular to the face of the limb. Now the index glass is fixed in its place by the maker, and generally remains perpendicular to the plane of the instrument, and, if it be correctly so, the horizon glass is adjusted by turning a small screw at the bottom of the frame in which it is set, till the reflected image passes exactly over the direct image.

To examine if the Index Glass be perpendicular to the Plane of the Instrument.—Bring the vernier to indicate about 45°, and look obliquely into this mirror, so as to view the sharp edge of the limb of the instrument by direct vision to the right hand, and by reflection to the left. If then the edge and its image appear as one continued arc of a circle, the index glass is correctly perpendicular to the plane of the instrument; but if the arc appears broken, the instrument must be sent to the maker to have the index glass adjusted.

To adjust the Line of Collimation.—1. Fix the telescope in its place and turn the eye-tube round, that the wires in the focus of the eye-glass may be parallel to the plane of the instrument. 2. Move the index till two objects, as the sun and moon, or the moon and a star, more than 90° distant from each other, are brought into contact at the wire of the diaphragm which is nearest the plane of the instrument. 3. Now fix the index, and altering slightly the position of the instrument, cause the objects to appear on the other wire; and if the contact still remain perfect, the line of collimation is in correct adjustment. If, however, the two objects appear to separate at the wire that is farther from the plane of the instrument, the object-end of the telescope inclines towards the plane of the instrument; but if they overlap, then the object-end of the telescope declines from the plane of the instrument. In either case, the correct adjustment is to be obtained by means of the two screws which fasten, to the up-and-down piece, the collar holding the telescope, tightening one screw and turning back the other, till, after a few trials, the contact remains perfect at both wires.

The instrument having been found by the preceding methods to be in perfect adjustment, set the index to zero, and if the direct and reflected images of any distant* object do not perfectly coincide, the arc through which the index has to be moved to bring them into perfect coincidence constitutes what is called the index error, which must be applied to all observed angles as a constant correction.

To determine the Index Error.—The most approved method is to measure the sun's diameter, both on the arc of the instrument, properly so called, to the left of the zero of the limb, and on the arc of excess to the right of the zero of the limb. For this purpose, firstly, clamp the index at about 30′ to the left of zero, and, looking at the sun, bring the reflected image of his upper limb into contact with the direct image of his lower limb, by turning the tangent screw, and set down the minutes and seconds denoted by the vernier; secondly, clamp the index at about 30′ to the right of zero, on the arc of excess, and, looking at the sun, bring the reflected image of his lower limb into contact with the direct image of his upper limb, by turning the tangent screw, and set down the minutes and seconds denoted by the vernier, underneath the reading before set down. Then half the sum of these two readings will be the correct diameter of the sun, and *half their difference will be the index error.* When the reading on the arc of excess is the greater of the two, the index error, thus found, must be added to all the readings of the instrument; and when the reading on the arc of excess is the less, the index error must be subtracted in all cases. To obtain the index error with the greatest accuracy, it is best to repeat the above operation several times, obtaining several readings on the arc of the instrument, and the same number on the arc of excess; and the difference of the sums of the readings in the two cases, divided by the whole number of readings, will be the index error; while the sum of all the readings, divided by their number, will be the sun's diameter.

* If the distance of the object from the observer be less than a quarter of a mile, there will be a sensible error due to parallax. See p. 58.

EXAMPLE.

	Readings on the Arc of Instrument.			Readings on the Arc of Excess.		
	35			29	25	
	35	5		29	35	
	35	10		29	20	
	105	15		88	20	
	88	20		105	15	
No. of readings	6) 16	55	Difference.	6) 193	35	Sum.
	2	49	Index error.	32	15·8	Sun's diam.

The readings on the arc of excess being less than those on the arc of the instrument, the index error, 2′ 49″, is to be subtracted from all the readings of the instrument.

NOTE.—In taking off the readings on the arc of excess, the vernier must be read backwards; that is, the division read off on the limb being the division next to the left of the zero of the vernier, the divisions of the vernier to be added must be reckoned from the other end of the vernier, to the division coinciding with a division upon the limb; or the reading of the vernier forwards, according to the usual method, may be subtracted from 10′, the limb being divided to 10′, and the remainder added to the reading of the division upon the limb next to the left of the zero of the vernier, as before.

The manner of observing with the sextant has been already explained. when treating of the pocket sextant.

CHAPTER II.

THE TRANSIT INSTRUMENT.

THE reflecting instruments which we have just described, from their portability, and the promptitude and facility with which they may be used in all situations and upon all occasions, are most useful instruments to the surveyor. The sextant, with an artificial horizon, and a good chronometer, forms, in fact, a complete observatory, with which the latitudes and longitudes of places may be determined to a great degree of accuracy; and to the navigator a reflecting instrument is indispensable; all other instruments requiring to be supported upon a stand perfectly at rest,* while the

* In observatories the instruments are supported by stone walls, or pillars, which pass below the floors, without touching them, or any

sextant and similar instruments are held in the hand, and perform their duty well on the deck of a rolling ship. In permanent observatories, however, the capital angular instruments are placed permanently in the plane of the meridian, and the measurements sought for by their aid are the exact times at which the observed objects pass the meridian, and their angular altitudes or zenith distances when upon the meridian. The instrument with which the first of these measurements are obtained, is called a *transit instrument*, *transit telescope*, or merely a *transit*. Transits of portable dimensions, besides their use in small or temporary observatories, are also found serviceable to the surveyor, for determining, with the greatest possible accuracy, the true north point, and thence setting out a line in any required direction; and to the scientific traveller, for determining the longitude of any place from astronomical observations, and for adjusting his time-keepers with greater accuracy than can be obtained by his sextant. The annexed figure represents a portable transit.

T T is a telescope formed of two parts, connected by a spherical centre-piece, into which are fitted the larger ends of two cones, the common axis of which is placed at right angles to the axis of the telescope, to serve as the horizontal axis of the instrument. The two small ends of these cones are ground into two perfectly equal cylinders, called *pivots*. The pivots rests upon angular bearings or Y's. The Y's are supported upon the standards E and W, of which E may be called the eastern, and W the western standard; and one of the Y's is fixed in a horizontal groove on the western standard, so that, by means of the screw S, one end of the axis may be pushed a little forwards or backwards, and a small motion in azimuth be thus communicated to the telescope.* The standards, E and W, are fixed by screws upon a brass circle

part of the building, and are consequently independent of any tremor communicated to the floor or walls of the buildings. It was considered that the passage of a railway through Greenwich Park would impair the observations at the Royal Observatory, by communicating a tremor to the ground.

* The large transits in permanent observatories have their Y's placed in two dove-tailed grooves, one horizontal, and the other vertical. By means of the latter one end of the axis may be raised or depressed; but in the portable transit the same object is attained by turning one of the foot screws upon which the entire instrument rests.

O O, and steadied by oblique braces, B, B, which spring from the cross-piece C.

On one end of the axis is fixed, so as to revolve with the axis, a vertical circle V V; and a double index bar.

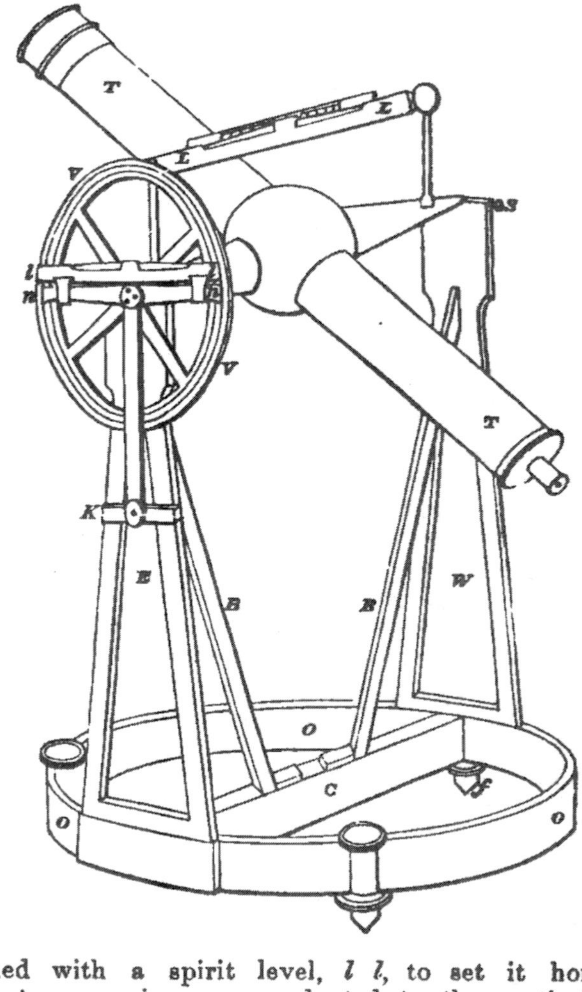

furnished with a spirit level, *l l*, to set it horizontal, carries two verniers, *n n*, adapted to the vertical circle, and showing the angle of elevation of the telescope. The index bar is fixed in its position by the clamping screw

K, and can be fixed upon either the eastern or western standard, at pleasure, while the telescope, with its attached circle, can also be lifted out of the Y's, and have its position reversed in them. The pivot which does not carry the vertical circle is pierced, and allows the light from a lamp to fall upon a plane speculum, fixed in the spherical centre piece on the axis of the telescope, and inclined to this axis at an angle of 45°. The light is thus thrown directly down the telescope, and illuminates the wires of the diaphragm, placed in the principal focus of the telescope. Of these wires, one is horizontal; and a vertical wire, intersecting it in the centre of the field of view, gives, by its intersection with it, the collimating point. There are also other vertical wires, arranged in pairs, equidistant from the central vertical wire, so that we have either three, five, or seven vertical wires, the most common number being five. The lamp has a contrivance for regulating the quantity of light thrown into the telescope, by turning a screw, so that the light from a small star may not be overpowered by the superior light of the lamp.

The requisites of a good instrument are—1stly, that the telescope be of the best quality, which is to be tested by the methods already given (vol. ii.); 2ndly, that the feet-screws act well and remain steady; 3rdly, that all the screws, by which the instrument is put together, are turned home, and remain so, after the instrument has been shaken by carriage; 4thly, that the length of the axis be just sufficient to reach from one Y to the other, without either friction or liberty; 5thly, that the lamp be held so as not to require adjustment for position; 6thly, that the screws for adjustment of the diaphragm and Y's be competent to give security of position to the parts adjusted by them; 7thly, that the metallic parts be free from flaws in casting, and that the pivots be formed of hard bell-metal, and incapable of rusting.

ADJUSTMENTS.

The principal adjustments of the transit are three:—

1st. To make the axis on which the telescope moves horizontal.

2nd. To make the line of collimation move in a great vertical circle, by setting it perpendicular to the horizontal axis.

3rd. To make it move in that vertical circle, which is the meridian.

To make the Axis Horizontal.—Apply to the pivots the large level, L L, which is supplied with the instrument for this purpose, and is constructed either to stand upon the pivots, in which case it is called a striding level, or of the form shown at page 10, in which case it is suspended from the pivots, and is called a hanging level. Bring the air-bubble to the centre of its run, by turning the foot screw *f*. Turn the level end for end, and if the air-bubble retains its position, the axis is horizontal; but if not, it must be brought back, half by the foot-screw, *f*, and half by turning the small screw at one end of the level. Repeat the operation till the bubble retains the same position in both positions of the level, and the axis will be horizontal.

To adjust the Line of Collimation in Azimuth.—Direct the telescope to some distant, small, and well-defined object, and bisect it by one extremity of the middle vertical wire, giving the telescope the azimuthal motion necessary for this purpose by turning the screw s. By elevating or depressing the telescope, examine whether the object is bisected by every part of the middle vertical wire; and if not, loosen the screws which hold the eye-end of the telescope in its place, and turn the end round very carefully till the error is moved. Lift the transit off the Y's, and reverse it, so that the end of the axis which was upon the eastern Y may now be upon the western, and *vice versâ;* and, if the object is still bisected by the central vertical wire, the collimation in azimuth is perfect, but if not, move the centre of the cross wires half way towards the object, by turning the small screws which hold the diaphragm, and if this half distance has been correctly estimated, the adjustment will be accomplished. Again bisect the object by the centre of the cross wires by turning the azimuthal screw s, and repeat the operation till the object is bisected by the centre of the cross wires in both positions of the instrument, and the adjustment will be known to be perfect.*

To adjust the Transit to the Meridian.—The line of collimation by reason of the previous adjustment describes a vertical circle, and, therefore, bisects the zenith, which is one point in the meridian; if, then, we can make it also bisect another point in the meridian, it will move entirely in the

* The horizontal motion given to the Y by the azimuthal screw s, forms, evidently, no part of the adjustment for collimation, but only enables us to examine if the adjustment has been made with sufficient exactness.

meridian. Compute from the tables in the Nautical Almanack the time of Polaris coming to the meridian, and at the computed time bisect the star by the middle vertical wire, and the transit will be very nearly adjusted to the meridian.

To make the great vertical circle described by the line of collimation more nearly coincident with the meridian, let the intervals between the successive passages of Polaris across the meridian be observed, as indicated by the instrument. Then, if the interval between the inferior and superior passage be equal to the interval between the superior and inferior, the adjustment to the meridian is perfect; but if the interval between the inferior and superior passage be less than the interval between the superior and inferior, the circle described by the line of collimation deviates to the eastward of the true meridian, from the zenith to the north point of the horizon, and to the westward, from the zenith to the south point of the horizon; while if the interval between the inferior and superior passage be the greater, the deviation is in the contrary directions.

Let δ be the observed difference of the intervals from twelve hours, or half the difference between the two intervals, in seconds, π the polar distance of the star Polaris, and L the latitude of the place; then, z representing, in time, the deviation from the meridian, the value of z will be given by the logarithmic formula,

$$\log. z = \log. \frac{\delta}{2} + \log. \sec. L + \log. \tan. \pi - 20$$

EXAMPLE.

Place of observation, Cambridge, latitude 52° 12′ 36″.
Polar distance of Polaris, 1° 39′ 25″·05.
Difference of intervals from 12^h, 7^m, 22^s = 442^s.

$\frac{\delta}{2} = 221$ .... log.	=	2·3443923
L = 52° 12′ 36″ .... log. sec.	=	10·2127030
π = 1° 39′ 25″·05 ... log. tan.	=	8·4513064
z = 10^s·195 .... log.	=	21·0084017

To determine the value of a revolution of the azimuthal screw s, the time* of passage of an equatorial star across the middle vertical wire must be noted one day; and then,

* The time here spoken of, and throughout the description of this instrument, unless otherwise expressly stated, is sidereal, and not mean time.

turning the screw, s, once round, the time of passage must be noted again the next day; and the difference of these times will be the equatorial value in time of a revolution of the screw. Suppose the difference thus observed to amount to two seconds, then the equatorial value of one complete revolution of the screw, s, is two seconds, and the value of the motion of the adjusting screw, thus obtained, must be reduced to the horizon, by increasing it in the ratio of radius to cosine of latitude, and may then be applied to correct the error of deviation, as found above.

A second method, founded on the same principles as the preceding, consists in observing the polar star, and another star, which crosses the meridian near the zenith of the place of observation. The time of passage of such a star, Capella, for instance, when near its superior transit, across the middle wire of the telescope, will differ but very little from the time of passing the true meridian, if the deviation of the instrument from the meridian be but small. Assume the two times to agree exactly, and the difference between the times of superior transit of Capella and Polaris will be the difference of the observed right ascensions of these two stars. From this difference subtract the difference of the computed, or catalogued, right ascensions of the two stars, and call the result D; and the deviation will be given by the formula,

$$\log. z = \log. D + \log. \sin. \pi + \log. \sec. (L + \pi);$$

π being the polar distance of Polaris, and L the latitude of the place of observation. From Capella not having been exactly on the meridian, when on the middle vertical wire, the value of D, as above obtained, is only an approximation to the error of the observed right ascension of Polaris, and the deviation computed from it will be only approximately correct; but, by repeating the operation, the adjustment mav be completely perfected.

D is actually the value of the sum of the errors of the observed right ascensions of Capella and Polaris, and hence the value of z will be correctly given, by so considering it, instead of supposing as above, that this error for Capella is zero. The true deviation then is given by the formula,

$$\log. z = \log. D + \log. \sin. \pi + \log. \sin. \pi' + \log. \text{cosec.} (\pi' - \pi) + \log. \sec. L;$$

π' being the polar distance of Capella.

Using this last formula, the method may be applied to

Polaris, and any star distant from the pole, or to any two stars differing from each other not less than 40° in declination. If, however, the transit of one star is observed above, and of the other below the pole, the formula will be

$$\log. z = \log. D + \log. \sin. \pi + \log. \sin. \pi' + \log. \text{cosec.} (\pi' + \pi) + \log. \sec. L.$$

Considerable advantage may be obtained by selecting two stars that differ but little in right ascension, as there is then the less probability of error from a change in the rate of the clock, or in the position of the instrument, on which account such methods are to be preferred in temporary observatories, where the stability of the instrument is not to be depended upon for any length of time.

In all the preceding formulæ, the deviation from the meridian is given in time; but, to convert it into angular measure, if desirable, we have only to multiply by 15, and the seconds of time will be converted into seconds of a degree.

When the instrument is, by any of the methods explained above, brought into the meridian, a distant mark may be set up in the plane of the meridian, by which the adjustment to the meridian may afterwards be tested.

METHOD OF OBSERVING WITH THE TRANSIT.

The adjustments having been completed, in making observations with the instrument, the instant of a star's passing the middle vertical wire will be the time of the star's transit; but the time of the star's passing all the five wires must be noted, and the mean of the times, taken as the time of transit, will be a more accurate result than the time observed at the middle wire only.

When the sun is the object observed, the time of the centre of his disc passing the middle wire is the time of transit; but, as it would be impossible to estimate the centre with accuracy, the time of both his limbs coming into contact with each wire in succession is to be noted, and a mean of all these times will be the time of transit required. This mean may be conveniently taken, by writing the observed times of contact of the first and second limbs underneath each other in the reverse order, when the sums of each pair will be nearly equal.*

* This is Dr. Pearson's method.

EXAMPLE.

	s.	s.	h.	m.	s.	s.	s.	
1826, Sept. 23	20·4	38·7	11	58	57·0	15·5	33·7	⊙ 1 Limb.
	42·3	24·0	12	1	5·7	47·2	28·7	⊙ 2 Limb.
	2·7	2·7	24	0	2·7	2·7	2·4	The sum = 13·2

The time of either limb passing the centre wire is recorded in full, but for the other wires, the seconds only are recorded, as the sums of the several pairs only differ by decimals of a second. Half the sum of the times at the middle gives, then, the correct time of transit as far as the seconds, and the decimals are found by removing the decimal point one place to the left in the sum 13·2, which is equivalent to dividing by 10. Then the time of transit, or mean of observations in the above example, is $12^{h}\ 0^{m}\ 1^{s}{\cdot}32$. This example is taken from observations made with a large transit; and if with a smaller instrument the sums of the several pairs of observations should differ by more than a second, it will be necessary to take the sums of both figures of the seconds, and the division by 10, performed as above, will give the last figure of the seconds, as well as the decimals.

In taking transits of the moon the luminous edge alone can be observed, from which the time of transit of the centre must be deduced by the aid of Lunar tables.

In observing the larger planets, one limb may be observed at the first, third, and fifth wires, and the other at the second and fourth, and the mean of these observations will give the transit of the planet's centre.

It will sometimes happen that from the state of weather, or from some other cause, a heavenly body may not have been observed at all the wires; but, if the declination of the body be known, an observation at any one of the wires may be reduced to the central wire, so as to give the time of transit, as deduced from this observation. If an observation be obtained at more than one wire, the mean of the times of passing the centre, as deduced from each wire observed, is to be taken as the time of transit. The reduction to the centre wire is given by the formula,

$$\text{R} = \text{V cosec. } \pi,$$

$$\text{or log. R} = \text{log. V} + \text{log. cosec. } \pi\,;$$

in which R represents the reduction, π the polar distance of the body observed, and v the equatorial interval from the wire at which the observation has been made to the central wire. The equatorial intervals for each side wire must, therefore, be carefully observed, and tabulated for the purpose of this reduction. The formula $R = v \operatorname{cosec.} \pi$ is only an approximate value of the reduction, and with large instruments capable of giving results within $0''\cdot05$, a further correction is necessary for bodies within $10°$ of the pole. The whole reduction in this case is given by the formula,

$$R' = \frac{1}{15} \sin.^{-1} (\operatorname{cosec.} \pi \sin. 15 v).$$

The time of any star's passage from one of the side wires to the centre wire being observed, the equatorial interval from that wire to the centre is obtained by multiplying the observed interval by the sine of the star's polar distance; and the equatorial intervals being deduced in this manner from a great many stars, the mean of the results may be considered as very correct values of the equatorial intervals required. No star very near the pole should, however, be taken for this purpose.

USE OF THE PORTABLE TRANSIT.

The large transits in permanent observatories are used to obtain, with the greatest possible accuracy, the right ascensions of the heavenly bodies, from which, and the meridian altitudes observed by a mural circle, an instrument consisting of a telescope attached to a large circle, and placed in the plane of the meridian, nearly all the data necessary for every astronomical computation are obtained. For such purposes the small portable transit is not adapted; but it is competent to determine the time to an accuracy of half a second, to determine the longitude by observations of the moon and moon culminating stars, and to determine the latitude by placing it at right angles to the meridian, or in the plane of the prime vertical.*

The transit of the sun's centre gives the apparent noon at the place of observation, and the mean time at apparent noon is found by subtracting or adding the equation of time, as

* The prime vertical is the great circle which passes through the zenith and the east and west points of the horizon.

found in the Nautical Almanack, to 24 hours.* The difference between the mean time, thus found, and the time of the sun's transit, as shown by a clock or chronometer, is the error of the clock or chronometer for mean time at the place of observation.

The time shown by a sidereal clock, when any heavenly body crosses the meridian, should coincide with the right ascension of that body, as given in the Nautical Almanack. The differencebetween the time shown by the sidereal clock, at the transit, and the right ascension of the body, taken from the almanack, will, therefore, be the error of the clock, +, or too fast, when the clock time is greater than the right ascension, —, or too slow, when it is less.

CHAPTER III.

THE PORTABLE ALTITUDE AND AZIMUTH INSTRUMENT.

THE bending of an unbraced telescope renders it unfit for the determination of altitudes; but by placing the telescope between two circles braced together, an instrument may be formed capable of observing both the meridian altitudes and times of transit of the heavenly bodies. The increased weight of the instrument, however, must now be prevented from producing flexure in the horizontal axis, and this has been very ingeniously accomplished by Troughton. By mounting the transit and altitude instrument, as Troughton's transit-circle may be called, upon a horizontal plate or circle, having an azimuthal motion round a vertical axis, an instrument is formed by which observations may be made either in or out of the meridian. When constructed of a portable size, the altitude and azimuth instrument may also be used in important surveying operations; for, in fact, it may be considered as a rather large theodolite of superior construction.

The altitude and azimuth instrument may be considered

* The astronomical day commences at noon, and contains 24 hours, the hours after midnight being called 13, 14, &c., and the day ends at the next noon. The equation of time is given in the Nautical Almanack for apparent noon at the meridian of Greenwich, and the correction to give the equation of time at any other meridian will be found by multiplying the difference for one hour, as given in the almanack, by the longitude of the place, estimated in time.

as consisting of three parts: 1, the tripod carrying the vertical axis about which the instrument turns; 2, the horizontal revolving plate carrying the vertical pillars, with their appendages; and 3, the vertical circles with the telescope.

The tripod, A A, is supported by three foot-screws, by which the vertical axis is brought into adjustment, and carries the lower horizontal plate, which is graduated to show the azimuths, or horizontal angles. The vertical axis is a solid metallic cone rising from the centre of the tripod to a height about equal to the radius of the horizontal circle.

To point out the graduation, upon the lower horizontal plate, or azimuth circle, the upper horizontal plate, or horizontal revolving plate, V V, carries an index, which denotes nearly the angle to be read off. The graduations upon the azimuth circle, as well as upon the vertical circle, are subdivided by reading microscopes, the construction and adjustments of which we shall presently explain. The reading microscopes of the azimuth circle are attached to the revolving plate, V V, which also carries two upright pillars. From the centre of the upper horizontal plate, V V, rises a hollow brass cone which just fits over, and moves smoothly upon the solid metallic vertical axis rising from the tripod stand. A horizontal brace connects the two upright pillars with one another and with the

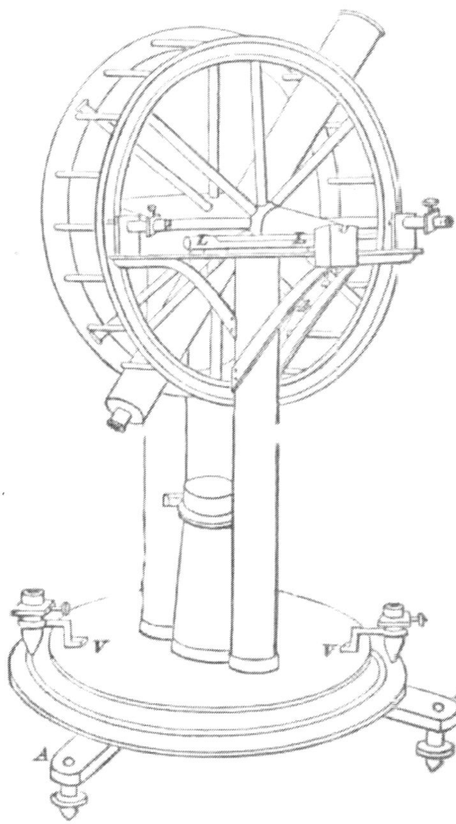

top of the hollow brass cone, and keeps the pillars firm, and parallel to one another. On the top of each pillar a gibbet piece is fixed, projecting beyond the pillar, and upon the extreme ends of these pieces are carried the Y's for supporting the pivots of the horizontal or transit axis. The Y's are each capable of being raised or lowered by turning a milled-headed screw. The top of one of the pillars carries a cross-piece, for supporting the two reading microscopes of the vertical circle; and to this cross-piece is attached the level, L L, by which the adjustment of the vertical axis is denoted.

The third portion of the instrument consists of the vertical circle and its telescope. This circle consists of two limbs firmly braced together, and preventing any tendency to flexure in the tube of the telescope, by affording it support at the opposite ends of a diameter. One of the limbs only is graduated, and the graduated side is called the face of the instrument, and the clamp and tangent screw, for giving a slow motion to the vertical circle, act upon the ungraduated limb, and are fixed to the vertical pillar on the side of that limb. The horizontal axis, which supports the telescope and vertical circle, is constructed exactly as the axis of the transit instrument already described; but, as it might press too heavily on the Y's from the increased load of the vertical circle, a spiral spring, fixed in the body of each pillar, presses up a friction roller against the conical axis with a force which is nearly a counterpoise to its weight. The adjustment of the horizontal axis is denoted by a striding level, as in the portable transit already described.

ADJUSTMENT.

Adjustments of the Vertical Axis.—Turn the instrument round till the level, L L, is over two of the foot-screws, and adjust the level, so that its bubble may retain the same position when the instrument is turned half round, and the level is again over the same foot-screws, but in the reverse position. The error at each trial is corrected, as nearly as can be judged, half by the foot-screws, and half by the adjusting screw of the level itself.

Next turn the instrument round 90° in azimuth, so that the level, L L, may be at right angles to its former positions, and bring the bubble to the same position as before, by turning the third foot-screw. Repeat the whole operation till the result is satisfactory.

Adjustment of the Horizontal Axis.—This adjustment is performed in the same manner as already described for the transit instrument (p. 135), with the single exception that one end of the axis is to be raised or lowered, if necessary, by the screw acting upon its Y, and not by moving a foot-screw, which would derange the previous adjustment.

Adjustment of the Circle to its Reading Microscopes.—This is performed by raising or lowering both the Y's equally, so as not to derange the previous adjustment, till the microscopes are directed to opposite points in its horizontal diameter.

Adjustment of Collimation in Azimuth.—Instead of taking the axis out of its bearings and turning it end for end, the whole instrument is turned round in azimuth; but in all other respects the method of performing this adjustment is the same as that already described for the transit instrument (p. 136).

Adjustment of Collimation in Altitude.—Point the telescope to a very distant object, or star, and, bisecting it by the cross wires, read off the angle upon the vertical circle denoted by the reading microscopes. Turn the instrument half round in azimuth, and, again bisecting the same object by the cross wires, read off the angle. One of these readings will be an altitude, and the other a zenith distance,* and their sum, therefore, when there is no error of collimation in altitude, will be 90°. If the sum is not 90°, half its difference from 90° will be the error of collimation in altitude, and this error being added to, or subtracted from, the observed angles, according as the sum of the readings is less or greater than 90°, will give the true zenith distance and altitude. The error of collimation in altitude may then be corrected by adjusting the microscopes to read the true zenith distance and altitude thus found, while the object is bisected by the cross wires of the telescope. The error of collimation of this and other astronomical instruments may also be found, or corrected, by the collimator.

* Both the horizontal and vertical circles are usually divided alike into four quadrants, and each quadrant graduated from 0° to 90°, proceeding in the same direction all round the circles.

METHOD OF OBSERVING WITH THE ALTITUDE AND AZIMUTH INSTRUMENT.

In using the altitude and azimuth instrument for astronomical purposes double observations should always be made, with the face first to the east, and then to the west, or *vice versâ*, or several observations may be made with the face to the east, and as many with the face to the west, and the mean of the results, reduced to the meridian, taken as the true results. The place for a meridian mark may be determined by the methods already explained when describing the transit instrument, or by observing the readings of the azimuthal circle, or noting the times when any celestial object has equal altitudes. Since the diaphragm of the telescope is furnished not only with the central horizontal wire, but with other horizontal wires at equal distances above and below it, so that there may be altogether either three, or five, or seven horizontal wires, the azimuths and times may be notified when the object observed is bisected by each of these wires. If a fixed star be the object observed, the mean of the times will give the time of the star's passing the meridian, and the mean of the azimuths will give the reading of the azimuth circle when the star was on the meridian, or the correction to be applied to the readings of the azimuth circle to give the true azimuths. If the sun be the body observed, a correction is necessary on account of the change of his declination during the intervals between the observations.

The correction for the time, as deduced from a pair of equal altitudes of the sun, is given by the formula—

$$\text{Correction} = \frac{\delta}{720} \times \frac{\frac{t}{2}}{\sin. 15^\circ \cdot \frac{t}{2}} \left(\tan. \text{D} \times \cos. 15^\circ . \frac{t}{2} - \tan. \text{L}\right)''$$

in which δ represents the variation in the sun's declination from the noon of the day preceding the observations to the noon of the day succeeding;

t represents the interval between the observations expressed in hours and decimals of an hour;

D represents the sun's declination at noon on the day on which the observations are made;

L represents the latitude of the place.

δ is to be reckoned positive when the sun's declination is increasing, and negative when it is decreasing.

The correction for azimuth is given by the formula—

$$\text{Correction} = \tfrac{1}{2}\,(D' - D)\ \text{sec. lat. cosec.}\ \frac{15}{2}\,(T' - T).$$

in which $D' - D$ represents the change of the sun's declination, and $T' - T$ represents the interval in time, } between the observations.

When the sun is advancing towards the north pole, this correction will carry the middle point towards the west of the approximate south point; but when he is approaching the south pole, it will carry the same point towards the east, and must be applied accordingly.

USE OF THE ALTITUDE AND AZIMUTH INSTRUMENT.

The altitude and azimuth instrument being adapted to observe the heavenly bodies in any part of the visible expanse of the heavens, its powers may be applied at any time, to determine the data from which the time, the latitude of the place of observation, or the declination of the body observed may be at once determined. We subjoin some of the formulæ, adapted to logarithmic computation, connecting the parts of what may be called the *astronomical triangle*, of which the angular points are—the pole, P; the zenith, Z; and the apparent place of the body observed, S.

Let PZ, the colatitude of the place, be represented by λ
PS, the polar distance of the body observed ...... ϖ
ZS, the zenith distance of the body observed ... z
ZPS, the hour angle from the meridian............ h
PZS, the azimuthal angle .... a

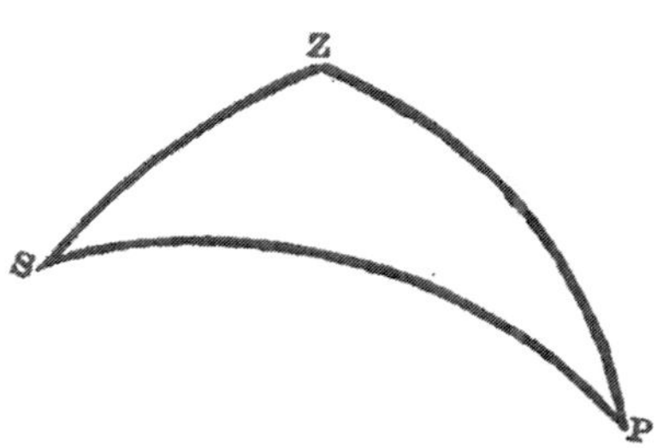

Then we have the following formulæ for determining the time, the latitude, and the declination of the body observed.

No.	Given.	Required.	Auxiliary Angles.	Formulæ.
1	z, π, λ	h	....	$\tan \frac{1}{2} h = \sqrt{\frac{\sin \frac{1}{2}(z + \pi - \lambda) \cdot \sin \frac{1}{2}(z + \lambda - \pi)}{\sin \frac{1}{2}(z + \pi + \lambda) \cdot \sin \frac{1}{2}(\pi + \lambda - z)}}$
2	π, λ, a	h	$\tan \phi = \frac{\cot a}{\cos \lambda}$	$\cos (h \sim \phi) = \frac{\cot \pi \cos \phi}{\cot \lambda}$
3	z, λ, a	h	$\cot \phi = \frac{\cot z}{\cos a}$	$\cot h = \frac{\cot a \sin (\lambda - \phi)}{\sin \phi}$
4	z, π, a	h	....	$\sin h = \frac{\sin z \sin a}{\sin \pi}$
5	z, π, a	λ	$\tan \phi = \cos a \tan z$	$\cos (\lambda \sim \phi) = \frac{\cos \pi \cos \phi}{\cos z}$
6	z, π, h	λ	$\tan \phi = \cos h \tan \pi$	$\cos (\lambda \sim \phi) = \frac{\cos z \cos \phi}{\cos \pi}$
7	z, a, h	λ	$\cot \phi = \frac{\cot z}{\cos a}$	$\sin (\lambda - \phi) = \frac{\cot h \sin \phi}{\cot a}$
8	π, a, h	λ	$\cot \phi = \frac{\cot \pi}{\cos h}$	$\sin (\lambda - \phi) = \frac{\cot a \sin \phi}{\cot h}$
9	z, λ, a	π	$\tan \phi = \cos a \tan z$	$\cos \pi = \frac{\cos z \cos (\lambda \sim \phi)}{\cos \phi}$
10	z, λ, h	π	$\tan \phi = \cos h \tan \lambda$	$\cos (\pi \sim \phi) = \frac{\cos z \cos \phi}{\cos \lambda}$
11	z, a, h	π	....	$\sin \pi = \frac{\sin a \sin z}{\sin h}$
12	λ, a, h	π	$\tan \phi = \frac{\cot a}{\cos \lambda}$	$\cot \pi = \frac{\cot \lambda \cos (h \sim \phi)}{\cos \phi}$

CHAPTER IV.

SUBSIDIARY INSTRUMENTS.

THE READING MICROSCOPE.

The first of the annexed figures represents a longitudinal section of this instrument, and the second represents the field of view, showing the magnified divisions of the limb of

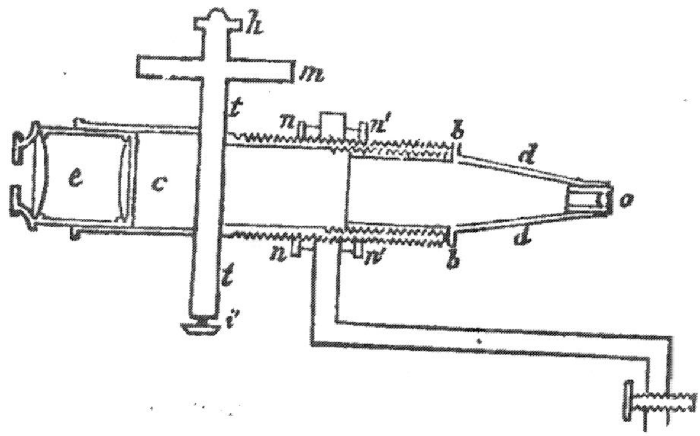

the instrument to which the microscope is applied, and the diaphragm, *d d*, of the microscope, with its comb, *c c*, and cross wires, *w w*. The diaphragm is contained in the box *t t*, and consists of two parts, moving one over the other: the comb, *c c*, which is moved by the screw, *i'*, at the bottom of the box, for the purpose of adjustment; and the cross wires *w w*, and index *i*, which are moved over the comb and the magnified image of the limb, by turning the milled head *h*. The micrometer head, *m*, is attached by friction to the screw turned by the milled head, so that by holding fast the milled head, the micrometer head can be turned round for adjustment.

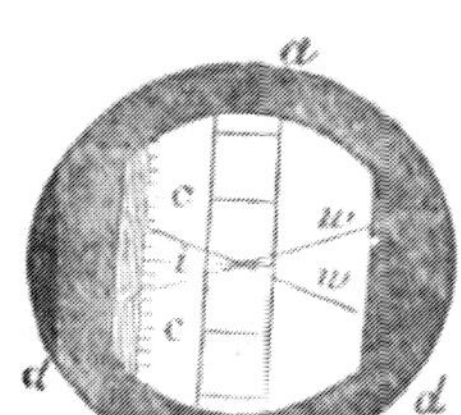

e is the eye-piece, which slides with friction into the cell *c*, so as to produce distinct vision of the spider's lines of the micrometer. The object-glass, *o*, is held by a conical piece, *d d*, which screws farther into or out of the body of the instrument, so as to produce distinct vision of the divided limb to be read by the microscope, and, when adjusted, is held firmly in its place by the nut *b b*. The microscope screws into a collar, so as to be capable of adjustment with respect to its distance from the divided limb, and, when so adjusted, is held firmly in its place by the nuts *n n*, *n′ n′*.

Adjustments of the Reading Microscope.—Screw the object-glass home. Insert the body of the microscope into the collar destined to receive it, and screw home the nuts *n n* and *n′ n′*. Make the diaphragm and spider-lines visible distinctly, by putting the eye-piece, *e*, the proper depth into the cell *c*. Then make the graduated limb also distinctly visible, without parallax, by turning the nuts, *n n*, and *n′ n′*, unscrewing one, and screwing up the other, till the desired end is attained.

Now bring the point of intersection of the spider-lines upon a stroke of the limb, and turn the micrometer head, *m*, to zero; then, turning the screw *h*, through five revolutions, if the point of intersection of the spider-lines has not moved over the whole of one of the divided spaces on the limb, the object lens must be screwed up, to diminish the power, by turning the cone *d d;* and if it has moved over more than one of the divided spaces, it must be unscrewed to increase the power, and then altering the position of the microscope, by turning the nuts *n n* and *n′ n′* till distinct vision of the limb is again obtained, the measure of the space moved over by five revolutions of the screw must be repeated as before. When, after repeated trials, the result is satisfactory, the three nuts *n n*, *n′ n′*, and *b b*, must be screwed tight home to render the adjustment permanent.

When the microscope has been thus adjusted for distance, the zero of the divisions on the limb must be brought to the point of intersection of the spider-lines, and the divided head, *m*, turned, till its zero is pointed to by its index, and then, if the zero on the comb, *c c*, be not covered exactly by the index *i*, the comb must be moved by turning the screw *i′*, which enters the bottom of the micrometer box, till its zero is covered by the index pin. The adjustments of the reading microscope will now be perfect; and the graduated limb to be read by it, being divided at every five minutes, the degree and nearest five minutes of an observed angle will be shown by the pointer or index to this graduated limb; while the number of complete revolutions, and the parts of a revolution of the screw *h*, in the order of the numbers upon the micrometer head, *m*, required, to bring the point of intersection of the spider-lines upon a division of the graduated limb, will be the number of minutes and seconds, respectively, to be added to the degrees and minutes shown by the index of the circle. The complete revolutions, or minutes, to be added, are shown by the number of teeth the index, *i*, has passed over from zero, and the parts of a revolution, or seconds and tenths to be added, are pointed out upon the micrometer head, *m*, by its index.

THE COLLIMATOR.

B B is a rectangular mahogany box, partly filled with mercury. F F is a float of cast-iron, partly immersed in the mercury. *b*, *b* are two iron bearing pieces, screwed to the bottom of the box by short iron screws; and each of these pieces has two vertical plates turned up, the inner one of which has a longitudinal slit in it, into which slits iron pivots, screwed into the sides of the float, are admitted. The use of these parts is to keep the sides of the float parallel to the sides of the box, and at an inch, or more, from contact with any part of the box, that the mercury may assume a flat surface. H and K are two holding pieces of metal cast along with the float, and are perforated, to receive each a socket. The socket at H receives an achromatic object-glass, and is adjustable by a screw for its focal distance, and the socket at K holds two cross wires; while another socket, let into the end of the box at L, carries a lense forming an eye-piece; so that the collimator is, in fact, an astronomical telescope, with a system of cross wires in the common focus of the object-glass and eye-lens. The inclination, as compared with the surface of the fluid, of the optical axis of this telescope, or of the line joining the centre of the object-glass and the intersection of the cross wires, can be modified by the addition of perforated pieces of iron, held steady by the vertical pin P, and by their weight depressing the end of the float. The mercury must be as pure as can be obtained, and particles of dust must be constantly excluded by a lid that covers

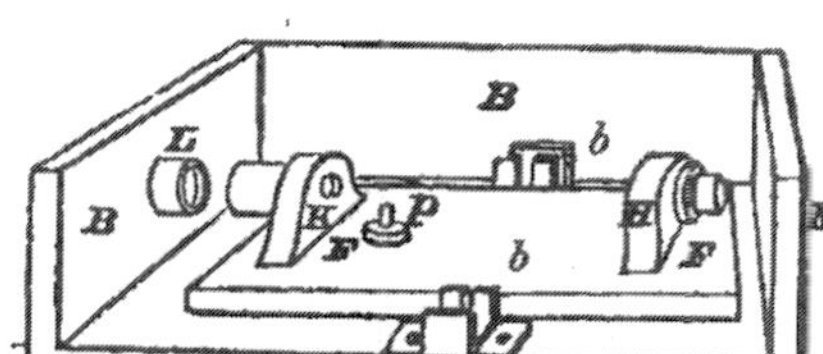

over the top of the box. At M is a circular hole, closed when the instrument is not in use, through which the telescope, of which the error of collimation is sought, is to be directed; and a lamp is placed behind the eye-lens at L, to illuminate the cross wires.

Use of the Collimator with an Altitude and Azimuth Instrument.—Place the collimator in the plane of the meridian on the south side of the observatory, and direct it so that the cross wires of the telescope of the altitude and azimuth instrument may be seen through it, in the centre of the field of view; then also will the cross wires of the collimator be seen through the telescope, in the centre of its field of view. Read off the altitude of the cross wires of the collimator, and then, turning the instrument half round in azimuth, observe again the cross wires of the collimator and read off the angle upon the vertical limb, which will now be a zenith distance. The difference between the sum of these readings and 90°, is the correction which is to be applied to the altitudes and zenith distances observed with the instrument.

Example.—The sun's meridian altitude had been observed on the 20th December, 1826, and the following determination of the error was made immediately after the observations were finished; viz.:—

	°	′	″
Before reversion the apparent altitude of the cross wires was	0	1	2·33
After reversion the apparent zenith distance of the cross wires was	89	58	10 ·33
Sum	89	59	12 ·66
Defect from 90° . . .			47 ·33
Correction of errors of collimation, &c. . .			23 ·66
Altitude of cross wires corrected $= \left\{ \begin{array}{r} 1' \ 2''\cdot33 \\ + \quad 23\ \cdot66 \end{array} \right\} =$		1	26 ·00
Zenith distance . . $= \left\{ \begin{array}{r} 89°\ 58'\ 10''\cdot33 \\ + \quad 23\ \cdot66 \end{array} \right\} =$	89	58	34 ·00
	90	0	0 ·00

The collimator may also be used for a meridian mark with the transit instrument. When used with a circle for measuring altitudes and zenith distances, which has no motion in azimuth, the collimator must be moved from the north to the south side of the observatory, and the mean of the observations in each of these two positions will give the correction for the errors of collimation, &c., as above.

CHAPTER V.

THE EQUATORIAL.

THE position of a body in the heavens is determined by its *right ascension* and *declination*, or its *right ascension* and *polar distance*, the polar distance being the complement of the declination. The right ascension is determined by the time of transit of the body over the meridian of the place of observation, being, in fact, identical with the sidereal time of this transit, and, consequently, equal to the mean solar time of the transit $\pm$ the equation of time $+$ the right ascension of the sun. The polar distance is also most simply and accurately determined from the observed altitude of the body when it crosses the meridian, being equal to the co-latitude plus the meridian zenith distance, when the observed body and the pole are on opposite sides of the zenith, and to the co-latitude minus the zenith distance, when the body is between the zenith and the pole. The formulæ at page 148 show how the hour angle and polar distance may be computed from observations made out of the meridian with the altitude and azimuth instrument, and from the hour angle the time of transit over the meridian, and, consequently, the right ascension is known.

However, neither the transit instrument, nor the altitude and azimuth instrument, is adapted for a prolonged view of a heavenly body; the transit being adjusted to the meridian, the object observed passes rapidly across and out of the field of view, and though by combined motion round the horizontal and vertical axes, the object may be followed with the telescope of the altitude and azimuth instrument, the operation is both inconvenient and imperfect, since it is impracticable so to combine the motions, as to keep the object in the same part of the field of view.

The equatorial supplies the means of prolonging an observation of a heavenly body to an unlimited extent; it also points out at once the right ascension and declination of an observed body; or can, without any computation, be directed to a celestial object, whose position is given, by merely setting the verniers or microscopes to read its known right ascension and declination.

The frontispiece to this volume represents a very elegant

and simple form of the equatorial. The instrument is mounted on a strong cast-iron pillar, P, which carries the bearings for the polar axis, A, with adjustments for setting it parallel to the earth's axis. The declination axis, B, revolves in a collar, which is fitted into a cast-iron socket, bolted on to the polar axis. The telescope, T, is fixed at right angles to one end of the declination axis, and balanced by the counterpoise, W, at the other end of the axis; and on this axis is the declination circle D. The polar axis carries the right ascension circle H, and can either be moved round by the hand, to set it to direct the telescope to any desired object, or made to revolve by the driving-wheel *w*, connected with a clock, the rate of which is governed by balls and fans, so as to have a continuous motion, instead of the intermittent motion of the ordinary time-keepers. The telescope has a fine adjustment for declination, worked by the handle *h*, and the intervention of a Hook's joint, carries a finder, F, for the more easy discovery of an object, and has a diagonal eye-piece, *d*, for looking at elevated objects.

ADJUSTMENTS OF THE EQUATORIAL.

Suppose the latitude of the place and the direction of the meridian to be approximately known, the instrument to show north polar distance, and the hour circle, when the sun is observed to read as an ordinary clock; and let the polar axis be placed nearly in the direction of the poles of the heavens: then the adjustments are—

1. *The elevation of the polar axis to the altitude of the pole.*

2. *The adjustment for collimation in declination, so that the indices of the declination circle point to 0, when the telescope points to the pole.*

3. *Adjustment of the polar axis to the plane of the meridian, so that its prolongation may coincide with the poles of the heavens.*

4. *Adjustment for collimation in right ascension, the line of sight being made perpendicular to the declination axis. This is similar to the adjustment for collimation in azimuth of the transit.*

5. *The setting of the declination axis exactly at right angles to the polar axis, if the construction of the instrument allow of this adjustment.*

6. *The adjustment of the right ascension or hour circle to*

read 0^h, *when the telescope is in the meridian of the place of observation.*

1. *Adjustment of Polar Axis in Altitude.*—Observe any known star, and note the north polar distance as indicated on the declination circle. Turn the instrument half round in right ascension, observe the same star again, and again note the reading of the declination circle. Take the mean of the two readings, correct it for refraction, and the result will be the polar distance indicated by the instrument. Compare this instrumental polar distance with the true polar distance given by the *Nautical Almanack*. If the star observed be above the pole, and the polar distance indicated by the instrument be greater than that given by the *Almanack*, the pole of the instrument is too low, and if the instrumental polar distance be less than the true, the pole of the instrument is too high. If the star observed be below the pole, an error of excess in the polar distance shows that the pole of the instrument is too high, and an error of defect, that it is too low. The position of the axis is to be corrected for elevation by raising or depressing one of its bearings.

2. *Adjustment for Collimation in Declination.*—Take half the difference of the readings of the declination circle in the two observations for index error, move the verniers or reading microscopes through this extent, and the polar distance read off will be the true instrumental polar distance. Several pairs of observation should be made, and the mean of the index errors indicated by them, taken as the real index error.

3. *Adjustment of the Polar Axis to the Plane of the Meridian.*—Turn the instrument round its polar axis, until the telescope is six hours from the meridian either way, and observe the north polar distance of a known star, not *very* near the pole, nor yet near the horizon. Correct for refraction, and compare the result with the true polar distance set down in the *Nautical Almanack*. Suppose the star to be east of the meridian, and the polar distance indicated by the instrument to be too great; then the pole of the instrument inclines to the west, and the upper bearing must be shifted east, or the lower bearing west. Several stars should be observed, and the mean of the corrections indicated be taken as the true amount through which the polar axis is to be moved; but there is no necessity for reversed observations, as in first adjustment.

The polar axis is now adjusted to the meridian, both in altitude and azimuth.

4. *Adjustment for Collimation in Right Ascension.*—Observe the transit of an equatorial star, note the time, and read the verniers of the hour circle: turn the instrument half way round on the polar axis, and observe again; then the difference of the times, and of the readings of the hour circle, should agree. If they do not, suppose the difference in the clock times to be eight seconds greater than the difference of the readings of the hour circle; then the first transit is four seconds too early, and the second transit is four seconds too late. In the first position, then, move the wires for four seconds, to follow the star in right ascension. If the difference of the clock times be less than the differences of the readings of the hour circle, the first transit is too late, and the second transit too early, and, in the second position, the wires must be moved to follow the star in right ascension.

5. *Adjustment of the Declination Axis exactly at right angles to the Polar Axis.*—This adjustment may be tested either astronomically or mechanically.

Astronomically.—Observe the transit of a star in reversed positions of the polar axis, exactly as for the preceding adjustment, only in this case the star, instead of being on the equator, should be not less than 45° from it. A defect in this adjustment will not produce any error in the time of transit of an equatorial star, the law of the error produced being that it varies as the tangent of the declination. As in the observations for the preceding adjustment, the interval between the two transits is increased or diminished by twice the error in the time of each transit. If, then, the clock interval be eight seconds greater than the difference of the readings of the right ascension, the first transit is four seconds too early, which shows that the telescope points to the east, and that, consequently, the west end of the declination axis is too elevated. At the second transit it is just as much depressed. If the declination of the star be exactly 45°, the amount by which the declination axis is out of adjustment, is equal to the error in the time of transit, converted into angular measure: thus if the transit be four seconds too early, the axis is out of adjustment $4 \times 15''$ or $1'$. the west end being too elevated.

If the star have any other declination δ, the amount by which the axis is out of adjustment is $\frac{4}{\tan \delta} \times 15''$.

Before making this adjustment, the previous adjustment is understood to have been perfected, no error of collimation being supposed to exist.

Mechanically.—The declination axis is to be placed exactly horizontal by testing it with a level, as in the cases of the transit instrument and altitude and azimuth instrument; and the reading of the hour circle to be noted. The polar axis is then to be turned half round, the declination axis again placed exactly horizontal, and the reading of the hour circle again taken. If the adjustment be perfect, the difference of the two readings will be exactly twelve hours; but if not, take half the amount by which this difference exceeds or falls short of twelve hours, convert it into angular measure, and call the result e; then the deviation of the axis is given by the equation $\sin d = \tan e \cot l$, in which d represents this deviation, and l the latitude of the place of observation. If the difference of the readings exceed twelve hours, the end of the declination axis, which in the course of the semi-revolution has passed above the polar axis, is too elevated, and *vice versâ*.

6. *Adjustment of the Verniers or Reading Microscopes of the Hour Circle, to read* 0, *when the Telescope is on the Meridian of the Place of Observation.*—If the observer have the means of getting the exact time, let him observe the time of transit of one or more stars not far from the equator, the polar axis of the instrument having been placed and clamped with the verniers of the right ascension or hour circle reading 0. Correct for error of clock. Then the reduced time of transit should be the same as the right ascension of the star, recorded in the *Nautical Almanack;* and if not, the verniers must be shifted, and the instrument again adjusted and clamped, until complete coincidence of the sidereal times of transit with the true right ascension be obtained. If the observer have not the means of getting the time with sufficient accuracy for the application of the preceding test, the declination axis must be placed horizontal by means of a level, and the hour angle indicated on the right ascension circle should then be 0, which the verniers or microscopes should be made to read without altering the position of the instrument.

For all the adjustments, except 8, the observations should be made with the telescope nearly in the meridian. The correction for refraction is, then, at once found from the ordinary tables for refraction, the zenith distance being the difference between the polar distance and the co-latitude for stars above the pole, and the sum of the polar distance and the co-latitude for stars below the pole.

USES OF THE EQUATORIAL.

However well the operations for testing and correcting the adjustments may be performed, the equatorial will not possess the accuracy of the transit and altitude and azimuth instruments. Its uses, however, are two-fold: 1, to observe phenomena, such as the eclipses of Jupiter's satellites, requiring prolonged observation; 2, to determine, in a rapid and simple manner, the place of a body in the heavens, by observations made at any time, this place being previously quite unknown.

METHOD OF OBSERVING WITH THE EQUATORIAL.

For the purpose of prolonged observation of a body, whose right ascension and declination are known, the instrument may be set at once to find it, and the hour circle being then put in connection with the driving-wheel, the clock will move the instrument in right ascension, so as to exactly follow the apparent motion of the heavenly body, caused by the revolution of the earth on its axis.

When used to determine the position of some stranger in the heavens, as a comet, the equatorial should be used merely as a differential instrument. The comet having been observed, and the time noted, let the instrument be clamped both in right ascension and declination, until a known body passes over the field of view, when its time of transit must also be noted and its instrumental declination observed. The observations, both of the comet and of the known body, will be equally effected both by refraction and by the imperfections of the adjustments, and, consequently, the differences, both in the right ascensions and declinations of the two bodies will be accurately determined. If there be no body of nearly the same declination as the comet sufficiently near in right ascension to pass across the field of view within a short time after the comet, the instrument may be clamped in right ascension only, and stars of both greater and lesser declination by nearly equal amounts may be observed, so that the mean of the corrections for refraction may be nearly the same as the correction for the comet.

INDEX TO VOL. III.

Altitude and azimuth instrument, description of, 142; adjustment of, 144; mode of observing with, 146; use of, 147.
Aneroid barometer, 104.
Artificial horizon, 52.
Astronomical triangle, 147; table of formulæ connecting the parts of, 148.

Barometer, standard, 98; the syphon, 99; the portable or mountain, 99; the wheel, 103; the aneroid, 104; application of, to determination of difference of altitude, 108; meteorological phenomena denoted by, 118.
Bench marks, 31.
Box sextant, 47; parallax of, 49; adjustments of, 51; mode of observing with, 51.

Chain, land, 2; Gunter's, 2; measuring with, 4; examination and correction of, by standard, 5; survey, 9.
Chords, protracting by, 75.
Clinometers, 7.
Collimation, line of, 15; adjustment of, for level and theodolite, 149; in altitude, adjustment for, 63, 145; in azimuth, adjustment for, 63, 136, 145.
Collimator, 151.
Compass, the prismatic, 39; surveying with, 43.
Cross staff, 53.

D'Abbadie's theodolite described, 66.

Edgeworth's stadiometer, 84.
Elliott's improved dumpy level, 22.
Equatorial, the, description of, 153; adjustment of, 154; mode of observing with, 158; uses of, 158.

Formulæ, for true altitude of barometer, 101; for determining height by the barometer, 110; connecting the parts of astronomical triangle, 148.

Gravatt, his adjustment of the level, 18; his level, 20.
Gunter's chain, 2.

Hadley's sextant, 127; adjustments of, 130.
Horizon, artificial, 52.
Hypsometrical instruments, preliminary remarks on, 92.
Hypsometry, barometric, 108; thermometric, 124.

Land chain, 2; surveying with, 9.
Level, the spirit, 10; criterion of a good, 11; the Y, 12; Gravatt's dumpy, 20; Elliott's improved dumpy, 22; the improved dumpy Y, 21; water, 33; reflecting, 34; improved reflecting, 35; draining and mining, 37.
Levelling, remarks on, 26; for section, 28; field-book, 29.
Levelling staves, 24.
Line ranger, the, 54.
Linear and square measure, tables of, 3.

Measuring tape, 9.
Mercurial thermometer, 96.
Meteorological phenomena denoted by barometer, 118.
Microscope, reading, 149.
Mountain barometer, 99.

Northings, southings, eastings, and westings, 75.

Offsetting staff, 8.
Omnimeter, 88.
Optical square, 53.

Parallax from misplacement of spider lines, detected and corrected, 15.
Plotting a section, 32; a survey, 74.
Poles, station, 9.
Portable barometer, 99; transit instrument, 132; altitude and azimuth instrument, 142.
Prismatic compass, the, 39; illustrations of its use, 43.
Protracting by chords, 75; by northings, southings, eastings, and westings, 75.

Ranger, the line, 54.
Ranging rods, 9; lines with the theodolite, 70.
Reading microscope, 149.

Sextant, box, 47; parallax of, 49; adjustments of, 49; mode of observing with, 51; Hadley's, 127; adjustments of, 130.
Spirit level, 10; criteria of a good, 11.
Square, optical, 53; measures, table of, 3.
Stadiometer, Edgeworth's, 84.
Stadium, the, for determining distances in rifle practice, 79.
Staff, levelling, 24; offsetting, 8; cross, 53.
Station poles and ranging rods, 9.
Surveying with chain only, 9; with prismatic compass, 43; with theodolite, 70; with the stadiometer, 86; with the omnimeter, 89.

Tables of linear and square measures, 3; for the reduction of inclined to horizontal distances, 6, 30; of corrections for capillarity, in the heights of the barometric column, 99; of feet in column of atmosphere equivalent in weight to one-tenth of an inch of mercury, at different temperatures, 109; of corrections for temperature to be applied to barometers, 112, 113; of factors and divisors for computation of heights by barometer, 114, 115; of altitudes and corresponding heights of barometer, 116, 117.
Tape, the measuring, 9.
Theodolite, the, 55; the transit, 60; Everest's, 64; Lister's, 66; D'Abbadie's, 66; measuring angles with the, 67; ranging lines with the, 70; traversing with the, 71.
Thermometer, the mercurial, 96.
Thermometric hypsometry, 124.
Transit instrument, 132; adjustments of, 135; method of observing with, 139; use of, 141; theodolite, 60.
Triangle, astronomical, 147; table of formulæ connecting the parts of, 148.

Wheel barometer, the, 103.
Whittley's slide-rule for determination of heights by the barometer, 118.

Y level, the, 12.

THE END.

A CATALOGUE OF BOOKS

INCLUDING MANY NEW AND STANDARD WORKS IN

ENGINEERING, MECHANICS, ARCHITECTURE, CARPENTRY, MINING, NAVAL ARCHITECTURE, SCIENCE, MATHEMATICS, THE INDUSTRIAL ARTS, MANUFACTURES, AGRICULTURE, GARDENING, ESTATE MANAGEMENT, &c.

PUBLISHED BY

CROSBY LOCKWOOD & CO.

CIVIL ENGINEERING, SURVEYING, etc.

The Water Supply of Cities and Towns.

A COMPREHENSIVE TREATISE on the WATER-SUPPLY OF CITIES AND TOWNS. By WILLIAM HUMBER, A-M. Inst. C.E., and M. Inst. M.E., Author of "Cast and Wrought Iron Bridge Construction," &c., &c. Illustrated with 50 Double Plates, 1 Single Plate, Coloured Frontispiece, and upwards of 250 Woodcuts, and containing 400 pages of Text. Imp. 4to, £6 6s. elegantly and substantially half-bound in morocco.

List of Contents.

I. Historical Sketch of some of the means that have been adopted for the Supply of Water to Cities and Towns.—II. Water and the Foreign Matter usually associated with it.—III. Rainfall and Evaporation.—IV. Springs and the water-bearing formations of various districts.—V. Measurement and Estimation of the flow of Water —VI. On the Selection of the Source of Supply.—VII. Wells.—VIII. Reservoirs.—IX. The Purification of Water.—X. Pumps. — XI. Pumping Machinery. — XII. Conduits.—XIII. Distribution of Water.—XIV. Meters, Service Pipes, and House Fittings.—XV. The Law and Economy of Water Works. XVI. Constant and Intermittent Supply.—XVII. Description of Plates. — Appendices, giving Tables of Rates of Supply, Velocities, &c. &c., together with Specifications of several Works illustrated, among which will be found Aberdeen, Bideford, Canterbury, Dundee Halifax, Lambeth, Rotherham, Dublin, and others.

"The most systematic and valuable work upon water supply hitherto produced in English, or in any other language. . . . Mr. Humber's work is characterised almost throughout by an exhaustiveness much more distinctive of French and German than of English technical treatises. —*Engineer.*

"We can congratulate Mr. Humber on having been able to give so large an amount of information on a subject so important as the water supply of cities and towns. The plates, fifty in number, are mostly drawings of executed works, and alone would have commanded the attention of every engineer whose practice may lie in this branch of the profession."—*Builder.*

Cast and Wrought Iron Bridge Construction.

A COMPLETE AND PRACTICAL TREATISE ON CAST AND WROUGHT IRON BRIDGE CONSTRUCTION, including Iron Foundations. In Three Parts—Theoretical, Practical, and Descriptive. By WILLIAM HUMBER, A-M. Inst. C.E., and M. Inst. M.E. Third Edition, Revised and much improved, with 115 Double Plates (20 of which now first appear in this edition), and numerous Additions to the Text. In Two Vols., imp. 4to, £6 16s. 6d. half-bound in morocco.

"A very valuable contribution to the standard literature of civil engineering. In addition to elevations, plans and sections, large scale details are given, which very much enhance the instructive worth of these illustrations. No engineer would willingly be without so valuable a fund of information."—*Civil Engineer and Architect's Journal.*

"Mr. Humber's stately volumes, lately issued—in which the most important bridges erected during the last five years, under the direction of the late Mr. Brunel, Sir W. Cubitt, Mr. Hawkshaw, Mr. Page, Mr. Fowler, Mr. Hemans, and others among our most eminent engineers, are drawn and specified in great detail."—*Engineer.*

HUMBER'S GREAT WORK ON MODERN ENGINEERING.

Complete in Four Volumes, imperial 4to, price £12 12s., half-morocco, each Volume sold separately as follows:—

A RECORD OF THE PROGRESS OF MODERN ENGINEERING. First Series. Comprising Civil, Mechanical, Marine, Hydraulic, Railway, Bridge, and other Engineering Works, &c. By William Humber, A-M. Inst. C.E., &c. Imp. 4to, with 36 Double Plates, drawn to a large scale, Photographic Portrait of John Hawkshaw, C.E., F.R.S., &c., and copious descriptive Letterpress, Specifications, &c., £3 3s. half-morocco.

List of the Plates and Diagrams.

Victoria Station and Roof, L. B. & S. C. R. (8 plates); Southport Pier (2 plates); Victoria Station and Roof, L. C. & D. and G. W. R. (6 plates); Roof of Cremorne Music Hall; Bridge over G. N. Railway; Roof of Station, Dutch Rhenish Rail (2 plates); Bridge over the Thames, West London Extension Railway (5 plates); Armour Plates: Suspension Bridge, Thames (4 plates); The Allen Engine; Suspension Bridge, Avon (3 plates); Underground Railway (3 plates).

"Handsomely lithographed and printed. It will find favour with many who desire to preserve in a permanent form copies of the plans and specifications prepared for the guidance of the contractors for many important engineering works."—*Engineer.*

HUMBER'S RECORD OF MODERN ENGINEERING. Second Series. Imp. 4to, with 36 Double Plates, Photographic Portrait of Robert Stephenson, C.E., M.P., F.R.S., &c., and copious descriptive Letterpress, Specifications, &c., £3 3s. half-morocco.

List of the Plates and Diagrams.

Birkenhead Docks, Low Water Basin (15 plates); Charing Cross Station Roof, C. C. Railway (3 plates); Digswell Viaduct, Great Northern Railway; Robbery Wood Viaduct, Great Northern Railway; Iron Permanent Way; Clydach Viaduct, Merthyr, Tredegar, and Abergavenny Railway; Ebbw Viaduct, Merthyr, Tredegar, and Abergavenny Railway; College Wood Viaduct, Cornwall Railway; Dublin Winter Palace Roof (3 plates); Bridge over the Thames, L. C. & D. Railway (6 plates); Albert Harbour, Greenock (4 plates).

"Mr. Humber has done the profession good and true service, by the fine collection of examples he has here brought before the profession and the public."—*Practical Mechanic's Journal.*

HUMBER'S RECORD OF MODERN ENGINEERING. Third Series. Imp. 4to, with 40 Double Plates, Photographic Portrait of J. R. M'Clean, Esq., late Pres. Inst. C.E., and copious descriptive Letterpress, Specifications, &c., £3 3s. half-morocco.

List of the Plates and Diagrams.

Main Drainage, Metropolis.—*North Side.*—Map showing Interception of Sewers; Middle Level Sewer (2 plates); Outfall Sewer, Bridge over River Lea (3 plates); Outfall Sewer, Bridge over Marsh Lane, North Woolwich Railway, and Bow and Barking Railway Junction; Outfall Sewer, Bridge over Bow and Barking Railway (3 plates); Outfall Sewer, Bridge over East London Waterworks' Feeder (2 plates); Outfall Sewer, Reservoir (2 plates); Outfall Sewer, Tumbling Bay and Outlet; Outfall Sewer, Penstocks. *South Side.*—Outfall Sewer, Bermondsey Branch (2 plates); Outfall Sewer, Reservoir and Outlet (4 plates); Outfall Sewer, Filth Hoist; Sections of Sewers (North and South Sides).

Thames Embankment.—Section of River Wall; Steamboat Pier, Westminster (2 plates), Landing Stairs between Charing Cross and Waterloo Bridges; York Gate (2 plates); Overflow and Outlet at Savoy Street Sewer (3 plates); Steamboat Pier, Waterloo Bridge (3 plates); Junction of Sewers, Plans and Sections; Gullies, Plans and Sections; Rolling Stock; Granite and Iron Forts.

"The drawings have a constantly increasing value, and whoever desires to possess clear representations of the two great works carried out by our Metropolitan Board wil obtain Mr. Humber's volume."—*Engineer.*

HUMBER'S RECORD OF MODERN ENGINEERING. Fourth Series. Imp. 4to, with 36 Double Plates, Photographic Portrait of John Fowler, Esq., late Pres. Inst. C.E., and copious descriptive Letterpress, Specifications, &c., £3 3s. half-morocco.

List of the Plates and Diagrams.

Abbey Mills Pumping Station, Main Drainage, Metropolis (4 plates); Barrow Docks (5 plates); Manquis Viaduct, Santiago and Valparaiso Railway (2 plates); Adam's Locomotive, St. Helen's Canal Railway (2 plates); Cannon Street Station Roof, Charing Cross Railway (3 plates); Road Bridge over the River Moka (2 plates); Telegraphic Apparatus for Mesopotamia; Viaduct over the River Wye, Midland Railway (3 plates); St. Germans Viaduct, Cornwall Railway (2 plates); Wrought-Iron Cylinder for Diving Bell; Millwall Docks (6 plates); Milroy's Patent Excavator; Metropolitan District Railway (6 plates); Harbours Ports, and Breakwaters (3 plates).

"We gladly welcome another year's issue of this valuable publication from the able pen of Mr. Humber. The accuracy and general excellence of this work are well known, while its usefulness in giving the measurements and details of some of the latest examples of engineering, as carried out by the most eminent men in the profession, cannot be too highly prized."—*Artizan.*

Trigonometrical Surveying.

AN OUTLINE OF THE METHOD OF CONDUCTING A TRIGONOMETRICAL SURVEY, for the Formation of Geographical and Topographical Maps and Plans, Military Reconnaissance, Levelling, &c., with Useful Problems, Formulæ, and Tables. By Lieut.-General FROME, R.E. Fourth Edition, Revised and partly Re-written by Captain CHARLES WARREN, R.E. With 19 Plates and 115 Woodcuts, royal 8vo, 16s. cloth.

Contents.—Chap. I. General Outline of the System of carrying on a Trigonometrical Survey.—Chap. II. Measurements of a Base Line.—Chap. III. Triangulation.—Chap. IV. Interior Filling-in of a Survey, either entirely or partially by Measurement.—Chap. V. Levelling and Contouring.—Chap. VI. Levelling by Barometer and Thermometric Hypsometer.—Chap. VII. Plotting, Penning-in, Copying, and Engraving Typographical Plans.—Chap. VIII. Modelling.—IX. Military Reconnaissance, and Hints on Sketching Ground, Colonial Surveying.—Chap. XI. Geodesical Operations connected with a Trigonometrical Survey.—Chap. XII. Practical Astronomy.—Problems, Tables, &c.

"The simple fact that a fourth edition has been called for is the best testimony to its merits. No words of praise from us can strengthen the position so well and so steadily maintained by this work. Captain Warren has revised the entire work, and made such additions as were necessary to bring every portion of the contents up to the present date."—*Broad Arrow.*

Oblique Bridges.

A PRACTICAL AND THEORETICAL ESSAY ON OBLIQUE BRIDGES. With 13 large Plates. By the late GEORGE WATSON BUCK, M.I.C.E. Third Edition, revised by his Son, J. H. WATSON BUCK, M.I.C.E.; and with the addition of Description to Diagrams for Facilitating the Construction of Oblique Bridges, by W. H. BARLOW, M.I.C.E. Royal 8vo, 12s. cloth.

"The standard text-book for all engineers regarding skew arches is Mr. Buck's treatise, and it would be impossible to consult a better."—*Engineer.*

"Mr. Buck's treatise is recognised as a standard text-book, and his treatment has divested the subject of many of the intricacies supposed to belong to it. As a guide to the engineer and architect, on a confessedly difficult subject, Mr. Buck's work is unsurpassed."—*Building News.*

Bridge Construction in Masonry, Timber and Iron.

EXAMPLES OF BRIDGE AND VIADUCT CONSTRUCTION OF MASONRY, TIMBER, AND IRON. Consisting of 46 Plates from the Contract Drawings or Admeasurement of select Works. By W. D. HASKOLL, C.E. Second Edition, with the addition of 554 Estimates, and the Practice of Setting out Works. Illustrated with 6 pages of Diagrams. Imp. 4to, £2 12s. 6d. half-morocco.

"A work of the present nature by a man of Mr. Haskoll's experience must prove invaluable. The tables of estimates will considerably enhance its value."—*Engineering.*

Earthwork.

EARTHWORK TABLES. Showing the Contents in Cubic Yards of Embankments, Cuttings, &c., of Heights or Depths up to an average of 80 feet. By JOSEPH BROADBENT, C.E., and FRANCIS CAMPIN, C.E. Crown 8vo, 5s. cloth.

"The way in which accuracy is attained, by a simple division of each cross section into three elements, two in which are constant and one variable, is ingenious."—*Athenæum.*

Barlow's Strength of Materials, enlarged.

A TREATISE ON THE STRENGTH OF MATERIALS; with Rules for Application in Architecture, the Construction of Suspension Bridges, Railways, &c. By PETER BARLOW, F.R.S. A New Edition, revised by his Sons, P. W. BARLOW, F.R.S., and W. H. BARLOW, F.R.S.; to which are added, Experiments by HODGKINSON, FAIRBAIRN, and KIRKALDY; and Formulæ for Calculating Girders, &c. Arranged and Edited by W. HUMBER, A-M. Inst. C.E. Demy 8vo, 400 pp., with 19 large Plates and numerous Woodcuts, 18s. cloth.

"Valuable alike to the student, tyro, and the experienced practitioner, it will always rank in future as it has hitherto done, as the standard treatise on that particular subject."—*Engineer.*

"A book which no engineer of any kind can afford to be without."—*Colliery Guardian.*

"There is no greater authority than Barlow."—*Building News.*

"The book is undoubtedly worthy of the highest commendation, and of an honourable place in the library of every engineer."—*Mining Journal.*

". . . As a scientific work of the first class, it deserves a foremost place on the bookshelves of every civil engineer and practical mechanic."—*English Mechanic.*

Survey Practice.

AID TO SURVEY PRACTICE, for Reference in Surveying, Levelling, Setting-out and in Route Surveys of Travellers by Land and Sea. With Tables, Illustrations, and Records. By LOWIS D'A. JACKSON, A.M.I.C.E., Author of "Hydraulic Manual," "Modern Metrology," &c. Large crown 8vo, 12s. 6d. cloth.

"Mr. Jackson has produced a valuable *vade-mecum* for the surveyor. We can recommend this book as containing an admirable supplement to the teaching of the accomplished surveyor."—*Athenæum.*

"We cannot recommend to the student who knows something of the mathematical principles of the subject a better course than to fortify his practice in the field under a competent surveyor with a study of Mr. Jackson's useful manual."—*Building News.*

"The author brings to his work a fortunate union of theory and practical experience which, aided by a clear and lucid style of writing, renders the book a very useful one."—*Builder.*

Levelling.

A TREATISE ON THE PRINCIPLES AND PRACTICE OF LEVELLING. Showing its Application to purposes of Railway and Civil Engineering, in the Construction of Roads; with Mr. TELFORD'S Rules for the same. By FREDERICK W. SIMMS, F.G.S., M. Inst. C.E. Seventh Edition, with the addition of LAW'S Practical Examples for Setting-out Railway Curves, and TRAUTWINE'S Field Practice of Laying-out Circular Curves. With 7 Plates and numerous Woodcuts, 8vo, 8s. 6d. cloth. *** TRAUTWINE on Curves, separate, 5s.

"The text-book on levelling in most of our engineering schools and colleges."—*Engineer.*

"The publishers have rendered a substantial service to the profession, especially to the younger members, by bringing out the present edition of Mr. Simms' useful work."—*Engineering.*

Tunnelling.

PRACTICAL TUNNELLING. Explaining in detail the Setting-out of the works, Shaft-sinking and Heading-driving, Ranging the Lines and Levelling underground, Sub-Excavating, Timbering, and the Construction of the Brickwork of Tunnels, with the amount of Labour required for, and the Cost of, the various portions of the work. By FREDERICK W. SIMMS, F.G.S., M. Inst. C.E. Third Edition, Revised and Extended by D. KINNEAR CLARK, M. Inst. C.E. Imp. 8vo, with 21 Folding Plates and numerous Wood Engravings, 30s. cloth.

"The estimation in which Mr. Simms' book on tunnelling has been held for over thirty years cannot be more truly expressed than in the words of the late Professor Rankine:—'The best source of information on the subject of tunnels is Mr. F. W. Simms' work on Practical Tunnelling.'"—*Architect.*

"It has been regarded from the first as a text-book of the subject. . . . . Mr. Clark has added immensely to the value of the book."—*Engineer.*

"The additional chapters by Mr. Clark, containing as they do numerous examples of modern practice, bring the book well up to date."—*Engineering.*

Statics, Graphic and Analytic.

GRAPHIC AND ANALYTIC STATICS, in Theory and Comparison: Their Practical Application to the Treatment of Stresses in Roofs, Solid Girders, Lattice, Bowstring and Suspension Bridges, Braced Iron Arches and Piers, and other Frameworks. To which is added a Chapter on Wind Pressures. By R. HUDSON GRAHAM, C.E. With numerous Examples, many taken from existing Structures. 8vo, 16s. cloth.

"Mr. Graham's book will find a place wherever graphic and analytic statics are used or studied. *Engineer.*

"This exhaustive treatise is admirably adapted for the architect and engineer, and will tend to wean the profession from a tedious and laboured mode of calculation. To prove the accuracy of the graphical demonstrations, the author compares them with the analytic formulæ given by Rankine."—*Building News.*

"The work is excellent from a practical point of view, and has evidently been prepared with much care. The directions for working are ample, and are illustrated by an abundance of well-selected examples. It is an excellent text-book for the practical draughtsman."—*Athenæum.*

Strains, Formulæ and Diagrams for Calculation of.

A HANDY BOOK FOR THE CALCULATION OF STRAINS IN GIRDERS AND SIMILAR STRUCTURES, AND THEIR STRENGTH. Consisting of Formulæ and Corresponding Diagrams, with numerous details for Practical Application, &c. By WILLIAM HUMBER, A-M. Inst. C.E., &c. Third Edition. Crown 8vo, nearly 100 Woodcuts and 3 Plates, 7s. 6d. cloth.

"The formulæ are neatly expressed, and the diagrams good."—*Athenæum.*

"We heartily commend this really *handy* book to our engineer and architect readers."—*English Mechanic.*

Hydraulic Tables.

HYDRAULIC TABLES, CO-EFFICIENTS, and FORMULÆ for finding the Discharge of Water from Orifices, Notches, Weirs, Pipes, and Rivers. With New Formulæ, Tables and General Information on Rainfall, Catchment-Basins, Drainage, Sewerage, Water Supply for Towns and Mill Power. By JOHN NEVILLE, Civil Engineer, M.R.I.A. Third Edition, carefully revised, with considerable Additions. Numerous Illustrations. Crown 8vo, 14*s.* cloth.

"Alike valuable to students and engineers in practice; its study will prevent the annoyance of avoidable failures, and assist them to select the readiest means of successfully carrying out any given work connected with hydraulic engineering."—*Mining Journal.*

"It is, of all English books on the subject, the one nearest to completion. . . . From the good arrangement of the matter, the clear explanations, and abundance of formulæ, the carefully calculated tables, and, above all, the thorough acquaintance with both theory and construction, which is displayed from first to last, the book will be found to be an acquisition."—*Architect.*

River Engineering.

RIVER BARS: The Causes of their Formation, and their Treatment by "Induced Tidal Scour." With a Description of the Successful Reduction by this Method of the Bar at Dublin. By I. J. MANN, Assist. Eng. to the Dublin Port and Docks Board. Royal 8vo, 7*s.* 6*d.* cloth.

"We recommend all interested in harbour works—and, indeed, those concerned in the improvement of rivers generally—to read Mr. Mann's interesting work on the treatment of river bars."—*Engineer.*

"The author's discussion on wave-action, currents, and scour is intelligent and interesting. . . a most valuable contribution to the history of this branch of engineering."—*Engineering and Mining Journal.*

Hydraulics.

HYDRAULIC MANUAL. Consisting of Working Tables and Explanatory Text. Intended as a Guide in Hydraulic Calculations and Field Operations. By LOWIS D'A. JACKSON. Fourth Edition, Rewritten and Enlarged. Large crown 8vo, 16*s.* cloth.

"The author has had a wide experience in hydraulic engineering, both in South America and in India, and has been a careful observer of the facts which have come under his notice, as well as a painstaking collector and critic of the results of the experiments of others; and from the great mass of material at his command he has constructed a manual which may be accepted as a trustworthy guide to this branch of the engineer's profession. We can heartily recommend this volume to all who desire to be acquainted with the latest development of this important subject."—*Engineering.*

"The most useful feature of this work is its freedom from what is superannuated, and its thorough adoption of recent experiments; the text is, in fact, in great part a short account of the great modern experiments."—*Nature.*

Tramways and their Working.

TRAMWAYS: THEIR CONSTRUCTION AND WORKING. Embracing a Comprehensive History of the System; with an exhaustive Analysis of the various Modes of Traction, including Horse-Power, Steam, Heated Water, and Compressed Air; a Description of the Varieties of Rolling Stock; and ample Details of Cost and Working Expenses: the Progress recently made in Tramway Construction, &c. &c. By D. KINNEAR CLARK, M. Inst. C.E. With over 200 Wood Engravings, and 13 Folding Plates. Two Vols., large crown 8vo, 30*s.* cloth.

"All interested in tramways must refer to it, as all railway engineers have turned to the author's work 'Railway Machinery.'"—*Engineer.*

"An exhaustive and practical work on tramways, in which the history of this kind of locomotion, and a description and cost of the various modes of laying tramways, are to be found."—*Building News.*

"The best form of rails, the best mode of construction, and the best mechanical appliances are so fairly indicated in the work under review, that any engineer about to construct a tramway will be enabled at once to obtain the practical information which will be of most service to him."—*Athenæum.*

Oblique Arches.

A PRACTICAL TREATISE ON THE CONSTRUCTION OF OBLIQUE ARCHES. By JOHN HART. Third Edition, with Plates. Imperial 8vo, 8*s.* cloth.

Strength of Girders.

GRAPHIC TABLE FOR FACILITATING THE COMPUTATION OF THE WEIGHTS OF WROUGHT IRON AND STEEL GIRDERS, &c., for Parliamentary and other Estimates. By J. H. WATSON BUCK, M. Inst. C.E. On a Sheet, 2*s.* 6*d.*

Tables for Setting-out Curves.

TABLES OF TANGENTIAL ANGLES AND MULTIPLES for Setting-out Curves from 5 to 200 Radius. By ALEXANDER BEAZELEY, M. Inst. C.E. Third Edition. Printed on 48 Cards, and sold in a cloth box, waistcoat-pocket size, 3*s.* 6*d.*

"Each table is printed on a small card, which, being placed on the theodolite, leaves the hands free to manipulate the instrument—no small advantage as regards the rapidity of work."—*Engineer*

"Very handy; a man may know that all his day's work must fall on two of these cards, which he puts into his own card-case, and leaves the rest behind."—*Athenæum.*

Engineering Fieldwork.

THE PRACTICE OF ENGINEERING FIELDWORK, applied to Land and Hydraulic, Hydrographic, and Submarine Surveying and Levelling. Second Edition, Revised, with considerable Additions, and a Supplement on Waterworks, Sewers, Sewage, and Irrigation. By W. DAVIS HASKOLL, C.E. Numerous Folding Plates. In One Volume, demy 8vo, £1 5*s.* cloth.

Large Tunnel Shafts.

THE CONSTRUCTION OF LARGE TUNNEL SHAFTS: A Practical and Theoretical Essay. By J. H. WATSON BUCK, M. Inst. C.E., Resident Engineer, London and North-Western Railway. Illustrated with Folding Plates, royal 8vo, 12*s.* cloth.

"Many of the methods given are of extreme practical value to the mason; and the observations on the form of arch, the rules for ordering the stone, and the construction of the templates will be found of considerable use. We commend the book to the engineering profession."—*Building News.*

"Will be regarded by civil engineers as of the utmost value, and calculated to save much time nd obviate many mistakes."—*Colliery Guardian.*

Field-Book for Engineers.

THE ENGINEER'S, MINING SURVEYOR'S, AND CONTRACTOR'S FIELD-BOOK. Consisting of a Series of Tables, with Rules, Explanations of Systems, and use of Theodolite for Traverse Surveying and Plotting the Work with minute accuracy by means of Straight Edge and Set Square only; Levelling with the Theodolite, Casting-out and Reducing Levels to Datum, and Plotting Sections in the ordinary manner; setting-out Curves with the Theodolite by Tangential Angles and Multiples, with Right and Left-hand Readings of the Instrument: Setting-out Curves without Theodolite, on the System of Tangential Angles by sets of Tangents and Offsets: and Earthwork Tables to 80 feet deep, calculated for every 6 inches in depth. By W. DAVIS HASKOLL, C.E. With numerous Woodcuts. Fourth Edition, Enlarged. Crown 8vo, 12*s.* cloth.

"The book is very handy, and the author might have added that the separate tables of sines and tangents to every minute will make it useful for many other purposes, the genuine traverse tables existing all the same."—*Athenæum.*

"Every person engaged in engineering field operations will estimate the importance of such a work and the amount of valuable time which will be saved by reference to a set of reliable tables prepared with the accuracy and fulness of those given in this volume."—*Railway News.*

Earthwork, Measurement and Calculation of.

A MANUAL ON EARTHWORK. By ALEX. J. S. GRAHAM, C.E. With numerous Diagrams. 18mo, 2*s.* 6*d.* cloth.

"A great amount of practical information, very admirably arranged, and available for rough estimates, as well as for the more exact calculations required in the engineer's and contractor's offices."—*Artisan.*

Strains.

THE STRAINS ON STRUCTURES OF IRONWORK; with Practical Remarks on Iron Construction. By F. W. SHEILDS, M. Inst. C.E. Second Edition, with 5 Plates. Royal 8vo, 5*s.* cloth.

"The student cannot find a better little book on this subject."—*Engineer.*

Strength of Cast Iron, etc.

A PRACTICAL ESSAY ON THE STRENGTH OF CAST IRON AND OTHER METALS. By THOMAS TREDGOLD, C.E. Fifth Edition, including HODGKINSON'S Experimental Researches. 8vo, 12*s.* cloth.

MECHANICS & MECHANICAL ENGINEERING.

Mechanic's Workshop Companion.

THE OPERATIVE MECHANIC'S WORKSHOP COMPANION, AND SCIENTIFIC GENTLEMAN'S PRACTICAL ASSISTANT. Comprising a great variety of the most useful Rules in Mechanical Science; with numerous Tables of Practical Data and Calculated Results. By W. TEMPLETON, Author of "The Engineer's Practical Assistant." Thirteenth Edition. With Tables for Operative Smiths, Millwrights, Engineers, &c., and Useful and Practical Rules in Hydraulics and Hydrodynamics, a variety of Experimental Results, and an extensive Table of Powers and Roots. 11 Plates. 12mo, 5s. bound.

"As a text-book of reference, in which mechanical and commercial demands are judiciously met, 'Templeton's Companion' stands unrivalled."—*Mechanic's Magazine.*

"It has met with great success in the engineering workshop, as we can testify; and there are a great many men who, in a great measure, owe their rise in life to this little book."—*Building News.*

Engineer's and Machinist's Assistant.

THE ENGINEER'S, MILLWRIGHT'S, and MACHINIST'S PRACTICAL ASSISTANT. Comprising a collection of Useful Tables, Rules and Data. Compiled and Arranged, with Original Matter, by WILLIAM TEMPLETON. Seventh Edition. Carefully revised, with Additions. 18mo, 2s. 6d. cloth. [*Just published.*

"With the utmost confidence we commend this book to the attention of our readers."—*Mechanic's Magazine.*

"A more suitable present to an apprentice to any of the mechanical trades could not possibly be made."—*Building News.*

Mechanics.

THE HANDBOOK OF MECHANICS. By DIONYSIUS LARDNER, D.C.L., formerly Professor of Natural Philosophy and Astronomy in University College, London. New Edition, Edited and considerably Enlarged by BENJAMIN LOEWY, F.R.A.S., &c. 378 Illustrations, post 8vo, 6s. cloth.

"The explanations throughout are studiously popular, and care has been taken to show the application of the various branches of physics to the industrial arts, and to the practical business of life."—*Mining Journal.*

Turning.

LATHE-WORK: A Practical Treatise on the Tools, Appliances, and Processes employed in the Art of Turning. By PAUL N. HASLUCK. Second Edition, thoroughly Revised, with a New Chapter on the Screw-cutting Lathe. Crown 8vo, 5s. cloth.

"Written by a man who knows, not only how work ought to be done, but who also knows how to do it, and how to convey his knowledge to others. To all turners this book would be valuable."—*Engineering.*

"We can safely recommend the work to young engineers. To the amateur it will simply be invaluable. To the student it will convey a great deal of useful information."—*Engineer.*

"A compact, succinct, and handy guide to lathe-work did not exist in our language until Mr. Hasluck, by the publication of this treatise, gave the turner a true *vade-mecum.*"—*House Decorator.*

Metal Turning.

THE METAL TURNER'S HANDBOOK. By PAUL N. HASLUCK. With over 100 Cuts. Crown 8vo, 1s. cloth.

Iron and Steel.

"IRON AND STEEL": A Work for the Forge, Foundry, Factory, and Office. Containing ready, useful, and trustworthy Information for Ironmasters and their Stock-takers; Managers of Bar, Rail, Plate, and Sheet Rolling Mills; Iron and Metal Founders; Iron Ship and Bridge Builders; Mechanical, Mining, and Consulting Engineers; Architects, Contractors, Builders, and Professional Draughtsmen. By CHARLES HOARE, Author of "The Slide Rule," &c. Eighth Edition, Revised throughout and considerably Enlarged. With folding Scales of "Foreign Measures compared with the English Foot," and "Fixed Scales of Squares, Cubes, and Roots, Areas, Decimal Equivalents, &c." Oblong 32mo, leather, elastic band, 6s.

For comprehensiveness the book has not its equal."—*Iron.*

One of the best of the pocket books, and a useful companion in other branches of work tha iron and steel."—*English Mechanic.*

Stone-working Machinery.

STONE-WORKING MACHINERY, and the Rapid and Economical Conversion of Stone. With Hints on the Arrangement and Management of Stone Works. By M. POWIS BALE, M.I.M.E., A.M.I.C.E. With numerous Illustrations. Large crown 8vo, 9s. cloth.

"The book should be in the hands of every mason or student of stone-work."—*Colliery Guardian.*

Engineer's Reference Book.

THE WORKS' MANAGER'S HANDBOOK OF MODERN RULES, TABLES, AND DATA. For Engineers, Millwrights, and Boiler Makers; Tool Makers, Machinists, and Metal Workers; Iron and Brass Founders, &c. By W. S. HUTTON, Civil and Mechanical Engineer. Second Edition, carefully revised, with Additions. In One handsome Volume, medium 8vo, price 15s. strongly bound. [*Just published.*

Contents:

Proportions and Rules for Modern Stationary and Locomotive Engines—Lancashire, Cornish, Vertical, Return Tube, and Portable Engine Boilers—Chimneys—Water-Wheels, Pumps, Pipes, Girders, Gearing, Shafting, and Millwork—Iron Foundry Work and Brass Work—Rules and Practical Data relating to the Strength and Weight of Metals and Materials—Iron and Brass Casting—Cutting Metals—Wheel Cutting—Screw Cutting—Wheel Gearing—Rope Gearing—Pulleys—Weight of Pulleys—Weight of Spur, Bevel, and Mortice Wheels—Warming and Ventilating—Weight of Iron, Steel, Brass, and various Metals and Materials.

Also, The Indicator and Indicator Diagrams—Various Memoranda for the Foundry and Workshop—Rules for the Weight of Castings—The New Patent Law and Costs of Patents—Legal Memoranda—List of French Words for English Engineering Terms — French and English Weights and Measures—And a variety of Modern Practical Information for Civil and Mechanical Engineers.

"The volume is an exceedingly useful one, brimful with engineers' notes, memoranda, and rules, and well worthy of being on every mechanical engineer's bookshelf. . . . There is valuable information on every page."—*Mechanical World.*

"The information is precisely that likely to be required in practice. . . . The work forms a desirable addition to the library, not only of the works manager, but of anyone connected with general engineering."—*Mining Journal.*

"A formidable mass of facts and figures, readily accessible through an elaborate index . . . . Such a volume will be found absolutely necessary as a book of reference in all sorts of 'works' connected with the metal trades. . . . Any ordinary foreman or workman can find all he wants in the crowded pages of this useful work."—*Ryland's Iron Trades Circular.*

Engineering Construction.

PATTERN-MAKING: A Practical Treatise, embracing the Main Types of Engineering Construction, and including Gearing, both Hand and Machine made, Engine Work, Sheaves and Pulleys, Pipes and Columns, Screws, Machine Parts, Pumps and Cocks, the Moulding of Patterns in Loam and Greensand, &c., together with the methods of Estimating the weight of Castings; to which is added an Appendix of Tables for Workshop Reference. By a FOREMAN PATTERN MAKER. With upwards of Three Hundred and Seventy Illustrations. Crown 8vo, 7s. 6d. cloth. [*Just published.*

Smith's Tables for Mechanics, etc.

TABLES, MEMORANDA, AND CALCULATED RESULTS, FOR MECHANICS, ENGINEERS, ARCHITECTS, BUILDERS, etc. Selected and Arranged by FRANCIS SMITH. Third Edition, Revised and Enlarged, 250 pp., waistcoat-pocket size, 1s. 6d. limp leather. [*Just published.*

"It would, perhaps, be as difficult to make a small pocket-book selection of notes and formulæ to suit ALL engineers as it would be to make a universal medicine; but Mr. Smith's waistcoat-pocket collection may be looked upon as a successful attempt."—*Engineer.*

"A veritable pocket treasury of knowledge."—*Iron.*

The High-Pressure Steam Engine.

THE HIGH-PRESSURE STEAM-ENGINE: An Exposition of its Comparative Merits and an Essay towards an Improved System of Construction. By Dr. ERNST ALBAN. Translated from the German, with Notes, by Dr. POLE, M. Inst. C.E., &c. With 28 Plates. 8vo, 16s. 6d. cloth.

"Goes thoroughly into the examination of the high-pressure engine, the boiler, and its appendages, and deserves a place in every scientific library."—*Steam Shipping Chronicle.*

Steam Boilers.

A TREATISE ON STEAM BOILERS: Their Strength, Construction, and Economical Working. By ROBERT WILSON, C.E., late Inspector, Manchester Steam Users' Association. Fifth Edition 12mo, 6s. cloth.

"The best treatise that has ever been published on steam boilers."—*Engineer.*

"The author shews himself perfect master of his subject, and we heartily recommend all employing steam power to possess themselves of the work."—*Ryland's Iron Trade Circular.*

Boiler Making.

THE BOILER-MAKER'S READY RECKONER. With Examples of Practical Geometry and Templating, for the Use of Platers, Smiths and Riveters. By JOHN COURTNEY, Edited by D. K. CLARK, M.I.C.E. Second Edition, revised, with Additions, 12mo, 5s. half-bound.

"A reliable guide to the working boiler-maker."—*Iron.*

"Boiler-makers will readily recognise the value of this volume. . . . The tables are clearly printed, and so arranged that they can be referred to with the greatest facility, so that it cannot be doubted that they will be generally appreciated and much used."—*Mining Journal.*

Steam Engine.

TEXT-BOOK ON THE STEAM ENGINE. By T. M. GOODEVE, M.A., Barrister-at-Law, Author of "The Elements of Mechanism," &c. Sixth Edition. With numerous Illustrations. Crown 8vo, 6s. cloth.

"Professor Goodeve has given us a treatise on the steam engine which will bear comparison with anything written by Huxley or Maxwell, and we can award it no higher praise."—*Engineer.*

Steam.

THE SAFE USE OF STEAM. Containing Rules for Unprofessional Steam-users. By an ENGINEER. Fifth Edition. Sewed, 6d.

"If steam-users would but learn this little book by heart, boiler explosions would become sensations by their rarity."—*English Mechanic.*

Coal and Speed Tables.

A POCKET BOOK OF COAL AND SPEED TABLES, for Engineers and Steam-users. By NELSON FOLEY, Author of "Boiler Construction." Pocket-size, 3s. 6d. cloth; 4s. leather.

Contents:

Table 1. Decimals of Tons in cwts., qrs., lbs.—Table 2. Consumption per Day per I.H.P.—Table 3. Consumption per Hour per I.H.P.—Table 4. Consumption per Day per sq. ft. grate.—Table 5. Consumption per Hour per sq. ft. grate.—Table 6. Stowage of Coal and Composition.—Table 7. Knots per Hour (Measured Mile).—Table 8. Knots per Hour (Cloch and Cumbrae).—Table 9. Decimal Equivalents of Time.—Table 10. Knots per Hour in ft. per min.—Table 11. Constants for I.H.P.—Table 12. Decimals of Cwts. in lbs.—Table 13. Measures of Weight; also Miscellaneous Results—Combustion, Air required, Stoking, &c.—Time and Speed Examples.

Fire Engineering.

FIRES, FIRE-ENGINES, AND FIRE-BRIGADES. With a History of Fire-Engines, their Construction, Use, and Management; Remarks on Fire-Proof Buildings, and the Preservation of Life from Fire; Statistics of the Fire Appliances in English Towns; Foreign Fire Systems; Hints on Fire Brigades, &c. &c. By CHARLES F. T. YOUNG, C.E. With numerous Illustrations, 544 pp., demy 8vo, £1 4s. cloth.

"To such of our readers as are interested in the subject of fires and fire apparatus, we can most heartily commend this book. It is really the only English work we now have upon the subject."—*Engineering.*

"It displays much evidence of careful research; and Mr. Young has put his facts neatly together. It is evident enough that his acquaintance with the practical details of the construction of steam fire engines, old and new, and the conditions with which it is necessary they should comply, is accurate and full."—*Engineer.*

Gas Lighting.

COMMON SENSE FOR GAS-USERS: A Catechism of Gas-Lighting for Householders, Gasfitters, Millowners, Architects, Engineers, etc. By ROBERT WILSON, C.E., Author of "A Treatise on Steam Boilers." Second Edition. Crown 8vo, sewed, with Folding Plates and Wood Engravings, 2s. 6d.

"All gas-users will decidedly benefit, both in pocket and comfort, if they will avail themselves of Mr. Wilson's counsels."—*Engineering.*

THE POPULAR WORKS OF MICHAEL REYNOLDS.

(*Known as* "THE ENGINE DRIVER'S FRIEND").

Locomotive-Engine Driving.

LOCOMOTIVE-ENGINE DRIVING: A Practical Manual for Engineers in charge of Locomotive Engines. By MICHAEL REYNOLDS, Member of the Society of Engineers, formerly Locomotive Inspector L. B. and S. C. R. Sixth Edition. Including a KEY TO THE LOCOMOTIVE ENGINE. With Illustrations and Portrait of Author. Crown 8vo, 4*s.* 6*d.* cloth.

"Mr. Reynolds has supplied a want, and has supplied it well. We can confidently recommend the book, not only to the practical driver, but to everyone who takes an interest in the performance of locomotive engines."—*The Engineer.*

"Mr. Reynolds has opened a new chapter in the literature of the day. Of the practical utility of Mr. Reynolds's book we have to speak in terms of warm commendation."—*Athenæum.*

"Evidently the work of one who knows his subject thoroughly."—*Railway Service Gazette.*

"Were the cautions and rules given in the book to become part of the every-day working of our engine-drivers, we might have fewer distressing accidents to deplore."—*Scotsman.*

The Engineer, Fireman, and Engine-Boy.

THE MODEL LOCOMOTIVE ENGINEER, FIREMAN, and ENGINE-BOY. Comprising a Historical Notice of the Pioneer Locomotive Engines and their Inventors, with a project for the establishment of Certificates of Qualification in the Running Service of Railways. By MICHAEL REYNOLDS, Author of "Locomotive-Engine Driving." With numerous Illustrations and a fine Portrait of George Stephenson. Crown 8vo, 4*s.* 6*d.* cloth.

"From the technical knowledge of the author it will appeal to the railway man of to-day more forcibly than anything written by Dr. Smiles. . . . The volume contains information of a technical kind, and facts that every driver should be familiar with."—*English Mechanic.*

"We should be glad to see this book in the possession of everyone in the kingdom who has ever laid, or is to lay, hands on a locomotive engine."—*Iron.*

"Replete with information and graphic detail for the railway man."—*Railway Service Gazette.*

Stationary Engine Driving.

STATIONARY ENGINE DRIVING: A Practical Manual for Engineers in charge of Stationary Engines. By MICHAEL REYNOLDS. Second Edition, Enlarged. With Plates and Woodcuts. Crown 8vo, 4*s.* 6*d.* cloth.

"The author is thoroughly acquainted with his subjects, and his advice on the various points treated is clear and practical. . . . He has produced a manual which is an exceedingly useful one for the class for whom it is specially intended."—*Engineering.*

"Our author leaves no stone unturned. He is determined that his readers shall not only know something about the stationary engine, but all about it."—*Engineer.*

"An engineman who has mastered the contents of Mr. Reynolds's book will require but little actual experience with boilers and engines before he can be trusted to look after them."—*English Mechanic.*

"The book should be possessed by every intelligent and ambitious engineman."—*Builder.*

Continuous Railway Brakes.

CONTINUOUS RAILWAY BRAKES: A Practical Treatise on the several Systems in Use in the United Kingdom; their Construction and Performance. With copious Illustrations and numerous Tables. By MICHAEL REYNOLDS. Large crown 8vo, 9*s.* cloth.

"A popular explanation of the different brakes. It will be of great assistance in forming public opinion, and will be studied with benefit by those who take an interest in the brake. . . . It is a clear exposition of the brake question as it relates to the rival systems."—*English Mechanic.*

"Written with sufficient technical detail to enable the principle and relative connection of the various parts of each particular brake to be readily grasped."—*Mechanical World.*

"May be recommended to all who desire to study the subject of continuous brakes."—*Iron.*

Engine-Driving Life.

ENGINE-DRIVING LIFE; or, Stirring Adventures and Incidents in the Lives of Locomotive-Engine Drivers. By MICHAEL REYNOLDS. Eighth Thousand. Crown 8vo, 2*s.* cloth.

"The book from first to last is perfectly fascinating. Wilkie Collins' most thrilling conceptions are thrown into the shade by true incidents, endless in their variety, related in every page."—*North British Mail.*

"Anyone who wishes to get a real insight into railway life cannot do better than read 'Engine-Driving Life' for himself; and if he once takes it up he will find that the author's enthusiasm and real love of the engine-driving profession will carry him on till he has read every page."—*Saturday Review.*

ARCHITECTURE, BUILDING, etc.

Construction.

THE SCIENCE OF BUILDING: An Elementary Treatise on the Principles of Construction. By E. WYNDHAM TARN, M.A. Architect. Second Edition, Revised, with 58 Engravings. Crown 8vo, 7*s.* 6*d.* cloth.

"A very valuable book, which we strongly recommend to all students."—*Builder.*

"No architectural student should be without this handbook of constructional knowledge."—*Architect.*

Villa Architecture.

A HANDY BOOK OF VILLA ARCHITECTURE: Being a Series of Designs for Villa Residences in various Styles. With Outline Specifications and Estimates. By C. WICKES, Architect, Author of "The Spires and Towers of England," &c. 30 Plates, 4to, half-morocco, £1 1*s.*

*** Also an Enlarged Edition of the above. 61 Plates, with Outline Specifications, Estimates, &c. £2 2*s.* half-morocco.

"The whole of the designs bear evidence of their being the work of an artistic architect, and they will prove very valuable and suggestive."—*Building News.*

Useful Text-Book for Architects.

THE ARCHITECT'S GUIDE: Being a Text-Book of Useful Information for Architects, Engineers, Surveyors, Contractors, Clerks of Works, &c. &c. By FREDERICK ROGERS, Architect, Author of "Specifications for Practical Architecture," &c. Second Edition, Revised and Enlarged. With numerous Illustrations. Crown 8vo, 6*s.* cloth.

"As a text-book of useful information for architects, engineers, surveyors, &c., it would be hard to find a handier or more complete little volume."—*Standard.*

"A young architect could hardly have a better guide-book."—*Timber Trades Journal.*

Taylor and Cresy's Rome.

THE ARCHITECTURAL ANTIQUITIES OF ROME. By the late G. L. TAYLOR, Esq., F.R.I.B.A., and EDWARD CRESY, Esq. New Edition, thoroughly revised by the Rev. ALEXANDER TAYLOR, M.A. (son of the late G. L. Taylor, Esq.), Fellow of Queen's College, Oxford, and Chaplain of Gray's Inn. Large folio, with 130 Plates, half-bound, £3 3*s.*

N.B.—This is the only book which gives on a large scale, and with the precision of architectural measurement, the principal Monuments of Ancient Rome in plan, elevation, and detail.

"Taylor and Cresy's work has from its first publication been ranked among those professional books which cannot be bettered. . . . It would be difficult to find examples of drawings, even among those of the most painstaking students of Gothic, more thoroughly worked out than are the one hundred and thirty plates in this volume."—*Architect.*

Drawing for Builders and Students in Architecture.

PRACTICAL RULES ON DRAWING, for the Operative Builder and Young Student in Architecture. By GEORGE PYNE. With 14 Plates, 4to, 7*s.* 6*d.* boards.

Specifications for Practical Architecture.

SPECIFICATIONS FOR PRACTICAL ARCHITECTURE: A Guide to the Architect, Engineer, Surveyor, and Builder; with an Essay on the Structure and Science of Modern Buildings. By FREDERICK ROGERS, Architect. With numerous Illustrations, demy 8vo, 15*s.* cloth.

*** A volume of specifications of a practical character being greatly required, and the old standard work of Alfred Bartholomew being out of print, the author, on the basis of that work, has produced the above.—*Extract from Preface.*

The House-Owner's Estimator.

THE HOUSE-OWNER'S ESTIMATOR; or, What will it Cost to Build, Alter, or Repair? A Price Book adapted to the Use of Unprofessional People, as well as for the Architectural Surveyor and Builder. By the late JAMES D. SIMON, A.R.I.B.A. Edited and Revised by FRANCIS T. W. MILLER, A.R.I.B.A. With numerous Illustrations. Third Edition, Revised. Crown 8vo, 3*s.* 6*d.* cloth.

"In two years it will repay its cost a hundred times over."—*Field.*

"A very handy book."—*English Mechanic.*

Civil Architecture.

THE DECORATIVE PART OF CIVIL ARCHITECTURE. By Sir WILLIAM CHAMBERS, F.R.S. With Illustrations, Notes, and an Examination of Grecian Architecture, by JOSEPH GWILT, F.S.A. Edited by W. H. LEEDS. 66 Plates, 4to, 21s. cloth.

Designing, Measuring, and Valuing.

THE STUDENT'S GUIDE to the PRACTICE of MEASURING AND VALUING ARTIFICERS' WORKS. Containing Directions for taking Dimensions, Abstracting the same, and bringing the Quantities into Bill, with Tables of Constants, and copious Memoranda for the Valuation of Labour and Materials in the respective Trades of Bricklayer and Slater, Carpenter and Joiner, Painter and Glazier, Paperhanger, &c. With 8 Plates and 63 Woodcuts. Originally edited by EDWARD DOBSON, Architect. Fifth Edition, Revised, with considerable Additions on Mensuration and Construction, and a New Chapter on Dilapidations, Repairs, and Contracts, by E. WYNDHAM TARN, M.A. Crown 8vo, 9s.

"Well fulfils the promise of its title-page, and we can thoroughly recommend it to the class for whose use it has been compiled. Mr. Tarn's additions and revisions have much increased the usefulness of the work, and have especially augmented its value to students."—*Engineering.*

"The work has been carefully revised and edited by Mr. E. Wyndham Tarn, M.A., and comprises several valuable additions on construction, mensuration, dilapidations and repairs, and other matters. . . . This edition will be found the most complete treatise on the principles of measuring and valuing artificers' work that has yet been published."—*Building News.*

"Might be read and studied with advantage by almost everyone having any connection with building operations, from the architect downwards."—*Building World.*

Pocket Estimator.

THE POCKET ESTIMATOR for the BUILDING TRADES. Being an Easy Method of Estimating the various parts of a Building collectively, more especially applied to Carpenters' and Joiners' work. By A. C. BEATON, Author of "Quantities and Measurements." Third Edition, carefully revised, 33 Woodcuts, leather, waistcoat-pocket size, 1s. 6d.

"Contains a good deal of information not easily to be obtained from the ordinary price books. The prices given are accurate, and up to date."—*Building News.*

"This book is as complete as it is possible to make it."—*Building and Engineering Times.*

Builder's and Surveyor's Pocket Technical Guide.

THE POCKET TECHNICAL GUIDE AND MEASURER FOR BUILDERS AND SURVEYORS. Containing a Complete Explanation of the Terms used in Building Construction, Memoranda for Reference, Technical Directions for Measuring Work in all the Building Trades, with a Treatise on the Measurement of Timber, Complete Specifications, &c., &c. By A. C. BEATON. Second Edition, with 19 Woodcuts, leather, waistcoat-pocket size, 1s. 6d.

"An exceedingly handy pocket companion, thoroughly reliable."—*Builder's Weekly Reporter.*

"This neat little compendium contains all that is requisite in carrying out contracts for excavating, tiling, bricklaying, paving, &c."—*British Trade Journal.*

Handbook of Specifications.

THE HANDBOOK OF SPECIFICATIONS; or, Practical Guide to the Architect, Engineer, Surveyor, and Builder, in drawing up Specifications and Contracts for Works and Constructions. Illustrated by Precedents of Buildings actually executed by eminent Architects and Engineers. By Professor T. L. DONALDSON, P.R.I.B.A., &c. New Edition, in One large Vol., 8vo, with upwards of 1,000 pages of Text, and 33 Plates, £1 11s. 6d. cloth.

"In this work forty-four specifications of executed works are given, including the specifications for parts of the new Houses of Parliament, by Sir Charles Barry, and for the new Royal Exchange, by Mr. Tite, M.P. The latter, in particular, is a very complete and remarkable document. It embodies, to a great extent, as Mr. Donaldson mentions, 'the bill of quantities with the description of the works.' . . . It is valuable as a record, and more valuable still as a book of precedents. . . . Suffice it to say that Donaldson's 'Handbook of Specifications' must be bought by all architects."—*Builder.*

Boiler and Factory Chimneys.

BOILER AND FACTORY CHIMNEYS: Their Draught-power and Stability; with a Chapter on Lightning-Conductors. By ROBERT WILSON, C.E., Author of "Treatise on Steam Boilers." Cr. 8vo, 3s. 6d. cloth.

"A most valuable book of its kind, full of useful information."—*Local Government Chronicle.*

DECORATIVE ARTS, etc.

Woods and Marbles (Imitation of).

SCHOOL OF PAINTING FOR THE IMITATION OF WOODS AND MARBLES, as Taught and Practised by A. R. VAN DER BURG and P. VAN DER BURG, Directors of the Rotterdam Painting Institution. Royal folio, 18½ by 12½ in., Illustrated with 24 full-size Coloured Plates; also 12 plain Plates, comprising 154 Figures, price £2 12s. 6d.

List of Contents

Introductory Chapter — Tools required for Wood Painting—Observations on the different species of Wood: Walnut —Observations on Marble in general—Tools required for Marble Painting—St. Remi Marble: Preparation of the Paints: Process of Working—Wood Graining: Preparation of Stiff and Flat Brushes: Sketching different Grains and Knots: Glazing of Wood—Ash: Painting of Ash—Breche (Breccia) Marble: Breche Violette: Process of Working—Maple: Process of Working—The different species of White Marble: Methods of Working: Painting White Marble with Lac-dye: Painting White Marble with Poppy-paint—Mahogany: Methods of Working—Yellow Sienna Marble Process of Working—Juniper: Characteristics of the Natural Wood: Method of Imitation—Vert de Mer Marble: Description of the Marble: Process of Working—Oak: Description of the varieties of Oak: Manipulation of Oak-painting: Tools employed: Method of Working—Waulsort Marble: Varieties of the Marble: Process of Working—The Painting of Iron with Red Lead: How to make Putty: Out-door Work: Varnishing: Priming and Varnishing Woods and Marbles: Painting in General: Ceilings and Walls: Gilding: Transparencies, Flags, &c.

List of Plates.

1. Various Tools required for Wood Painting—2, 3. Walnut: Preliminary Stages of Graining and Finished Specimen—4. Tools used for Marble Painting and Method of Manipulation—5, 6. St. Remi Marble: Earlier Operations and Finished Specimen—7. Methods of Sketching different Grains, Knots, &c.—8, 9. Ash: Preliminary Stages and Finished Specimen—10. Methods of Sketching Marble Grains—11, 12. Breche Marble: Preliminary Stages of Working and Finished Specimen—13. Maple: Methods of Producing the different Grains—14, 15. Bird's-eye Maple: Preliminary Stages and Finished Specimen—16. Methods of Sketching the different Species of White Marble—17, 18. White Marble: Preliminary Stages of Process and Finished Specimen—19. Mahogany: Specimens of various Grains and Methods of Manipulation—20, 21. Mahogany: Earlier Stages and Finished Specimen—22, 23, 24. Sienna Marble: Varieties of Grain, Preliminary Stages and Finished Specimen—25, 26, 27. Juniper Wood: Methods of producing Grain, &c.: Preliminary Stages and Finished Specimen—28, 29, 30. Vert de Mer Marble: Varieties of Grain and Methods of Working Unfinished and Finished Specimens—31, 32, 33. Oak: Varieties of Grain, Tools Employed, and Methods of Manipulation, Preliminary Stages and Finished Specimen—34, 35, 36. Waulsort Marble: Varieties of Grain, Unfinished and Finished Specimens.

"A comprehensive guide to the art. The explanations of the processes, the manipulation and management of the colours, and the beautifully executed plates will not be the least valuable to the student who aims at making his work a faithful transcript of nature."—*Building News.*

"Students and novices are fortunate who are able to become the possessors of *so noble a work.*"—*Architect.*

Colour.

A GRAMMAR OF COLOURING. Applied to Decorative Painting and the Arts. By GEORGE FIELD. New Edition, adapted to the use of the Ornamental Painter and Designer. By ELLIS A. DAVIDSON. With New Coloured Diagrams and Engravings. 12mo, 3s. 6d. cloth boards.

"The book is a most useful *resume* of the properties of pigments."—*Builder.*

Glass Painting.

GLASS STAINING AND THE ART OF PAINTING ON GLASS. From the German of Dr. GESSERT and EMANUEL OTTO FROMBERG. With an Appendix on THE ART OF ENAMELLING. 12mo, 2s. 6d. cloth limp

Decoration.

ELEMENTARY DECORATION. A Guide to the Simpler Forms of Everyday Art, as applied to the Interior and Exterior Decoration of Dwelling Houses, &c. By JAMES W. FACEY, Jun. With 68 Cuts. 12mo, 2s. cloth limp.

"As a technical guide-book to the decorative painter it will be found reliable."—*Building News.*

House Painting, etc.

HOUSE PAINTING, GRAINING, MARBLING, AND SIGN WRITING, A Practical Manual of. By ELLIS A. DAVIDSON. Fourth Edition. With Coloured Plates and Wood Engravings. 12mo, 6s. cloth boards.

"A mass of information, of use to the amateur and of value to the practical man."—*English Mechanic.*

"Simply invaluable to the youngster entering upon this particular calling, and highly serviceable to the man who is practising it."—*Furniture Gazette.*

DELAMOTTE'S WORKS on ILLUMINATION & ALPHABETS.

A PRIMER OF THE ART OF ILLUMINATION, for the Use of Beginners: with a Rudimentary Treatise on the Art, Practical Directions for its exercise, and numerous Examples taken from Illuminated MSS., printed in Gold and Colours. By F. DELAMOTTE. Small 4to, 9s. elegantly bound, cloth antique.

". . . . The examples of ancient MSS. recommended to the student, which, with much good sense, the author chooses from collections accessible to all, are selected with judgment and knowledge, as well as taste."—*Athenæum.*

ORNAMENTAL ALPHABETS, Ancient and Mediæval, from the Eighth Century, with Numerals; including Gothic, Church-Text, large and small, German, Italian, Arabesque, Initials for Illumination, Monograms, Crosses, &c., &c., for the use of Architectural and Engineering Draughtsmen, Missal Painters, Masons, Decorative Painters, Lithographers, Engravers, Carvers, &c., &c. Collected and Engraved by F. DELAMOTTE, and printed in Colours. New and Cheaper Edition. Royal 8vo, oblong, 2s. 6d. ornamental boards.

"For those who insert enamelled sentences round gilded chalices, who blazon shop legends over shop-doors, who letter church walls with pithy sentences from the Decalogue, this book will be useful."—*Athenæum.*

EXAMPLES OF MODERN ALPHABETS, Plain and Ornamental; including German, Old English, Saxon, Italic, Perspective, Greek, Hebrew, Court Hand, Engrossing, Tuscan, Riband, Gothic, Rustic, and Arabesque; with several Original Designs, and an Analysis of the Roman and Old English Alphabets, large and small, and Numerals, for the use of Draughtsmen, Surveyors, Masons, Decorative Painters, Lithographers, Engravers, Carvers, &c. Collected and Engraved by F. DELAMOTTE, and printed in Colours. New and Cheaper Edition. Royal 8vo, oblong, 2s. 6d. ornamental boards.

"There is comprised in it every possible shape into which the letters of the alphabet and numerals can be formed, and the talent which has been expended in the conception of the various plain and ornamental letters is wonderful."—*Standard.*

"This volume has borne the test of practical use, and is invariably spoken well of by those who possess it."—*Jeweller and Metal-worker.*

MEDIÆVAL ALPHABETS AND INITIALS FOR ILLUMINATORS. By F. DELAMOTTE, Illuminator, Designer, and Engraver on Wood. Containing 21 Plates and Illuminated Title, printed in Gold and Colours. With an Introduction by J. WILLIS BROOKS. Small 4to, 6s. cloth gilt.

"A volume in which the letters of the alphabet come forth glorified in gilding and all the colours of the prism interwoven and intertwined and intermingled."—*Sun.*

THE EMBROIDERER'S BOOK OF DESIGN. Containing Initials, Emblems, Cyphers, Monograms, Ornamental Borders, Ecclesiastical Devices, Mediæval and Modern Alphabets, and National Emblems. Collected by F. DELAMOTTE, and printed in Colours. Oblong royal 8vo, 1s. 6d., ornamental wrapper.

"The designs are practical—not designs that cannot be done in embroidery; and the book will no doubt be found exceedingly useful in the lady's work-room and in schools."—*Newcastle Courant.*

"The book will be of great assistance to ladies and young children who are endowed with the art of plying the needle in this most ornamental and useful pretty work."—*East Anglian Times.*

Wood Carving.

INSTRUCTIONS IN WOOD-CARVING, for Amateurs; with Hints on Design. By A LADY. With Ten large Plates, 2s. 6d. in emblematic wrapper.

"The handicraft of the wood-carver, so well as a book can impart it, may be learnt from 'A Lady's' publication."—*Athenæum.*

"The directions given are plain and easily understood."—*English Mechanic.*

Letter Painting.

THE ART OF LETTER PAINTING MADE EASY. By JAMES GREIG BADENOCH. With 12 full-page Engravings of Examples, 1s. cloth limp.

"The system is a simple one, but quite original, and well worth the careful attention of letter painters. It can be easily mastered and remembered."—*Building News.*

CARPENTRY, TIMBER, etc.

Tredgold's Carpentry, Enlarged by E. W. Tarn.

THE ELEMENTARY PRINCIPLES OF CARPENTRY. A Treatise on the Pressure and Equilibrium of Timber Framing, the Resistance of Timber, and the Construction of Floors, Arches, Bridges, Roofs, Uniting Iron and Stone with Timber, &c. To which is added an Essay on the Nature and Properties of Timber, &c., with Descriptions of the kinds of Wood used in Building; also numerous Tables of the Scantlings of Timber for different purposes, the Specific Gravities of Materials, &c. By THOMAS TREDGOLD, C.E. With an Appendix of Specimens of Various Roofs of Iron and Stone, Illustrated. Sixth Edition, thoroughly revised and considerably enlarged by E. WYNDHAM TARN, M.A., Author of "The Science of Building," &c. With 61 Plates, Portrait of the Author, and several Woodcuts. In one large vol., 4to, price £1 5s. cloth. [*Just published.*

"Ought to be in every architect's and every builder's library."—*Builder.*

"Tredgold's 'Elementary Principles of Carpentry' is, without doubt, the standard English authority upon the subject. Mr. Tarn, by supplementing the text of the author, adds much to the work, and makes it an indispensable addition to the library of the student, the architect, and the engineer."—*Building News.*

Woodworking Machinery.

WOODWORKING MACHINERY: Its Rise, Progress, and Construction. With Hints on the Management of Saw Mills and the Economical Conversion of Timber. Illustrated with Examples of Recent Designs by leading English, French, and American Engineers. By M. POWIS BALE, A.M. Inst. C.E., M.I.M.E. Large crown 8vo, 12s. 6d. cloth.

"Mr. Bale is evidently an expert on the subject, and he has collected so much information that his book is all-sufficient for builders and others engaged in the conversion of timber."—*Architect.*

"The most comprehensive compendium of wood-working machinery we have seen. The author is a thorough master of his subject."—*Building News.*

"Converters of timber, wholesale joiners, the members generally of the building trades, architects, and owners of and dealers in timber, owe Mr. Bale tribute to the value of the cost of his book. It is sound from end to end, well designed and well executed."—*Iron.*

"The appearance of this book at the present time will, we should think, give a considerable impetus to the onward march of the machinist engaged in the designing and manufacture of wood-working machines. It should be in the office of every wood-working factory."—*English Mechanic.*

Saw Mills.

SAW MILLS: Their Arrangement and Management, and the Economical Conversion of Timber. (Being a Companion Volume to "Woodworking Machinery.") By M. POWIS BALE, A.M. Inst. C.E., M.I.M.E. With numerous Illustrations. Crown 8vo, 10s. 6d. cloth.

"The author is favourably known by his former work on 'Woodworking Machinery,' of which we were able to speak approvingly. This is a companion volume, in which the *administration* of a large sawing establishment is discussed, and the subject examined from a financial standpoint. Hence the size, shape, order, and disposition of saw-mills and the like are gone into in detail, and the course of the timber is traced from its reception to its delivery in its converted state. We could not desire a more complete or practical treatise."—*Builder.*

"We highly recommend Mr. Bale's work to the attention and perusal of all those who are engaged in the art of wood conversion, or who are about building or remodelling saw-mills on improved principles."—*Building News.*

"Will be found of much value by that special class of readers for whose information it is designed. We recommend the book to those about to construct or to manage saw-mills."—*Athenæum.*

Underwood and Woodland Tables.

TABLES FOR PLANTING AND VALUING UNDERWOOD AND WOODLAND; also Lineal, Superficial, Cubical, Wages, Marketing, and Decimal Tables. Together with Tables for Converting Land-measure from one denomination to another, and Instructions for Measuring Round Timber. By RICHARD HORTON. 12mo, 2s. strongly bound in leather.

Carpentering.

THE CARPENTER'S NEW GUIDE; or, Book of Lines for Carpenters; comprising all the Elementary Principles essential for acquiring a knowledge of Carpentry. Founded on the late PETER NICHOLSON'S Standard Work. A New Edition, revised by ARTHUR ASHPITEL, F.S.A. Together with Practical Rules on Drawing, by GEORGE PYNE. With 74 Plates, 4to, £1 1s. cloth.

Timber Merchant's Companion.

THE TIMBER MERCHANT'S AND BUILDER'S COMPANION. Containing New and Copious Tables of the Reduced Weight and Measurement of Deals and Battens, of all sizes, from One to a Thousand Pieces, and the relative Price that each size bears per Lineal Foot to any given Price per Petersburg Standard Hundred; the Price per Cube Foot of Square Timber to any given Price per Load of 50 Feet; the proportionate Value of Deals and Battens by the Standard, to Square Timber by the Load of 50 Feet; the readiest mode of ascertaining the Price of Scantling per Lineal Foot of any size, to any given Figure per Cube Foot. Also a variety of other valuable information. By WILLIAM DOWSING, Timber Merchant. Third Edition, Revised and Corrected. Crown 8vo, 3s., cloth.

"Everything is as concise and clear as it can possibly be made. There can be no doubt that every timber merchant and builder ought to possess it."—*Hull Advertiser.*

"We are glad to see a third edition of these admirable tables, which for correctness and simplicity of arrangement leave nothing to be desired."—*Timber Trades Journal.*

Practical Timber Merchant.

THE PRACTICAL TIMBER MERCHANT. Being a Guide for the use of Building Contractors, Surveyors, Builders, &c., comprising useful Tables for all purposes connected with the Timber Trade, Marks of Wood, Essay on the Strength of Timber, Remarks on the Growth of Timber, &c. By W. RICHARDSON. Fcap. 8vo, 3s. 6d. cloth.

"This handy manual contains much valuable information for the use of timber merchants, builders, foresters, and all others connected with the growth, sale, and manufacture of timber."—*Journal of Forestry.*

"To timber merchants or users this compact treatise will be found very useful."—*Illustrated Carpenter.*

Timber Freight Book.

THE TIMBER MERCHANT'S, SAW MILLER'S, AND IMPORTER'S FREIGHT BOOK AND ASSISTANT. Comprising Rules, Tables, and Memoranda relating to the Timber Trade. By WILLIAM RICHARDSON, Timber Broker; together with a Chapter on "Speeds of Saw Mill Machinery," by M. POWIS BALE, M.I.M.E., &c. 12mo, 3s. 6d. cloth.

Tables for Packing-Case Makers.

PACKING-CASE TABLES; showing the number of Superficial Feet in Boxes or Packing-Cases, from six inches square and upwards. By W. RICHARDSON, Timber Broker. Second Edition. Oblong 4to, 3s. 6d. cloth.

"Will save much labour and calculation to makers and users of packing-cases."—*Grocer.*

"Invaluable labour-saving tables."—*Ironmonger*

Superficial Measurement.

THE TRADESMAN'S GUIDE TO SUPERFICIAL MEASUREMENT. Tables calculated from 1 to 200 inches in length, by 1 to 108 inches in breadth. For the use of Architects, Surveyors, Engineers, Timber Merchants, Builders, &c. By JAMES HAWKINGS. Third Edition. Fcap., 3s. 6d. cloth.

Forestry.

THE ELEMENTS OF FORESTRY. Designed to afford Information concerning the Planting and Care of Forest Trees for Ornament or Profit, with Suggestions upon the Creation and Care of Woodlands. By F. B. HOUGH. Large crown 8vo, 10s. cloth.

Timber Importer's Guide.

THE TIMBER IMPORTER'S, TIMBER MERCHANT'S AND BUILDER'S STANDARD GUIDE. By RICHARD E. GRANDY. Comprising an Analysis of Deal Standards, Home and Foreign, with Comparative Values and Tabular Arrangements for fixing Nett Landed Cost on Baltic and North American Deals, including all intermediate Expenses, Freight, Insurance, &c., &c.; together with copious Information for the Retailer and Builder. Second Edition, carefully revised and corrected. 12mo, 3s. 6d. cloth boards.

"Everything it pretends to be: built up gradually, it leads one from a forest to a treenail, and throws in, as a makeweight, a host of material concerning bricks, columns, cisterns, &c."—*English Mechanic.*

MINING AND MINING INDUSTRIES.

Mining in the United Kingdom.

BRITISH MINING: A Treatise on the History, Discovery, Practical Development, and Future Prospects of Metalliferous Mines in the United Kingdom. By ROBERT HUNT, F.R.S., Keeper of Mining Records; Editor of "Ure's Dictionary of Arts, Manufactures, and Mines," &c. Upwards of 950 pp., with 230 Illustrations. Super-royal 8vo, £3 3s. cloth.

Synopsis of Contents:

BOOK I. HISTORICAL SKETCH OF BRITISH MINING:—Chap. I. Mining Previous to the Roman Invasion.—II. Mining during the Roman Occupation.—III. Mining to the Eighteenth Century.—IV. Mining for Tin and Copper to the End of the Eighteenth Century.—V. Mining for Lead, Silver, &c., to the End of the Eighteenth Century.—VI. Gold, Plumbago, Iron Ore, and Sundries to the End of the Eighteenth Century

BOOK II. ON THE FORMATION OF METALLIFEROUS DEPOSITS.—Chap. I. The Rocks of Mining Districts, and the Distribution of Metalliferous Deposits.—II. Mechanics of Mineral Lodes, Faults, Cross Courses, &c.—III. The Laws relating to Mineral Deposits.—IV. Remarkable Phenomena observed in Metalliferous Ore Deposits.

BOOK III. PRACTICAL MINING.—Chap. I. Discovery of Mineral Lodes and the Opening of Mines.—II. Practical Operations for the Extraction of Metalliferous Ores.—III. Ventilation and Drainage of Mines, &c.—IV. Dressing Metalliferous Ores; Preparation for Smelter.—V. Discovery and Extraction of Iron Ores from Veins and other Deposits.

BOOK IV. THE FUTURE PROSPECTS OF BRITISH MINING.—Chap. I. Summary, Examination of the Probable Exhaustion of Metalliferous Minerals.—II. On the Limits of the Metalliferous Zone.—III. The Occurrence of Ores at Great Depths or in New Districts.—IV. Improvements and Economy in Working British Mines.—V. General Summary and Conclusion—Appendix—Glossary of Terms.

"One of the most valuable works of reference of modern times. Mr. Hunt, as keeper of mining records of the United Kingdom, has had opportunities for such a task not enjoyed by anyone else, and has evidently made the most of them. . . . The language and style adopted are good, and the treatment of the various subjects laborious, conscientious, and scientific."—*Engineering.*

"A mass of information not elsewhere available, and of the greatest value to those who may be interested in our great mineral industries."—*Engineer.*

"A sound, business-like collection of interesting facts. . . . The amount of information Mr. Hunt has brought together is enormous. . . . The volume appears likely to convey more instruction upon the subject than any work hitherto published."—*Mining Journal.*

"The work will be for the mining industry what Dr. Percy's celebrated treatise has been for the metallurgical—a book that cannot with advantage be omitted from the library."—*Iron and Coal Trades Review.*

"Probably no one in this country was better qualified than Mr. Hunt for undertaking such a work. Brought into frequent and close association during a long life-time with the principal guardians of our mineral and metallurgical industries, he enjoyed a position exceptionally favourable for collecting the necessary information. The use which he has made of his opportunities is sufficiently attested by the dense mass of information crowded into the handsome volume which has just been published. . . . In placing before the reader a sketch of the present position of British Mining, Mr. Hunt treats his subject so fully and illustrates it so amply that this section really forms a little treatise on practical mining. . . . The book is, in fact, a treasure-house of statistical information on mining subjects, and we know of no other work embodying so great a mass of matter of this kind. Were this the only merit of Mr. Hunt's volume it would be sufficient to render it indispensable in the library of everyone interested in the development of the mining and metallurgical industries of this country."—*Athenæum.*

"The literature of mining has hitherto possessed no work approaching in importance to that which has just been published. There is much in Mr. Hunt's valuable work that every shareholder in a mine should read with close attention. The entire subject of practical mining—from the first search for the lode to the latest stages of dressing the ore—is dealt with in a masterly manner."—*Academy.*

Coal and Iron.

THE COAL AND IRON INDUSTRIES OF THE UNITED KINGDOM. Comprising a Description of the Coal Fields, and of the Principal Seams of Coal, with Returns of their Produce and its Distribution, and Analyses of Special Varieties. Also an Account of the occurrence of Iron Ores in Veins or Seams; Analyses of each Variety; and a History of the Rise and Progress of Pig Iron Manufacture since the year 1740, exhibiting the Economies introduced in the Blast Furnaces for its Production and Improvement. By RICHARD MEADE, Assistant Keeper of Mining Records. With Maps of the Coal Fields and Ironstone Deposits of the United Kingdom. 8vo, £1 8s. cloth.

"The book is one which must find a place on the shelves of all interested in coal and iron production, and in the iron, steel, and other metallurgical industries."—*Engineer.*

"Of this book we may unreservedly say that it is the best of its class which we have ever met. . . . A book of reference which no one engaged in the iron or coal trades should omit from his library."—*Iron and Coal Trades' Review.*

"An exhaustive treatise and a valuable work of reference. . . . The amount of labour which the preparation of the work has involved must have been enormous."—*Mining Journal.*

Metalliferous Minerals and Mining.

TREATISE ON METALLIFEROUS MINERALS AND MINING. By D. C. DAVIES, F.G.S., Mining Engineer, &c., Author of "A Treatise on Slate and Slate Quarrying." Illustrated with numerous Wood Engravings. Second Edition, carefully Revised. Crown 8vo, 12s. 6d. cloth.

"Neither the practical miner nor the general reader interested in mines, can have a better book for his companion and his guide."—*Mining Journal.*

"A book that will not only be useful to the geologist, the practical miner, and the metallurgist; but also very interesting to the general public."—*Iron.*

"As a history of the present state of mining throughout the world this book has a real value, and it supplies an actual want, for no such information has hitherto been brought together within such limited space."—*Athenæum.*

Earthy Minerals and Mining.

A TREATISE ON EARTHY AND OTHER MINERALS AND MINING. By D. C. DAVIES, F.G.S. Uniform with, and forming a Companion Volume to, the same Author's "Metalliferous Minerals and Mining." With 76 Wood Engravings. Crown 8vo, 12s. 6d. cloth.

Summary of Contents :

PART I. Silica—Alumina—Lime—Magnesia—Glucina—Zirconia—Thoria—with some of their combinations. PART II. Chloride of Sodium (Common Salt)—Nitrate of Soda—Borax—Baryta—Gypsum—Alum Shale—Phosphate of Lime. PART III. Carbon and Compounds of Carbon—Sulphur. PART IV. Arsenic—Cobalt—Molybdenum—Antimony and Manganese—Classified List of Mineral Substances.

"It is essentially a practical work, intended primarily for the use of practical men. . . . We do not remember to have met with any English work on mining matters that contains the same amount of information packed in equally convenient form."—*Academy.*

Underground Pumping Machinery.

MINE DRAINAGE. Being a Complete and Practical Treatise on Direct-Acting Underground Steam Pumping Machinery, with a Description of a large number of the best known Engines, their General Utility and the Special Sphere of their Action, the Mode of their Application, and their merits compared with other forms of Pumping Machinery. By STEPHEN MICHELL. 8vo, 15s. cloth.

"Will be highly esteemed by colliery owners and lessees, mining engineers, and students generally who require to be acquainted with the best means of securing the drainage of mines. It is a most valuable work, and stands almost alone in the literature of steam pumping machinery."—*Colliery Guardian.*

Mining Tools.

A MANUAL OF MINING TOOLS. For the Use of Mine Managers, Agents, Students, &c. By WILLIAM MORGANS, Lecturer on Practical Mining at the Bristol School of Mines. 12mo, 3s. cloth boards.

ATLAS OF ENGRAVINGS to Illustrate the above, containing 235 Illustrations of Mining Tools, drawn to scale. 4to, 6s. cloth boards.

"Students in the science of mining, and overmen, captains, managers, and viewers may gain practical knowledge and useful hints by the study of Mr. Morgans' manual."—*Colliery Guardian.*

Coal Mining.

COAL AND COAL MINING: A Rudimentary Treatise on. By WARINGTON W. SMYTH, M.A., F.R.S., &c., Chief Inspector of the Mines of the Crown. New Edition, Revised and Corrected. With numerous Illustrations. 12mo, 4s. cloth boards.

"As an outline is given of every known coal-field in this and other countries, as well as of the principal methods of working, the book will doubtless interest a very large number of readers."—*Mining Journal.*

Mining Surveying.

THE MINERAL SURVEYOR AND VALUER'S COMPLETE GUIDE. By WILLIAM LINTERN, Mining and Civil Engineer. With Four Plates of Diagrams, Plans, &c. 12mo, 4s. cloth boards.

"Contains much valuable information given in a small compass, and which, as far as we have tested it, is thoroughly trustworthy."—*Iron and Coal Trades Review.*

Subterraneous Surveying.

SUBTERRANEOUS SURVEYING, Elementary and Practical Treatise on; with and without the Magnetic Needle. By THOMAS FENWICK, Surveyor of Mines, and THOMAS BAKER, C.E. Illustrated. 12mo, 3s. cloth.

NAVAL ARCHITECTURE, NAVIGATION, etc.

Pocket-Book for Naval Architects and Shipbuilders.

THE NAVAL ARCHITECT'S AND SHIPBUILDER'S POCKET-BOOK of Formulæ, Rules, and Tables, and Marine Engineer's and Surveyor's Handy Book of Reference. By CLEMENT MACKROW, Member of the Institution of Naval Architects, Naval Draughtsman. Third Edition, Revised. With numerous Diagrams, &c. Fcap., 12s. 6d. strongly bound in leather.

"Should be used by all who are engaged in the construction or design of vessels. . . . Will be found to contain the most useful tables and formulæ required by shipbuilders, carefully collected from the best authorities, and put together in a popular and simple form."—*Engineer.*

"The professional shipbuilder has now, in a convenient and accessible form, reliable data for solving many of the numerous problems that present themselves in the course of his work."—*Iron.*

"There is scarcely a subject on which a naval architect or shipbuilder can require to refresh his memory which will not be found within the covers of Mr. Mackrow's book."—*English Mechanic.*

Pocket-Book for Marine Engineers.

A POCKET-BOOK OF USEFUL TABLES AND FORMULÆ FOR MARINE ENGINEERS. By FRANK PROCTOR, A.I.N.A. Third Edition. Royal 32mo, leather, gilt edges, with strap, 4s.

"We recommend it to our readers as going far to supply a long-felt want."—*Naval Science.*

"A most useful companion to all marine engineers."—*United Service Gazette.*

Grantham's Iron Shipbuilding.

ON IRON SHIPBUILDING. With Practical Examples and Details. By JOHN GRANTHAM, M. Inst. C.E., &c. Fifth Edition. Imp. 4to, boards, enlarged to 40 Plates, including the latest Examples. Together with separate Text, also considerably enlarged. 12mo, cloth limp, price £2 2s. complete.

"Mr. Grantham's work is of great interest. . . . It is also valuable as a record of the progress of iron shipbuilding. . . . It will, we are confident, command an extensive circulation among shipbuilders in general. . . . The text-book on which the examination in iron shipbuilding of candidates for promotion in the dockyards will be mainly based."—*Engineering.*

Lighthouses.

EUROPEAN LIGHTHOUSE SYSTEMS. Being a Report of a Tour of Inspection made in 1873. By Major GEORGE H. ELLIOT, Corps of Engineers, U.S.A. Illustrated by 51 Engravings and 31 Woodcuts. 8vo, 21s. cloth.

Navigation (Practical), with Tables.

PRACTICAL NAVIGATION. Consisting of the Sailor's Sea-Book, by JAMES GREENWOOD and W. H. ROSSER; together with the requisite Mathematical and Nautical Tables for the Working of the Problems, by HENRY LAW, C.E., and Professor J. R. YOUNG. Illustrated. 12mo, 7s. strongly half-bound.

Storms.

STORMS: Their Nature, Classification, and Laws; with the Means of Predicting them by their Embodiments, the Clouds. By WILLIAM BLASIUS. With Coloured Plates and numerous Wood Engravings. Crown 8vo, 10s. 6d. cloth.

"A very readable book. . . . The fresh facts contained in its pages, collected with evident care, form a useful repository to meteorologists in the study of atmospherical disturbances. . . . The book will repay perusal as being the production of one who gives evidence of acute observation."—*Nature.*

The following books on NAVAL ARCHITECTURE, etc., *are published in* WEALE'S RUDIMENTARY SERIES.

MASTING, MAST-MAKING, AND RIGGING OF SHIPS. By ROBERT KIPPING, N.A. Fourteenth Edition. 12mo, 2s. 6d. cloth boards.

SAILS AND SAIL-MAKING. Tenth Edition, Enlarged, with an Appendix. By ROBERT KIPPING, N.A. Illustrated. 12mo, 3s. cloth boards.

NAVAL ARCHITECTURE. By JAMES PEAKE. Fourth Edition, with Plates and Diagrams. 12mo, 4s. cloth boards.

NATURAL PHILOSOPHY AND SCIENCE.

Electricity.

A MANUAL OF ELECTRICITY: Including Galvanism, Magnetism, Dia-Magnetism, Electro-Dynamics, Magno-Electricity, and the Electric Telegraph. By HENRY M. NOAD, Ph.D., F.R.S., F.C.S. Fourth Edition. With 500 Woodcuts. 8vo, £1 4*s.* cloth.

"The accounts given of electricity and galvanism are not only complete in a scientific sense, but, which is a rarer thing, are popular and interesting."—*Lancet.*

"Among the numerous writers on the attractive and fascinating subject of electricity, the author of the present volume has occupied our best attention. It is worthy of a place in the library of every public institution, and we have no doubt it will be deservedly patronised by the scientific community."—*Mining Journal.*

Text Book of Electricity.

THE STUDENT'S TEXT-BOOK OF ELECTRICITY. By HENRY M. NOAD, Ph.D., F.R.S., F.C.S. New Edition, carefully Revised. With an Introduction and Additional Chapters, by W. H. PREECE, M.I.C.E., Vice-President of the Society of Telegraph Engineers, &c. With 470 Illustrations. Crown 8vo, 12*s.* 6*d.* cloth.

"The original plan of this book has been carefully adhered to so as to make it a reflex of the existing state of electrical science, adapted for students. . . . Discovery seems to have progressed with marvellous strides; nevertheless it has now apparently ceased, and practical applications have commenced their career; and it is to give a faithful account of these that this fresh edition of Dr. Noad's valuable text-book is launched forth."—*Extract from Introduction by W. H. Preece, Esq.*

"We can recommend Dr. Noad's book for clear style, great range of subject, a good index, and a plethora of woodcuts. Such collections as the present are indispensable."—*Athenæum.*

"An admirable text-book for every student—beginner or advanced—of electricity."—*Engineering.*

"A most elaborate compilation of the facts of electricity and magnetism."—*Popular Science Review.*

"Dr. Noad's text-book has earned for itself the reputation of a truly scientific manual for the student of electricity, and we gladly hail this new amended edition, which brings it once more to the front. Mr. Preece as reviser, with the assistance of Mr. H. R. Kempe and Mr. J. P. Edwards, has added all the practical results of recent invention and research to the admirable theoretical expositions of the author, so that the book is about as complete and advanced as it is possible for any book to be within the limits of a text-book."—*Telegraphic Journal.*

Dr. Lardner's School Handbooks.

NATURAL PHILOSOPHY FOR SCHOOLS. By Dr. LARDNER. 328 Illustrations. Sixth Edition. One Vol., 3*s.* 6*d.* cloth.

"A very convenient class-book for junior students in private schools. It is intended to convey, in clear and precise terms, general notions of all the principal divisions of Physical Science."—*British Quarterly Review.*

ANIMAL PHYSIOLOGY FOR SCHOOLS. By Dr. LARDNER. With 190 Illustrations. Second Edition. One Vol., 3*s.* 6*d.* cloth.

"Clearly written, well arranged, and excellently illustrated."—*Gardener's Chronicle.*

Dr. Lardner's Electric Telegraph.

THE ELECTRIC TELEGRAPH. By Dr. LARDNER. New Edition. Revised and Re-written by E. B. BRIGHT, F.R.A.S. 140 Illustrations. Small 8vo, 2*s.* 6*d.* cloth.

"One of the most readable books extant on the Electric Telegraph."—*English Mechanic.*

Field Fortification.

A TREATISE ON FIELD FORTIFICATION, THE ATTACK OF FORTRESSES, MILITARY MINING, AND RECONNOITRING. By Colonel I. S. MACAULAY, late Professor of Fortification in the R.M.A., Woolwich. Sixth Edition, crown 8vo, cloth, with separate Atlas of 12 Plates, 12*s.* complete.

Lightning.

THE ACTION OF LIGHTNING, and the Means of Defending Life and Property from its Effects. By Major A. PARNELL, R.E. 12mo, 7*s.* 6*d.* cloth.

"Major Parnell has written an original work on a scientific subject of unusual interest; and he has prefaced his arguments by a patient and almost exhaustive citation of the best writers on the subject in the English language."—*Athenæum.*

Geology and Genesis.

THE TWIN RECORDS OF CREATION; or, Geology and Genesis: their Perfect Harmony and Wonderful Concord. By GEORGE W. VICTOR LE VAUX. Numerous Illustrations. Fcap. 8vo, 5s. cloth.

"A valuable contribution to the evidences of revelation, and disposes very conclusively of the arguments of those who would set God's Works against God's Word. No real difficulty is shirked, and no sophistry is left unexposed."—*The Rock.*

"The remarkable peculiarity of this author is that he combines an unbounded admiration of science with an unbounded admiration of the Written Record. The two impulses are balanced to a nicety; and the consequence is that difficulties which to minds less evenly poised would be serious find immediate solutions of the happiest kinds."—*London Review.*

The Blowpipe.

THE BLOWPIPE IN CHEMISTRY, MINERALOGY, AND GEOLOGY. Containing all known Methods of Anhydrous Analysis, many Working Examples, and Instructions for Making Apparatus. By Lieut.-Colonel W. A. ROSS, R.A., F.G.S. With 120 Illustrations. Crown 8vo, 3s. 6d. cloth.

Contents:

CHAP. I. Manufacture of Blowing Apparatus. II. Construction of Pyrological Lamps.—III. Supports and Apparatus.—IV. On Auxiliary Apparatus.—V. On the Structure and Management of Pyrocones.—VI. On Pyrological Reagents.—VII. On Reagents and Simple Mineral Analyses.—VIII. First Operations: Aluminium-plate Reactions of Metals and Alloys.—IX. On Pyrological Mineralogy and a New Specific Gravitometer.—X. Inner Calcium-pyroborate Balls, "Chemical Water," &c.—XI. Rationale of Outer and Inner Ball Formation in Boric Acid before the Blowpipe.—XII. Frieberg Qualitative Examples.

"The student who goes conscientiously through the course of experimentation here laid down will gain a better insight into inorganic chemistry and mineralogy than if he had 'got up' any of the best text-books of the day, and passed any number of examinations in their contents."—*Chemical News.*

The Military Sciences.

AIDE-MEMOIRE TO THE MILITARY SCIENCES. Framed from Contributions of Officers and others connected with the different Services. Originally edited by a Committee of the Corps of Royal Engineers. Second Edition, most carefully revised by an Officer of the Corps, with many Additions; containing nearly 350 Engravings and many hundred Woodcuts. Three Vols., royal 8vo, extra cloth boards, and lettered, £4 10s.

"A compendious encyclopædia of military knowledge, to which we are greatly indebted."—*Edinburgh Review.*

"The most comprehensive work of reference to the military and collateral sciences.'—*Volunteer Service Gazette.*

Astronomy.

ASTRONOMY. By the late Rev. ROBERT MAIN, M.A., F.R.S., formerly Radcliffe Observer at Oxford. Third Edition, Revised and Corrected to the present time, by WILLIAM THYNNE LYNN, B.A., F.R.A.S., formerly of the Royal Observatory, Greenwich. 12mo, 2s. cloth limp.

"A sound and simple treatise, very carefully edited, and a capital book for beginners."—*Knowledge.*

"The present edition seems to have been carefully and accurately brought down to the requirements of the present time by Mr. Lynn."—*Educational Times.*

Geology.

GEOLOGY, Physical and Historical. Consisting of "Physical Geology," which sets forth the leading Principles of the Science; and "Historical Geology," which treats of the Mineral and Organic Conditions of the Earth at each successive epoch, especial reference being made to the British Series of Rocks. By RALPH TATE, A.L.S., F.G.S., &c., &c. With over 250 Illustrations. 12mo, 5s. cloth boards.

"The fulness of the matter has elevated the book into a manual. Its information is exhaustive and well arranged."—*School Board Chronicle.*

Conchology.

MANUAL OF THE MOLLUSCA: A Treatise on Recent and Fossil Shells. By Dr. S. P. WOODWARD, A.L.S. With Appendix by RALPH TATE, A.L.S., F.G.S. With numerous Plates and 300 Woodcuts. Handsomely bound in cloth boards, 7s. 6d.

"A most valuable storehouse of conchological and geological information."—*Hardwicke's Science Gossip.*

Dr. LARDNER'S HANDBOOKS of NATURAL PHILOSOPHY.

**** The following five volumes, though each is complete in itself, and to be purchased separately, form* A Complete Course of Natural Philosophy. *The style is studiously popular. It has been the author's aim to supply Manuals for the Student, the Engineer, the Artisan, and the superior classes in Schools.*

THE HANDBOOK OF MECHANICS. Enlarged and almost re-written by Benjamin Loewy, F.R.A.S. With 378 Illustrations. Post 8vo, 6s. cloth.

"The perspicuity of the original has been retained, and chapters which had become obsolete have been replaced by others of more modern character. The explanations throughout are studiously popular, and care has been taken to show the application of the various branches of physics to the industrial arts, and to the practical business of life."—*Mining Journal.*

"Mr. Loewy has carefully revised the book, and brought it up to modern requirements."—*Nature.*

"Natural philosophy has had few exponents more able or better skilled in the art of popularising the subject than Dr. Lardner; and Mr. Loewy is doing good service in fitting this treatise and the others of the series, for use at the present time."—*Scotsman.*

THE HANDBOOK OF HYDROSTATICS AND PNEUMATICS. New Edition, Revised and Enlarged by Benjamin Loewy, F.R.A.S. With 236 Illustrations. Post 8vo, 5s. cloth.

"For those 'who desire to attain an accurate knowledge of physical science without the profound methods of mathematical investigation,' this work is not merely intended, but well adapted."—*Chemical News.*

"The volume before us has been carefully edited, augmented to nearly twice the bulk of the former edition, and all the most recent matter has been added. . . . It is a valuable text-book."—*Nature.*

"Candidates for pass examinations will find it, we think, specially suited to their requirements."—*English Mechanic.*

THE HANDBOOK OF HEAT. Edited and almost entirely re written by Benjamin Loewy, F.R.A.S., &c. 117 Illustrations. Post 8vo, 6s. cloth.

"The style is always clear and precise, and conveys instruction without leaving any cloudiness or lurking doubts behind."—*Engineering.*

"A most exhaustive book on the subject on which it treats, and is so arranged that it can be understood by all who desire to attain an accurate knowledge of physical science. . . . . Mr. Loewy has included all the latest discoveries in the varied laws and effects of heat."—*Standard.*

"A complete and handy text-book for the use of students and general readers."—*English Mechanic.*

THE HANDBOOK OF OPTICS. By Dionysius Lardner, D.C.L., formerly Professor of Natural Philosophy and Astronomy in University College, London. New Edition. Edited by T. Olver Harding, B.A. Lond., of University College, London. With 298 Illustrations. Small 8vo, 448 pages, 5s. cloth.

"Written by one of the ablest English scientific writers, beautifully and elaborately illustrated."—*Mechanics' Magazine.*

THE HANDBOOK OF ELECTRICITY, MAGNETISM, AND ACOUSTICS. By Dr. Lardner. New Edition. Edited by George Carey Foster, B.A., F.C.S. With 400 Illustrations. Small 8vo, 5s. cloth.

"The book could not have been entrusted to anyone better calculated to preserve the terse and lucid style of Lardner, while correcting his errors and bringing up his work to the present state of cientific knowledge."—*Popular Science Review.*

Dr. Lardner's Handbook of Astronomy.

THE HANDBOOK OF ASTRONOMY. Forming a Companion to the "Handbook of Natural Philosophy." By Dionysius Lardner, D.C.L., formerly Professor of Natural Philosophy and Astronomy in University College, London. Fourth Edition. Revised and Edited by Edwin Dunkin, F.R.A.S., Royal Observatory, Greenwich. With 38 Plates and upwards of 100 Woodcuts. In One Vol., small 8vo, 550 pages, 9s. 6d. cloth.

"Probably no other book contains the same amount of information in so compendious and well-arranged a form—certainly none at the price at which this is offered to the public."—*Athenæum.*

"We can do no other than pronounce this work a most valuable manual of astronomy, and we strongly recommend it to all who wish to acquire a general—but at the same time correct—acquaintance with this sublime science."—*Quarterly Journal of Science.*

"One of the most deservedly popular books on the subject . . . We would recommend not only the student of the elementary principles of the science, but he who aims at mastering the higher and mathematical branches of astronomy, not to be without this work beside him."—*Practical Magazine.*

DR. LARDNER'S MUSEUM OF SCIENCE AND ART.

THE MUSEUM OF SCIENCE AND ART. Edited by DIONYSIUS LARDNER, D.C.L., formerly Professor of Natural Philosophy and Astronomy in University College, London. With upwards of 1,200 Engravings on Wood. In 6 Double Volumes, £1 1s., in a new and elegant cloth binding; or handsomely bound in half-morocco, 31s. 6d.

Contents:

The Planets: Are they Inhabited Worlds?—Weather Prognostics — Popular Fallacies in Questions of Physical Science—Latitudes and Longitudes — Lunar Influences — Meteoric Stones and Shooting Stars—Railway Accidents—Light—Common Things: Air—Locomotion in the United States—Cometary Influences—Common Things: Water—The Potter's Art—Common Things: Fire — Locomotion and Transport, their Influence and Progress—The Moon — Common Things: The Earth — The Electric Telegraph — Terrestrial Heat — The Sun—Earthquakes and Volcanoes—Barometer, Safety Lamp, and Whitworth's Micrometric Apparatus—Steam—The Steam Engine—The Eye — The Atmosphere — Time — Common Things: Pumps—Common Things: Spectacles, the Kaleidoscope — Clocks and Watches — Microscopic Drawing and Engraving—Locomotive — Thermometer — New Planets: Leverrier and Adams's Planet—Magnitude and Minuteness—Common Things: The Almanack—Optical Images—How to observe the Heavens — Common Things: The Looking-glass — Stellar Universe—The Tides—Colour—Common Things: Man—Magnifying Glasses—Instinct and Intelligence—The Solar Microscope—The Camera Lucida—The Magic Lantern—The Camera Obscura—The Microscope—The White Ants: Their Manners and Habits—The Surface of the Earth, or First Notions of Geography—Science and Poetry—The Bee—Steam Navigation — Electro-Motive Power — Thunder, Lightning, and the Aurora Borealis—The Printing Press—The Crust of the Earth—Comets—The Stereoscope—The Pre-Adamite Earth—Eclipses—Sound.

Opinions of the Press.

"This series, besides affording popular but sound instruction on scientific subjects, with which the humblest man in the country ought to be acquainted, also undertakes that teaching of 'Common Things' which every well-wisher of his kind is anxious to promote. Many thousand copies of this serviceable publication have been printed, in the belief and hope that the desire for instruction and improvement widely prevails; and we have no fear that such enlightened faith will meet with disappointment."—*Times.*

"A cheap and interesting publication, alike informing and attractive. The papers combine subjects of importance and great scientific knowledge, considerable inductive powers, and a popular style of treatment."—*Spectator.*

"The 'Museum of Science and Art' is the most valuable contribution that has ever been made to the Scientific Instruction of every class of society."—Sir DAVID BREWSTER, in the *North British Review.*

"Whether we consider the liberality and beauty of the illustrations, the charm of the writing, or the durable interest of the matter, we must express our belief that there is hardly to be found among the new books one that would be welcomed by people of so many ages and classes as a valuable present."—*Examiner.*

*** *Separate books formed from the above, suitable for Workmen's Libraries, Science Classes, &c.*

Common Things Explained. Containing Air, Earth, Fire, Water, Time, Man, the Eye, Locomotion, Colour, Clocks and Watches, &c. 233 Illustrations, cloth gilt, 5s.

The Microscope. Containing Optical Images, Magnifying Glasses, Origin and Description of the Microscope, Microscopic Objects, the Solar Microscope, Microscopic Drawing and Engraving, &c. 147 Illustrations, cloth gilt, 2s.

Popular Geology. Containing Earthquakes and Volcanoes, the Crust of the Earth, &c. 201 Illustrations, cloth gilt, 2s. 6d.

Popular Physics. Containing Magnitude and Minuteness, the Atmosphere, Meteoric Stones, Popular Fallacies, Weather Prognostics, the Thermometer, the Barometer, Sound, &c. 85 Illustrations, cloth gilt, 2s 6d.

Steam and its Uses. Including the Steam Engine, the Locomotive, and Steam Navigation. 89 Illustrations, cloth gilt, 2s.

Popular Astronomy. Containing How to observe the Heavens—The Earth, Sun, Moon, Planets, Light, Comets, Eclipses, Astronomical Influences, &c. 182 Illustrations, 4s. 6d.

The Bee and White Ants: Their Manners and Habits. With Illustrations of Animal Instinct and Intelligence. 135 Illustrations, cloth gilt, 2s.

The Electric Telegraph Popularised. To render intelligible to all who can Read, irrespective of any previous Scientific Acquirements, the various forms of Telegraphy in Actual Operation. 100 Illustrations, cloth gilt, 1s, 6d.

MATHEMATICS, GEOMETRY, TABLES, etc.

Practical Mathematics.

MATHEMATICS FOR PRACTICAL MEN. Being a Common-place Book of Pure and Mixed Mathematics. Designed chiefly for the Use of Civil Engineers, Architects, and Surveyors. Part I. PURE MATHEMATICS: comprising Arithmetic, Algebra, Geometry, Mensuration, Trigonometry, Conic Sections, Properties of Curves. Part II. MIXED MATHEMATICS: comprising Mechanics in general Statics, Dynamics, Hydrostatics, Hydrodynamics, Pneumatics, Mechanical Agents, Strength of Materials. With an Appendix of copious Logarithmic and other Tables. By OLINTHUS GREGORY, LL.D., F.R.A.S., Enlarged by HENRY LAW, C.E. 4th Edition, carefully Revised by J. R. YOUNG, formerly Professor of Mathematics, Belfast College. With 13 Plates, 8vo, £1 1s. cloth.

"The engineer or architect will here find ready to his hand rules for solving nearly every mathematical difficulty that may arise in his practice. The rules are in all cases explained by means of examples, in which every step of the process is clearly worked out."—*Builder.*

"One of the most serviceable books for practical mechanics. . . . Professor Young has modernised the notation throughout, introduced a few paragraphs here and there, and corrected the numerous typographical errors which had escaped the eyes of the former Editor. The book is now as complete as it is possible to make it. It is an instructive book for the student, and a Text-book for him who, having once mastered the subjects it treats of, needs occasionally to refresh his memory upon them."—*Building News.*

"As a standard work on mathematics it has not been excelled."—*Artisan.*

Metrical Units and Systems, etc.

MODERN METROLOGY: A Manual of the Metrical Units and Systems of the Present Century. With an Appendix containing a proposed English System. By LOWIS D'A. JACKSON, A.M. Inst. C.E., Author of "Aid to Survey Practice," &c. Large crown 8vo, 12s. 6d. cloth.

"The author has brought together much valuable and interesting information. . . . The main object of the work appears to be the discussion of a remedy for the evils of the complex systems of weights and measures which are unfortunately still in use in this country. To provide such a remedy is a serious task, and one well worthy of the attention of a great statesman such as Mr. Gladstone, to whom the work is inscribed. . . . We cannot but recommend the work to the consideration of all interested in the practical reform of our weights and measures."—*Nature.*

"For the large collection of measures brought together, the author supplies the French scientific equivalents, in terms of the metrical system. He also gives throughout a double set of English equivalents, the first of which he calls the 'Commercial Measures,' and the second the 'Scientific Value.' . . . There is no doubt as to the great amount of labour that must have been devoted to the calculation; and the juxtaposition of the two values affords an easy means of checking the work by comparison."—*Athenæum.*

"For exhaustive tables of equivalent weights and measures of all sorts, and for clear demonstrations of the effects of the various systems that have been proposed or adopted, Mr. Jackson's treatise is without a rival."—*Academy.*

The Metric System.

A SERIES OF METRIC TABLES, in which the British Standard Measures and Weights are compared with those of the Metric System at present in Use on the Continent. By C. H. DOWLING, C.E. Second Edition, Revised and Enlarged. 8vo, 10s. 6d. strongly bound.

"Their accuracy has been certified by Professor Airy, the Astronomer-Royal."—*Builder.*

"Mr. Dowling's Tables, which are well put together, come just in time as a ready-reckoner for the conversion of one system into the other."—*Athenæum*

Geometry for the Architect, Engineer, etc.

PRACTICAL GEOMETRY, for the Architect, Engineer and Mechanic. Giving Rules for the Delineation and Application of various Geometrical Lines, Figures and Curves. By E. W. TARN, M.A., Architect, Author of "The Science of Building," &c. Second Edition. With Appendices on Diagrams of Strains and Isometrical Projection. With 172 Illustrations, demy 8vo, 9s. cloth.

"No book with the same objects in view has ever been published in which the clearness of the rules laid down and the illustrative diagrams have been so satisfactory."—*Scotsman.*

"This is a manual for the practical man, whether architect, engineer, or mechanic. . . . The object of the author being to avoid all abstruse formulæ or complicated methods, and to enable persons with but a moderate knowledge of geometry to work out the problems required."—*English Mechanic.*

Comprehensive Weight Calculator.

THE WEIGHT CALCULATOR. Being a Series of Tables upon a New and Comprehensive Plan, exhibiting at One Reference the exact Value of any Weight from 1 lb. to 15 tons, at 300 Progressive Rates, from *1d.* to *168s.* per cwt., and containing 186,000 Direct Answers, which, with their Combinations, consisting of a single addition (mostly to be performed at sight), will afford an aggregate of 10,266,000 Answers; the whole being calculated and designed to ensure correctness and promote despatch. By HENRY HARBEN, Accountant, Sheffield, Author of "The Discount Guide." An entirely New Edition, carefully revised. Royal 8vo, strongly half-bound, £1 5s.

"A practical and useful work of reference for men of business generally; it is the best of the kind we have seen. We have frequently been asked if such a work as this could be obtained, and therefore refer to it with pleasure."—*Ironmonger.*

"Of priceless value to business men. Its accuracy and completeness have secured for it a eputation which renders it quite unnecessary for us to say one word in its praise. It is a necessary book in all mercantile offices."—*Sheffield Independent.*

"An enormous amount of labour must have been bestowed upon the preparation of the tables, but the result is a series which gives the answers sought more readily than any of its predecessors." *Mining Journal.*

Comprehensive Discount Guide.

THE DISCOUNT GUIDE. Comprising several Series of Tables for the use of Merchants, Manufacturers, Ironmongers, and others, by which may be ascertained the exact Profit arising from any mode of using Discounts, either in the Purchase or Sale of Goods, and the method of either Altering a Rate of Discount or Advancing a Price, so as to produce, by one operation, a sum that will realise any required profit after allowing one or more Discounts: to which are added Tables of Profit or Advance from $1\frac{1}{4}$ to 90 per cent., Tables of Discount from $1\frac{1}{4}$ to $98\frac{3}{4}$ per cent., and Tables of Commission, &c., from $\frac{1}{8}$ to 10 per cent. By HENRY HARBEN, Accountant, Author of "The Weight Calculator." New Edition, carefully Revised and Corrected. Demy 8vo, 544 pp. half-bound, £1 5s.

"All these tables are well arranged and clearly printed; and the collection will be found of great value in those businesses for which the book has been especially compiled."—*Engineering.*

"A book such as this can only be appreciated by business men, to whom the saving of time means saving of money. We have the high authority of Professor J. R. Young that the tables throughout the work are constructed upon strictly accurate principles. The work is a model of typographical clearness, and must prove of great value to merchants, manufacturers, and general traders."—*British Trade Journal.*

"Much time and labour will be saved by the use of this book—even to the most expert arithmetician. The great practical utility of the 'Discount Guide' will no doubt make it a standard work of reference."—*London Commercial Record.*

Iron and Metal Trades' Calculator.

THE IRON AND METAL TRADES' COMPANION. Being a Calculator, containing a Series of Tables upon a New and Comprehensive Plan, for expeditiously ascertaining the Value of any Goods bought or sold by Weight, from 1s. per cwt. to 112s. per cwt., and from one farthing per pound to one shilling per pound. Each Table extends from one pound to 100 tons. To which are appended Rules on Decimals, Square and Cube Root, Mensuration of Superficies and Solids, &c.; also Tables of Weights of Materials, and other Useful Memoranda. By THOMAS DOWNIE. Strongly bound in leather, 396 pp., 9s.

"A most useful set of tables, and will supply a want, for nothing like them before existed."—*Building News.*

"Will save the possessor the trouble of making numerous intricate calculations. Although specially adapted to the iron and metal trades, the tables contained in this handy little companion will be found useful in every other business in which merchandise is bought and sold by weight."—*Railway News.*

Practical Geometry.

THE GEOMETRY OF COMPASSES; or, Problems Resolved by the mere Description of Circles and the use of Coloured Diagrams and Symbols. By OLIVER BYRNE. Coloured Plates. Crown 8vo, 3s. 6d. cloth.

"A very useful work, which will be valued by all who are pursuing the useful and fascinating study of geometry."—*Iron.*

"The treatise is a good one, and remarkable—like all Mr. Byrne's contributions to the science of geometry—for the lucid character of its teaching."—*Building News.*

"The problems in this useful book are cleverly worked out by beautifully coloured and uncoloured diagrams. It will no doubt be extensively circulated. We highly recommend it."—*Cambridge Express.*

INDUSTRIAL AND USEFUL ARTS.

Soap-making.

THE ART OF SOAP-MAKING: A Practical Handbook of the Manufacture of Hard and Soft Soaps, Toilet Soaps, &c. Including many New Processes, and a Chapter on the Recovery of Glycerine from Waste Leys. By ALEXANDER WATT, Author of "Electro-Metallurgy Practically Treated," &c. With numerous Illustrations. Crown 8vo, 9s. cloth.

"The work will prove very useful, not merely to the technological student, but to the practical soapboiler who wishes to understand the theory of his art."—*Chemical News.*

"It is really an excellent example of a technical manual, entering, as it does, thoroughly and exhaustively both into the theory and practice of soap manufacture. The book is well and honestly done, and deserves the considerable circulation with which it will doubtless meet."—*Knowledge.*

"Mr. Watt's book is a thoroughly practical treatise on an art which has almost no literature in our language. We congratulate the author on the success of his endeavour to fill a void in English technical literature."—*Nature.*

". . . . Clearly and concisely written, and appears to be comprehensive and complete."—*Engineering.*

Leather Manufacture.

THE ART OF LEATHER MANUFACTURE. Being a Practical Handbook, in which the Operations of Tanning, Currying, and Leather Dressing are fully Described, and the Principles of Tanning Explained, and many Recent Processes introduced; as also Methods for the Estimation of Tannin, and a Description of the Arts of Glue Boiling, Gut Dressing, &c. By ALEXANDER WATT, Author of "Soap-Making," "Electro-Metallurgy," &c. With numerous Illustrations. Crown 8vo, 12s. 6d. cloth. [*Just published.*

"Every item of use and interest to the leather trade has been touched upon, and the descriptions and explanations of the various processes are exhaustively given."—*Tanners' and Curriers' Journal.*

"A most lucid and readable book upon difficult and intricate subjects. Every known process of tanning, from the most primitive to the most recent, is fully and accurately described."—*Scottish Leather Trader.*

Boot and Shoe Making.

THE ART OF BOOT AND SHOE-MAKING. A Practical Handbook, including Measurement, Last-Fitting, Cutting-Out, Closing and Making, with a Description of the most approved Machinery employed. By JOHN B. LENO, late Editor of *St. Crispin*, and *The Boot and Shoe-Maker*. With numerous Illustrations. Crown 8vo, 5s. cloth. [*Just published.*

"A very complete account of the art and science of bootmaking, which includes all that need be said about leathers and other materials, as well as about hand-tools and the various machines that have latterly been introduced to supplement or supersede the old-fashioned handiwork."—*Weekly Dispatch.*

Dentistry.

MECHANICAL DENTISTRY: A Practical Treatise on the Construction of the various kinds of Artificial Dentures. Comprising also Useful Formulæ, Tables and Receipts for Gold Plate, Clasps, Solders, &c. &c. By CHARLES HUNTER. Second Edition, Revised. With upwards of 100 Wood Engravings. Crown 8vo, 7s. 6d. cloth.

"The work is very practical."—*Monthly Review of Dental Surgery.*

"An authoritative treatise. . . . We can strongly recommend Mr. Hunter's treatise to all students preparing for the profession of dentistry, as well as to every mechanical dentist."—*Dublin Journal of Medical Science.*

"A work in a concise form that few could read without gaining information from."—*British Journal of Dental Science.*

"The best book on the subject with which we are acquainted."—*Medical Press and Circular.*

Brewing.

A HANDBOOK FOR YOUNG BREWERS. By HERBERT EDWARDS WRIGHT, B.A. Crown 8vo, 3s. 6d. cloth.

"This little volume, containing such a large amount of good sense in so small a compass, ought to recommend itself to every brewery pupil, and many who have passed that stage."—*Brewers' Guardian.*

"The book is very clearly written, and the author has successfully brought his scientific knowledge to bear upon the various processes and details of brewing. To the young student of brewing the reading of such a book as this is calculated to do good, for it will lead him to inquire not only what is to be done, but why it should be done."—*Brewer.*

Electroplating, etc.

ELECTROPLATING: A Practical Handbook. By J. W. URQUHART, C.E. With numerous Illustrations. Crown 8vo, 5*s.* cloth.

"The volume is without a rival in its particular sphere, and the lucid style in which it is written commends it to those amateurs and experimental electrotypers who have but slight, if any, knowledge of the processes of the art to which they turn their attention."—*Design and Work.*

"The information given appears to be based on direct personal knowledge. . . . Its science is sound and the style is always clear."—*Athenæum.*

Electrotyping, etc.

ELECTROTYPING: The Reproduction and Multiplication of Printing Surfaces and Works of Art by the Electro-deposition of Metals. By J. W. URQUHART, C.E. Crown 8vo, 5*s.* cloth.

"The book is so thoroughly practical that it assumes to start with an entire ignorance on the reader's part of electricity. He is, therefore, conducted through its leading laws, then through the metals used by electrotypers, the apparatus, and the depositing processes, up to the final preparation of the work."—*Art Journal.*

"In this work the author enters systematically and thoroughly into every department of the process, in the style of one who combines theory with practice. We can recommend this treatise, not merely to amateurs, but to those actually engaged in the trade."—*Chemical News.*

Electric Lighting.

ELECTRIC LIGHT: Its Production and Use. Embodying Plain Directions for the Treatment of Voltaic Batteries, Electric Lamps, and Dynamo-Electric Machines. By J. W. URQUHART, C.E., Author of "Electroplating: A Practical Handbook." Edited by F. C. WEBB, M.I.C.E., M.S.T.E. Second Edition, revised, with large Additions and 128 Illusts. 7*s.* 6*d.* cloth.

"The book is by far the best that we have yet met with on the subject."—*Athenæum.*

"It is the only work at present available which gives, in language intelligible for the most part to the ordinary reader, a general but concise history of the means which have been adopted up to the present time in producing the electric light. . . . A chapter on the comparative cost of the electric light and gas contains much valuable and interesting information."—*Metropolitan.*

"The book contains a general account of the means adopted in producing the electric light, not only as obtained from voltaic or galvanic batteries, but treats at length of the dynamo-electric machine in several of its forms. . . . An important addition to the literature of the electric light. Students of the subject should not fail to read it."—*Colliery Guardian.*

Electro-Metallurgy.

ELECTRO-METALLURGY; Practically Treated. By ALEXANDER WATT, F.R.S.S.A. Eighth Edition, Revised, with Additional Matter and Illustrations, including the most recent Processes. 12mo, 3*s.* 6*d.* cloth boards.

"From this book both amateur and artisan may learn everything necessary for the successful prosecution of electroplating."—*Iron.*

"A practical treatise for the use of those who desire to work in the art of electro-deposition as a business."—*English Mechanic.*

Silversmiths' Work.

THE SILVERSMITH'S HANDBOOK. Containing full Instructions for the Alloying and Working of Silver, including the different modes of Refining and Melting the Metal, its Solders, the Preparation of Imitation Alloys, Methods of Manipulation, Prevention of Waste, Instructions for Improving and Finishing the Surface of the Work, together with other useful Information and Memoranda. By GEORGE E. GEE, Jeweller, &c. Second Edition, Revised, with numerous Illustrations. 12mo, 3*s.* 6*d.* cloth boards. [*Just published.*

"The chief merit of the work is its practical character. . . . The workers in the trade will speedily discover its merits when they sit down to study it."—*English Mechanic.*

"This work forms a valuable sequel to the author's 'Goldsmith's Handbook,' and supplies a want long felt in the silver trade."—*Silversmiths' Trade Journal.*

Goldsmiths' Work.

THE GOLDSMITH'S HANDBOOK. Containing full Instructions in the Art of Alloying, Melting, Reducing, Colouring, Collecting and Refining. The processes of Manipulation, Recovery of Waste, Chemical and Physical Properties of Gold, with a New System of Mixing its Alloys; Solders, Enamels, and other useful Rules and Recipes, &c. By GEORGE E. GEE. Second Edition, considerably enlarged. 12mo, 3*s.* 6*d.* cloth boards.

"A good, sound, technical educator, and will be generally accepted as an authority. It gives full particulars for mixing alloys and enamels, is essentially a book for the workshop, and exactly fulfils the purpose intended."—*Horological Journal.*

"The best work yet printed on its subject for a reasonable price. We have no doubt that it will speedily become a standard book which few will care to do without."—*Jeweller and Metalworker.*

CHEMICAL MANUFACTURES & COMMERCE.

Alkali Trade, Manufacture of Sulphuric Acid, etc.

A MANUAL OF THE ALKALI TRADE, including the Manufacture of Sulphuric Acid, Sulphate of Soda, and Bleaching Powder. By JOHN LOMAS, Alkali Manufacturer, Newcastle-upon-Tyne and London. With 232 Illustrations and Working Drawings, and containing 386 pages of Text. Super-royal 8vo, £2 12s. 6d. cloth.

*** This work provides (1) a Complete Handbook for intending Alkali and Sulphuric Acid Manufacturers, and for those already in the field who desire to improve their plant, or to become practically acquainted with the latest processes and developments of the trade: (2) a Handy Volume which Manufacturers can put into the hands of their Managers and Foremen as a useful guide in their daily rounds of duty.

Synopsis of Contents:

Chap. I. Choice of Site and General Plan of Works.—II. Sulphuric Acid.—III. Recovery of the Nitrogen Compounds, and Treatment of Small Pyrites.—IV. The Salt Cake Process.—V. Legislation upon the Noxious Vapours Question.—VI. The Hargreaves' and Jones' Processes.—VII. The Balling Process.—VIII. Lixiviation and Salting Down.—IX. Carbonating or Finishing.—X. Soda Crystals.—XI. Refined Alkali.—XII. Caustic Soda.—XIII. Bi-carbonate of Soda.—XIV. Bleaching Powder.—XV. Utilisation of Tank Waste.—XVI. General Remarks—Four Appendices treating of Yields, Sulphuric Acid Calculations, Anemometers, and Foreign Legislation upon the Noxious Vapours Question.

"The author has given the fullest, most practical, and, to all concerned in the alkali trade, most valuable mass of information that, to our knowledge, has been published in any language."—*Engineer.*

"This book is written by a manufacturer for manufacturers. The working details of the most approved forms of apparatus are given, and these are accompanied by no less than 232 wood engravings, all of which may be used for the purposes of construction. Every step in the manufacture is very fully described in this manual, and each improvement explained. Everything which tends to introduce economy into the technical details of this trade receives the fullest attention."—*Athenæum.*

"The author is not one of those clever compilers who, on short notice, will 'read up' any conceivable subject, but a practical man in the best sense of the word. We find here not merely a sound and luminous explanation of the chemical principles of the trade, but a notice of numerous matters which have a most important bearing on the successful conduct of alkali works, but which are generally overlooked by even the most experienced technological authors."—*Chemical Review.*

Commercial Chemical Analysis.

THE COMMERCIAL HANDBOOK OF CHEMICAL ANALYSIS; or, Practical Instructions for the determination of the Intrinsic or Commercial Value of Substances used in Manufactures, in Trades, and in the Arts. By A. NORMANDY, Author of "Practical Introduction to Rose's Chemistry," and Editor of Rose's "Treatise on Chemical Analysis." New Edition, Enlarged and to a great extent re-written, by HENRY M. NOAD, Ph.D., F.R.S. With numerous Illustrations. Crown 8vo, 12s. 6d. cloth.

"We recommend this book to the careful perusal of everyone; it may be truly affirmed to be of universal interest, and we strongly recommend it to our readers as a guide, alike indispensable to the housewife as to the pharmaceutical practitioner."—*Medical Times.*

"Essential to the analysts appointed under the new Act. The most recent results are given, and the work is well edited and carefully written."—*Nature.*

Dye-Wares and Colours.

THE MANUAL OF COLOURS AND DYE-WARES: Their Properties, Applications, Valuation, Impurities, and Sophistications. For the use of Dyers, Printers, Drysalters, Brokers, &c. By J. W. SLATER. Second Edition, Revised and greatly Enlarged. Crown 8vo, 7s. 6d. cloth.

*** This book contains a description of about Six Hundred Colours, Chemicals, and Drugs used in the Tinctorial Arts, and their Sources, Applications, and possible Impurities.

"A complete encyclopædia of the *materia tinctoria*. The information given respecting each article is full and precise, and the methods of determining the value of articles such as these, so liable to sophistication, are given with clearness, and are practical as well as valuable."—*Chemist and Druggist.*

"Practical dyers, &c., will welcome the work in its improved form. There is no other work in the language which covers precisely the same ground. To technological students preparing for examinations in dyeing and printing it will prove exceedingly useful."—*Chemical News.*

AGRICULTURE, LAND MANAGEMENT, etc.

Youatt and Burn's Complete Grazier.

THE COMPLETE GRAZIER, and FARMER'S and CATTLE-BREEDER'S ASSISTANT. A Compendium of Husbandry; especially in the departments connected with the Breeding, Rearing, Feeding, and General Management of Stock; the Management of the Dairy, &c. With Directions for the Culture and Management of Grass Land, of Grain and Root Crops, the Arrangement of Farm Offices, the use of Implements and Machines, and on Draining, Irrigation, Warping, &c.; and the Application and Relative Value of Manures. By WILLIAM YOUATT, Esq., V.S. Twelfth Edition, very considerably enlarged and brought up to the present requirements of Agricultural Practice by ROBERT SCOTT BURN, Author of "Outlines of Modern Farming," "Landed Estates Management," "Farm Management," "The Lessons of My Farm," &c. One large 8vo Volume, 860 pp., with 244 Illustrations, £1 1s. half-bound.

"The standard and text-book with the farmer and grazier."—*Farmers' Magazine.*

"A treatise which will remain a standard work on the subject as long as British agriculture endures."—*Mark Lane Express* (First Notice).

"The book deals with all departments of agriculture, and contains an immense amount of valuable information. It is, in fact, an encyclopædia of agriculture put into readable form, and it is the only work equally comprehensive brought down to present date. It is excellently printed on thick paper, and strongly bound, and deserves a place in the library of every agriculturist."—*Mark Lane Express* (Second Notice).

"Of great value to the farmer, more especially to the young cattle breeder and feeder. . . . This esteemed work is well worthy of a place in the libraries of agriculturists."—*North British Agriculturist.*

"A valuable repertory of intelligence for all who make agriculture a pursuit, and especially for those who aim at keeping pace with the improvements of the age. . . . The new matter is of so valuable a nature that the volume is now almost entitled to be considered as a distinct work."—*Bell's Messenger.*

Modern Farming.

OUTLINES OF MODERN FARMING. By R. SCOTT BURN. Soils, Manures, and Crops—Farming and Farming Economy—Cattle, Sheep, and Horses—Management of the Dairy, Pigs and Poultry—Utilisation of Town-Sewage, Irrigation, &c. Sixth Edition. In One Vol., 1,250 pp., half-bound, profusely Illustrated, 12s.

"The aim of the author has been to make his work at once comprehensive and trustworthy, and in this aim he has succeeded to a degree which entitles him to much credit."—*Morning Advertiser.*

"Eminently calculated to enlighten the agricultural community on the varied subjects of which it treats, and hence it should find a place in every farmer's library."—*City Press.*

"No farmer should be without this book."—*Banbury Guardian.*

Agricultural Engineering.

THE COMPLETE TEXT-BOOK OF FARM ENGINEERING. Comprising Practical Treatises on Draining and Embanking; Irrigation and Water Supply; Farm Roads, Fences, and Gates; Farm Buildings, Barn Implements, and Machines; Field Implements and Machines; and Agricultural Surveying, Levelling, &c. By Prof. JOHN SCOTT, Editor of the *Farmers' Gazette,* late Professor of Agriculture and Rural Economy at the Royal Agricultural College, Cirencester, &c., &c. In One Vol., 1,150 pages, wih 600 Illustrations, 12s. half-bound. [*Just published.*

"A copy of this work should be treasured up in every library where the owner thereof is in any way connected with land."—*Farm and Home.*

"Written with great care, as well as with knowledge and ability. The author has done his work well; we have found him a very trustworthy guide wherever we have tested his statements. The volume will be of great value to agricultural students, and we have much pleasure in recommending it."—*Mark Lane Express.*

"For a young agriculturist we know of no handy volume so likely to be more usefully studied." —*Bell's Weekly Messenger.*

Amateur Farming.

THE LESSONS OF MY FARM: A Book for Amateur Agriculturists. Being an Introduction to Farm Practice. By ROBERT SCOTT BURN. With numerous Illustrations.

[*A New and Enlarged Edition in preparation.*

A Text-Book of English Agriculture.

THE FIELDS OF GREAT BRITAIN: A Text-Book of Agriculture, adapted to the Syllabus of the Science and Art Department. For Elementary and Advanced Students. By HUGH CLEMENTS (Board of Trade). 18mo, 2s. 6d. cloth.

"A most comprehensive volume, giving a mass of information."—*Agricultural Economist.*

"It is a long time since we have seen a book which has pleased us more, or which contains such a vast and useful fund of knowledge."—*Educational Times.*

Agricultural Data.

NOTE BOOK of AGRICULTURAL FACTS and FIGURES, for Farmers and Farm Students. By PRIMROSE MCCONNELL, Fellow of the Highland and Agricultural Society; late Professor of Agriculture, Glasgow Veterinary College. Royal 32mo oblong, leather, with strap, 4s.

"It is full of very valuable information. Farmers' sons and other youths who wish to become farmers at home or abroad, might, even before their school education is completed, become familiar with the facts and figures furnished in this interesting and valuable little book."—*Aberdeen Free Press.*

Hudson's Land Valuer's Pocket-Book.

THE LAND VALUER'S BEST ASSISTANT: Being Tables on a very much Improved Plan, for Calculating the Value of Estates. With Tables for reducing Scotch, Irish, and Provincial Customary Acres to Statute Measure, &c. By R. HUDSON, C.E. New Edition. Royal 32mo, leather, gilt edges, elastic band, 4s.

"This new edition includes tables for ascertaining the value of leases for any term of years; and for showing how to lay out plots of ground of certain acres in forms, square, round, &c., with valuable rules for ascertaining the probable worth of standing timber to any amount; and is of incalculable value to the country gentleman and professional man."—*Farmers' Journal.*

Ewart's Land Improver's Pocket-Book.

THE LAND IMPROVER'S POCKET-BOOK OF FORMULÆ, TABLES and MEMORANDA required in any Computation relating to the Permanent Improvement of Landed Property. By JOHN EWART, Land Surveyor and Agricultural Engineer. Royal 32mo, oblong, leather, gilt edges, with elastic band, 4s.

"A compendious and handy little volume."—*Spectator.*

Complete Agricultural Surveyor's Pocket-Book.

THE LAND VALUER'S AND LAND IMPROVER'S COMPLETE POCKET-BOOK. Consisting of the above Two Works bound together. Leather, gilt edges, with strap, 7s. 6d.

"We consider Hudson's book to be the best ready-reckoner on matters relating to the valuation of land and crops we have ever seen, and its combination with Mr. Ewart's work greatly enhances the value and usefulness of the latter-mentioned. . . . It is most useful as a manual for reference."—*North of England Farmer.*

Potato Culture.

POTATOES: How to Grow and Show them. A Practical Guide to the Cultivation and General Treatment of the Potato. By JAMES PINK. With Illustrations. Second Edition. Crown 8vo, 2s. cloth.

"A well-written little volume. The author gives good practical instructions under both divisions of his subject."—*Agricultural Gazette.*

GARDENING, FLORICULTURE, etc.

Early Fruits, Flowers and Vegetables.

THE FORCING GARDEN: or, How to Grow Early Fruits, Flowers, and Vegetables. With Plans, and Estimates for Building Glasshouses, Pits and Frames. Containing also Original Plans for Double Glazing, a New Method of Growing the Gooseberry under Glass, &c., &c., and on Ventilation, Protecting Vine Borders, &c. With Illustrations. By SAMUEL WOOD. Crown 8vo, 3s. 6d. cloth.

"A good book, and fairly fills a place that was in some degree vacant. The book is written with great care, and contains a great deal of valuable teaching."—*Gardeners' Magazine.*

"Mr. Wood's book is an original and exhaustive answer to the question 'How to Grow Early Fruits, Flowers and Vegetables?'"—*Land and Water.*

Good Gardening.

A PLAIN GUIDE TO GOOD GARDENING; or, How to Grow Vegetables, Fruits, and Flowers. With Practical Notes on Soils, Manures, Seeds, Planting, Laying-out of Gardens and Grounds, &c. By S. WOOD. With numerous Illustrations. Third Edition. Crown 8vo, 5s. cloth.

"A very good book, and one to be highly recommended as a practical guide. The practical directions are excellent."—*Athenæum.*

"May be recommended to young gardeners, cottagers, and specially to amateurs, for the plain, simple, and trustworthy information it gives on common matters too often neglected."—*Gardeners' Chronicle.*

Gainful Gardening.

MULTUM-IN-PARVO GARDENING; or, How to make One Acre of Land produce £620 a-year by the Cultivation of Fruits and Vegetables; also, How to Grow Flowers in Three Glass Houses, so as to realise £176 per annum clear Profit. By SAMUEL WOOD, Author of "Good Gardening," &c. Fourth Edition, revised. With Wood Engravings. Crown 8vo. 2s. cloth.

"We are bound to recommend it as not only suited to the case of the amateur and gentleman's gardener, but to the market grower."—*Gardeners' Magazine.*

"Of all the practical guides to the amateur, as well as being invaluable to most gardeners, Mr Wood's book is the most accurate and concise."—*Horticultural Record.*

Gardening for Ladies.

THE LADIES' MULTUM-IN-PARVO FLOWER GARDEN, and Amateur's Complete Guide. With Illustrations. By SAMUEL WOOD. Crown 8vo, 3s. 6d. cloth.

"This volume contains a good deal of sound, common-sense instruction."—*Florist.*

"Full of shrewd hints and useful instructions, based on a lifetime of experience."—*Scotsman.*

Receipts for Gardeners.

GARDEN RECEIPTS. Edited by CHARLES W. QUIN. 12mo, 1s. 6d. cloth limp.

"A useful and handy book, containing a good deal of valuable information."—*Athenæum.*

Kitchen Gardening.

THE KITCHEN AND MARKET GARDEN. By Contributors to "The Garden." Compiled by C. W. SHAW, Editor of "Gardening Illustrated." 12mo, 3s. 6d. cloth boards.

"The most valuable compendium of kitchen and market-garden work published."—*Farmer.*

Cottage Gardening.

COTTAGE GARDENING; or, Flowers, Fruits, and Vegetables for Small Gardens. By E. HOBDAY. 12mo, 1s. 6d. cloth limp.

"Definite instructions as to the cultivation of small gardens."—*Scotsman.*

"Contains much useful information at a small charge."—*Glasgow Herald.*

AUCTIONEERING, ESTATE AGENCY, etc.

Auctioneer's Assistant.

THE APPRAISER, AUCTIONEER, BROKER, HOUSE AND ESTATE AGENT AND VALUER'S POCKET ASSISTANT, for the Valuation for Purchase, Sale, or Renewal of Leases, Annuities and Reversions, and of property generally; with Prices for Inventories, &c. By JOHN WHEELER, Valuer, &c. Fifth Edition, Re-written and greatly Extended by C. NORRIS, Surveyor, Valuer, &c. Royal 32mo, 5s. cloth.

"A neat and concise book of reference, containing an admirable and clearly-arranged list of prices for inventories, and a very practical guide to determine the value of furniture, &c."—*Standard.*

"Cram full of valuable information of practical value. It is a trustworthy and compendious guide to all sorts of valuation."—*Insurance Agent.*

Auctioneering.

AUCTIONEERS: Their Duties and Liabilities. By ROBERT SQUIBBS, Auctioneer. Demy 8vo, 10s. 6d. cloth.

"The position and duties of auctioneers are treated compendiously and clearly."—*Builder.*

"Every auctioneer ought to possess a copy of this excellent work."—*Ironmonger.*

"Of great value to the profession. . . . We readily welcome this book."—*Estates Gazette.*

House Property.

HANDBOOK OF HOUSE PROPERTY : A Popular and Practical Guide to the Purchase, Mortgage, Tenancy, and Compulsory Sale of Houses and Land. By E. L. TARBUCK, Architect and Surveyor. Third Edition, 12mo, 3s. 6d. cloth.

"The advice is thoroughly practical."—*Law Journal.*

"This is a well-written and thoughtful work. We commend the work to the careful study of all interested in questions affecting houses and land."—*Land Agents' Record.*

Inwood's Estate Tables.

TABLES FOR THE PURCHASING OF ESTATES, Freehold, Copyhold, or Leasehold; Annuities, Advowsons, &c., and for the Renewing of Leases held under Cathedral Churches, Colleges, or other Corporate bodies, for Terms of Years certain, and for Lives; also for Valuing Reversionary Estates, Deferred Annuities, Next Presentations, &c.: together with SMART'S Five Tables of Compound Interest, and an Extension of the same to Lower and Intermediate Rates. By W. INWOOD. 22nd Edition, with considerable Additions, and new and valuable Tables of Logarithms for the more Difficult Computations of the Interest of Money, Discount, Annuities, &c., by M. FEDOR THOMAN, of the Société Crédit Mobilier of Paris. 12mo, 8s. cloth.

"Those interested in the purchase and sale of estates, and in the adjustment of compensation cases, as well as in transactions in annuities, life insurances, &c., will find the present edition of eminent service."—*Engineering.*

"'Inwood's Tables' still maintain a most enviable reputation. The new issue has been enriched by large additional contributions by M. Fedor Thoman, whose carefully arranged Tables cannot fail to be of the utmost utility."—*Mining Journal.*

How to Invest.

HINTS FOR INVESTORS: Being an Explanation of the Mode of Transacting Business on the Stock Exchange. To which are added Comments on the Fluctuations and Table of Quarterly Average prices of Consols since 1759. Also a Copy of the London Daily Stock and Share List. By WALTER M. PLAYFORD, Sworn Broker. Crown 8vo, 2s. cloth.

"A clearly-written book, by one who evidently knows the sort of information which the investor is likely to want."—*Lloyd's News.*

"An invaluable guide to investors and speculators."—*Bullionist.*

A Complete Epitome of the Laws of this Country.

EVERY MAN'S OWN LAWYER: A Handy-book of the Principles of Law and Equity. By A BARRISTER. Twenty-second Edition. Carefully revised and brought down to the end of the last Session, including Summaries of the Latest Statute Laws. With Notes and References to the Authorities. Crown 8vo, price 6s. 8d. (saved at every consultation), strongly bound in cloth.

Comprising THE RIGHTS AND WRONGS OF INDIVIDUALS—MERCANTILE AND COMMERCIAL LAW—CRIMINAL LAW—PARISH LAW—COUNTY COURT LAW—GAME AND FISHERY LAWS—POOR MEN'S LAWSUITS—THE LAWS OF BANKRUPTCY—BETS AND WAGERS—CHEQUES, BILLS, AND NOTES—CONTRACTS AND AGREEMENTS—COPYRIGHT—ELECTIONS AND REGISTRATION—INSURANCE—LIBEL AND SLANDER—MARRIAGE AND DIVORCE—MERCHANT SHIPPING—MORTGAGES—SETTLEMENTS—STOCK EXCHANGE PRACTICE—TRADE MARKS AND PATENTS—TRESPASS—NUISANCES, &c.—TRANSFER OF LAND, &c.—WARRANTY—WILLS AND AGREEMENTS, &c. &c.

Also, Law for Landlord and Tenant—Master and Servant—Workmen and Apprentices—Heirs—Devisees and Legatees—Husband and Wife—Executors and Trustees—Guardian and Ward—Married Women and Infants—Partners and Agents—Lender, Borrower and Sureties—Debtor and Creditor—Purchaser and Vendor—Companies and Associations—Friendly Societies—Clergymen—Churchwardens—Medical Practitioners, &c.—Bankers—Farmers—Contractors—Stock and Share Brokers—Sportsmen—Gamekeepers—Farriers and Horse Dealers—Auctioneers—House Agents—Innkeepers, &c.—Bakers—Millers, &c.—Pawnbrokers—Surveyors—Railways and Carriers—Constables—Seamen—Soldiers, &c. &c.

Opinions of the Press.

"*No Englishman ought to be without this book.* . . . Any person perfectly uninformed on legal matters, who may require sound information on unknown law points, will, by reference to this book, acquire the necessary information, and thus on many occasions save the expense and loss of time of a visit to a lawyer."—*Engineer.*

"It is a complete code of English Law, written in plain language, which all can understand. . . . Should be in the hands of every business man, and all who wish to abolish lawyers' bills."—*Weekly Times.*

"A useful and concise epitome of the law, compiled with considerable care."—*Law Magazine.*

"What it professes to be—a complete epitome of the laws of this country, thoroughly intelligible to non-professional readers. The book is a handy one to have in readiness when some knotty point requires ready solution."—*Bell's Life.*

J. OGDEN AND CO., PRINTERS, 172, ST. JOHN STREET, E.C.

WEALE'S RUDIMENTARY SCIENTIFIC SERIES.

*** The volumes of this Series are freely Illustrated with Woodcuts, or otherwise, where requisite. Throughout the following List it must be understood that the books are bound in limp cloth, unless otherwise stated; *but the volumes marked with a ‡ may also be had strongly bound in cloth boards for 6d. extra.*

N.B.—In ordering from this List it is recommended, as a means of facilitating business and obviating error, to quote the numbers affixed to the volumes, as well as the titles and prices.

CIVIL ENGINEERING, SURVEYING, ETC.

No.

31. *WELLS AND WELL-SINKING.* By JOHN GEO. SWINDELL, A.R.I.B.A., and G. R. BURNELL, C.E. Revised Edition. With a New Appendix on the Qualities of Water. Illustrated. 2s.

35. *THE BLASTING AND QUARRYING OF STONE*, for Building and other Purposes. With Remarks on the Blowing up of Bridges. By Gen. Sir JOHN BURGOYNE, Bart., K.C.B. Illustrated. 1s. 6d.

43. *TUBULAR, AND OTHER IRON GIRDER BRIDGES*, particularly describing the Britannia and Conway Tubular Bridges. By G. DRYSDALE DEMPSEY, C.E. Fourth Edition. 2s.

44. *FOUNDATIONS AND CONCRETE WORKS*, with Practical Remarks on Footings, Sand, Concrete, Béton, Pile-driving, Caissons, and Cofferdams, &c. By E. DOBSON. Fifth Edition. 1s. 6d.

60. *LAND AND ENGINEERING SURVEYING.* By T. BAKER, C.E. New Edition, revised by EDWARD NUGENT, C.E. 2s.‡

80*. *EMBANKING LANDS FROM THE SEA.* With examples and Particulars of actual Embankments, &c. By J. WIGGINS, F.G.S. 2s.

81. *WATER WORKS*, for the Supply of Cities and Towns. With a Description of the Principal Geological Formations of England as influencing Supplies of Water; and Details of Engines and Pumping Machinery for raising Water. By SAMUEL HUGHES, F.G.S., C.E. New Edition. 4s.‡

118. *CIVIL ENGINEERING IN NORTH AMERICA*, a Sketch of. By DAVID STEVENSON, F.R.S.E., &c. Plates and Diagrams. 3s.

167. *IRON BRIDGES, GIRDERS, ROOFS, AND OTHER WORKS.* By FRANCIS CAMPIN, C.E. 2s. 6d.‡

197. *ROADS AND STREETS (THE CONSTRUCTION OF).* By HENRY LAW, C.E., revised and enlarged by D. K. CLARK, C.E., including pavements of Stone, Wood, Asphalte, &c. 4s. 6d.‡

203. *SANITARY WORK IN THE SMALLER TOWNS AND IN VILLAGES.* By C. SLAGG, A.M.I.C.E. Revised Edition. 3s.‡

212. *GAS-WORKS, THEIR CONSTRUCTION AND ARRANGEMENT*; and the Manufacture and Distribution of Coal Gas. Originally written by SAMUEL HUGHES, C.E. Re-written and enlarged by WILLIAM RICHARDS, C.E. Seventh Edition, with important additions. 5s. 6d.‡

213. *PIONEER ENGINEERING.* A Treatise on the Engineering Operations connected with the Settlement of Waste Lands in New Countries. By EDWARD DOBSON, Assoc. Inst. C.E. 4s. 6d.‡

216. *MATERIALS AND CONSTRUCTION;* A Theoretical and Practical Treatise on the Strains, Designing, and Erection of Works of Construction. By FRANCIS CAMPIN, C.E. Second Edition, revised. 3s.‡

219. *CIVIL ENGINEERING.* By HENRY LAW, M.Inst. C.E. Including HYDRAULIC ENGINEERING by GEO. R. BURNELL, M.Inst. C.E. Seventh Edition, revised, with large additions by D. KINNEAR CLARK, M.Inst. C.E. 6s. 6d., Cloth boards, 7s. 6d.

☞ *The ‡ indicates that these vols. may be had strongly bound at 6d. extra.*

MECHANICAL ENGINEERING, ETC.

33. *CRANES*, the Construction of, and other Machinery for Raising Heavy Bodies. By JOSEPH GLYNN, F.R.S. Illustrated. 1s. 6d.

34. *THE STEAM ENGINE*. By Dr. LARDNER. Illustrated. 1s. 6d.

59. *STEAM BOILERS:* their Construction and Management. By R. ARMSTRONG, C.E. Illustrated. 1s. 6d.

82. *THE POWER OF WATER*, as applied to drive Flour Mills, and to give motion to Turbines, &c. By JOSEPH GLYNN, F.R.S. 2s.‡

98. *PRACTICAL MECHANISM*, the Elements of; and Machine Tools. By T. BAKER, C.E. With Additions by J. NASMYTH, C.E. 2s. ‡

139. *THE STEAM ENGINE*, a Treatise on the Mathematical Theory of, with Rules and Examples for Practical Men. By T. BAKER, C.E. 1s. 6d.

164. *MODERN WORKSHOP PRACTICE*, as applied to Marine, Land, and Locomotive Engines, Floating Docks, Dredging Machines, Bridges, Cranes, Ship-building, &c., &c. By J. G. WINTON. Illustrated. 3s.‡

165. *IRON AND HEAT*, exhibiting the Principles concerned in the Construction of Iron Beams, Pillars, and Bridge Girders, and the Action of Heat in the Smelting Furnace. By J. ARMOUR, C.E. 2s. 6d.‡

166. *POWER IN MOTION:* Horse-Power, Toothed-Wheel Gearing, Long and Short Driving Bands, and Angular Forces. By J. ARMOUR, 2s. 6d.‡

171. *THE WORKMAN'S MANUAL OF ENGINEERING* DRAWING. By J. MAXTON. 5th Edn. With 7 Plates and 350 Cuts. 3s. 6d.‡

190. *STEAM AND THE STEAM ENGINE*, Stationary and Portable. By JOHN SEWELL and D. K. CLARK, M.I.C.E. 3s. 6d.‡

200. *FUEL*, its Combustion and Economy. By C. W. WILLIAMS, With Recent Practice in the Combustion and Economy of Fuel—Coal, Coke, Wood, Peat, Petroleum, &c.—by D. K. CLARK, M.I.C.E. 3s. 6d.‡

202. *LOCOMOTIVE ENGINES*. By G. D. DEMPSEY, C.E.; with large additions by D. KINNEAR CLARK, M.I.C.E. 3s.‡

211. *THE BOILERMAKER'S ASSISTANT* in Drawing, Templating, and Calculating Boiler and Tank Work. By JOHN COURTNEY, Practical Boiler Maker. Edited by D. K. CLARK, C.E. 100 Illustrations. 2s.

217. *SEWING MACHINERY:* Its Construction, History, &c., with full Technical Directions for Adjusting, &c. By J. W. URQUHART, C.E. 2s.‡

223. *MECHANICAL ENGINEERING*. Comprising Metallurgy, Moulding, Casting, Forging, Tools, Workshop Machinery, Manufacture of the Steam Engine, &c. By FRANCIS CAMPIN, C.E. 2s. 6d.‡

236. *DETAILS OF MACHINERY*. Comprising Instructions for the Execution of various Works in Iron. By FRANCIS CAMPIN, C.E. 3s.‡

237. *THE SMITHY AND FORGE;* including the Farrier's Art and Coach Smithing. By W. J. E. CRANE. Illustrated. 2s. 6d.‡

238. *THE SHEET-METAL WORKER'S GUIDE;* a Practical Handbook for Tinsmiths, Coppersmiths, Zincworkers, &c. With 94 Diagrams and Working Patterns. By W. J. E. CRANE. 1s. 6d.

251. *STEAM AND MACHINERY MANAGEMENT:* with Hints on Construction and Selection. By M. POWIS BALE, M.I.M.E. 2s. 6d.‡

254. *THE BOILERMAKER'S READY-RECKONER*. With Examples of Practical Geometry and Templating. By JOHN COURTNEY. Edited by D. K. CLARK, M.Inst.C.E. 4s., limp; 5s., half-bound.

255. *LOCOMOTIVE ENGINE-DRIVING*. A Practical Manual for Engineers in charge of Locomotive Engines. By MICHAEL REYNOLDS, M.S.E. Seventh Edition. 3s. 6d., limp; 4s. 6d. cloth boards. [*Just published*.

256. *STATIONARY ENGINE-DRIVING*. A Practical Manual for Engineers in charge of Stationary Engines. By MICHAEL REYNOLDS, M.S.E. Third Edition. 3s. 6d. limp; 4s. 6d. cloth boards. [*Just published*.

☞ *The ‡ indicates that these vols. may be had strongly bound at 6d. extra.*

MINING, METALLURGY, ETC.

4. *MINERALOGY*, Rudiments of; a concise View of the General Properties of Minerals. By A. RAMSAY, F.G.S., F.R.G.S., &c. Third Edition, revised and enlarged. Illustrated. 3s. 6d.‡ [*Just published.*

117. *SUBTERRANEOUS SURVEYING*, Elementary and Practical Treatise on, with and without the Magnetic Needle. By THOMAS FENWICK, Surveyor of Mines, and THOMAS BAKER, C.E. Illustrated. 2s. 6d.‡

133. *METALLURGY OF COPPER*; an Introduction to the Methods of Seeking, Mining, and Assaying Copper, and Manufacturing its Alloys. By ROBERT H. LAMBORN, Ph.D. Woodcuts. 2s. 6d.‡

135. *ELECTRO-METALLURGY*; Practically Treated. By ALEXANDER WATT, F.R.S.S.A. Eighth Edition, revised, with additional Matter and Illustrations, including the most recent Processes. 3s.‡

172. *MINING TOOLS*, Manual of. For the Use of Mine Managers, Agents, Students, &c. By WILLIAM MORGANS. 2s. 6d.‡

72*. *MINING TOOLS, ATLAS* of Engravings to Illustrate the above, containing 235 Illustrations, drawn to Scale. 4to. 4s. 6d.; cloth boards, 6s.

176. *METALLURGY OF IRON.* Containing History of Iron Manufacture. Methods of Assay, and Analyses of Iron Ores, Processes of Manufacture of Iron and Steel, &c. By H. BAUERMAN, F.G.S. Fifth Edition, revised and enlarged. 5s.‡

180. *COAL AND COAL MINING.* By WARINGTON W. SMYTH, M.A., F.R.S. Sixth Edition, revised. 3s. 6d.‡ [*Just published.*

195. *THE MINERAL SURVEYOR AND VALUER'S COMPLETE GUIDE*, with new Traverse Tables, and Descriptions of Improved Instruments; also the Correct Principles of Laying out and Valuing Mineral Properties. By WILLIAM LINTERN, Mining and Civil Engineer. 3s. 6d.‡

214. *SLATE AND SLATE QUARRYING*, Scientific, Practical, and Commercial. By D. C. DAVIES, F.G.S., Mining Engineer, &c. 3s.‡

220. *MAGNETIC SURVEYING, AND ANGULAR SURVEYING*, with Records of the Peculiarities of Needle Disturbances. Compiled from the Results of carefully made Experiments. By W. LINTERN. 2s.

ARCHITECTURE, BUILDING, ETC.

16. *ARCHITECTURE—ORDERS*—The Orders and their Æsthetic Principles. By W. H. LEEDS. Illustrated. 1s. 6d.

17. *ARCHITECTURE—STYLES*—The History and Description of the Styles of Architecture of Various Countries, from the Earliest to the Present Period. By T. TALBOT BURY, F.R.I.B.A., &c. Illustrated. 2s.
*** ORDERS AND STYLES OF ARCHITECTURE, *in One Vol.*, 3s. 6d.

18. *ARCHITECTURE—DESIGN*—The Principles of Design in Architecture, as deducible from Nature and exemplified in the Works of the Greek and Gothic Architects. By E. L. GARBETT, Architect. Illustrated. 2s.6d.

*** *The three preceding Works, in One handsome Vol., half bound, entitled* "MODERN ARCHITECTURE," *price* 6s.

22. *THE ART OF BUILDING*, Rudiments of. General Principles of Construction, Materials used in Building, Strength and Use of Materials, Working Drawings, Specifications, and Estimates. By E. DOBSON, 2s.‡

25. *MASONRY AND STONECUTTING*: Rudimentary Treatise on the Principles of Masonic Projection and their application to Construction. By EDWARD DOBSON, M.R.I.B.A., &c. 2s. 6d.‡

42. *COTTAGE BUILDING.* By C. BRUCE ALLEN, Architect. Ninth Edition, revised and enlarged. Numerous Illustrations. 1s. 6d.

45. *LIMES, CEMENTS, MORTARS, CONCRETES, MASTICS*, PLASTERING, &c. By G. R. BURNELL, C.E. Twelfth Edition. 1s. 6d.

☞ *The ‡ indicates that these vols. may be had strongly bound at 6d. extra.*

Architecture, Building, etc., *continued.*

57. *WARMING AND VENTILATION.* An Exposition of the General Principles as applied to Domestic and Public Buildings, Mines, Lighthouses, Ships, &c. By C. TOMLINSON, F.R.S., &c. Illustrated. 3s.

111. *ARCHES, PIERS, BUTTRESSES, &c.:* Experimental Essays on the Principles of Construction. By W. BLAND. Illustrated. 1s. 6d.

116. *THE ACOUSTICS OF PUBLIC BUILDINGS;* or, The Principles of the Science of Sound applied to the purposes of the Architect and Builder. By T. ROGER SMITH, M.R.I.B.A., Architect. Illustrated. 1s. 6d.

127. *ARCHITECTURAL MODELLING IN PAPER,* the Art of. By T. A. RICHARDSON, Architect. Illustrated. 1s. 6d.

128. *VITRUVIUS—THE ARCHITECTURE OF MARCUS VITRUVIUS POLLO.* In Ten Books. Translated from the Latin by JOSEPH GWILT, F.S.A., F.R.A.S. With 23 Plates. 5s.

130. *GRECIAN ARCHITECTURE,* An Inquiry into the Principles of Beauty in; with an Historical View of the Rise and Progress of the Art in Greece. By the EARL OF ABERDEEN. 1s.

*** *The two preceding Works in One handsome Vol., half bound, entitled* "ANCIENT ARCHITECTURE," *price* 6s.

132. *THE ERECTION OF DWELLING-HOUSES.* Illustrated by a Perspective View, Plans, Elevations, and Sections of a pair of Semi-detached Villas, with the Specification, Quantities, and Estimates, &c. By S. H. BROOKS. New Edition, with Plates. 2s. 6d.‡

156. *QUANTITIES AND MEASUREMENTS,* How to Calculate and Take them in Bricklayers', Masons', Plasterers', Plumbers', Painters', Paper-hangers', Gilders', Smiths', Carpenters', and Joiners' Work. By A. C. BEATON, Architect and Surveyor. New and Enlarged Edition. Illus. 1s. 6d.

175. *LOCKWOOD & CO.'S BUILDER'S AND CONTRACTOR'S* PRICE BOOK, containing the latest Prices of all kinds of Builders' Materials and Labour, and of all Trades connected with Building, &c., &c. Edited by F. T. W. MILLER, Architect. Published annually. 3s. 6d.; half bound, 4s.

182. *CARPENTRY AND JOINERY*—THE ELEMENTARY PRINCIPLES OF CARPENTRY. Chiefly composed from the Standard Work of THOMAS TREDGOLD, C.E. With Additions, Alterations, and Corrections from the Works of the most Recent Authorities, and a TREATISE ON JOINERY by E. WYNDHAM TARN, M.A. Fourth Edition, Revised and extended, with numerous Illustrations. 3s. 6d.‡ [*Just published.*

182*. *CARPENTRY AND JOINERY. ATLAS* of 35 Plates to accompany the above. With Descriptive Letterpress. 4to. 6s.; cloth, 7s. 6d.

185. *THE COMPLETE MEASURER;* the Measurement of Boards, Glass, &c.; Unequal-sided, Square-sided, Octagonal-sided, Round Timber and Stone, and Standing Timber, &c. By RICHARD HORTON. Fifth Edition. 4s.; strongly bound in leather, 5s.

187. *HINTS TO YOUNG ARCHITECTS.* By G. WIGHTWICK. New Edition. By G. H. GUILLAUME. Illustrated. 3s. 6d.‡

188. *HOUSE PAINTING, GRAINING, MARBLING, AND SIGN WRITING:* a Practical Manual of, with a Course of Elementary Drawing for House-Painters, Sign-Writers, &c., and a Collection of Useful Receipts. By ELLIS A. DAVIDSON. Fourth Edition. With Coloured Plates. 5s. cloth limp; 6s. cloth boards.

189. *THE RUDIMENTS OF PRACTICAL BRICKLAYING.* In Six Sections: General Principles; Arch Drawing, Cutting, and Setting; Pointing; Paving, Tiling, Materials; Slating and Plastering; Practical Geometry, Mensuration, &c. By ADAM HAMMOND. Fifth Edition. 1s. 6d.

191. *PLUMBING.* A Text-Book to the Practice of the Art or Craft of the Plumber. With Chapters upon House Drainage. Fourth Edition. With 330 Illustrations. By W. P. BUCHAN. 3s. 6d.‡

☞ *The ‡ indicates that these vols. may be had strongly bound at 6d. extra.*

Architecture, Building, etc., *continued.*

192. *THE TIMBER IMPORTER'S, TIMBER MERCHANT'S,* and BUILDER'S STANDARD GUIDE. By RICHARD E. GRANDY. Second Edition, Revised. 3s.‡

206. *A BOOK ON BUILDING, Civil and Ecclesiastical,* including CHURCH RESTORATION. With the Theory of Domes and the Great Pyramid, &c. By Sir EDMUND BECKETT, Bart., LL.D., Q.C., F.R.A.S. 4s. 6d.‡

226. *THE JOINTS MADE AND USED BY BUILDERS* in the Construction of various kinds of Engineering and Architectural Works. By WYVILL J. CHRISTY, Architect. With upwards of 160 Engravings on Wood. 3s.‡

228. *THE CONSTRUCTION OF ROOFS OF WOOD AND IRON* By E. WYNDHAM TARN, M.A., Architect. Second Edition, revised. 1s. 6d.

229. *ELEMENTARY DECORATION:* as applied to the Interior and Exterior Decoration of Dwelling-Houses, &c. By JAMES W. FACEY, Jun. Illustrated with Sixty-eight explanatory Engravings. 2s.

230. *HANDRAILING* (A Practical Treatise on). Showing New and Simple Methods for finding the Pitch of the Plank, Drawing the Moulds, Bevelling, Jointing-up, and Squaring the Wreath. By GEORGE COLLINGS. Illustrated with Plates and Diagrams. 1s. 6d.

247. *BUILDING ESTATES:* a Rudimentary Treatise on the Development, Sale, Purchase, and General Management of Building Land, including the Formation of Streets and Sewers, and the Requirements of Sanitary Authorities. By FOWLER MAITLAND, Surveyor. Illustrated. 2s.

248. *PORTLAND CEMENT FOR USERS.* By HENRY FAIJA, Assoc. M. Inst. C.E. Second Edition, corrected. Illustrated. 2s.

252. *BRICKWORK:* a Practical Treatise, embodying the General and Higher Principles of Bricklaying, Cutting and Setting, &c. By F. WALKER, Certificated by the Science and Art Department in Building Construction, Solid Geometry, &c. 1s. 6d.

253. *THE TIMBER MERCHANT'S, SAW-MILLER'S, AND* IMPORTER'S FREIGHT-BOOK AND ASSISTANT. By WM. RICHARDSON. With a Chapter on Speeds of Saw-Mill Machinery, &c. By M. POWIS BALE, A.M.Inst.C.E., and a London Price List for Timber and Deal Sawing, &c., 1884. 3s.‡ [*Just published.*

SHIPBUILDING, NAVIGATION, MARINE ENGINEERING, ETC.

51. *NAVAL ARCHITECTURE.* An Exposition of the Elementary Principles of the Science, and their Practical Application to Naval Construction. By J. PEAKE. Fifth Edition, with Plates and Diagrams. 3s. 6d.‡

53*. *SHIPS FOR OCEAN AND RIVER SERVICE,* Elementary and Practical Principles of the Construction of. By H. A. SOMMERFELDT, Surveyor of the Royal Norwegian Navy. With an Appendix. 1s. 6d.

53**. *AN ATLAS OF ENGRAVINGS* to Illustrate the above. Twelve large folding plates. Royal 4to, cloth. 7s. 6d.

54. *MASTING, MAST-MAKING, AND RIGGING OF SHIPS,* Rudimentary Treatise on. Also Tables of Spars, Rigging, Blocks; Chain, Wire, and Hemp Ropes, &c., relative to every class of vessels. By ROBERT KIPPING, N.A. Fifteenth Edition. Illustrated. 2s.‡

54*. *IRON SHIP-BUILDING.* With Practical Examples and Details for the Use of Ship Owners and Ship Builders. By JOHN GRANTHAM, Consulting Engineer and Naval Architect. 5th Edition, with Additions. 4s.

54**. *AN ATLAS OF FORTY PLATES* to Illustrate the above. Fifth Edition. 4to, boards. 38s.

55. *THE SAILOR'S SEA BOOK:* a Rudimentary Treatise on Navigation. By JAMES GREENWOOD, B.A. With numerous Woodcuts and Coloured Plates. New and enlarged edition. By W. H. ROSSER. 2s. 6d.‡

☞ *The ‡ indicates that these vols. may be had strongly bound at 6d. extra.*

Shipbuilding, Navigation, Marine Engineering, etc., *cont.*

83*bis*. *THE FORMS OF SHIPS AND BOATS:* Hints, Experimentally Derived, on some of the Principles regulating Shipbuilding. By W. BLAND. Seventh Edition, revised, with numerous Illustrations and Models. 1s. 6d.

99. *NAVIGATION AND NAUTICAL ASTRONOMY*, in Theory and Practice. By Prof. J. R. YOUNG. New Edition. 2s. 6d.

106. *SHIPS' ANCHORS*, a Treatise on. By G. COTSELL, N.A. 1s. 6d.

149. *SAILS AND SAIL-MAKING*, an Elementary Treatise on. With Draughting, and the Centre of Effort of the Sails. Also, Weights and Sizes of Ropes: Masting, Rigging, and Sails of Steam Vessels, &c., &c. Eleventh Edition. By ROBERT KIPPING, N.A., Sailmaker. Illustrated. 2s. 6d.‡

155. *THE ENGINEER'S GUIDE TO THE ROYAL AND* MERCANTILE NAVIES. By a PRACTICAL ENGINEER. Revised by D. F. M'CARTHY, late of the Ordnance Survey Office, Southampton. 3s.

55 & 204. *PRACTICAL NAVIGATION*. Consisting of The Sailor's Sea-Book. By JAMES GREENWOOD and W. H. ROSSER. Together with the requisite Mathematical and Nautical Tables for the Working of the Problems. By HENRY LAW, C.E., and Professor J. R. YOUNG. With numerous Wood Engravings and Coloured Plates. 7s. Strongly half-bound.

AGRICULTURE, GARDENING, ETC.

61*. *A COMPLETE READY RECKONER FOR THE ADMEA-* SUREMENT OF LAND, &c. By A. ARMAN. Second Edition, revised and extended by C. NORRIS, Surveyor, Valuer, &c. 2s.

131. *MILLER'S, MERCHANT'S, AND FARMER'S READY* RECKONER. With approximate values of Millstones, Millwork, &c. 1s.

140. *SOILS, MANURES, AND CROPS*. (Vol. 1. OUTLINES OF MODERN FARMING.) By R. SCOTT BURN. Woodcuts. 2s.

141. *FARMING & FARMING ECONOMY*, Notes, Historical and Practical, on. (Vol. 2. OUTLINES OF MODERN FARMING.) By R. SCOTT BURN. 3s.

142. *STOCK; CATTLE, SHEEP, AND HORSES*. (Vol. 3. OUTLINES OF MODERN FARMING.) By R. SCOTT BURN. Woodcuts. 2s. 6d.

145. *DAIRY, PIGS, AND POULTRY*, Management of the. By R. SCOTT BURN. With Notes on the Diseases of Stock. (Vol. 4. OUTLINES OF MODERN FARMING.) Woodcuts. 2s.

146. *UTILIZATION OF SEWAGE, IRRIGATION, AND* RECLAMATION OF WASTE LAND. (Vol. 5. OUTLINES OF MODERN FARMING.) By R. SCOTT BURN. Woodcuts. 2s. 6d.

*** *Nos. 140-1-2-5-6, in One Vol., handsomely half-bound, entitled* "OUTLINES OF MODERN FARMING." By ROBERT SCOTT BURN. *Price* 12s.

177. *FRUIT TREES*, The Scientific and Profitable Culture of. From the French of DU BREUIL. Revised by GEO. GLENNY. 187 Woodcuts. 3s. 6d.‡

198. *SHEEP; THE HISTORY, STRUCTURE, ECONOMY, AND* DISEASES OF. By W. C. SPOONER, M.R.V.C., &c. Fourth Edition, enlarged, including Specimens of New and Improved Breeds. 3s. 6d.‡

201. *KITCHEN GARDENING MADE EASY*. By GEORGE M. F. GLENNY. Illustrated. 1s. 6d.‡

207. *OUTLINES OF FARM MANAGEMENT, and the Organization of Farm Labour:* Treating of the General Work of the Farm; Field and Live Stock; Contract Work; Labour, &c. By R. SCOTT BURN. 2s. 6d.‡

208. *OUTLINES OF LANDED ESTATES MANAGEMENT:* Treating of the Varieties of Lands, Methods of Farming, Farm Buildings, Irrigation, Drainage, &c. By R. SCOTT BURN. 2s. 6d.‡

*** *Nos. 207 & 208 in One Vol., handsomely half-bound, entitled* "OUTLINES OF LANDED ESTATES AND FARM MANAGEMENT." By R. SCOTT BURN. *Price* 6s.

☞ *The ‡ indicates that these vols. may be had strongly bound at 6d. extra.*

Agriculture, Gardening, etc., *continued.*

209. *THE TREE PLANTER AND PLANT PROPAGATOR.* A Practical Manual on the Propagation of Forest Trees, Fruit Trees, Flowering Shrubs, Flowering Plants, &c. By SAMUEL WOOD. 2s.‡

210. *THE TREE PRUNER.* A Practical Manual on the Pruning of Fruit Trees, including also their Training and Renovation; also the Pruning of Shrubs, Climbers, and Flowering Plants. By SAMUEL WOOD. 2s.‡

*** *Nos.* 209 *&* 210 *in One Vol., handsomely half-bound, entitled* "THE TREE PLANTER, PROPAGATOR, AND PRUNER." By SAMUEL WOOD. *Price* 5*s*.

218. *THE HAY AND STRAW MEASURER:* Being New Tables for the Use of Auctioneers, Valuers, Farmers, Hay and Straw Dealers, &c. By JOHN STEELE. Fourth Edition. 2s.

222. *SUBURBAN FARMING.* The Laying-out and Cultivation of Farms, adapted to the Produce of Milk, Butter, and Cheese, Eggs, Poultry, and Pigs. By Prof. JOHN DONALDSON and R. SCOTT BURN. 3s. 6d.‡

231. *THE ART OF GRAFTING AND BUDDING.* By CHARLES BALTET. With Illustrations. 2s. 6d.‡

232. *COTTAGE GARDENING;* or, Flowers, Fruits, and Vegetables for Small Gardens. By E. HOBDAY. 1s. 6d.

233. *GARDEN RECEIPTS.* Edited by CHARLES W. QUIN. 1s. 6d.

234. *THE KITCHEN AND MARKET GARDEN.* Compiled by C. W. SHAW, Editor of "Gardening Illustrated." 3s.‡

239. *DRAINING AND EMBANKING.* A Practical Treatise, embodying the most recent experience in the Application of Improved Methods. By JOHN SCOTT, late Professor of Agriculture and Rural Economy at the Royal Agricultural College, Cirencester. With 68 Illustrations. 1s. 6d.

240. *IRRIGATION AND WATER SUPPLY.* A Treatise on Water Meadows, Sewage Irrigation, Warping, &c.; on the Construction of Wells, Ponds, and Reservoirs; and on Raising Water by Machinery for Agricultural and Domestic Purposes. By Prof. JOHN SCOTT. With 34 Illus. 1s. 6d.

241. *FARM ROADS, FENCES, AND GATES.* A Practical Treatise on the Roads, Tramways, and Waterways of the Farm; the Principles of Enclosures; and the different kinds of Fences, Gates, and Stiles. By Professor JOHN SCOTT. With 75 Illustrations. 1s. 6d.

242. *FARM BUILDINGS.* A Practical Treatise on the Buildings necessary for various kinds of Farms, their Arrangement and Construction, including Plans and Estimates. By Prof. JOHN SCOTT. With 105 Illus. 2s.

243. *BARN IMPLEMENTS AND MACHINES.* A Practical Treatise on the Application of Power to the Operations of Agriculture; and on various Machines used in the Threshing-barn, in the Stock-yard, and in the Dairy, &c. By Prof. J. SCOTT. With 123 Illustrations. 2s.

244. *FIELD IMPLEMENTS AND MACHINES.* A Practical Treatise on the Varieties now in use, with Principles and Details of Construction, their Points of Excellence, and Management. By Professor JOHN SCOTT. With 138 Illustrations. 2s.

245. *AGRICULTURAL SURVEYING.* A Practical Treatise on Land Surveying, Levelling, and Setting-out; and on Measuring and Estimating Quantities, Weights, and Values of Materials, Produce, Stock, &c. By Prof. JOHN SCOTT. With 62 Illustrations. 1s. 6d.

*** *Nos.* 239 *to* 245 *in One Vol., handsomely half-bound, entitled* "THE COMPLETE TEXT-BOOK OF FARM ENGINEERING." By Professor JOHN SCOTT. *Price* 12*s*.

250. *MEAT PRODUCTION.* A Manual for Producers, Distributors, &c. By JOHN EWART. 2s. 6d.‡

☞ *The ‡ indicates that these vols. may be had strongly bound at 6d. extra.*

MATHEMATICS, ARITHMETIC, ETC.

32. *MATHEMATICAL INSTRUMENTS*, a Treatise on; in which their Construction and the Methods of Testing, Adjusting, and Using them are concisely Explained. By J. F. HEATHER, M.A., of the Royal Military Academy, Woolwich. Original Edition, in 1 vol., Illustrated. 1s. 6d.

*** *In ordering the above, be careful to say, "Original Edition" (No. 32), to distinguish it from the Enlarged Edition in 3 vols. (Nos. 168-9-70.)*

76. *DESCRIPTIVE GEOMETRY*, an Elementary Treatise on; with a Theory of Shadows and of Perspective, extracted from the French of G. MONGE. To which is added, a description of the Principles and Practice of Isometrical Projection. By J. F. HEATHER, M.A. With 14 Plates. 2s.

178. *PRACTICAL PLANE GEOMETRY*: giving the Simplest Modes of Constructing Figures contained in one Plane and Geometrical Construction of the Ground. By J. F. HEATHER, M.A. With 215 Woodcuts. 2s.

83. *COMMERCIAL BOOK-KEEPING*. With Commercial Phrases and Forms in English, French, Italian, and German. By JAMES HADDON, M.A., Arithmetical Master of King's College School, London. 1s. 6d.

84. *ARITHMETIC*, a Rudimentary Treatise on: with full Explanations of its Theoretical Principles, and numerous Examples for Practice. By Professor J. R. YOUNG. Tenth Edition, corrected. 1s. 6d.

84*. A KEY to the above, containing Solutions in full to the Exercises, together with Comments, Explanations, and Improved Processes, for the Use of Teachers and Unassisted Learners. By J. R. YOUNG. 1s. 6d.

85. *EQUATIONAL ARITHMETIC*, applied to Questions of Interest, Annuities, Life Assurance, and General Commerce; with various Tables by which all Calculations may be greatly facilitated. By W. HIPSLEY. 2s.

86. *ALGEBRA*, the Elements of. By JAMES HADDON, M.A. With Appendix, containing miscellaneous Investigations, and a Collection of Problems in various parts of Algebra. 2s.

86*. A KEY AND COMPANION to the above Book, forming an extensive repository of Solved Examples and Problems in Illustration of the various Expedients necessary in Algebraical Operations. By J. R. YOUNG. 1s. 6d.

88. *EUCLID*, THE ELEMENTS OF: with many additional Propositions
89. and Explanatory Notes: to which is prefixed, an Introductory Essay on Logic. By HENRY LAW, C.E. 2s. 6d.‡

*** *Sold also separately, viz.:—*

88. EUCLID, The First Three Books. By HENRY LAW, C.E. 1s. 6d.
89. EUCLID, Books 4, 5, 6, 11, 12. By HENRY LAW, C.E. 1s. 6d.

90. *ANALYTICAL GEOMETRY AND CONIC SECTIONS*, By JAMES HANN. A New Edition, by Professor J. R. YOUNG. 2s.‡

91. *PLANE TRIGONOMETRY*, the Elements of. By JAMES HANN, formerly Mathematical Master of King's College, London. 1s. 6d.

92. *SPHERICAL TRIGONOMETRY*, the Elements of. By JAMES HANN. Revised by CHARLES H. DOWLING, C.E. 1s.

*** *Or with "The Elements of Plane Trigonometry," in One Volume, 2s. 6d.*

93. *MENSURATION AND MEASURING*. With the Mensuration and Levelling of Land for the Purposes of Modern Engineering. By T. BAKER, C.E. New Edition by E. NUGENT, C.E. Illustrated. 1s. 6d.

101. *DIFFERENTIAL CALCULUS*, Elements of the. By W. S. B. WOOLHOUSE, F.R.A.S., &c. 1s. 6d.

102. *INTEGRAL CALCULUS*, Rudimentary Treatise on the. By HOMERSHAM COX, B.A. Illustrated. 1s.

105. *MNEMONICAL LESSONS*.—GEOMETRY, ALGEBRA, AND TRIGONOMETRY, in Easy Mnemonical Lessons. By the Rev. THOMAS PENYNGTON KIRKMAN, M.A. 1s. 6d.

136. *ARITHMETIC*, Rudimentary, for the Use of Schools and Self-Instruction. By JAMES HADDON, M.A. Revised by A. ARMAN. 1s. 6d.

137. A KEY TO HADDON'S RUDIMENTARY ARITHMETIC. By A. ARMAN. 1s. 6d.

☞ *The ‡ indicates that these vols. may be had strongly bound at 6d. extra.*

Mathematics, Arithmetic, etc., *continued.*

168. *DRAWING AND MEASURING INSTRUMENTS.* Including—I. Instruments employed in Geometrical and Mechanical Drawing, and in the Construction, Copying, and Measurement of Maps and Plans. II. Instruments used for the purposes of Accurate Measurement, and for Arithmetical Computations. By J. F. HEATHER, M.A. Illustrated. 1s. 6d.

169. *OPTICAL INSTRUMENTS.* Including (more especially) Telescopes, Microscopes, and Apparatus for producing copies of Maps and Plans by Photography. By J. F. HEATHER, M.A. Illustrated. 1s. 6d.

170. *SURVEYING AND ASTRONOMICAL INSTRUMENTS.* Including—I. Instruments Used for Determining the Geometrical Features of a portion of Ground. II. Instruments Employed in Astronomical Observations. By J. F. HEATHER, M.A. Illustrated. 1s. 6d.

*** *The above three volumes form an enlargement of the Author's original work, "Mathematical Instruments."* (*See No.* 32 *in the Series.*)

168.
169.
170. *MATHEMATICAL INSTRUMENTS.* By J. F. HEATHER, M.A. Enlarged Edition, for the most part entirely re-written. The 3 Parts as above, in One thick Volume. With numerous Illustrations. 4s. 6d.‡

158. *THE SLIDE RULE, AND HOW TO USE IT;* containing full, easy, and simple Instructions to perform all Business Calculations with unexampled rapidity and accuracy. By CHARLES HOARE, C.E. With a Slide Rule in tuck of cover. 2s. 6d.‡

196. *THEORY OF COMPOUND INTEREST AND ANNUITIES;* with Tables of Logarithms for the more Difficult Computations of Interest, Discount, Annuities, &c. By FÉDOR THOMAN. 4s.‡

199. *THE COMPENDIOUS CALCULATOR;* or, Easy and Concise Methods of Performing the various Arithmetical Operations required in Commercial and Business Transactions; together with Useful Tables. By D. O'GORMAN. Twenty-sixth Edition, carefully revised by C. NORRIS. 3s., cloth limp; 3s. 6d., strongly half-bound in leather.

204. *MATHEMATICAL TABLES,* for Trigonometrical, Astronomical, and Nautical Calculations; to which is prefixed a Treatise on Logarithms. By HENRY LAW, C.E. Together with a Series of Tables for Navigation and Nautical Astronomy. By Prof. J. R. YOUNG. New Edition. 4s.‡

204*. *LOGARITHMS.* With Mathematical Tables for Trigonometrical, Astronomical, and Nautical Calculations. By HENRY LAW, M.Inst.C.E. New and Revised Edition. (Forming part of the above Work). 3s.

221. *MEASURES, WEIGHTS, AND MONEYS OF ALL NATIONS,* and an Analysis of the Christian, Hebrew, and Mahometan Calendars. By W. S. B. WOOLHOUSE, F.R.A.S., F.S.S. Sixth Edition. 2s.‡

227. *MATHEMATICS AS APPLIED TO THE CONSTRUCTIVE ARTS.* Illustrating the various processes of Mathematical Investigation, by means of Arithmetical and Simple Algebraical Equations and Practical Examples. By FRANCIS CAMPIN, C.E. Second Edition. 3s.‡

PHYSICAL SCIENCE, NATURAL PHILOSOPHY, ETC.

1. *CHEMISTRY.* By Professor GEORGE FOWNES, F.R.S. With an Appendix on the Application of Chemistry to Agriculture. 1s.

2. *NATURAL PHILOSOPHY,* Introduction to the Study of. By C. TOMLINSON. Woodcuts. 1s. 6d.

6. *MECHANICS,* Rudimentary Treatise on. By CHARLES TOMLINSON. Illustrated. 1s. 6d.

7. *ELECTRICITY;* showing the General Principles of Electrical Science, and the purposes to which it has been applied. By Sir W. SNOW HARRIS, F.R.S., &c. With Additions by R. SABINE, C.E., F.S.A. 1s. 6d.

7*. *GALVANISM.* By Sir W. SNOW HARRIS. New Edition by ROBERT SABINE, C.E., F.S.A. 1s. 6d.

8. *MAGNETISM;* being a concise Exposition of the General Principles of Magnetical Science. By Sir W. SNOW HARRIS. New Edition, revised by H. M. NOAD, Ph.D. With 165 Woodcuts. 3s. 6d.‡

☞ *The ‡ indicates that these vols. may be had strongly bound at 6d. extra.*

Physical Science, Natural Philosophy, etc., *continued.*

11. *THE ELECTRIC TELEGRAPH;* its History and Progress; with Descriptions of some of the Apparatus. By R. SABINE, C.E., F.S.A. 3s.

12. *PNEUMATICS,* for the Use of Beginners. By CHARLES TOMLINSON. Illustrated. 1s. 6d.

72. *MANUAL OF THE MOLLUSCA;* a Treatise on Recent and Fossil Shells. By Dr. S. P. WOODWARD, A.L.S. Fourth Edition. With Appendix by RALPH TATE, A.L.S., F.G.S. With numerous Plates and 300 Woodcuts. 6s. 6d. Cloth boards, 7s. 6d.

96. *ASTRONOMY.* By the late Rev. ROBERT MAIN, M.A. Third Edition, by WILLIAM THYNNE LYNN, B.A., F.R.A.S. 2s.

97. *STATICS AND DYNAMICS,* the Principles and Practice of; embracing also a clear development of Hydrostatics, Hydrodynamics, and Central Forces. By T. BAKER, C.E. 1s. 6d.

138. *TELEGRAPH,* Handbook of the; a Guide to Candidates for Employment in the Telegraph Service. By R. BOND. Fourth Edition. Including Questions on Magnetism, Electricity, and Practical Telegraphy, by W. McGREGOR. 3s.‡

173. *PHYSICAL GEOLOGY,* partly based on Major-General PORTLOCK'S "Rudiments of Geology." By RALPH TATE, A.L.S., &c. Woodcuts. 2s.

174. *HISTORICAL GEOLOGY,* partly based on Major-General PORTLOCK'S "Rudiments." By RALPH TATE, A.L.S., &c. Woodcuts. 2s. 6d.

173 & 174. *RUDIMENTARY TREATISE ON GEOLOGY,* Physical and Historical. Partly based on Major-General PORTLOCK'S "Rudiments of Geology." By RALPH TATE, A.L.S., F.G.S., &c. In One Volume. 4s. 6d.‡

183 & 184. *ANIMAL PHYSICS,* Handbook of. By Dr. LARDNER, D.C.L., formerly Professor of Natural Philosophy and Astronomy in University College, Lond. With 520 Illustrations. In One Vol. 7s. 6d., cloth boards.

*** *Sold also in Two Parts, as follows:—*

183. ANIMAL PHYSICS. By Dr. LARDNER. Part I., Chapters I.—VII. 4s.

184. ANIMAL PHYSICS. By Dr. LARDNER. Part II., Chapters VIII.—XVIII. 3s.

FINE ARTS.

20. *PERSPECTIVE FOR BEGINNERS.* Adapted to Young Students and Amateurs in Architecture, Painting, &c. By GEORGE PYNE. 2s.

40 *GLASS STAINING, AND THE ART OF PAINTING ON GLASS.* From the German of Dr. GESSERT and EMANUEL OTTO FROMBERG. With an Appendix on THE ART OF ENAMELLING. 2s. 6d.

69. *MUSIC,* A Rudimentary and Practical Treatise on. With numerous Examples. By CHARLES CHILD SPENCER. 2s. 6d.

71. *PIANOFORTE,* The Art of Playing the. With numerous Exercises & Lessons from the Best Masters. By CHARLES CHILD SPENCER. 1s. 6d.

69-71. *MUSIC & THE PIANOFORTE.* In one vol. Half bound, 5s.

181. *PAINTING POPULARLY EXPLAINED,* including Fresco, Oil, Mosaic, Water Colour, Water-Glass, Tempera, Encaustic, Miniature, Painting on Ivory, Vellum, Pottery, Enamel, Glass, &c. With Historical Sketches of the Progress of the Art by THOMAS JOHN GULLICK, assisted by JOHN TIMBS, F.S.A. Fifth Edition, revised and enlarged. 5s.‡

186. *A GRAMMAR OF COLOURING,* applied to Decorative Painting and the Arts. By GEORGE FIELD. New Edition, enlarged and adapted to the Use of the Ornamental Painter and Designer. By ELLIS A. DAVIDSON. With two new Coloured Diagrams, &c. 3s.‡

246. *A DICTIONARY OF PAINTERS, AND HANDBOOK FOR PICTURE AMATEURS;* including Methods of Painting, Cleaning, Relining and Restoring, Schools of Painting, &c. With Notes on the Copyists and Imitators of each Master. By PHILIPPE DARYL. 2s. 6d.‡

☞ *The ‡ indicates that these vols. may be had strongly bound at 6d. extra.*

INDUSTRIAL AND USEFUL ARTS.

23. *BRICKS AND TILES*, Rudimentary Treatise on the Manufacture of. By E. DOBSON, M.R.I.B.A. Illustrated, 3s.‡

67. *CLOCKS, WATCHES, AND BELLS*, a Rudimentary Treatise on. By Sir EDMUND BECKETT, LL.D., Q.C. Seventh Edition, revised and enlarged. 4s. 6d. limp; 5s. 6d. cloth boards.

83**. *CONSTRUCTION OF DOOR LOCKS*. Compiled from the Papers of A. C. HOBBS, and Edited by CHARLES TOMLINSON, F.R.S. With Additions by ROBERT MALLET, M.I.C.E. Illus. 2s. 6d.

162. *THE BRASS FOUNDER'S MANUAL;* Instructions for Modelling, Pattern-Making, Moulding, Turning, Filing, Burnishing, Bronzing, &c. With copious Receipts, &c. By WALTER GRAHAM. 2s.‡

205. *THE ART OF LETTER PAINTING MADE EASY*. By J. G. BADENOCH. Illustrated with 12 full-page Engravings of Examples. 1s.

215. *THE GOLDSMITH'S HANDBOOK*, containing full Instructions for the Alloying and Working of Gold. By GEORGE E. GEE. 3s.‡

224. *COACH BUILDING*, A Practical Treatise, Historical and Descriptive. By J. W. BURGESS. 2s. 6d.‡

225. *THE SILVERSMITH'S HANDBOOK*, containing full Instructions for the Alloying and Working of Silver. By GEORGE E. GEE. 3s.‡

235. *PRACTICAL ORGAN BUILDING*. By W. E. DICKSON, M.A., Precentor of Ely Cathedral. Illustrated. 2s. 6d.‡

249. *THE HALL-MARKING OF JEWELLERY PRACTICALLY CONSIDERED*. By GEORGE E. GEE. 3s.‡

MISCELLANEOUS VOLUMES.

36. *A DICTIONARY OF TERMS used in ARCHITECTURE, BUILDING, ENGINEERING, MINING, METALLURGY, ARCHÆOLOGY, the FINE ARTS, &c.* By JOHN WEALE. Fifth Edition. Revised by ROBERT HUNT, F.R.S. Illustrated. 5s. limp; 6s. cloth boards.

50. *THE LAW OF CONTRACTS FOR WORKS AND SERVICES*. By DAVID GIBBONS. Third Edition, enlarged. 3s.‡

112. *MANUAL OF DOMESTIC MEDICINE*. By R. GOODING, B.A., M.D. Intended as a Family Guide in all Cases of Accident and Emergency. Third Edition. 2s.‡

112*. *MANAGEMENT OF HEALTH*. A Manual of Home and Personal Hygiene. By the Rev. JAMES BAIRD, B.A. 1s.

150. *LOGIC*, Pure and Applied. By S. H. EMMENS. 1s. 6d.

153. *SELECTIONS FROM LOCKE'S ESSAYS ON THE HUMAN UNDERSTANDING*. With Notes by S. H. EMMENS. 2s.

154. *GENERAL HINTS TO EMIGRANTS*. Notices of the various Fields for Emigration, Hints on Outfits, Useful Recipes, &c. 2s.

157. *THE EMIGRANT'S GUIDE TO NATAL*. By ROBERT JAMES MANN, F.R.A.S., F.M.S. Second Edition. Map. 2s.

193. *HANDBOOK OF FIELD FORTIFICATION*, intended for the Guidance of Officers Preparing for Promotion. By Major W. W. KNOLLYS, F.R.G.S. With 163 Woodcuts. 3s.‡

194. *THE HOUSE MANAGER:* Being a Guide to Housekeeping. Practical Cookery, Pickling and Preserving, Household Work, Dairy Management, the Table and Dessert, Cellarage of Wines, Home-brewing and Wine-making, the Boudoir and Dressing-room, Travelling, Stable Economy, Gardening Operations, &c. By AN OLD HOUSEKEEPER. 3s. 6d.‡

194, 112, & 112*. *HOUSE BOOK (The)*. Comprising:—I. THE HOUSE MANAGER. By an OLD HOUSEKEEPER. II. DOMESTIC MEDICINE. By RALPH GOODING, M.D. III. MANAGEMENT OF HEALTH. By JAMES BAIRD. In One Vol., strongly half-bound, 6s.

☞ *The ‡ indicates that these vols. may be had strongly bound at 6d. extra.*

EDUCATIONAL AND CLASSICAL SERIES.

HISTORY.

1. **England, Outlines of the History of**; more especially with reference to the Origin and Progress of the English Constitution. By WILLIAM DOUGLAS HAMILTON, F.S.A., of Her Majesty's Public Record Office. 4th Edition, revised. 5s.; cloth boards, 6s.

5. **Greece, Outlines of the History of**; in connection with the Rise of the Arts and Civilization in Europe. By W. DOUGLAS HAMILTON, of University College, London, and EDWARD LEVIEN, M.A., of Balliol College, Oxford. 2s. 6d.; cloth boards, 3s. 6d.

7. **Rome, Outlines of the History of**: from the Earliest Period to the Christian Era and the Commencement of the Decline of the Empire. By EDWARD LEVIEN, of Balliol College, Oxford. Map, 2s. 6d.; cl. bds. 3s. 6d.

9. **Chronology of History, Art, Literature, and Progress**, from the Creation of the World to the Conclusion of the Franco-German War. The Continuation by W. D. HAMILTON, F.S.A. 3s.; cloth boards, 3s. 6d.

50. **Dates and Events in English History**, for the use of Candidates in Public and Private Examinations. By the Rev. E. RAND. 1s.

ENGLISH LANGUAGE AND MISCELLANEOUS.

11. **Grammar of the English Tongue**, Spoken and Written. With an Introduction to the Study of Comparative Philology. By HYDE CLARKE, D.C.L. Fourth Edition. 1s. 6d.

11*. **Philology**: Handbook of the Comparative Philology of English, Anglo-Saxon, Frisian, Flemish or Dutch, Low or Platt Dutch, High Dutch or German, Danish, Swedish, Icelandic, Latin, Italian, French, Spanish, and Portuguese Tongues. By HYDE CLARKE, D.C.L. 1s.

12. **Dictionary of the English Language**, as Spoken and Written. Containing above 100,000 Words. By HYDE CLARKE, D.C.L. 3s. 6d.; cloth boards, 4s. 6d.; complete with the GRAMMAR, cloth bds., 5s. 6d.

48. **Composition and Punctuation**, familiarly Explained for those who have neglected the Study of Grammar. By JUSTIN BRENAN. 17th Edition. 1s. 6d.

49. **Derivative Spelling-Book**: Giving the Origin of Every Word from the Greek, Latin, Saxon, German, Teutonic, Dutch, French, Spanish, and other Languages; with their present Acceptation and Pronunciation. By J. ROWBOTHAM, F.R.A.S. Improved Edition. 1s. 6d.

51. **The Art of Extempore Speaking**: Hints for the Pulpit, the Senate, and the Bar. By M. BAUTAIN, Vicar-General and Professor at the Sorbonne. Translated from the French. 7th Edition, carefully corrected. 2s. 6d.

52. **Mining and Quarrying**, with the Sciences connected therewith. First Book of, for Schools. By J. H. COLLINS, F.G.S., Lecturer to the Miners' Association of Cornwall and Devon. 1s.

53. **Places and Facts in Political and Physical Geography**, for Candidates in Examinations. By the Rev. EDGAR RAND, B.A. 1s.

54. **Analytical Chemistry, Qualitative and Quantitative, a Course of.** To which is prefixed, a Brief Treatise upon Modern Chemical Nomenclature and Notation. By WM. W. PINK and GEORGE E. WEBSTER. 2s.

THE SCHOOL MANAGERS' SERIES OF READING BOOKS,

Edited by the Rev. A. R. GRANT, Rector of Hitcham, and Honorary Canon of Ely; formerly H.M. Inspector of Schools.

INTRODUCTORY PRIMER, 3d.

	s.	d.		s.	d.
FIRST STANDARD	0	6	FOURTH STANDARD	1	2
SECOND "	0	10	FIFTH "	1	6
THIRD "	1	0	SIXTH "	1	6

LESSONS FROM THE BIBLE. Part I. Old Testament. 1s.

LESSONS FROM THE BIBLE. Part II. New Testament, to which is added THE GEOGRAPHY OF THE BIBLE, for very young Children. By Rev. C. THORNTON FORSTER. 1s. 2d. *** Or the Two Parts in One Volume. 2s.

FRENCH.

24. **French Grammar.** With Complete and Concise Rules on the Genders of French Nouns. By G. L. STRAUSS, Ph.D. 1s. 6d.

25. **French-English Dictionary.** Comprising a large number of New Terms used in Engineering, Mining, &c. By ALFRED ELWES. 1s. 6d.

26. **English-French Dictionary.** By ALFRED ELWES. 2s.

25,26. **French Dictionary** (as above). Complete, in One Vol., 3s.; cloth boards, 3s. 6d. *** Or with the GRAMMAR, cloth boards, 4s. 6d.

47. **French and English Phrase Book:** containing Introductory Lessons, with Translations, several Vocabularies of Words, a Collection of suitable Phrases, and Easy Familiar Dialogues. 1s. 6d.

GERMAN.

39. **German Grammar.** Adapted for English Students, from Heyse's Theoretical and Practical Grammar, by Dr. G. L. STRAUSS. 1s. 6d.

40. **German Reader:** A Series of Extracts, carefully culled from the most approved Authors of Germany; with Notes, Philological and Explanatory. By G. L. STRAUSS, Ph.D. 1s.

41-43. **German Triglot Dictionary.** By N. E. S. A. HAMILTON. In Three Parts. Part I. German-French-English. Part II. English-German-French. Part III. French-German-English. 3s., or cloth boards, 4s.

41-43 & 39. **German Triglot Dictionary** (as above), together with German Grammar (No. 39), in One Volume, cloth boards, 5s.

ITALIAN.

27. **Italian Grammar,** arranged in Twenty Lessons, with a Course of Exercises. By ALFRED ELWES. 1s. 6d.

28. **Italian Triglot Dictionary,** wherein the Genders of all the Italian and French Nouns are carefully noted down. By ALFRED ELWES. Vol. 1. Italian-English-French. 2s. 6d.

30. **Italian Triglot Dictionary.** By A. ELWES. Vol. 2. English-French-Italian. 2s. 6d.

32. **Italian Triglot Dictionary.** By ALFRED ELWES. Vol. 3. French-Italian-English. 2s. 6d.

28,30, 32. **Italian Triglot Dictionary** (as above). In One Vol., 7s. 6d. Cloth boards.

SPANISH AND PORTUGUESE.

34. **Spanish Grammar,** in a Simple and Practical Form. With a Course of Exercises. By ALFRED ELWES. 1s. 6d.

35. **Spanish-English and English-Spanish Dictionary.** Including a large number of Technical Terms used in Mining, Engineering, &c., with the proper Accents and the Gender of every Noun. By ALFRED ELWES. 4s.; cloth boards, 5s. *** Or with the GRAMMAR, cloth boards, 6s.

55. **Portuguese Grammar,** in a Simple and Practical Form. With a Course of Exercises. By ALFRED ELWES. 1s. 6d.

56. **Portuguese-English and English-Portuguese Dictionary.** Including a large number of Technical Terms used in Mining, Engineering, &c., with the proper Accents and the Gender of every Noun. By ALFRED ELWES. 5s.; cloth boards, 6s. *** Or with the GRAMMAR, cloth boards, 7s. [*Just published.*

HEBREW.

46*. **Hebrew Grammar.** By Dr. BRESSLAU. 1s. 6d.

44. **Hebrew and English Dictionary,** Biblical and Rabbinical; containing the Hebrew and Chaldee Roots of the Old Testament Post-Rabbinical Writings. By Dr. BRESSLAU. 6s.

46. **English and Hebrew Dictionary.** By Dr. BRESSLAU. 3s.

44,46. 46*. **Hebrew Dictionary** (as above), in Two Vols., complete, with the GRAMMAR, cloth boards, 12s.

LATIN.

19. **Latin Grammar.** Containing the Inflections and Elementary Principles of Translation and Construction. By the Rev. THOMAS GOODWIN, M.A., Head Master of the Greenwich Proprietary School. 1s.

20. **Latin-English Dictionary.** By the Rev. THOMAS GOODWIN, M.A. 2s.

22. **English-Latin Dictionary;** together with an Appendix of French and Italian Words which have their origin from the Latin. By the Rev. THOMAS GOODWIN, M.A. 1s. 6d.

20,22. **Latin Dictionary** (as above). Complete in One Vol., 3s. 6d. cloth boards, 4s. 6d. *** Or with the GRAMMAR, cloth boards, 5s. 6d.

LATIN CLASSICS. With Explanatory Notes in English.

1. **Latin Delectus.** Containing Extracts from Classical Authors, with Genealogical Vocabularies and Explanatory Notes, by H. YOUNG. 1s. 6d.

2. **Cæsaris** Commentarii de Bello Gallico. Notes, and a Geographical Register for the Use of Schools, by H. YOUNG. 2s.

3. **Cornelius Nepos.** With Notes. By H. YOUNG. 1s.

4. **Virgilii** Maronis Bucolica et Georgica. With Notes on the Bucolics by W. RUSHTON, M.A., and on the Georgics by H. YOUNG. 1s. 6d.

5. **Virgilii** Maronis Æneis. With Notes, Critical and Explanatory, by H. YOUNG. New Edition, revised and improved With copious Additional Notes by Rev. T. H. L. LEARY, D.C.L., formerly Scholar of Brasenose College, Oxford. 3s.

5* ——— Part 1. Books i.—vi., 1s. 6d.

5** ——— Part 2. Books vii.—xii., 2s.

6. **Horace;** Odes, Epode, and Carmen Sæculare. Notes by H. YOUNG. 1s. 6d.

7. **Horace;** Satires, Epistles, and Ars Poetica. Notes by W. BROWNRIGG SMITH, M.A., F.R.G.S. 1s. 6d.

8. **Sallustii** Crispi Catalina et Bellum Jugurthinum. Notes, Critical and Explanatory, by W. M. DONNE, B.A., Trin. Coll., Cam. 1s. 6d.

9. **Terentii** Andria et Heautontimorumenos. With Notes, Critical and Explanatory, by the Rev. JAMES DAVIES, M.A. 1s. 6d.

10. **Terentii** Adelphi, Hecyra, Phormio. Edited, with Notes, Critical and Explanatory, by the Rev. JAMES DAVIES, M.A. 2s.

11. **Terentii** Eunuchus, Comœdia. Notes, by Rev. J. DAVIES, M.A. 1s. 6d.

12. **Ciceronis** Oratio pro Sexto Roscio Amerino. Edited, with an Introduction, Analysis, and Notes, Explanatory and Critical, by the Rev. JAMES DAVIES, M.A. 1s. 6d.

13. **Ciceronis** Orationes in Catilinam, Verrem, et pro Archia. With Introduction, Analysis, and Notes, Explanatory and Critical, by Rev. T. H. L. LEARY, D.C.L. formerly Scholar of Brasenose College, Oxford. 1s. 6d.

14. **Ciceronis** Cato Major, Lælius, Brutus, sive de Senectute, de Amicitia, de Claris Oratoribus Dialogi. With Notes by W. BROWNRIGG SMITH, M.A., F.R.G.S. 2s.

16. **Livy:** History of Rome. Notes by H. YOUNG and W. B. SMITH, M.A. Part 1. Books i., ii., 1s. 6d.

16*. ——— Part 2. Books iii., iv., v., 1s. 6d.

17. ——— Part 3. Books xxi., xxii., 1s. 6d.

19. **Latin Verse Selections,** from Catullus, Tibullus, Propertius, and Ovid. Notes by W. B. DONNE, M.A., Trinity College, Cambridge. 2s.

20. **Latin Prose Selections,** from Varro, Columella, Vitruvius, Seneca, Quintilian, Florus, Velleius Paterculus, Valerius Maximus Suetonius, Apuleius, &c. Notes by W. B. DONNE, M.A. 2s.

21. **Juvenalis** Satiræ. With Prolegomena and Notes by T. H. S. ESCOTT, B.A., Lecturer on Logic at King's College, London. 2s.

GREEK.

14. **Greek Grammar,** in accordance with the Principles and Philological Researches of the most eminent Scholars of our own day. By HANS CLAUDE HAMILTON. 1s. 6d.

15,17. **Greek Lexicon.** Containing all the Words in General Use, with their Significations, Inflections, and Doubtful Quantities. By HENRY R. HAMILTON. Vol. 1. Greek-English, 2s. 6d.; Vol. 2. English-Greek, 2s. Or the Two Vols. in One, 4s. 6d.: cloth boards, 5s.

14,15. 17. **Greek Lexicon** (as above). Complete, with the GRAMMAR, in One Vol., cloth boards, 6s.

GREEK CLASSICS. With Explanatory Notes in English.

1. **Greek Delectus.** Containing Extracts from Classical Authors, with Genealogical Vocabularies and Explanatory Notes, by H. YOUNG. New Edition, with an improved and enlarged Supplementary Vocabulary, by JOHN HUTCHISON, M.A., of the High School, Glasgow. 1s. 6d.

2, 3. **Xenophon's Anabasis;** or, The Retreat of the Ten Thousand. Notes and a Geographical Register, by H. YOUNG. Part 1. Books i. to iii., 1s. Part 2. Books iv. to vii., 1s.

4. **Lucian's Select Dialogues.** The Text carefully revised, with Grammatical and Explanatory Notes, by H. YOUNG. 1s. 6d.

5-12. **Homer, The Works of.** According to the Text of BAEUMLEIN. With Notes, Critical and Explanatory, drawn from the best and latest Authorities, with Preliminary Observations and Appendices, by T. H. L. LEARY, M.A., D.C.L.

THE ILIAD:	Part 1. Books i. to vi., 1s. 6d.	Part 3. Books xiii. to xviii., 1s. 6d.
	Part 2. Books vii. to xii., 1s. 6d.	Part 4. Books xix. to xxiv., 1s. 6d.
THE ODYSSEY:	Part 1. Books i. to vi., 1s. 6d	Part 3. Books xiii. to xviii., 1s. 6d.
	Part 2. Books vii. to xii., 1s. 6d.	Part 4. Books xix. to xxiv., and Hymns, 2s.

13. **Plato's Dialogues:** The Apology of Socrates, the Crito, and the Phædo. From the Text of C. F. HERMANN. Edited with Notes, Critical and Explanatory, by the Rev. JAMES DAVIES, M.A. 2s.

14-17. **Herodotus,** The History of, chiefly after the Text of GAISFORD. With Preliminary Observations and Appendices, and Notes, Critical and Explanatory, by T. H. L. LEARY, M.A., D.C.L.
Part 1. Books i., ii. (The Clio and Euterpe), 2s.
Part 2. Books iii., iv. (The Thalia and Melpomene), 2s.
Part 3. Books v.-vii. (The Terpsichore, Erato, and Polymnia), 2s.
Part 4. Books viii., ix. (The Urania and Calliope) and Index, 1s. 6d.

18. **Sophocles:** Œdipus Tyrannus. Notes by H. YOUNG. 1s.

20. **Sophocles:** Antigone. From the Text of DINDORF. Notes, Critical and Explanatory, by the Rev. JOHN MILNER, B.A. 2s.

23. **Euripides:** Hecuba and Medea. Chiefly from the Text of DINDORF. With Notes, Critical and Explanatory, by W. BROWNRIGG SMITH, M.A., F.R.G.S. 1s. 6d.

26. **Euripides:** Alcestis. Chiefly from the Text of DINDORF. With Notes, Critical and Explanatory, by JOHN MILNER, B.A. 1s. 6d.

30. **Æschylus:** Prometheus Vinctus: The Prometheus Bound. From the Text of DINDORF. Edited, with English Notes, Critical and Explanatory, by the Rev. JAMES DAVIES, M.A. 1s.

32. **Æschylus:** Septem Contra Thebes: The Seven against Thebes. From the Text of DINDORF. Edited, with English Notes, Critical and Explanatory, by the Rev. JAMES DAVIES, M.A. 1s.

40. **Aristophanes:** Acharnians. Chiefly from the Text of C. H. WEISE. With Notes, by C. S. T. TOWNSHEND, M.A. 1s. 6d.

41. **Thucydides:** History of the Peloponnesian War. Notes by H. YOUNG. Book 1. 1s. 6d.

42. **Xenophon's Panegyric on Agesilaus.** Notes and Introduction by LL. F. W. JEWITT. 1s. 6d.

43. **Demosthenes.** The Oration on the Crown and the Philippics. With English Notes. By Rev. T. H. L. LEARY, D.C.L., formerly Scholar of Brasenose College, Oxford. 1s. 6d.